# SOIL ORGANIC MATTER
## IN
# SUSTAINABLE AGRICULTURE

# Advances in Agroecology
Series Editor: Clive A. Edwards

# SOIL ORGANIC MATTER
## IN
# SUSTAINABLE AGRICULTURE

### Edited by
### Fred Magdoff
### Ray R. Weil

**CRC Press**
Taylor & Francis Group
Boca Raton London New York

CRC Press is an imprint of the
Taylor & Francis Group, an **informa** business

CRC Press
Taylor & Francis Group
6000 Broken Sound Parkway NW, Suite 300
Boca Raton, FL 33487-2742

First issued in paperback 2019

© 2004 by Taylor Francis Group, LLC
CRC Press is an imprint of Taylor & Francis Group, an Informa business

No claim to original U.S. Government works

ISBN-13: 978-0-8493-1294-6 (hbk)
ISBN-13: 978-0-367-39423-3 (pbk)

Library of Congress Card Number 2004043574

+--------------------------------------------------------------------+
| **Library of Congress Cataloging-in-Publication Data**             |
|                                                                    |
|   Soil organic matter in sustainable agriculture / edited by Fred Magdoff and |
|   Ray R. Weil.                                                     |
|       p. cm. -- (Advances in agroecology)                          |
|   Includes bibliographical references and index.                   |
|   ISBN 0-8493-1294-9                                               |
|     1. Humus. 2. Soil ecology. 3. Sustainable agriculture. I. Magdoff, Fred, 1942- II. Weil, |
|   Ray R. III. Series                                               |
|                                                                    |
|   S592.8.S674 2004                                                 |
|   631.4'17--dc22                                        2004043574 |
+--------------------------------------------------------------------+

**Visit the Taylor & Francis Web site at**
**http://www.taylorandfrancis.com**

**and the CRC Press Web site at**
**http://www.crcpress.com**

# Preface

During the past two centuries, scientists, farmers, and agricultural educators have tended to alternate their views of soil organic matter (SOM) between the extremes of great appreciation and low esteem. As an early 20th century bulletin explained, organic matter was "once extolled as the essential soil ingredient, the bright particular star in the firmament of the plant grower..." before it "...fell like Lucifer..." as a result of the findings of 19th century agricultural chemists that most of the plant structure (C, that is) originated in the atmosphere (Hills et al., 1908). In the early 20th century, soil organic matter was once again viewed as critically important, only to become considered close to irrelevant to agriculture following the availability of cheap fertilizers, especially N, following World War II.

The paradigm of industrial agriculture — which developed following World War II and continues even now to dominate agriculture — views soil and plant constraints that arise while growing crops as individual problems that are usually considered to be unrelated to one another. The industrial agriculture mindset believes the best way to deal with these individual problems is to try to resolve each one separately. The remedies, according to this view, lie in the application of specific (usually purchased) inputs: fertilizers and lime for soils low in nutrients; pesticides for crops threatened by diseases, nematodes, insects, and weeds; extra irrigation water for soils with declining water holding capacity; deep tillage for overly compacted soils, etc. The development, promotion, and implementation of the industrial agriculture model has been facilitated by the specialized training that many agricultural scientists and extension specialists receive — causing them to concentrate efforts in their disciplines (and frequently subdisciplines) with little ongoing active professional engagement with those in other disciplines.

A renewed appreciation of soil organic matter — in all its astonishing complexity — began toward the end of the 20th century and continues to this day. Scientists have come to appreciate SOM's profound influence on almost all soil properties — such as structure (and hence on water infiltration and storage, susceptibility to surface runoff and erosion), cation exchange capacity, nutrient availability, buffering (pH, nutrient availability), color, and plant pest pressure. Additionally, new tools have become available to help explore SOM's chemical and physical properties as well as the diversity of soil life. With its critical role in so many soil properties and processes, organic matter can provide an integrating concept for understanding and promoting soil health and soil quality. Viewing healthy (high quality) soils as a major goal fosters a whole-system, preventive approach to agricultural management. In contrast to the industrial model, this approach aims to enhance soil properties so as to make the field ecosystem more self regulating, self sufficient, resistant to degradation, and resilient. Rather than look for *post*-facto "band-aid" solutions, the preventive approach emphasizes management to prevent chemical, biological (pest), and physical problems from developing in the first place.

In addition to an immense volume of scientific journal articles dealing with SOM, a number of books have been published over the last decade that deal with both scientific and practical aspects of SOM management. In this book, we have brought together scientific reviews concerning issues that are key for practical SOM management. We have included evaluations of the various types of organic constituents in the soil — the living organisms, the relatively fresh residues, and well decomposed substances (for which we have reserved the use of the term *humus*). The health (quality) of a soil is strongly and positively affected by soil organic matter (Weil and Magdoff, Chapter 1) and the various practices that enhance SOM (Magdoff and Weil, Chapter 2). There are chapters that focus on the effects of soil and crop management practices on soil organisms (Kennedy et al.,

Chapter 10) and organic matter gains and losses and the significance of various SOM fractions (Franzluebbers, Chapter 8; Wander, Chapter 3; Chen et al., Chapter 4). Chapters also discuss the contributions to soil quality and crop growth by fungi (Nichols and Wright, Chapter 6) and earthworms (Edwards and Arancon, Chapter 11).

Soil organic matter and its management to promote soil health goes hand-in-hand with the emergence of an ecologically based approach to soil and crop management that stresses prevention of imbalances leading to soil and crop problems down the road (Magdoff and Weil, Chapter 2). This approach is especially apparent with regard to the role of SOM and its management in the development of soils suppressive to plant diseases (Stone et al., Chapter 5) and the relation of SOM management to supplying nutrients to crops (Seiter and Horwath, Chapter 9). It also recognizes the potential effects arising from the interactions between belowground food webs (based on SOM and its management) and aboveground food webs that influence crop health and productivity (Phelan, Chapter 7).

We hope this book will provide essential scientific background and pose challenging questions relevant to both scientists and students as they strive to better understand SOM and develop improved soil and crop management systems.

**Fred Magdoff**
**Ray R. Weil**

## REFERENCE

Hills, J.L., C.H. Jones, and C. Cutler. 1908. Soil deterioration and soil humus. Section VIII, in *Vermont Agricultural Experiment Station Bulletin* 135, pp. 142–177. College of Agriculture, Burlington, VT.

# Editors

**Fred Magdoff, Ph.D.,** is a Professor of Soil Science in the Department of Plant and Soil Science at the University of Vermont, Burlington. Dr. Magdoff received his B.A. from Oberlin College and his M.S. and Ph.D. from Cornell University. He was the Plant and Soil Science Department Chair for 8 years at the University of Vermont, Burlington. He is a member of the National Small Farm Commission, and he is the coordinator for the 12-state Northeast region of the U.S. Department of Agriculture's Sustainable Agriculture Research and Education Program. He is a Fellow of the American Society of Agronomy.

Dr. Magdoff's area of specialty is soil fertility and management. He has worked on problems of sodic and saline soils, acid soils, use of manures and sewage sludges, phosphorus soil tests, and nutrient cycling. He developed the first reliable soil test for nitrogen availability to corn for the humid regions of the U.S. This test, called the presidedress nitrate test (PSNT) and the spring or late spring nitrate test, is now used throughout much of the Northeast, Mid-Atlantic, and Midwestern states, as well as eastern Canada. It has also been adopted for use with a number of vegetable crops.

Dr. Magdoff has oriented his outreach activities to explaining how to apply ecological principles to agricultural production. His book *Building Soils for Better Crops* (2000; co-authored with Harold van Es) describes an ecologically based approach that explains how to work with and enhance the inherent built-in strengths of plant-soil systems. He is also interested in political and economic issues surrounding agriculture. He is the senior editor of *Hungry for Profit: The Agribusiness Threat to Farmers, Food, and the Environment* (2000).

**Ray R. Weil, Ph.D.,** is a Professor of Soil Science in the Department of Natural Resource Sciences and Landscape Architecture at the University of Maryland, College Park. He earned his B.S. at Michigan State University, his M.S. at Purdue University, and his Ph.D. at Virginia Tech University. Dr. Weil began his career in the early 1970s as a farm manager for a 200-hectare organic farming training center. He has become a leader in researching and promoting sustainable agricultural systems in both industrial and developing countries. His research focuses on organic matter management for enhanced soil quality and nutrient cycling for water quality and sustainability. During the past decade much of his research has benefited from collaboration with dozens of local innovative farmers.

Dr. Weil also has an active university teaching program that includes five undergraduate and graduate courses. He is co-author of the widely used textbook *The Nature and Properties of Soils.* Dr. Weil is a Fellow of both the Soil Science Society of America and the American Society of Agronomy. He has twice been awarded a Fulbright Fellowship to support his work in developing countries.

Dr. Weil lives with his wife, Trish, in Hyattsville, Maryland, where he enjoys gardening, helping the town protect its urban forest, and commuting by bicycle to the University of Maryland in nearby College Park.

# Contributors

**Norman Q. Arancon**
Soil Ecology Laboratory
The Ohio State University
Columbus, Ohio

**Tsila Aviad**
Department of Soil and Water Sciences
Faculty of Agricultural, Food and
  Environmental Quality Services
Hebrew University of Jerusalem
Rehovot, Israel

**Yona Chen**
Department of Soil and Water Sciences
Faculty of Agricultural, Food and
  Environmental Quality Services
Hebrew University of Jerusalem
Rehovot, Israel

**Heather M. Darby**
Department of Plant and Soil Science
University of Vermont
Burlington, Vermont

**Maria De Nobili**
Universita Degli Studi de Udine
Departimento Di Produzione Vegetale,
  E. Techonologie Agrarie
Udine, Italy

**Clive A. Edwards**
Soil Ecology Laboratory
The Ohio State University
Columbus, Ohio

**Alan J. Franzluebbers**
USDA-Agricultural Research Service
Watkinsville, Georgia

**William R. Horwath**
Department of Land, Air and Water Research
University of California, Davis
Davis, California

**Ann C. Kennedy**
USDA-Agricultural Research Service
Washington State University
Pullman, Washington

**Fred Magdoff**
Department of Plant and Soil Science
University of Vermont
Burlington, Vermont

**Kristine A. Nichols**
USDA-Agricultural Research Service
North Great Plains Research Laboratory
Mandan, North Dakota

**P. Larry Phelan**
Department of Entomology
The Ohio State University
Wooster, Ohio

**Steven J. Scheuerell**
Department of Botany and Plant Pathology
Oregon State University
Corvallis, Oregon

**William F. Schillinger**
Department of Crop and Soil Sciences
Washington State University
Lind, Washington

**Stefan Seiter**
Department of Agricultural Sciences
Linn-Benton Community College
Albany, Oregon

**Alexandra G. Stone**
Department of Horticulture
Oregon State University
Corvallis, Oregon

**Tami L. Stubbs**
Department of Crop and Soil Sciences
Washington State University
Pullman, Washington

**Michelle Wander**
Department of Natural Resources
  and Environmental Sciences
University of Illinois
Urbana, Illinois

**Ray R. Weil**
Department of Natural Resource Sciences
  and Landscape Architecture
University of Maryland
College Park, Maryland

**Sara F. Wright**
USDA-Agricultural Research Service
Sustainable Agricultural Systems Laboratory
Beltsville, Maryland

# Contents

# 1 Significance of Soil Organic Matter to Soil Quality and Health

*Ray R. Weil and Fred Magdoff*

## CONTENTS

## SOIL QUALITY, SOIL HEALTH, AND ECOSYSTEM FUNCTIONS

Soil scientists have developed the concept of soil quality (SQ) to describe the fitness of soils to perform particular ecosystem functions (Karlen et al., 2001). SQ is usually defined as the capacity of the soil to carry out ecological functions that support terrestrial communities (including agroecosystems and humans), resist erosion, and reduce negative impacts on associated air and water resources (Karlen et al., 1997). The soil quality concept addresses the associations among soil management practices, observable soil characteristics, soil processes, and the performance of soil ecosystem functions (Lewandowski et al., 1999). Soil quality can be influenced by many properties inherent to a particular soil and reflective of the environmental factors affecting long-term soil formation, such as steepness of slope, depth of solum, and soil texture. Soil quality can also reflect the condition of a soil resulting from the alteration of certain soil properties by management practices. Aggregate stability, infiltration capacity, nitrogen mineralization potential, and biodiversity of soil organisms are among the soil properties that are substantially altered by management. The term *soil health* is often used to describe those aspects of soil quality that reflect the condition of the soil as expressed by management-sensitive properties (Islam and Weil, 2000). The health of a soil refers not only to its lack of degradation or contamination, but also to its overall fitness for carrying out ecosystem functions and responding to environmental stresses (Lewandowski et al., 1999).

The many ecosystem processes mediated by soils can be grouped into five fundamental, though somewhat overlapping, functions (Brady and Weil, 2002): (1) promotion of plant growth; (2) biogeochemical cycling of elements (especially carbon and mineral nutrients elements); (3) provision of habitat for soil organisms; (4) partitioning, storage, translocation, and decontamination of water; and (5) support and protection of human structures and artifacts. Although soil organic matter (SOM) can be a negative factor with regard to the fitness of soil as a medium for constructing roads and buildings, it is usually a major positive factor in determining a soil's capacity to perform most of the other functions listed.

Specific soil functions important in promoting plant growth include (1) provision of mineral nutrients available to plant roots in time, space, and form; (2) retention of water in sufficient quantities and with appropriate potential energy to be available for root uptake on an almost continuous basis; (3) provision of a network of interconnected pores sufficient to provide pathways of low physical resistance to root growth and meet plant root needs by supplying oxygen and removing carbon dioxide and toxic gases; (4) support of plant growth-promoting soil organisms; and (5) provision of sufficient rooting depth and physical support for optimal plant growth. These functions, with the possible exception of the last, are greatly influenced by the increase in organic matter content of soils. High levels of SOM are associated with reduced erosion and runoff; enhanced soil aggregation and nutrient cycling; and improved infiltration, movement, and retention of water (Greenland and Szabolcs, 1994; Woomer and Swift, 1994). Areas that need more research to elucidate the relationship between SOM and soil quality are (1) the role of organic compounds as chelating agents that control the availability and toxicity of micronutrients to plants and microorganisms; (2) the role of soluble or easily oxidizable C as fuel for microbial biomass; and (3) the passage of SOM and its chemical energy through the trophic levels of the soil food web that cycles C and nutrients.

## SOIL QUALITY INDICATORS, PERCEPTIONS, AND INDICES

Many soil ecosystem functions are difficult to measure directly; therefore, SQ must often be inferred from easily measurable soil properties, the soil quality indicators (Acton and Padbury, 1993). Because of the multiple and complex functions associated with SQ, its assessment necessitates the integration of chemical, physical, and biological soil properties. Of particular interest are properties that can serve as early and sensitive indicators of ecosystem stress or changes in soil productivity. Given the pervasive role of organic matter in promoting soil ecosystem functions, it is not surprising that researchers have found SOM-related properties to be important indicators of SQ (Larson and Pierce, 1991; Arshad and Coen, 1992; Gregorich et al., 1994; Kennedy and Papendick, 1995; Wander and Bollero, 1999; Islam and Weil, 2000). Of the 13 SQ indicator properties suggested in 14 recent papers on SQ indices, the two most commonly included are pH (13 of 14) and SOM (12 of 14; Popp et al., 2002). Dumansky (1994) concluded that "soil organic matter is emerging as a key indicator for assessing sustainability" of land management systems.

Organic matter content and related soil properties can also largely account for farmers' perception of SQ. For example, in the mid-Atlantic region of the U.S., in a study of pairs of similar soils with similar pedological characteristics but different management histories, 88% of the 32 farmers interviewed cited SOM as a soil property indicative of relative SQ (Gruver and Weil, 1997). Organic-matter-related soil properties, such as total organic carbon, microbial biomass C, extractable carbohydrates, and macroaggregate stability, exhibited significantly higher values in the soils perceived by the farmers as being in better condition or in a state of higher SQ. Similarly, in the highlands of Kenya (Murage et al., 2000), 12 smallholder farmers identified paired fields as either productive or nonproductive. Total organic C, various labile fractions of organic C, and microbial biomass C were significantly higher in the soils perceived to be more productive. In the Kenyan setting, the lower-quality unproductive soils were also more acidic and lower in available plant nutrients, because the resource-limited farmers had applied inadequate levels of fertilizers, organic amendments, and crops residues. In the U.S., application of mineral fertilizer and limestone is so routine that researchers studying farmer perceptions of SQ (Liebig and Doran, 1999; Gruver and Weil, 1997) have found that farmer-identified high- and low-quality soils differ little with regard to levels of pH or available nutrients, leaving SOM-related properties as the main differentiating factors. Traditional indigenous knowledge of subsistence farmers is also often a factor in perceived SOM content when classifying the fertility or potential productivity of different soils in their local landscapes (Corbeels et al., 2000).

To assess overall SQ from an agricultural viewpoint, researchers have attempted to integrate numerous SQ indicator properties into a single index by which SQ under various management systems can be compared (Karlen et al., 1997). Glover et al. (2000) tested a systematic method for rating SQ in a study of apple orchards in Washington managed by conventional, organic, and integrated approaches. Their system focused on the capacity of soils to perform four specific ecosystem functions: accommodate water entry, accommodate water movement and availability, resist surface structure degradation, and sustain fruit production and quality. Each of these functions was given an equal (0.25) weightage and was assessed by averaging scores given for two levels of indicator properties. The two properties most closely related to SOM [soil organic content (SOC) and aggregate stability] together accounted for 41% of the total SQ scores. However, despite progress in developing SQ indices useful for assessing agricultural management practices, Sojka and Upchurch (1999) warn that it will probably not be possible to develop a single SQ index that applies to all five basic soil functions.

## NATURE AND COMPOSITION OF SOM

SOM is arguably the most complex and least understood component of soils. The content of organic matter in soils ranges from ca. 2 g/kg in some desert soils to more that 800 g/kg in some Histosols; however, cultivated mineral soils usually contain 10 to 40 g/kg in the A horizon. The major portion of SOM usually consists of a conglomeration of relatively recalcitrant organic molecules termed *humus* (see Chapter 4). Although the chemistry of humus is under active investigation and several phenolic-rich models of humus have been proposed (Schnitzer, 1995; Stevenson, 1994), the chemical nature of this large portion of SOM is highly variable and not yet well understood. However, advances in instrumentation in the 1980s and 1990s (Hatcher et al., 2001) have allowed identification and quantification of the major types of chemical groups present in a complex organic mixture. Solid-state CPMAAS $^{13}$C-NMR is at present routinely applied to quantify the alkyl, $N$-alkyl, $O$-alkyl, acetal, aromatic, phenolic, and carboxylic groups in various fractions of SOM (Oades, 1995; Guant et al., 2001).

A considerable advance in the chemical characterization of the recalcitrant portion of SOM came with the recent discovery of glomalin, a large (~90 kDa) iron-containing glycoprotein molecule synthesized by the hyphae of arbuscular mycorrhizal fungi (see Chapter 6). Glomalin, which appears to contain only 10 to 35% C, resists acids and enzymes and therefore accumulates to rather high levels in many soils (2 to 35 g/kg). Apparently, large amounts of glomalin are present in SOM fractions defined operationally as particulate organic matter, humic acid, fulvic acid, and humin.

Other works have suggested that the charred plant material resulting from incomplete burning might constitute an important part of what is often considered humus. Ponomarenko and Anderson (2001) combined high-energy UV photooxidation with micromorphology and electron microscopy to define and estimate the char portion of SOM in the A horizons of black prairie soils. Char was found to be highly recalcitrant, high in aromatic groups, and a constituent of all particle sizes, from sand to clay size.

Litter and detritus, the dead, decaying remains of plants and animals, comprise a significant proportion of the organic matter in (or on) soils. The chemistry of the litter and detritus is usually closely related to that of the original tissues, the greatest portion being derived from lignin and polysaccharides. Kögel-Knabner (2002) has reviewed the structural chemistry of litter and some of the alterations that occur as it decomposes. Finally, a small but biologically active portion of SOM consists of easily oxidized, often relatively soluble compounds derived from litter, such as sugars and amino acids, as well as a wide array of biochemicals synthesized by microorganisms or contributed by plant roots. The chemistry of this microbially available bioactive fraction is complex and poorly understood. In addition to dead remains, soils contain myriads of living organisms. The microorganisms, constituting the soil microbial biomass, are especially significant.

Because SOM is nonhomogenous and not well defined chemically, it cannot be measured directly as a specific substance. Instead, methods used routinely for its analysis generally measure some proxy for SOM, such as total organic C, total C, or weight loss on ignition. Unfortunately, the methods used to determine SOM vary from lab to lab. Each method in common use has its own limitations and inaccuracies, and the estimate of SOM obtained by one method can differ significantly from that obtained by one of the alternative methods. Making comparisons between SOM values obtained by different soil test labs or by different researchers can be fraught with ambiguity. However, most soil test labs routinely analyze for SOM, and therefore information on SOM content is widely available and used for many purposes. Therefore, it is important that users appreciate the differences among the methods in common use.

A common method for estimating SOM is the determination of total organic C by wet oxidation with chromate in strong acid (Walkley and Black, 1947). When SOM is reported by this method, the value determined for organic C must be multiplied by a standard factor designed to account for the method's incomplete oxidation of SOC (traditionally assumed to be 77%; however, for a range of tropical Alfisols, Inceptisols, and Vertisols, Olayinka et al. (1998) found the value to be close to 100%) and then by another factor based on the questionable assumption that the SOM always contains 58% C. Although most labs use the same standard factors for these purposes, the factors are certainly not equally appropriate for all soils.

A second commonly used method estimates total SOM from the weight lost when a dry soil sample is ignited at a temperature high enough to volatilize all organic material. Various loss on ignition (LOI) methods use different durations and temperatures for the ignition, but usually give values of SOM ca. 25% higher than those by the Walkley–Black method. They might greatly overestimate SOM if free carbonates are present and high temperatures used. Comparison of SOM values obtained from different labs is made more uncertain because some labs report raw weight LOI values, whereas others report values corrected to more closely match wet oxidation methods of SOC analysis by empirical regression equations (Magdoff, 1996). Nonetheless, Gajda et al. (2001) found LOI to be a useful measure for monitoring changes in the size of the SOM pool of soils under different management systems in the U.S. Great Plains.

Charcoal can be an important component of soil carbon in fire-dependent ecosystems and where farmers practice slash-and-burn land clearing or burn crop residues after harvest. This relatively inert material can account for 0 to 50% of the total C in soils and is generally uncorrelated with SOM contents (Lal et al., 2001). Charcoal is largely excluded from SOM estimated by wet digestion, but is largely included by most LOI methods.

# LEVELS OF SOC ACCUMULATION IN RELATION TO GLOBAL GREENHOUSE EFFECTS

During the fertilizer heydays of the 1950s, 1960s, and 1970s, when it appeared to most agriculturalists that the solution to almost all soil problems could be purchased in a bag of mineral salts, research on SOM languished on the scientific back burner. The past two decades have seen a renaissance in SOM research and a renewed emphasis on increasing the levels of organic matter in agricultural soils (Weil, 2001). In the more industrialized nations, enhanced SOM levels are now seen as a primary tool for overcoming soil problems other than simple nutrient supply. Alleviation of soil compaction, improved soil workability, efficient water use, suppression of plant pests and diseases, resistance to soil erosion by wind and water, and resistance to surface crusting are issues at least partially addressed by SOM management, even when fertilizers can maintain adequate supplies of nutrients. In less industrialized countries where most farmers still have little access to commercial fertilizers, SOM management is increasingly seen as an essential part of affordable, practical solutions to many nutrient supply problems.

At the same time, concerns were emerging about global climate change driven by increasing atmospheric content of so-called greenhouse gases, especially $CO_2$. In the Kyoto Protocol, most of the international community agreed to aim at reducing total greenhouse gas emissions by 2008 to 2012 to levels 5.3% below those of 1990. Critics (including the U.S. government) have charged that such reductions will place devastating economic burdens on industrial nations because they are forced to reduce their energy consumption and switch to more expensive energy technologies. However, as recently shown by DeLeo et al. (2001), the costs of the proposed emission reductions will be largely offset by lowered costs related to the environmental, human health, material, and agricultural damages caused by unchecked greenhouse gas emissions. Also contributing to the economic feasibility of meeting the proposed emission reduction targets is the role that soil management might play. Managing soils to increase the stocks of C stored as SOM can be expected to reduce the rate of increase in atmospheric $CO_2$ and to improve SQ, thus resulting in additional benefits beyond greenhouse gas reductions.

By the 1990s, an appreciation of the central role of soils in the global C cycle had engendered new interest in increasing the levels of SOM. Soil plays a major role in the Earth's system of self-regulation that has created and maintains the environmental conditions necessary for the survival of plants and animals on this planet. SOM represents by far the largest pool of C actively turning over in the global C cycle. Reicosky et al. (2000) and Lal et al. (2001) reviewed the literature and summarized current estimates for the sizes of global C pools, including 720–750 Gt in the atmosphere and 550–835 Gt in the plant biomass. In contrast, SOM is estimated to contain 1200 to 2200 Gt of C. Soils in dry regions can contain an additional 700 to 946 Gt of C as carbonates in petrocalcic horizons.

As the level of $CO_2$ rises in the atmosphere, many plants, including crops, respond with increased net productivity, some of which is shunted to the soil via increased root growth and activity (Islam et al., 1999) and results in more aboveground residues as well. The resulting sequestration of C in SOM might counteract, to some degree, the rise in C stored in the atmosphere as $CO_2$. This interplay between soil C and atmospheric $CO_2$ is one possible feedback control on global warming.

The amount of organic matter in a soil is largely a function of environmental and inherent soil factors that humans cannot control. However, human management of soils, including the decision to convert land in natural vegetation to agricultural use, can produce significant alterations in soil SOM levels. Most of the increase in atmospheric $CO_2$ during the past 150 years was caused by a combination of fossil fuel burning and reductions in SOM C stocks. As discussed later, the primary cause for the decline in global SOM during this period was the widespread conversion of natural forests and grasslands to agricultural cultivation (Wood et al., 2000).

Assuming 750 Gt of C ($750 \times 10^{12}$ kg) in atmospheric $CO_2$, there is ca. 1.5 kg C as atmospheric $CO_2$ per square meter of the Earth's surface. This is equivalent to the C content in the upper 15 cm of a meter square of a soil with a SOM content of ca. 13 mg SOM/kg. According to the calculations of Wood et al. (2000), the pool of SOC in agricultural soils averages ca. 10.2 kg $C/m^2$ to a depth of 1 m, or close to seven times the content of the atmosphere above a field. They report that soils in North America, Asia, and Europe are considerably richer in SOC (12.2, 12.6, and 14.6 kg $C/m^2$, respectively) than in Sub-Saharan Africa (7.7 kg $C/m^2$). Globally, agricultural soils account for only under one fourth of the SOM C pool (Wood et al., 2000); however, these soils represent the portion of the SOM pool that is most readily amenable to management. Therefore, agricultural soil management has to be considered as a serious option for helping balance anthropogenic greenhouse gas emissions. Batjes (1999) estimates that on average, 0.58 to 0.80 Pg C/year can be sequestered if all agricultural soils are restored to near their native SOM levels, and that this amount of sequestration will be equivalent to 9 to 12% of the annual anthropogenic $CO_2$-C production.

Table 1.1 illustrates the magnitude of $CO_2$ that can be released or sequestered with changes in land use (Arrouays et al., 2001), using France as an example. Mean C levels in the upper 30 cm of soil ranged from 1.5 kg $C/m^2$ in Lithosols under vineyards to 34.7 kg $C/m^2$ for Histosols under moor vegetation (extremes not shown in Table 1.1). For a given soil order, soils with the smallest C pools were in permanent crops such as orchards and vineyards, which in France usually have a long history of clean cultivation without cover crops. Forests, pastures and natural grasslands were nearly always much higher in SOC than was arable land. For all of France, the SOC stocks in the upper 30 cm were estimated as 3.1 Pg, ca. 0.44% the global total SOC estimated by Batjes (1996) for this depth. Annual C emissions in France are estimated to be 100 Tg, of which 4.5 come from agriculture. On this basis, Arrouays et al. (2001) calculate that the relative increase of SOC of 0.13%/year will entirely mitigate French agricultural C emissions. Of course, this rate of SOC increase is difficult to achieve and cannot continue indefinitely by using the same practices.

## INFLUENCES ON SOM ACCUMULATION

### ENVIRONMENTAL INFLUENCES

Geography exercises a major influence over the levels of organic matter found in soils. Although vegetation and parent material make important contributions to geographic SOM variations in some cases, the principal geographic factors involved are climatic, namely temperature and precipitation.

**TABLE 1.1**
**Mean C stocks in the Upper 30 cm of Soils in France for Selected Land Use — Soil Order Combinations**

| FAO Soil Order[a] | Arable Land (kg C/m²) | Orchards and Vineyards (kg C/m²) | Forested Land (kg C/m²) | Natural Grasslands (kg C/m²) |
|---|---|---|---|---|
| Cambisols (Inceptisols) | 5.1 | 3.0 | 7.2 | 7.7 |
| Fluvisols (alluvial Entisols/Inceptisols) | 5.1 | 2.7 | 10.2 | 8.0 |
| Luvisols (Alfisols) | 4.4 | 2.9 | 8.4 | 6.5 |
| Histosols | 17.0 | NA | 20.5 | 26.7 |
| Andosols | 8.2 | NA | 9.5 | 9.3 |
| Podzols (Spodosols) | 4.7 | NA | 7.3 | 8.0 |

[a] Equivalent U.S. soil taxonomy orders in parentheses.
*Source*: Data from Arrouays, D. et al. 2001. *Soil Use Manage.* 17:7–11.

Organic matter accumulation in soils is maximum when the difference between annual plant productivity and annual decomposition is highest. Low mean annual temperatures tend to slow decomposition much more than productivity, and therefore the highest SOM levels are generally found at extreme latitudes short of the permafrost regions near the poles. For this reason, Spodosols and other types of soils under boreal forests contain very high levels of organic matter in the surface layer. In the tropics, elevation rather than latitude tends to influence mean annual temperatures, and therefore soils of highlands and mountains (such as humic suborders or great groups of Ultisols and Inceptisols) tend to contain the highest C stocks.

The influence of precipitation can also be seen in both temperate and tropical zones. Within limits, more rainfall tends to increase plant growth more than it does decomposition; therefore, SOM tends to be positively correlated with annual precipitation. Under very wet conditions usually associated with a shallow water table, decomposition can become limited by oxygen availability. Under anaerobic conditions, decomposition is both slow and incomplete, even with warm temperatures. Histosols, which occur mainly in wetlands, account for more than 20% of global SOC stocks even though they cover less than 2% of the land surface.

The combined influences of temperature and precipitation are dramatically illustrated in the maps developed by Kern (1994) showing the geography of SOM in U.S. surface soils. Jenny (1941) earlier observed the changes in SOC levels along climosequences in the North American Great Plains region, such as the increasing SOC levels found as one travels north to cooler climates from Texas to Minnesota, or the increases seen in going east to wetter climates from Colorado to Iowa. The same factors, especially low temperatures associated with high elevations, also partially account for the much higher SOC content found in Ethiopian (20.2–47.3 g/kg) than Nigerian (12.0–24.0 g/kg) forested soils (Spaccini et al., 2001). Recent research on geographic C distribution in soils should dispel the myth that soils in tropical regions are low in SOC. The range of SOC contents found within the tropics is far greater than the range between temperate and tropical sites. It has been observed (Theng et al., 1989; Weil, 1984) that SOC levels soils are highest (>30 g/kg) where the ratio of mean annual temperature (in °C) to annual precipitation × 0.01 (in mm) is less than 1; as this ratio reaches 3 or more, SOC declines to very low levels.

Elevation is only one example of how topography affects SOM accumulation. Slope aspect, steepness, and landscape position are also important factors, often accounting for large differences in SOM accumulation within small distances. Wetland Histosols are the most extreme example of the influence of topography on SOM, as these soils occur mainly in the lowest landscape positions where accumulation of runoff water and interaction with groundwater (or tides in coastal areas) lead to extreme accumulations of 150 to 500 g C/kg in surface layers that might be meters thick.

More subtle differences occur between soils that are similar in slope steepness and landscape position, but differ in slope aspect. Because slopes facing away from the equator (north slopes in the Northern Hemisphere) receive less solar energy to evaporate water, they experience a wetter effective climate and are therefore commonly covered by soils that are wetter, deeper, and richer in SOC than their counterparts on the other side of the hill or mountain. The effect is less pronounced where slopes are less steep, but can still be significant over large areas. For example, Brejda et al. (2001) observed that aspect had a significant influence on SOC in the 0- to 10-cm layer of soils sampled in the Central High Plains (Figure 1.1), but not in the Northern Mississippi Valley Loess Hills or in the Palouse and Nez Perce Prairies. Other studies (e.g., Pikul and Aase, 1994) have documented major translocations of SOC from shoulder slopes to footslopes by erosion and sedimentation of cultivated land.

In addition to these environmental factors, SOC levels are also markedly influenced by inherent properties of the soils themselves. Soil texture is particularly important in this regard. If environmental factors are similar, finer-textured soils tend to accumulate higher amounts of organic C (Figure 1.2). Nichols (1984) found that among soils in the southern Great Plains region of the U.S., SOM levels were more closely related to soil clay content than to annual rainfall. There are several reasons for this relationship. High clay and silt contents provide the capacity to hold more water

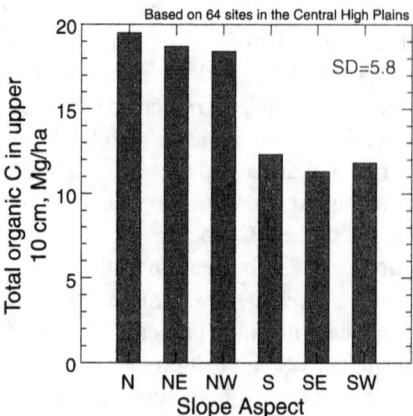

**FIGURE 1.1** Total soil organic C content of surface soils in the U.S. Central High Plains region as influenced by slope aspect. A significant effect of slope aspect was found in some regions but not in others. (From Brejda, J.J. et al. 2001. *Soil Sci. Soc. Am. J.* 65:842–849. With permission.)

and nutrients for more plant biomass production while inhibiting the free circulation of air that would stimulate rapid aerobic decomposition. The fine soil particles also provide large amounts of mineral surfaces that bind chemically with organic compounds and come together to form aggregates that physically protect organic matter from microbial attack (Oades, 1995). Coarse-textured soils, in contrast, have much less surface area to bind organic matter, produce less plant biomass, and annually lose a higher percentage of their SOM content to microbial respiration. Therefore, it is probably impractical to attempt to raise SOM in sandy soils to the levels that typify well-managed soils of silt loam or clay loam textures.

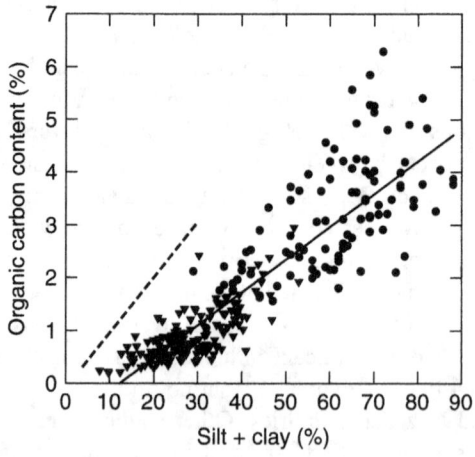

**FIGURE 1.2** Soils high in silt + clay tend to contain high levels of total soil organic C. The data shown are for surface soils in well drained to moderately well drained, tilled maize fields in subhumid regions of Malawi (triangles) and Honduras (circles). Sites varied in such factors as years under tillage, elevation, and temperature. The thick line represents linear regression, $N = 279$. The dashed line represents adequate SOM level for stable structure according to Pieri, C.J.M.G. 1992. *Fertility of Soils: A Future for Farming in the West African Savannah.* Springer-Verlag, Berlin. (Data from Brady, N.C., and R.R. Weil. 2002. *The Nature and Properties of Soils*, 13th ed. Prentice-Hall, Upper Saddle River, NJ. With permission.)

## SOIL MANAGEMENT PRACTICES THAT INFLUENCE THE BALANCE BETWEEN C GAINS AND LOSSES

Organic matter accumulates in soils to the degree that C inputs exceed C outputs, or as stated in Chapter 2, "gains exceed losses." The previous section describes the influence of environmental factors, but human management practices also exert considerable influence on this balance. Except where dry desert soils are brought under irrigated crop production, the shift in land use from natural vegetation to agriculture generally results in a marked decline in SOM levels. Compared with natural vegetation, agricultural systems nearly always cause less C to be added to the soil and more to be lost. In their study of adjacent forested and cultivated soils in eight agroecosystems from Ethiopian highlands and Nigerian lowlands, Spaccini et al. (2001) observed that "in all agroecosystems, SOC content was two to four times higher in the forested than the cultivated soils." The lower input of C is usually not due to reduced net primary productivity under agriculture, but rather to human appropriation and removal of much of the aboveground plant biomass produced. In addition, agricultural practices, especially intensive tillage, accelerate the loss of C from the soil by microbial respiration and erosion. Globally, ca. 1.7 billion ha of cropland is estimated to currently contain a stock of approximately 170 Pg of C. Past losses of SOM from this cropland are estimated to have contributed ca. 50 Pg of C to the atmospheric $CO_2$ pool (Paustian et al., 2000). Largely because of the loss of SOM, ca. 22% of global agricultural land is considered to be chemically degraded (Wood et al., 2000). This past loss of SOM represents a potential for improved soil management to mitigate a finite but significant portion of anthropogenic $CO_2$ emissions.

Allowing some cropland to return to natural forest or grassland vegetation is one way by which the farmers can increase the stock of C in SOM and mitigate some $CO_2$ emissions. In the humid lowland tropics, restoration of SOM levels in soils can be a relatively slow process. About 20 to 30% of the SOM stock is generally lost during only 2 years of cropping, but the restoration of this lost SOM by fallowing in natural vegetation is likely to take up to 35 years (Syers and Craswell, 1995). The Conservation Reserve Program in the U.S., which financially compensates farmers for the opportunity costs incurred by allowing a portion of their land to go out of harvestable production, has succeeded in sequestering significant amounts of C in the SOM of former cropland in the Central and Western regions of the U.S. (Gewin et al., 1999; Karlen et al., 1999). However, the gains can be quickly reversed if the land is once again cultivated, especially if by conventional tillage.

If farmers are to be compensated for sequestering C as a measure for mitigating $CO_2$ emissions, there will likely be a need for computer models that can predict, within an appropriate range of conditions, the amount of C sequestered because of improved management practices. Molina and Smith (1998) reviewed 24 such models and concluded that modelers still had to do much work to refine their models and even the best models needed local calibration to be effective in predicting long-term changes in SOC. Rickman et al. (2001) recently developed a computer-based model that predicts tillage and crop rotation effects on SOM trends by using data on soil properties and crop biomass production that are readily available from such sources as the C-factor files in the widely used RUSLE (Revised Universal Soil Loss Equation) program. They calibrated their model by using 60 years of SOM observations at Pendleton, OR, and then successfully validated the model at two sites in Kentucky and Wisconsin. The Natural Resources Conservation Service of the United States Department of Agriculture (USDA) has developed a soil conditioning index (USDA-NRCS, 2001) that attempts to predict long-term trends in soil organic C concentration by using information on climate (mean annual temperature and precipitation), erosion (as predicted by the RUSLE), and mean annual organic matter additions (mainly plant residues associated with various cropping systems). This index has had better success in predicting SOC losses and gains in long-term experiments in humid regions than in drier regions of the U.S. (Hubbs et al., 2002).

# SOM INFLUENCES ON SQ INDICATOR PROPERTIES AND FUNCTIONS

Most arable soils contain only 2 to 4% organic matter by weight, yet very little about these soils is not significantly influenced by the organic matter in them. Organic matter provides much of the soil's capacity to store nutrients and water. It plays a critical role in the formation and stabilization of soil structure, which in turn produces good tilth and drainage and resistance to erosion. It can not only carry and make available nitrogen, sulfur, and phosphorus, but also improve the availability of nearly all nutrients, whether applied as fertilizer or weathered from minerals. It promotes the health of the soil ecosystem and stimulates organisms that cycle C and protect plants from disease. Humic fractions of SOM can directly influence plants through enhanced nutrient uptake (see Chapter 4), effects on intermediary metabolism, and, according to some authors, possibly some hormone-like reactions (Nardi et al., 2002). However, knowledge of the mechanisms involved is quite limited and the importance of these direct effects in field soils is still quite uncertain. This chapter therefore focuses on impacts of SOM on the physical, chemical, and biological properties of soil, many of which indirectly influence plant growth.

## PHYSICAL PROPERTIES INFLUENCED BY SOM

To a remarkable degree, increased organic matter can counteract the ill effects of too much clay or too much sand. Increasing the SOM content usually increases total porosity and therefore decreases bulk density. Within a limited range of SOM contents, the relationship for a given soil is nearly linear (Weil and Kroontje, 1979). However, across a wider range of SOM, the relationship between these two variables is likely to be curvilinear (Figure 1.3), because at very high levels of SOM, additional OM has little further effect on soil aggregation and influences bulk density mainly because of its low particle density (Franzluebbers et al., 2001).

*Friability* is a term that describes the tendency of soil clods to easily crumble into their constituent natural aggregates. Highly friable soils are easy to cultivate or to sow with a no-till seeder. Soil management can have pronounced effects on friability. Among surface soils in the southern wheat belt of New South Wales, Australia, uncultivated woodland soils are most friable (mean friability index $k = 0.66$) and intensively tilled soils are the least friable (mean $k = 0.05$; Macks et al., 1996). Soils under direct drilling (no till) have friability similar to that from woodland sites (mean $k = 0.55$). Friability is significantly related to organic carbon and a number of soil structural properties, including aggregate stability and bulk density.

For many soil functions, the voids between solid particles are as important as the particles themselves. Relatively large biopores (channels from earthworms and decayed roots) or interaggregate pores are especially important for enabling root growth, free water drainage, and aeration.

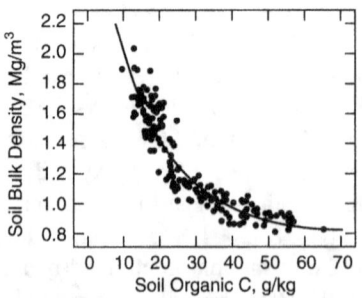

FIGURE 1.3 Relationship between bulk density and organic C in the upper 2 cm of a sandy loam soil (Hapludults) under various Bermudagrass management practices during a 5-year period in Georgia. (From Franzluebbers, A.J. et al. 2001. *Soil Sci. Soc. Am. J.* 65:8334–8341. With permission.)

The configurations of smaller pores strongly influence water retention and microbial activity in soils, although much more is known about pore interactions with the former than with the latter (Juma, 1993).

## Soil Aggregation

The aggregation of soil mineral particles into granular structure and the stabilization of these granules against slaking by water are sensitive indicators of SQ (McQuaid and Olson, 1998). The degree to which a soil is organized into water-stable aggregates influences many soil ecosystem functions, including the accumulation of SOM (Tisdall and Oades, 1982). Cropping systems that promote SOM accumulation usually also promote soil aggregation (Figure 1.4). The soil properties that contribute to the formation and stabilization of macroaggregates include soil texture, clay mineralogy, exchangeable cations, Fe and Al oxides, and $CaCO_3$, as well as SOM (Le Bissonnais, 1996). In three sites from semiarid to subhumid Spain, microaggregate stability positively correlated with clay content, whereas macroaggregates stability correlated mainly with SOM. However, below a threshold SOM content, aggregate stability was more closely associated with the soil carbonate content (Boix-Fayos et al., 2001).

A portion of the organic matter in soils is sorbed on the surfaces of the fine mineral particles (clay and silt), where it is protected from further microbial decomposition. Once the surface capacity of the silt and clay to bind organic matter is exceeded, additional organic matter accumulating in the soil is more likely to be involved with aggregate stabilization. Therefore, there is usually only a weak relationship between total SOM and aggregate stability over a wide range of different soils. However, the relationship between aggregation and SOM can be considerably stronger if SOM content is expressed as a proportion of the fine soil particles. (See Figure 1.5 for soils containing 11 to 93% sand.) This concept is of special importance in determining the level of SOM that is adequate for structural stabilization in coarse-textured soils in the tropics. Based on 495 West African soil samples (67 to 97% sand) for which structural condition was recorded, Pieri (1992) suggested that if the mass of SOM exceeds 9% of the mass of the fine mineral fraction (silt + clay) the SOM content is likely to be sufficient to assure structural stabilization and protect the soil from

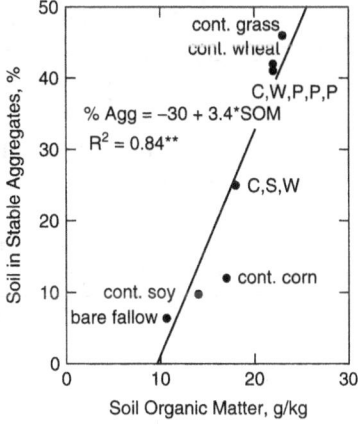

**FIGURE 1.4** Relationship between water-stable soil aggregates and level of soil organic matter after 20 years of various crop rotations. The aggregate stability was measured in the second year of conventional tillage corn production following the 20-year rotation period on a Beltsville silt loam (Typic Fragiudults) in Maryland. C: corn; P: perennial hay; W: wheat; S: soybean. (Plotted from data in Strickling, E. 1975. In *North East Branch American Society of Agronomy Abstracts*. Agronomy Society of America, Madison, WI, pp. 20–29.)

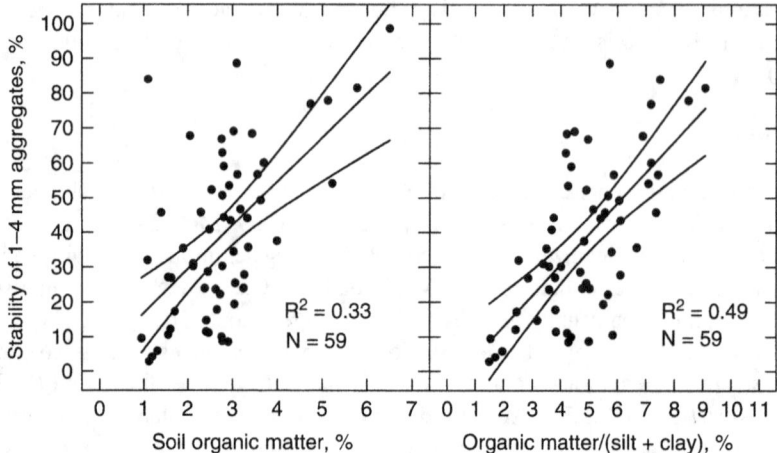

**FIGURE 1.5** Soil organic matter as a percentage of the silt + clay is often more closely related to structural stability than is the level of SOM as a percentage of the whole soil mass. Data are spring samples from 59 farm fields that had not had a recent sod crop. (From unpublished data of J. Gruver and R. Weil, University of Maryland. With permission.)

physical degradation. If this critical value is less than 5%, serious degradation is expected in these soils. If we risk extrapolating this relationship beyond the range of Pieri's dataset, the implication is that a soil with 50% (silt + clay) should have >4.5% SOM and a soil with 80% (silt + clay) should have >7.2% SOM for organic matter to have its maximum effect on soil structure.

Soil aggregates vary in size from microaggregates (0.02 to 0.25 mm in diameter) to macroaggregates (up to 10 mm in diameter), with the larger aggregates probably comprising clusters of smaller aggregates. The formation and stabilization of macroaggregates (>0.25 mm) seem to be especially dependent on SOM, except in certain highly weathered soils in which Fe and Al oxides provide the main agent that binds particles together into aggregates (Six et al., 2000a). The larger aggregates are especially important because they physically protect particulate organic matter from decomposition and provide large interaggregate pores for water, air, and root movement into and through surface soils. In moderately weathered (2:1 clay-dominated) soils, organic materials help bind microaggregates together into macroaggregates. Thus, the interaction between soil aggregates and organic mater is two way: organic matter stabilizes soil aggregates and soil aggregates stabilize SOM (Six et al., 2000b).

Although it is clear that higher SOM levels are associated with high levels of aggregate stability in most soils, increased SOM level is not the only mechanism by which aggregation is enhanced by such agricultural practices as the growing of pasture grasses and legumes. In eight pairs of forested and cultivated soils in Africa, cultivation consistently reduced the size and stability of aggregates and reduced levels of total SOM and carbohydrates (Spaccini et al., 2001). However, these measures of SOM did not account for most of the variability in soil aggregation in the study.

The type as well as quantity of organic material applied to soil can influence the stabilization of soil aggregates. For example, 1% dry weight of composted dairy manure, small grain straw, grass clippings, and fresh dairy manure were added to three different soils — loamy sand, silt loam, and clay. Compost had no significant effect whereas grass clippings and dairy manure had a significant positive effect on water-stable aggregates in the range of 0.25- to 4.0-mm diameter (Figure 1.6). Almost all the increase in water-stable aggregates occurred in the 2- to 4-mm size range. These results are consistent with the hypothesis that the active fractions of SOM derived from fresh, easily decomposable materials help bind smaller aggregates into larger ones, partly by supporting increased fungal growth. Anecdotal reports that compost increases aggregation after

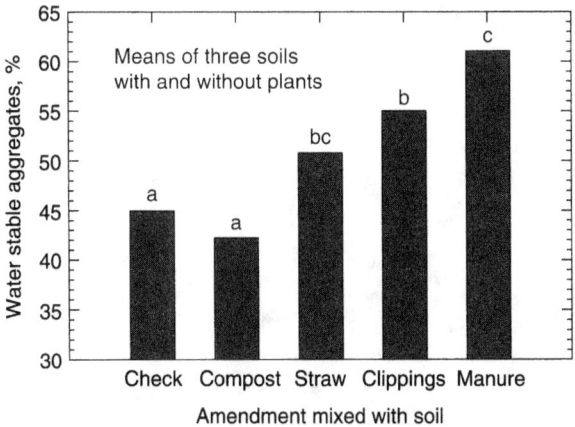

**FIGURE 1.6** Water-stable aggregates (0.25 to 4 mm) as percentage of whole soil mass following incubation with dairy manure compost, small grain straw, grass clippings, or fresh dairy manure. The values are means for three soils (sand, loam and clay) incubated with and without growing plants (the presence of growing plants had no significant effect). (From unpublished data of F. Magdoff. With permission.)

years of application might be a result of the stimulation of soil organisms by compost application and the passage of compost-containing soil through the digestive systems of soil animals such as earthworms (see Chapter 11). However, the relatively nonlabile organic compounds in compost do not appear have the same short-term effect on water-stable aggregation as do the more labile compounds resulting from fresh plant or animal materials.

Cover crops grown in the winter off-season in temperate regions are well known to improved soil aggregation. The improvement in aggregation is often related to increases in SOM but differs among cover crop species grown, regardless of effects on total SOM. Soil aggregation varies seasonally, but cover crops or mulches can prevent most of the decreased aggregate stability observed on bare soils from fall to spring (Hermawan and Bomke, 1997).

Figure 1.7 (Chan et al. 2001) also suggests that SOM plays a greater role in aggregation for some soils than for others. The figure illustrates linear relationships between SOM and aggregate stability in two soils dominated by 2:1 clays and a lack of relationship in a more weathered soil in which a high level of aggregate stability is likely because of binding by 1:1 clays and Fe oxides. The data also suggest that some types of vegetation are more effective than others at promoting aggregation, irrespective of their relative effects on SOM levels. Five pasture treatments used for 3 to 4 years for restoration of SOM and structure to these three degraded Alfisols were compared. The combination of lucerne and lovegrass resulted in levels of aggregate stability much higher than what would be predicted from the SOM–aggregation relationship derived from the other treatments (Figure 1.7).

A similar conclusion was drawn by Haynes and Beare (1997), who compared the soil aggregating effects of four grass species and two legumes grown in pots. For the four grasses, aggregate stability was related to the amount of root biomass produced (and probably to a proportional amount of rhizodeposition of organic binding agents). However, the soil in which the two legumes (white clover and lupin) were grown ended up with very high levels of aggregate stability despite the low root biomass produced by these species. The authors suggested that the rhizosphere microbial community associated with legumes might differ from that associated with grasses, possibly in relation to mycorrhizal fungi colonization, in a manner that enables more aggregate stabilization from a given amount of root biomass.

Including such legumes as lucerne (*Medicago sativa*) and medic (*M. scutella*) in wheat rotations on Vertisols in Australia enhanced soil aggregate stability with little effect on total SOM (Blair and Crocker, 2000). Differences in ability to enhance soil aggregation have also

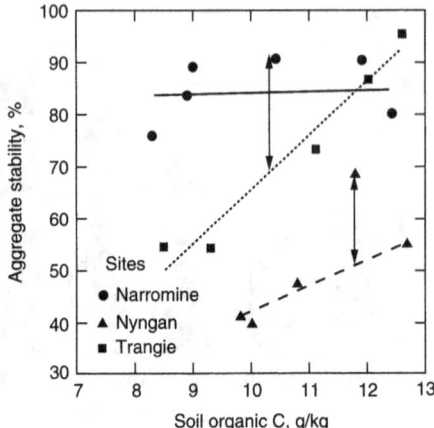

**FIGURE 1.7** Relationships between stability of macroaggregates (>0.25 mm) and soil organic matter content of degraded red Alfisols at three sites in New South Wales, Australia. Various levels of SOC were established by growing perennial pasture species, including ryegrass (*Lolium rigidum*), barrel medic (*Medicago truncatulata*), Consol lovegrass (*Eragrostis curvula*) and lucerne (*Medicago sativa*) or bare fallow kept vegetation-free with glyphosate. Lines represent regressions excluding the lucerne + lovegrass treatment, whose deviation from the trend lines are indicated by double headed arrows. (Drawn from data in Chan, K.Y. et al. 2001. *Aust. J. Exp. Agric.* 41:507–514.)

been found among grass species and cultivars used for pastures or hay. Carter et al. (1994) reported that the mean weighted diameter of aggregates in a Spodosol surface soil in Canada was higher under tall fescue (*Festuca arindinacea*) cultivars and timothy (*Phleum pratense*) cv. Farol than with orchard grass (*Dactylis glomerata*) and timothy cv. Champ. Differences among plant species and cultivars might be associated with their role as host for mycorrhiza. Fungi, especially those in mycorrhizal associations with plants, play important roles in soil aggregation (Chapter 6). The role of fungi might be most important in acid soils that are subjected to little, if any, physical disturbance.

Addition of organic amendments can have a pronounced influence on the water stability of soil aggregates and therefore on the resistance of soil to dispersion and surface crusting. Increased SOM seems to be most important in stabilizing soil structure for soils low in clay. For example, Bresson et al. (2001) worked in central France with a loess soil whose low clay content (<15%) and low SOM (15 mg/kg) limited cohesiveness and rendered the soil prone to physical degradation. Amending this low-clay soil with enough composted municipal solid waste (27.5% C, C/N = 24) to increase the SOC from 7.3 to 11.2 g C/kg greatly reduced the destruction of structure and pore continuity by raindrop impact. Structural aggregation in very sandy low-clay soils responds to even quite small differences in SOM.

Pieri (1992) reviewed structure stability data for West African soils (67 to 97% sand) and concluded that the critical concentration of SOM needed to stabilize soil structure depends on the soil texture (see previous discussion). As observed by Loveland and Webb (2003), "there is a widespread perception that the SOM…needs to be kept above a minimum level," 2.5% being most commonly mentioned "in order to prevent, or at least minimize, decline in a range of soil properties," such as structure, ease of cultivation, and water retention. In a study of tropical forests and converted pasture sites in Costa Rica, Krishnaswamy and Richter (2002) also found that aggregate stability in these low-activity clay soils increased with increasing SOC up to ca. 2% SOC (3.5% SOM) and then leveled off. Popp et al. (2002) concluded that a SOC level of 2.0% was sufficient for optimal SQ in their SQ index calculations. Nonetheless, in examining soil structure and SOM data for a large number of soils in England and Wales, researchers (Loveland and Webb, 2003) found little or no evidence for a threshold value of SOM in these soils. However, they did not report on the ratio of SOM to the fine mineral fraction (silt + clay) that Pieri (1992) showed to be so important in West Africa.

In finer-textured soils, earthworm activity can alter the effects of SOM on structural stability. The burrowing and ingesting activity of earthworms can have either a stabilizing or a disrupting effect on soil aggregates, depending on the texture and mineralogy of the soil (Schrader and Zhang, 1997). Physical disruption of soil during its passage through the earthworm gut might destabilize aggregates, but organic compounds added during the passage act to stabilize the aggregates. On balance, earthworm activity usually enhances aggregate stability.

## Soil Water Availability

The water regime in soils is influenced by SOM in several ways. First, organic matter increases the soil's capacity to hold plant available water, defined as the difference between the water content at field capacity and that held at the permanent wilting point (Figure 1.8; Hudson, 1994). It does so both by direct absorption of water and by enhancing the formation and stabilization of aggregates containing an abundance of pores that hold water under moderate tensions. However, as important to the provision of ample water for plant growth is the capacity of the soil to absorb water as it impacts from rain or irrigation. When, because of structural properties at the soil surface, the rate of water infiltration into the soil surface is lower than the rate of rainfall, a portion of the rain is lost as surface runoff. The effect on the supply of water available for plants growing in that soil is similar to a significant reduction in rainfall.

On cropland, residue management provides one of the greatest opportunities to positively influence soil infiltration rate and hydraulic conductivity and thereby increase the amount of water entering the soil to be stored for plant use. Although the return of crop residues is not likely to maintain SOM and related SQ properties at the levels seen under natural vegetation, residue return usually results in marked increases in soil water entry and storage compared with the levels observed where residues are burned or removed. (See Figure 1.9 for an example from Australia.) The relative effects of management on these soil functional properties might be much more marked than effects on the total SOC content (Whitbread et al., 2000). In a rolling landscape of the Canadian prairie in Saskatchewan, Elliott and Efetha (1999) found that maintaining crop residues on the soil surface by using zero tillage for 11 years resulted in a 42% increase in mean infiltration rate over that by

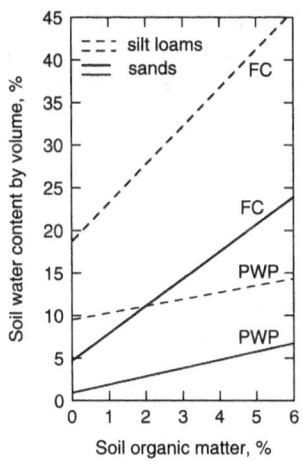

**FIGURE 1.8** Influence of SOM percentage on soil water content at field capacity (FC) and at permanent wilting point (PWP) for 20 sand-textured soils of Florida and 18 silt-loam-textured soils from the Midwest U.S. (From data in Hudson, B.D. 1994. *J. Soil Water Conserv.* 49:189–194.)

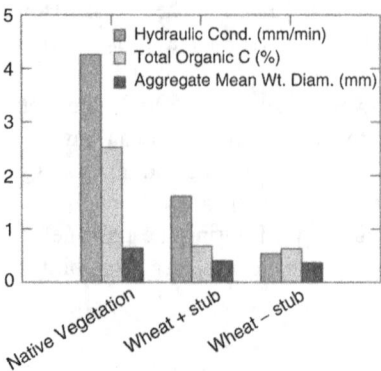

**FIGURE 1.9** The effect of organic matter management on soil organic C and two functional soil quality properties in a degraded Ferric Luvisol (Red Earth) soil in New South Wales. (Drawn from data in Whitbread, A.M. et al. 2000. *Soil Tillage Res.* 54:77–89.)

conventional tillage in a continuously cropped cereal or summer fallow system. The increases in infiltration were most pronounced on shoulder slope positions.

On the other hand, in some sandy soils or peats, SOM can lead to pronounced water repellency, especially where severe soil drying or heating (as by fires) occurs. When portions of the SOM volatilize and coat soil particles with hydrophobic compounds, soils exhibit a delay in absorbing water, leading to slow, erratic, and spatially variable rates of infiltration (Doerr et al., 2000). Water repellency also encourages preferential flow of water and solutes, especially finger flow (Dekker et al., 2001).

There are several different ways by which building SOM can reduce land degradation, erosion, and subsequent water pollution. The first and most obvious is the protection from raindrop impact that organic materials provide if they are maintained on the soil surface. In natural systems and in untilled agricultural soils (pastures and no-till cropland), accumulation of SOM occurs most rapidly at the soil surface, largely because much of the input of plant residues occurs there. The accumulation of surface residues particularly enhances the erosion resistance and water infiltration functions of the soil. Research suggests that surface placement of plant residues can maximize these functions while also significantly contributing to the nutrient cycling functions (Kayuki and Wortmann, 2001). Fortunately, the reduction in soil loss is not linearly related to the amount of residue cover on the soil surface; rather, soil loss falls off sharply with the first increments of organic residues maintained on the soil surface. Thus, very substantial reductions in erosion result from relatively small OM additions as surface cover. The data of Wildner (2000) in Figure 1.10 are typical of the results obtained in many studies worldwide. In addition to protecting the soil if maintained as surface mulch, organic matter additions can also increase soil resistance to erosion when the amendments are incorporated into the soil, because of their enhancement of structure at and near the soil surface (Table 1.2; Bresson et al., 2001).

Like organic mulches, plastic film mulches commonly used in vegetable production conserve water and inhibit weed growth. However, plastic mulch as normally applied does not behave like organic mulch with regard to controlling loss of sediment and dissolved chemicals in surface runoff. As typically installed, plastic mulch covers only 50 to 75% of the soil surface, causing runoff water to concentrate on the soil areas exposed in crop interrows. Rice et al. (2002) compared tomato production by using such plasticulture with that by using organic matter (residues from a hairy vetch winter cover crop) mulch and found that losses of soil and copper [from $Cu(OH)_2$ used as a fungicide-bactericide] were four to six times higher with the former. The runoff losses of copper from plastic mulched tomato fields has been linked to the decline of shellfish in the Chesapeake

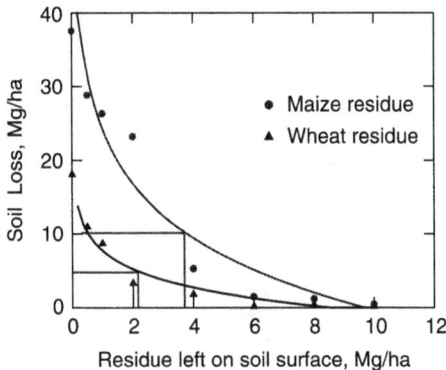

**FIGURE 1.10** Relationship between amount of organic residues maintained on the soil surface and the loss of soil by erosion. (From Wildner, L.D.P. 2000. In *Manual on Integrated Soil Management and Conservation Practices*. FAO Land and Water Bulletin No. 8, Food and Agriculture Organization of the United Nations, Rome, Italy, pp. 87–90. With permission.)

---

**TABLE 1.2**
**Generation of Runoff and Loss of Soil as Influenced by Incorporated Compost Amendments That Altered Soil Surface Structure on Loess Soils Prone to Physical Degradation Due to Their Low Clay (<15%) and SOM (<1.5%) Contents**

| Measured Attribute | Control Soil | Amended Soil |
| --- | --- | --- |
| Cumulative rain when runoff starts (mm) | 2.5 (0.1) | 9.2 (2.4) |
| Sediment concentration in incipient runoff (g/L) | 36.4 (9.7) | 11.0 (1.7) |
| Soil loss (g) | 54.6 (24) | 18.3 (9.1) |

*Note:* Mean with standard error in parentheses.
*Source:* Data from Bresson, L.M. et al. 2001. *Soil Sci. Soc. Am. J.* 65:1804–1811. With permission.

---

Bay watershed (Dietrich et al., 2001). It appears that building SOM with organic mulches from killed cover crops is a practice that is likely to improve both SQ and water quality.

Other forms of organic matter additions have been observed to result in similar win–win situations. For example, adding a moderate amount of manure to soil can actually reduce P losses in runoff, even though the manure itself carries significant quantities of P. This was the case in Wisconsin experiments (on silt loam Typic Argiudolls), especially when manure addition was combined with no-till soil management (Bundy et al., 2001). Total P concentrations in fall runoff from chisel-plowed, shallow-tilled, and no-till cropland were in the range of 3 to 4.5 mg/L, except for no-till plots with manure application, which had a significantly lower level (1.8 mg/L) of total P concentration in runoff (Figure 1.11). Likewise, manure application reduced the total P loading most dramatically on the no-till plots. Apparently, the nutrients and organic matter provided by the manure were sufficient to increase crop residue production and improve surface soil structural condition. SOM also enhances aggregation and stability of aggregates, leading to higher infiltration rates for both natural rainfall and irrigation waters (see below for a discussion).

A further important physical effect of adding organic matter to soils as surface mulch is the moderation of soil temperature fluctuations. In summer, soil temperature highs are reduced by organic surface mulch to considerably below air temperature highs, but in winter the reverse might be true as soils under an organic mulch are generally warmer than bare soils when air temperatures are low (Unger, 1978). Organic matter on the soil surface is much more effective at moderating

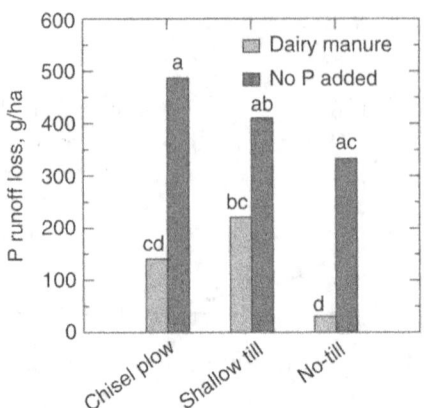

**FIGURE 1.11** Effect of spring applied dairy manure (12.3 Mg/ha dry wt containing 64 kg total P) and tillage on total P losses (g/ha) in runoff in September from a Wisconsin silt loam (Typic Argiudolls). (Redrawn from Bundy, L.G. et al. 2001. *J. Environ. Qual.* 30:1822–1828.)

soil temperature than plastic mulches. In summer, soil temperatures under black or clear plastic can become high enough to limit root growth, even if the soil is kept moist (Tindall et al., 1991). If organic matter is incorporated and humified, it reduces the thermal conductivity and heat capacity of soils. However, the increased soil water content associated with SOM-enhanced water-holding capacity and aggregation results in increased thermal conductivity and heat capacity of soils. The balance between these opposing direct and indirect effects determines the net effect of increased SOM on soil temperature regime.

In summary, SOM content influences the physical condition of soils in a multitude of ways, sometimes even overshadowing the effects of such fundamental soil properties as particle size and mineralogy.

## CHEMICAL SOIL PROPERTIES INFLUENCED BY ORGANIC MATTER

### Nutrient Storage and Release

Nearly all the nitrogen and large proportions of the phosphorus and sulfur found in soils occur as constituents of SOM. The SOM serves as both the principal long-term storage medium and as the primary short-term source of these and other nutrients. Contrary to the assumption that fertilizers feed the plant not the soil, most of the N taken up by crops comes from organic pools that cycle through the microbial biomass. Numerous studies that use [15]N and other isotopes to trace the fate of fertilizer N have shown that current-year fertilizer is the source of only 10 to 50% of the N taken up by such N-demanding crops as corn, unless the system is overwhelmed by excessively large applications of inorganic N (Omay et al., 1998; Reddy and Reddy, 1993). An example of such a study is illustrated in Figure 1.12, which shows that even when 168 kg N/ha was applied as inorganic fertilizer, unlabeled N from mainly organic soil pools accounted for 70% of the N taken up by continuously grown corn. Corn N uptake from soil pools increased further when corn was grown in rotation with soybean.

Crop residues serve as an important source of plant nutrients released though mineralization. Litter quality (content of lignin and polyphenol and C:N ratio) and method of placement (surface mulch or incorporated) largely control the rate of mineralization of nutrients for use by following crops. The C and N in litter largely move into biologically active pools of SOM (discussed in the next section). Moderate rates of fertilizer N in the range that effectively stimulates crop growth also stimulates the production of increased crop residues (litter) and the transfer of C

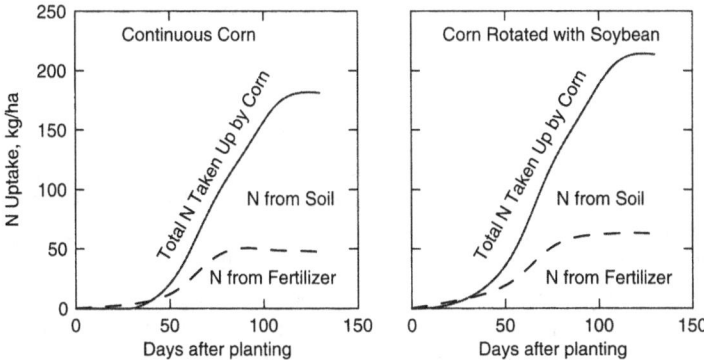

**FIGURE 1.12** Uptake of N by corn from applied fertilizer and from soil organic matter pools (a) corn after corn and (b) corn after soybean. Fertilizer N was applied at 168 kg/ha to each crop. (Drawn from data in Omay, A. et al. 1998. *Soil Sci. Soc. Am. J.* 62:1596–1603.)

and N into the biologically active SOM pools (e.g., Lal, 1997). Nitrogen in such biologically active pools as POM and biomass often correlates closely with short-term N mineralization (e.g., Gajda et al., 2001).

A diversity of organic matter inputs can increase the pool of mineralizable N ($N_{min}$). In Michigan (on Typic Hapludalfs), Sanchez et al. (2001) compared plots under continuous corn monocropping fertilized with only mineral NPK to plots under a more diverse system consisting of a corn–corn–soybean–wheat rotation with clover cover crops and composted manure added. After 70 d of incubation, net N mineralized in the rotation system soil was 90% higher than that in the monocrop soil. However, because adding an extra 2.5 Mg/ha of clover residues allowed the monocrop soil $N_{min}$ to equal that of the rotation soil, they concluded that "regardless of its history, the soil microbial biomass reacts to the input of easily decomposed compounds like those in clover." In addition, the rapid N release from these residues suggested that "N release may coincide with crop growth." Furthermore, the authors speculated that living crop roots would alter these results "by adding an energy source and exposing materials to decomposition." They concluded that a diversity of organic matter inputs can increase the pool of mineralizable N.

Research suggests that surface placement of plant residues can maximize the erosion-resistance and water infiltration functions while also significantly contributing to the nutrient cycling functions. For example, Kayuki and Wortmann (2001) used litterbags either buried at a 45° angle or pinned to the soil surface in maize fields to study nutrient mineralization from residues of four species of plants that occur abundantly around farm fields in Uganda. At the end of 16 weeks, 75% of the N in the incorporated materials was released compared with only 50% of that in the surface litter materials, but application method did not affect maize yields. High-quality organic residues (low lignin and phenols) were as effective as fertilizer at increasing maize yield and bean yields, whether applied as surface mulch or incorporated into the soil. They concluded that "plant materials such as those studied are best when surface applied, thereby reducing labor and improving soil cover until a crop canopy is established."

Although decomposer microorganisms carry out the bulk of decomposition activity that ultimately releases plant-available nutrients from residues, predatory organisms (especially protozoa and nematodes) in higher trophic levels can substantially hasten the process. One study (Ferris et al., 1998) used columns of sandy soil amended with ground alfalfa and cellulose in varying proportions to give residue C:N ratios ranging from 10 to 40. Well-established bacterial communities populated all the columns. Some soil columns were inoculated with bacteria-feeding nematodes (*Cephalobus persegnis*) and some were kept free of nematodes. By feeding on bacteria, the nematodes nearly doubled the amount of mineral N released from the added residues over a 21-day period. Without nematodes, much of the N was tied up in the bacterial biomass, but as the nematodes preyed on the bacteria, they excreted excess N into the soil solution as $NH_4^+$. Enhanced mineralization in the presence of bacteria-feeding nematodes allowed the maintenance of a given level of mineral N with residues of a higher C:N ratio than that required without the nematodes.

Composting is a practical means for storing organic residues and stabilizing their nitrogen for later use as a soil fertility amendment. Because of the low lability of its nitrogen, large amounts of compost might be needed initially to supply enough of this and other nutrients for optimal crop growth material. Beneficial effects on soil physical and chemical properties can make the use of heavy compost applications worthwhile. On the other hand, these benefits are not always apparent in the short term. For example, compared with synthetic NPK fertilizers, addition of 43 Mg dry matter/ha of composts over 2 years had no effect on penetration resistance, CEC, VAM colonization, or soil mineral N, but the composts did increase, or prevent a decrease in, pH (6.5 vs. 5.9) and gave slightly lower electrical conductivity (0.15 vs. 0.26 dS/m; Carpenter-Boggs et al., 2000). Disease suppression and other beneficial biological effects of composts and compost extracts have also been widely reported (see Chapter 5), largely because decomposable organic amendments can serve as an excellent food base for beneficial soil microorganisms (Hoitink and Boehm, 1999). In using heavy application of compost, more phosphorus is likely to be applied than the plant can take up, and therefore this element is likely to accumulate in the soil. The water quality impacts of potential P losses in runoff must therefore be taken into consideration (see Chapter 2).

## Microbial Enhancement of Nutrient Availability

The main contribution soil organisms make to plant nutrition is through cycling nutrients contained in crop residue and other organic amendments (see previous discussion). As organisms in the soil food web use stored photosynthetic energy and nutrients in organic residue, mineralization of N, P, S, and other elements occurs, rendering them more available to plants. However, there are a number of other ways by which soil organisms can enhance the availability of nutrients to plants:

- Biological $N_2$ fixation can make important contributions to both the short- and long-term nitrogen status of plants. For example, the mututalist *Rhizobium* (in legume root nodules), free-living *Azotobacter* and *Clostridia*, and those that form associations with nonagricultural plants enhance N availability.
- Arbuscular mycorrhizal fungi form mutualist associations with most crop plants, greatly enhancing the effective surface area of the root systems, and thus contact with soil solids (see Chapter 6). The greater surface area produced by mycorrhizal hyphae promotes uptake of water as well as P and other nutrients.
- Siderophores, low-molecular-weight compounds that chelate Fe and increase its availability to plants, are produced in low-Fe soils by a variety of microorganisms such as *Agrobacterium*, *Bacillus*, *Pseudonomonas*, and *Rhizobium* (Neilands, 1996).
- A wide variety of soil microorganisms are capable of enhancing P solubility in artificial media, and some have been shown to increase P availability in soils (Kucey et al., 1989). One fungus, *Penicillium bialji*, has been commercialized for this purpose.

## Cation Exchange Capacity

The colloidal fraction, which consists of both clay and humified organic matter, is recognized as the seat of chemical activity in soils, including the capacity for ion exchange (Brady and Weil, 2002). The relative contribution of clay and SOM to soil cation exchange capacity (CEC) is largely determined by the amount of SOM, the amount and mineralogy of clay, and the soil pH. For soils low in clay, SOM can be responsible for essentially all of a soil's CEC. Even for clay soils, SOM can contribute a significant fraction of the CEC. For example, illitic and chloritic clays have CECs ca. 20 cmol (+)/kg of clay. A soil with 50% of such clays (500 g/kg) and 5% SOM and a pH of

6.5 will have a CEC of ca. 20 cmol (+)/kg, with approximately half originating from the clay and half from the SOM.

Within the pH range of 5 to 7.5 of most agricultural soils, the CEC of SOM results mainly from carboxyl groups that have lost $H^+$ and thus gained a negative charge. In very acid soils (ca. pH 4) much of the potential CEC of SOM is not expressed because the soil's pH is far below the $pK_a$ values of most of the individual carboxyl groups of humus. Under these very acid conditions, $H^+$ is held tightly by SOM's acid groups. When acid soils are limed and the pH increases, neutralization of $H^+$ occurs from organic acid groups with increasingly higher $pK_a$ values. This removal and neutralization of $H^+$ leave organic acid sites with negative charges that have become exchange sites. The CEC of humus per unit mass is much higher and increases more with increased pH than that of most clay minerals. The CEC of humus from Wisconsin soils increased linearly with increasing pH of the saturating solution used to measure the CEC, from ca. 90 cmol (+)/kg at pH 4.0 to 220 cmol (+)/kg at pH 8.0 (Helling et al., 1964). More recent studies that used soils from other parts of the world support the characterization by Helling et al. (1964) of the CEC of soil humus. For example, regression of CEC against total SOC (mineral-associated C + particulate C) for southern Brazilian Ultisols with pH 5.0 to 5.3 gave a slope of ca. 0.26 cmol(+)/g C (Figure 1.13), which is equivalent to 130 cmol (+)/kg of SOM (assuming SOM to be 50% C), a value consistent with that given by Helling et al. (1964), who predicted ca. 140 cmol(+)/kg of SOM at pH 5.3. For tropical forest soils in Costa Rica (Krishnaswamy and Richter, 2002), CEC measured at pH 8.2 increased by ca. 0.5 cmol(+)/g SOC [equivalent to 250 cmol(+)/kg SOM], about the same level of CEC per kg of SOM reported by Helling et al. (1964) for cation exchange at pH 8.

## Sorption of Organic Compounds

SOM plays an important role in sorbing organic compounds as well as inorganic cations. Organic compounds that commonly carry a net positive charge, and are therefore cations, include quaternary N pesticides (e.g., diquat and paraquat). Basic compounds such as atrazine tend to become positively

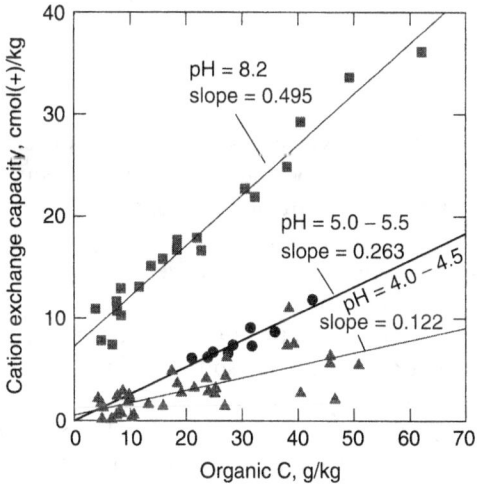

**FIGURE 1.13** Increased cation exchange capacity (CEC) of soil resulting from increased organic C concentration at various pH levels. The data for soils at pH 5.0 to 5.5 are from Bayer et al. (2001), the different SOC levels being associated with differences in soil depth (ranging from 2.5 to 17.5 cm) and cropping systems under no-tillage management. Soil texture varied little and CEC was measured at the pH of the soil (pH 5.0 to 5.5). The data for CEC measured at pH 8.2 and at the pH of soils (4.0 to 4.5) are from Krishnaswamy, J., and D.D. Richter. 2002. *Soil Sci. Soc. Am. J.* 66:244–253, in whose study the differences in SOC were associated mainly with land use (forest or pasture) and soil depth (either 0–10 or 10–30 cm). The slopes of the regression lines reflect the higher CEC of humus at higher pH levels.

charged by protonation in low-pH environments. The levels of SOM and pH also influence the sorption of carboxylic acid herbicides, which tend to sorb more strongly on lipophilic organic matter groups when they are not ionized (at low pH) than when they are present as anions (Magdoff, 1996). The importance of SOM for the sorption of pesticides is illustrated by the fact that 10 times more smectite clay than SOM was necessary to retain 50% of a PCB herbicide (Strek and Weber, 1982). In the same study, the use of peroxide to remove the SOM fraction from a sandy soil reduced the soil's capacity to sorb the herbicide by a factor of eight.

Retention of organic compounds in soils is commonly quantified by comparing the amount of the compound found in solution to that sorbed by the soil after coming to equilibrium, the ratio of the two pools being termed the *distribution coefficient* ($K_d$). Because many organic compounds sorb more strongly to SOM than to clay, the mobilities of herbicides and pesticides are commonly compared by calculating the $K_{oc}$ ($K_{oc} = K_d \times 100/\%SOC$). The $K_{oc}$ is especially useful for relative ranking of pesticide mobility, because even though the $K_{oc}$ for a compound varies from soil to soil, it varies substantially less than $K_d$ does.

Although SOM can sorb many organic pollutants, organic matter from soils can itself be a pollutant. Chlorination of organic compounds by reactions similar to those used to produce certain carcinogenic and toxic pesticides occurs in soils and sediments by both biotic and abiotic processes. *In situ* x-ray spectroscopy data (Myneni, 2002) indicate "that natural organic matter in soils, sediments, and natural waters contain stable, less volatile organic compounds with chlorinated phenolic and aliphatic groups as the principal Cl forms." The chlorine is incorporated during the humification, and the resulting chlorinated compounds influence the cycling of Cl and other elements. SOM, chlorinated or not, can move to waterways and have health implications on humans and the ecosystem. For example, scientists initially presumed that differences in water quality (especially biological oxygen demand from dissolved organic matter) in various reservoirs were due to management of the watershed. However, research on 16 watersheds in Southern Australia (Oades, 1995) showed that the major factor was capacity of the A horizon soil to retain organic matter. Highly sorptive soils prevented organic matter from entering the aqueous phase and leaching through the profile to the groundwater. The adsorption capacity of soils for organic C was related to the clay content and surface area of the soils.

## Anion Sorption

Phosphorus deficiency is commonly associated with acid soils because of the strong sorption or fixation of inorganic phosphorus to iron and aluminum oxide surfaces under low-pH conditions. High P fixation is a major constraint to crop production in ca. 5% of the world's agricultural soils and in 13 to 14% of soils in regions with warm humid and subhumid climates (Wood et al., 2000). Increasing SOM not only provides a source of P from mineralization but also can reduce the capacity of acid soils to lock up P by fixation. Organic soil amendments, such as digested sewage sludge or vegetative mulches, reduce the sorption of P in soils and increase the equilibrium P concentration in the soil solution ($EPC_0$) with no inorganic P added. In one study (Sui and Thompson, 2000), the $EPC_0$ nearly doubled from 1.2 to 2.2 mg/L. In contrast, another anion, boron, is usually more strongly adsorbed and therefore less available to the plant in soils with higher SOM levels, especially under high pH conditions (Yermiyahu et al., 1995).

## Metal Mobility

SOM also affects the mobility of heavy metals that may (e.g., Zn) or may not (e.g., Cd) serve as plant nutrients. Addition of organic matter to soil can either decrease or increase metal availability, solubility, and plant uptake. Insoluble organic matter usually forms insoluble organometal complexes or sorbs metal ions, making them less available for plant uptake or leaching (Sauve et al.,

1998). However, many organic amendments have a soluble C component or produce soluble decomposition products, and the soluble organic matter can increase metal solubility by forming soluble organometal complexes (Alvarez et al., 1999). The influence of SOM on metal mobility can also be modified by the solution pH (Yoo and James, 2002).

In addition to these effects, SOM affects the mobility of some metals, because decaying organic matter stimulates reducing conditions as microorganisms deplete the soil oxygen supply and reduce other electron acceptors. Among plant nutrients, manganese (Weil, 2000) and iron (Childers et al., 2002) are especially prone to increased solubility under reduced conditions. In contrast, decomposable organic matter (such as manure) can reduce the solubility and toxicity of chromium by reducing the toxic Cr(VI) to the nontoxic, much more immobile Cr(III) form (Losi et al., 1994).

## Soil pH Buffering and Amelioration

SOM exerts a major influence on the pH buffering of surface soils, both because it contributes much of the soil CEC and because of the dissociation of weak acid functional groups on the SOM molecules (Magdoff and Bartlett, 1985). Buffering of soil pH by organic matter and the variation of SOM's CEC with pH changes are two sides of the same coin. Carboxyl sites that lose (or gain) $H^+$ as the pH increases (or decreases) in the process of $H^+$ neutralization (or acidification) gain (or lose) negative charges capable of holding cations in exchangeable form. This pH buffering is one reason to favor management practices that increase SOM. (For a discussion of strategies and practices, see Chapter 2.)

Aluminum toxicity is a major limitation for productivity on acid soils in large areas of the more humid regions of the world. It is estimated to be a major constraint on 17% of agricultural soils worldwide, on 42% in the humid to subhumid tropics, and on 25% in the humid to subhumid subtropics (Wood et al., 2000). Organic matter, in the form of chicken manure applied to the soil surface, is effective in reducing Al saturation in subsoils horizons in an Ultisol, a desirable effect not easily achieved by using agricultural limestone, which is relatively immobile in soils. The manure probably forms organic Ca-complexes that move Ca down profile with percolating water, thus ameliorating the acid subsoil layers (Hue and Licudine, 1999). Organic matter is credited with ameliorating acidity by other mechanisms too. At pH levels between 3.5 and 5.5, adding 10% humified organic matter (peat) to soil greatly reduced exchangeable aluminum by binding the aluminum ions in nonexchangeable forms (Hargrove and Thomas, 1981). Complexation of aluminum ions by SOM is another important mechanism to ameliorate aluminum toxicity (Parfitt et al., 1999). The stabilization of organic matter by interaction with amorphous aluminum oxides and allophane is also important in retarding the decomposition of SOM and in maintaining SQ in regions with warm climates or volcanic parent materials. Some soil-testing laboratories recognize the protection against Al toxicity afforded by SOM by suggesting lower target pH ranges for soils with higher SOM contents.

## Growth-Regulating Substances

The significance of the rhizosphere population of organisms and the production of plant growth-regulating substances in the rhizosphere has been reviewed by Arshad and Frankenberger (1998) and Bowen and Rovira (1999). Although most of these substances stimulate the growth of plants, some substances reduce plant growth. Briefly, there is an intimate interaction between plant roots and microorganisms in the rhizosphere (see discussion in Chapter 10). Root exudates provide microorganisms with nutrients as well as cues that might induce the production of specific compounds. The variety of microbially produced substances that enhance plant growth include auxins, gibberellins, cytokinins, ethylene, abscisic acid, and siderophores (Arshad and Frankenberger, 1998).

## BIOLOGICAL PROPERTIES

SOM is both a raw material for and a product of biological activity in soils. Increased SOM can enhance the diversity of the microbial community as well as its biomass. The diversity of life in the soil is staggering: "Because of their small size, prokaryote diversity in a 100-cm$^3$ soil sample can be compared to the regional diversity of macroorganisms." (Torsvik et al., 2002). The diversity of microbial species and functional groups can play an important role in the accumulation of SOM (see Chapter 10). The influence of SOM extends to arbuscular mycorrhizal fungi (Hodge et al., 2001; Hamel et al., 1997) and beneficial rhizosphere bacteria (Gilbert et al., 1994). Rhizodeposition, the release of organic material by plant roots, can be especially important as a source of SOM, serving as substrate for rhizosphere bacteria (Haynes and Beare, 1997). The amount of C so deposited in the rhizosphere is estimated as ca. 40% of the net C photosynthetically assimilated by growing plants (e.g., Bottner et al., 1999). Bioremediation studies have shown that soil microbial communities stimulated by rhizodeposition can more efficiently break down organic contaminants, including xenobiotics (e.g., Reynolds et al., 1999).

An increase in the study of the microbial biomass in soils during the past two decades has led to major advances in the understanding of soil systems and especially C cycling in these systems. Microbial biomass is at present seen to exert a controlling influence on the dynamics of SOM and the availability of many nutrients. The critical roles that microbial biomass performs in the soil system can be summarized as follows (modified from Dalal, 1998):

- A labile source of carbon, nitrogen, phosphorus, and sulfur
- An immediate sink of carbon, nitrogen, phosphorus, and sulfur
- An agent of nutrient transformation and pesticide degradation
- An agent of mineral weathering and soil formation
- Includes organisms, especially in association with plant roots, that form and stabilize soil aggregates
- Includes organisms that are antagonistic to plant pathogens and parasitic nematodes
- Produces plant growth regulators

Both the turnover rate and the size of the microbial biomass are important in determining microbial activity (Theng et al., 1989). Depending on climatic and other variables, turnover rates of microbial biomass carbon ($C_{mic}$) usually range from 0.5 to 5 years, as compared to >20 years for the bulk of SOC (Dalal, 1998). Because of this rapid turnover, environmental factors, such as tillage, erosion, or cropping sequence, cause more rapid and greater proportional changes in biomass C than in total SOM. Therefore, $C_{mic}$ (by methods that measure either active or total biomass) is often a more sensitive indicator of SOM flux and change in SQ than is total SOC (Islam and Weil, 2000). Further advances in understanding SOM dynamics are likely to result from the use of stable isotopes ($^{13}$C and $^{15}$N) in microbial biomass C studies (Syers and Craswell, 1995). Microbial biomass C is an important parameter in at least two models (CENTURY and Rothamsted) widely used to predict SOM dynamics (Jenkinson, 1991; Parton et al., 1988; Rees et al., 2001).

The size of the living community in the soil, the microbial biomass, is generally positively related to the SOM level; rarely is $C_{mic}$ less than 1% or more than 5% of the total SOC (Carter, 2001; Dalal, 1998). Carbon input to agricultural soils from roots, residues, and amendments usually ranges from 1 to 15 Mg/ha/year, maintaining surface soil organic C stocks ranging from 5 to 50 Mg/ha and microbial biomass C stocks ranging from 0.05 to 2.5 Mg/ha. Duxbury et al. (1989) estimate that N flux through the soil microbial biomass has a 10-fold range from ca. 35 to 350 kg N/ha/year. In many soils, these fluxes account for the majority of the N made available to plants. In the North American Great Plains, Gajda et al. (2001) found a close relationship between soil microbial biomass N (kg/ha) and potentially mineralizable N (kg/ha) estimated by anaerobic incubation (anaerobic NH$_4$ release = $-4.71 + 0.94 \times$ biomass $N$, $R^2 = 0.84$).

Increases in SOM are usually associated with similar increases in microbial biomass, because the SOM provides the principal substrates for the microorganisms. However, the two parameters are only loosely linked, giving rise to the four- to fivefold variation commonly observed in the ratio $C_{mic}$:SOC. This microbial quotient, or the $C_{mic}$ as a percentage of SOC, is considered by some to be an indicator of health or stress of the soil microbial community, a healthy, low-stress community being able to sustain a relatively high level of $C_{mic}$ with a given level of SOC. Lavahun et al. (1996) found that the microbial quotient declined with soil depth and also with cultivation (Figure 1.14). In the surface15 cm, the lower microbial quotient in soil under cultivation as compared with that in soil under permanent grass was paralleled by similarly lower levels of SOC. The microbial quotient, but not the SOC, was considerably lower in the cultivated soil. Swift and Woomer (1993) cite data (from Sparling, 1992) for a Kairanga soil in New Zealand in which the microbial quotient, but not the SOC level, showed a relatively consistent decline and leveling off during 12 years of corn cultivation. They suggested that the microbial quotient provides a generally more stable index of SQ change over time than does SOC (see Figure 1.15). However, this is not consistently true: in the other two soils studied by Sparling (1992), the trends for the microbial quotient were as variable as those for SOC (Figure 1.15) and there was little consistency in the microbial quotient among the three soils used in the study.

Another biological quotient often influenced by changes in SOM is the ratio of basal respiration to $C_{mic}$, sometimes termed the maintenance respiration rate or respiratory quotient, $qCO_2$. Increased $qCO_2$ might reflect a stressed soil microbial community in which a relatively large amount of energy is exerted simply on maintenance rather than growth (Anderson and Domsch, 1990). Larger $qCO_2$ values can also indicate a system more dominated by bacteria, which are less efficient than fungi in converting SOC into biomass. Islam and Weil (2000) reported that among mid-Atlantic agricultural soils, those in which SOC was severely depleted by intensive tillage and row cropping had relatively low levels of $C_{mic}$ and very high $qCO_2$ values.

Carbon dioxide from microbial respiration is a major contributor to the increase of this greenhouse gas in the global atmosphere; soil microbial activity also produces other greenhouse gases. Among the most important of these gases are methane ($CH_4$) and nitrogen oxides (NOx). The processes that produce of $CH_4$ and NOx in soils require the combination of anaerobic conditions (usually waterlogged soils) and the availability of decomposable organic matter.

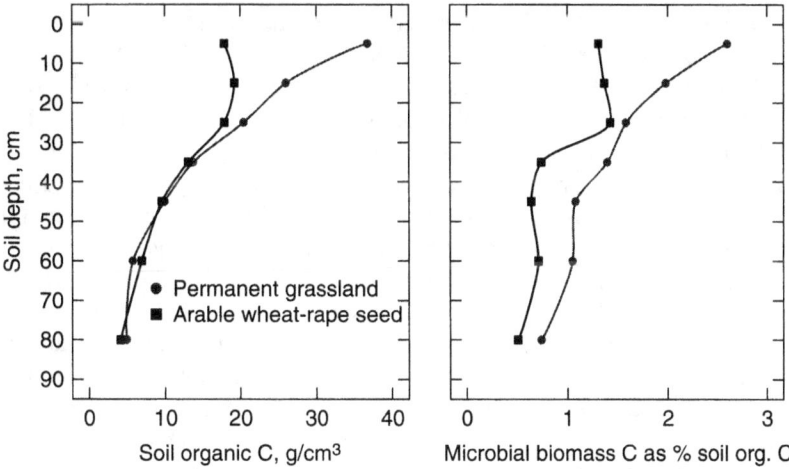

**FIGURE 1.14** Distribution of total organic C and the proportion accounted for by microbial biomass in German Luvisols under two types of management. (Drawn from data in Lavahun, M.F.E. et al. 1996. *Biol. Fertil. Soils* 23:38–42.)

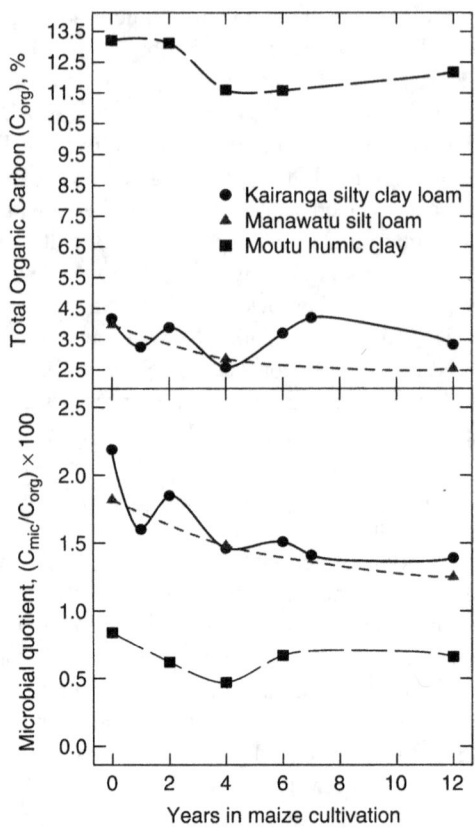

**FIGURE 1.15** Effect of years under continuous maize cultivation on the changes in total organic C and microbial biomass C as a percentage of TOC in three soils. (Redrawn from Sparling, G. 1992. *Aust. J. Agric. Res.* 30:195–207.)

Agriculture contributes to about one third of $CH_4$ emissions globally, and in the U.S. it is second only to landfills as a source of this greenhouse gas (Reicosky et al., 2000). The principal agricultural sources of $CH_4$ are flooded rice fields, livestock, animal manure storage, and biomass burning. Methane emissions from flooded rice paddies might increase over a 10-fold range with increasing levels of SOM (Neue et al., 1994). Paddies to which rice residues are returned also emit several fold more $CH_4$ than those from which residues are removed or burned (Neue et al., 1994; Redeker et al., 2000). In a California rice study (Redeker et al., 2000), $CH_4$ emissions were highest during the period of most active rice growth in mid-season. Very little $CH_4$ was released if no rice was planted. Moderate amounts were released if rice was planted but rice residues from the previous crop had been burned off. The highest amounts of $CH_4$ were released where rice was planted and the straw from the previous crop had been incorporated into the soil. Although total SOM certainly affects $CH_4$ emission levels, dissolved organic C, especially from root exudation, might be responsible for stimulating much of the methanogenic activity (see Figure 1.16; Lu et al., 2000).

Denitrification is another microbial process influenced greatly by the level of SOM. The four principal environmental conditions required for rapid loss of N by this pathway are anaerobic conditions, warm temperatures, presence of nitrate, and a supply of microbially oxidizable carbon. Wet periods during a warm growing season tend to cause the most rapid denitrification in soils that have been amended with organic materials, especially legume cover crop residues or animal manures. When nitrate and temperature are not limiting, the denitrification rate can vary over three orders of magnitude, depending on levels of water and carbon in the soil (Strong and Fillery, 2002). In a study (Weil et al., 1990) that used transects of wells to monitor groundwater under fertilized

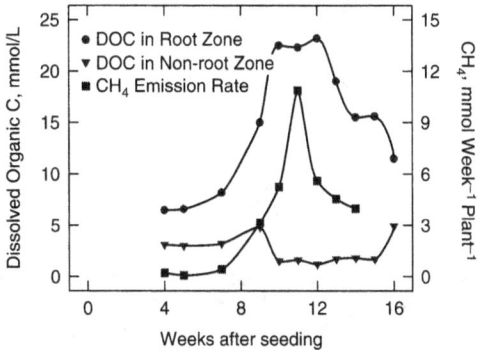

**FIGURE 1.16** Dissolved organic C and methane emissions in a clay soil flooded and planted to rice. Root exudates provided most of the DOC, which in turn provided the substrate for microbial methane production. The root zone was isolated by a 24-μm mesh nylon bag and the DOC and dissolved $CH_4$ were sampled by a ceramic suction lysimeter. (Data from Lu, Y. et al. 2000. *Soil Sci. Soc. Am. J.* 64:2011–2017.)

and irrigated sandy soils, localized buried layers of organic-matter-rich soil stimulated denitrification in shallow groundwater and resulted in the water sampled from one well being far lower in nitrate than that sampled from all the other wells. This well was located in a low-lying area that had been filled in with surface soil during land leveling several years earlier, resulting in the groundwater coming in contact with high OM soil ca. 1 m below the current surface. Given how dramatically SOM can stimulate denitrification in wet soils, McCarl and Schneider (2001) suggest that efforts to sequester C in agricultural soils should take into account the effects of the added C on denitrification, noting that 1 kg of NOx exerts many times greater effects on global temperature forcing than does 1 kg of $CO_2$.

SOM provides the underlying substrate not only for microbial metabolism but also for the entire soil food web. Macrofauna (mainly earthworms, ants, termites), mesofauna (mainly arthropods), and microfauna (nematodes, protozoa, rotifers, tardigrades) play important, if largely indirect, roles in the cycling of C and nutrients through SOM. Various faunal groups (e.g., nematodes, protozoa, mites, collembola, earthworms) help regulate microbial activity by grazing bacterial colonies and fungal hyphae. In addition, the larger species that ingest SOM physically shred plant residues in a manner that enhances microbial access to plant residues. Surface mulches of organic materials greatly stimulate faunal populations, especially earthworms (Syers and Springett, 1984) and termites (De Bruyn and Conacher, 1990), either by providing a litter habitat or by maintaining moist, moderate temperature soil conditions and substrate for their microbial diet. Addition of livestock manure can similarly stimulate populations of dung beetles. Plant residues can stimulate populations of certain species of termites and ants, especially in subtropical and tropical region soils. These invertebrates play important roles as soil ecosystem engineers (Lavelle et al., 1997), because their burrowing activities have major long-lasting effects on the soil. Their incorporation of organic materials enhances nutrient cycling and reduces nutrient losses. Their burrows increase the porosity, water infiltration rate, and drainage of soils (Larink and Schrader, 2000). As the invertebrates respond to increased SOM levels, these effects can be considered as indirect influences of SOM.

Much is known about how the incorporation of various organic materials into soils affects soil biology and nutrient cycling (Hulugalle et al., 1996; Lewandowski, 2002; Sanchez et al., 2001). However, much less is known about the effects of organic materials applied to the surface of the soil as mulch. Immobilization of N is usually observed after addition of material with a high C to N ratio. However, this might not be the case when more complex food web changes occur at the level of microbial feeding fauna (Forge et al., 2003). Various kinds of mulches were maintained by applying large amounts of organic materials between rows of apple trees. A shredded paper

**TABLE 1.3**
**Effects of Organic Mulch Treatments on Soil Fauna and Nutrient Fluxes in an Apple Orchard**

| Treatments Applied as Mulch between Tree Rows | Root Parasites[a] Nematodes 100 g⁻¹ | Bacterivorous Organisms | | Microbial Biomass | | |
|---|---|---|---|---|---|---|
| | | 1000s Protozoa MPN g⁻¹ | Nematodes 100 g⁻¹ | C Consumed (mg/kg/ year) | N Flux (mg/ kg/ year) | P Flux (mg/kg/ year) |
| Paper | 45 c | 107 c | 3402 bc | 1345 b | 266 b | 67 b |
| Alfalfa | 66 bc | 135 c | 2551 bcd | 1416 b | 282 b | 71 b |
| Biosolids | 569 a | 255ab | 3291 bc | 1838 ab | 364ab | 92 ab |
| Biosolids + paper | 69 bc | 278 a | 5209 a | 3094 a | 617 a | 458 a |
| Mun. compost + paper | 45 bc | 164 bc | 3691 b | 1935ab | 384ab | 247 ab |
| Black plastic | 266 ab | 101 c | 1041 d | 490 b | 97 b | 25 b |
| Control (herbicide) | 513 a | 72 c | 2080 cd | 825 b | 161 b | 41 b |

[a] *Pratylenchus penetrans.*
*Note*: In a column, means followed by the same lowercase letter do not differ statistically.
*Source:* Data from Forge, T.A. et al. 2003. *Appl. Soil Ecol.* 22:39–54.

mulch (C:N ratio of 183:1) did not reduce N in the tree leaves, and the amount of N released through the microbial biomass was estimated at 240% of the N applied in the paper (Forge et al., 2003). The mulches had pronounced effects on the population of bacterivorous protozoa and nematodes, resulting in marked differences in predation on microbial biomass and fluxes of N and P (Table 1.3). In another example of mulch effects on the food web, maize stover applied as mulch in Uganda increased nesting by predatory ants, which decreased termite damage to the maize plants (Sekamatte et al., 2001).

### Influence of Organic Residues and SOM Management on Crop Pests

One of the reasons for the renewed focus on SOM and its management is the many reports of the beneficial effects of organic residues and improved management on reducing damage caused by crop pests. The complex effects of organic residues on the soil food web and the corresponding effects of soil organisms on plant growth are further elucidated in Chapter 5, Chapter 10, and Chapter 11. In addition, SOM management influences plant susceptibility to insect pests (Chapter 7) as well as weed seed populations (Liebman and Davis, 2000). For a discussion of the influence of SOM management practices on crop pests, see Chapter 2.

The physical, chemical, and biological effects of SOM are complex and intertwined. The addition of organic materials to a soil usually leads to a cascade of cause-and-effect relationships that produce a series of changes in soil properties and ultimately on how a soil interacts with the environment (Figure 1.17).

## POOLS OF SOIL ORGANIC CARBON

One of the major advances of the past few decades in our understanding of SOM was the realization that SOM is not all the same type of substance, neither in composition nor behavior. Although their exact chemical and physical nature remains to be described, several pools of

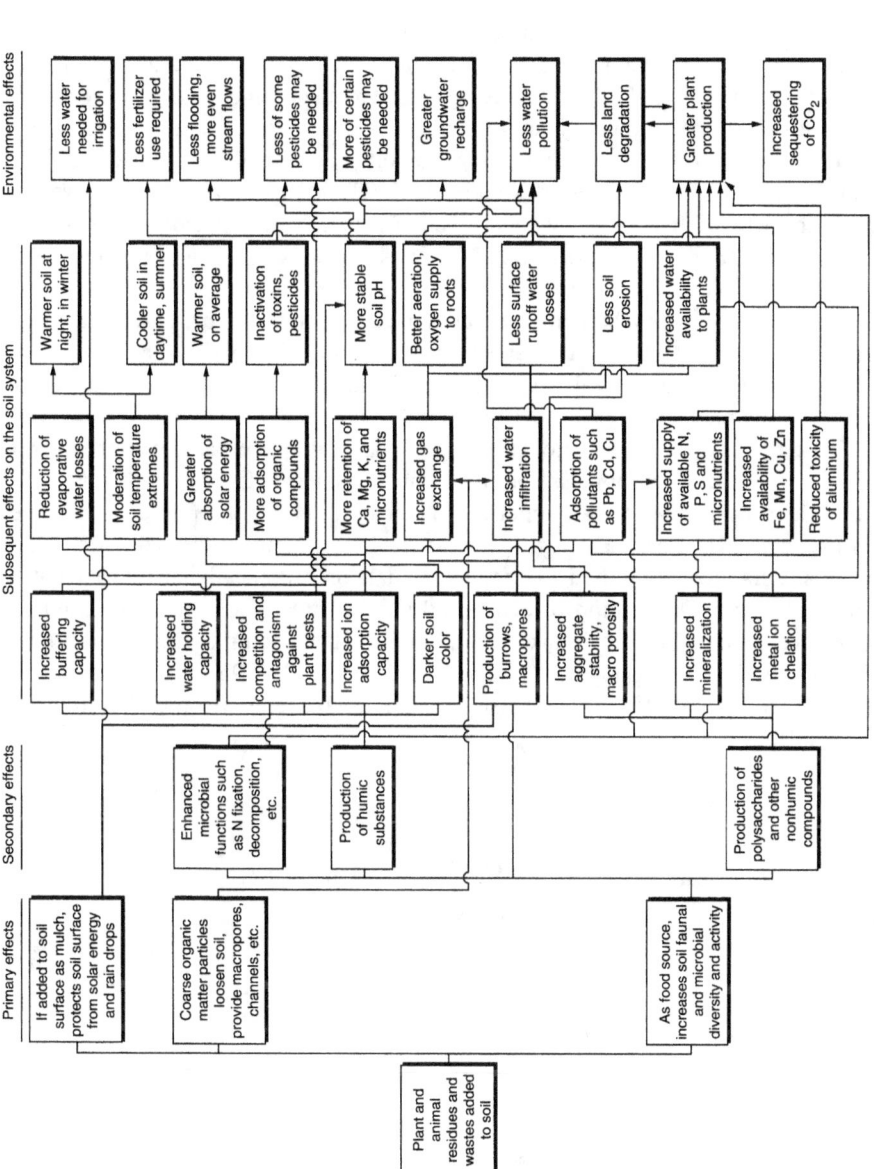

**FIGURE 1.17** An illustration of how addition of organic materials to a soil sets off a far-reaching, complex web of interrelated changes, some more direct than others, that alter the soil's properties, behaviors and environmental impacts. The bold line illustrates one cause and effect chain of alterations that might result from applying organic residues to the soil surface as mulch. (From Brady, N.C., and R.R. Weil. 2002. *The Nature and Properties of Soils*, 13th ed. Prentice-Hall, Upper Saddle River, NJ. With permission.)

organic C can be defined by their turnover rates and availability to microbial processes (Figure 1.18). Models designed to predict SOM dynamics have made use of this concept, incorporating at least two contrasting pools: a passive or stable pool and a labile or active pool (Duxbury et al., 1989). A relatively small portion of SOM with a half-life measured in months or a few years accounts for most of the biological activity in soils and plays a particularly important role in maintaining SQ (Weil, 1992). The living biomass itself is often considered to be part of this active pool.

The labile or active pool of SOM also seems largely responsible for building soil aggregate structure, micronutrient chelation, and nutrient mineralization (Blair and Crocker, 2000; Gunapala and Scow, 1998; Tisdall and Oades, 1982) and is very sensitive to management changes. According to Loveland and Webb (2003), a "growing body of evidence suggests that the 'active' fraction of SOM is a more important factor in controlling change in soil properties than the total SOM." A simple chemical method recently developed by Weil et al. (2003) promises to provide a convenient means of estimating the active C fraction for SQ assessment. Active C by this method is more closely related to microbial activity (Figure 1.19; Weil et al., 2000) and is more sensitive to soil management (Figure 1.20) than is total SOC.

In contrast, a passive pool, with a half-life measured in centuries, is composed of recalcitrant compounds that are quite resistant to decay and prone to accumulate over time. It therefore represents the bulk of the C in most soils. Attempts to chemically or physically separate and characterize these various pools are ongoing. The work of Lemaitre et al. (1995) can be taken as an example, where their "objective was to determine a pool of labile organic matter which is intermediate between the microbial biomass and the humic substances. Until now this pool was just a concept described by mathematical models." Bottner et al. (1999) proposed to fit a five-pool decomposition model (labile and recalcitrant plant residues, labile microbial metabolites, microbial biomass, and stabilized humified compounds) to data on isotopically labeled and native SOM of several arable soils. Delineation of active and passive SOC pools must be based not only on the organic substances' chemical make-up and lability, but also on the location of these substances between or within soil aggregates (Figure 1.21; Whitbread, 1995), because this might control their physical accessibility to microorganisms of various sizes and therefore their turnover rates (Six et al., 2002).

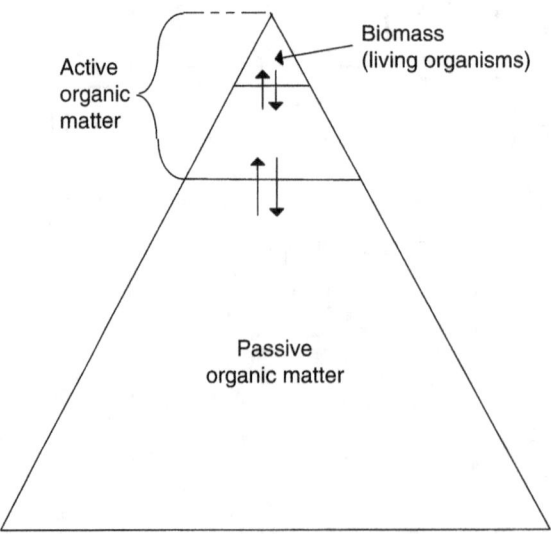

**FIGURE 1.18** The total organic matter in a soil includes microbially active and passive pools, the biomass being part of the active pool. (Figure courtesy of R. Weil.)

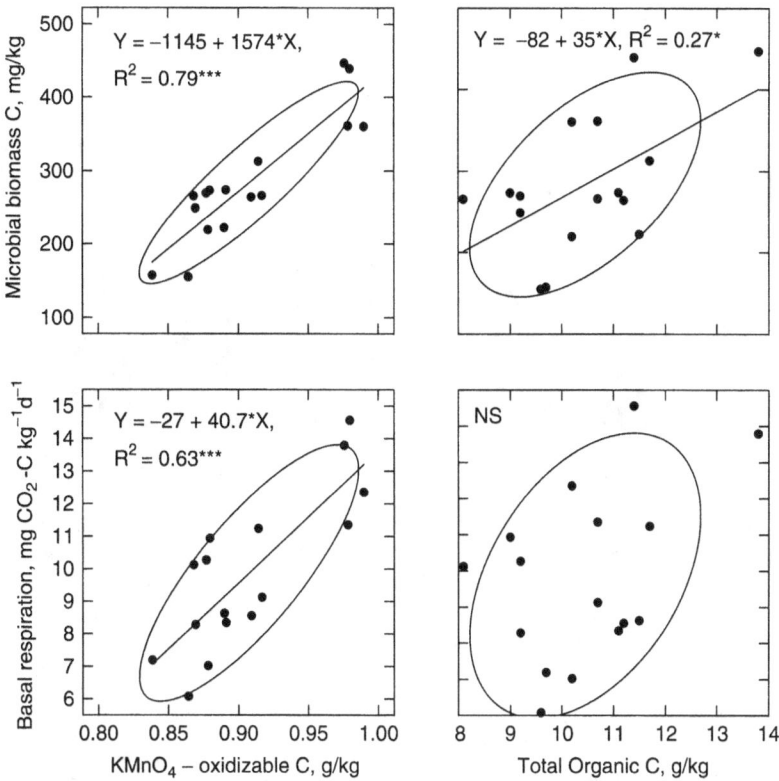

**FIGURE 1.19** Relationships between active or total soil organic carbon and soil microbial biomass C (upper) or soil basal respiration rate (lower). The KMnO4 oxidizable (active) C was a much better predictor of microbial soil functions than total SOC (by LECO high-temperature combustion). Data are for the upper 7.5 cm of Codorus silt loam soil in soybean–wheat agroecosystems exposed to various levels of atmospheric $CO_2$ in open-top chambers for the previous 5 years. Ellipses represent regions in which 90% of the cases are expected to occur. (From Weil, R.R. et al. 2000. In *Agronomy Abstracts for the Annual Meetings of the Soil Science Society of America*, 5–9 November 2000, Minneapolis, MN, p. 47. With permission.)

The active pool of SOC provides the fuel that drives the soil food web. The type and diversity of plants and organic residues added to a soil can influence the type and diversity of organisms that make up the soil community, and vice versa (Bradford et al., 2002; Chapter 6, this book). A diversity of species at many trophic levels is thought to be important for a healthy plant–soil–animal system, including efficient nutrient cycling, disease suppression, and plant pest resistance. Active C is available for microbial assimilation of inorganic N (immobilization) and is likely an important factor in the ability of natural ecosystems to cycle N efficiently with little loss. DeLuca and Keeney (1993) showed that the ratio of soluble or active C to inorganic N was a much more sensitive indicator of the status of N cycling efficiency than was the ratio of total C to total N. They also calculated that in the 11 cultivated Iowa fields they studied, the available active C did not provide enough energy for microbial assimilation and reduction of the $NO_3^-$ present, whereas there was more than enough energy in the active C of the paired prairie soils to assimilate all the $NO_3^-$ in those systems.

Jannsen (1984) developed an early model to separate, by decomposition rates, what he called young and old SOM. He used this model to illustrate the effects of long-term fertilization with mineral N, green manure, and animal manure on the mineralization of soil N. After 25 years, the differences in N mineralization were more closely related to the amounts of young SOM than to

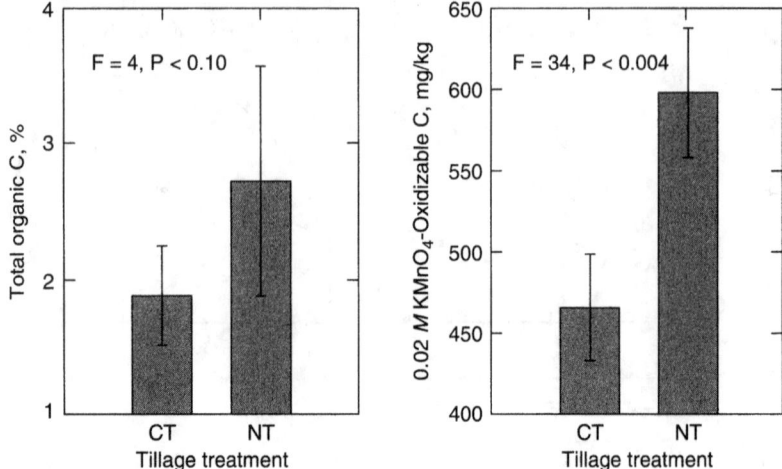

**FIGURE 1.20** Comparative sensitivity of total organic C and active C determined by oxidation with dilute $KMnO_4$ to long-term tillage treatments in a replicated wheat-based rotation experiment at Mandan, ND. The total C levels were not statistically significantly different between the treatments, but a highly significant difference was observed in the active C levels. The soil was a Wilton silt loam (fine-silty, mixed, superactive frigid Pachic Haplustoll). (From Weil, R.R. et al. 2003. *Am. J. Altern. Agric.* 18:3–17. With permission.)

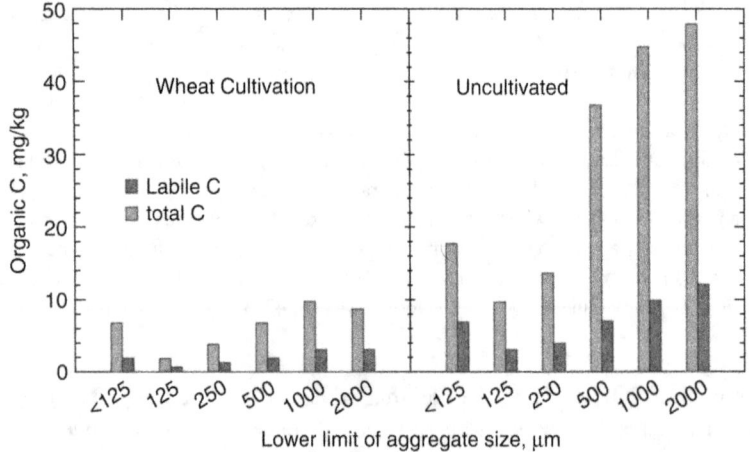

**FIGURE 1.21** Distribution of organic carbon (total and active) among various size classes of water-stable aggregates in soil from adjacent plots under long-term wheat cultivation or uncultivated perennial vegetation in Australia. (From Whitbread, A. 1995. In Lefroy, R. et al. (Eds.), *Soil Organic Matter Management for Sustainable Agriculture: A Workshop.* ACIAR Proceedings No. 56, Australian Center for International Agricultural Research, Canberra, pp. 124–130. With permission.)

the total N contents of the soils (Table 1.4). A more recent two-compartment SOC computer model has been developed by Andrén and Kätterer (2001). Their introductory carbon balance model (ICBM) attempts to describe the global diversity of soil carbon dynamics by using a model with only two state variables and five parameters. The same authors derived calibration parameter values for the model from a 35-year experiment with arable crops on a clay soil in central Sweden (Andrén and Kätterer, 1997). They then showed that the model could predict effects of changed inputs, climate, initial pools, and litter quality on soil carbon pools over a medium-term period (30 years). Wander (Chapter 3) provides a more detailed discussion of the pools of SOC and the fractionation schemes aimed at helping scientists better understand them.

**TABLE 1.4**
**Effects of 25 Years of Applying Three Types of N Amendments on Organic Matter Related Soil Properties in the Netherlands**

| Soil Property | Fertility Management Practice | | |
|---|---|---|---|
| | MinN | MinN + GM | MinN + GM +AnMan |
| BD (g/cm³) | 1.5 | 1.45 | 1.35 |
| Total SOM (Mg/ha) | 78.4 | 81.9 | 86.9 |
| Young SOM (Mg/ha) | 6.2 | 11.0 | 15.2 |
| Old SOM (Mg/ha) | 72.2 | 70.9 | 71.7 |
| Total N (Mg/ha) | 4.0 | 4.4 | 4.6 |
| Mineralized N from young SOM (Mg/ha) | 34 | 74 | 108 |
| Mineralized N from old SOM (Mg/ha) | 26 | 26 | 26 |

*Note:* MinN: N fertilizer; GM: green manure, AnMan: animal manure and pasture ley.
*Source:* Data selected from Jannsen, B.H. 1984. *Plant Soil* 76:297–304.

## RELATIONSHIP OF SOM WITH SOIL PRODUCTIVITY AND CROP YIELDS

Research during the past 50 years has placed much emphasis on the importance of mineral nutrients for crop productivity, with notably less research on the crop productivity effects of SOM management. Pieri (1992) noted that in semiarid regions, roles of SOM in increasing crop yields can include (1) promotion of root development, which can improve efficiency of water use from soil; (2) stabilization of soil structure, which improves permeability and the infiltration of rainwater into the soil; (3) mineralization of nitrogen (and presumably P and S), which affects plant nutrition; and (4) contribution to soil chemical properties such as CEC and acidity. He reported that in West Africa, "regular application of farmyard manure to sandy soils has greatly improved water supply to crops...not due to changes in hydraulic conductivity or water retention capacity, but entirely due to stimulation of root growth and activity."

Although the beneficial effects of SOM management on soil functions have been well researched (Karlen et al., 1992; Seybold et al., 1996), it is more difficult to demonstrate the influence of SOM on crop yields (Strickling, 1975; Lucas et al., 1977). One reason for this difficulty is that SOM levels are usually related to climate, topography, and soil texture. Where crop productivity is compared across widely differing soils, the yield effects of these environmental factors tend to obscure or be confounded with those due to the SOM levels themselves.

In India (Kanchikerimath and Singh, 2001), linear correlations were reported between 26-year average yields for the three crops and the final SOM level in experimental plots. In the Kenyan highlands on Humic Nitosols, Kapkiyai et al. (1998) showed some relationships between crop yields and SOM when maize stover retention, cattle manure application (10 t/ha/year), and fertilizer inputs (120 and 23 kg/ha annually for N and P, respectively) were varied in a maize–bean system. Particulate organic matter (POM) was correlated with yield for the no-fertilizer treatments ($r = 0.82$) but not for fertilized treatments. Crop yields were significantly affected by manure and fertilizer addition ($P < 0.001$) but not stover retention ($P = 0.13$). In Honduras, Stine and Weil (2002) found that soil C parameters were highly predictive of macroaggregate stability and soil porosity across different tillage systems (Figure 1.22). Macroaggregate stability and soil C (particularly active C) were highly correlated with crop productivity across tillage systems. They

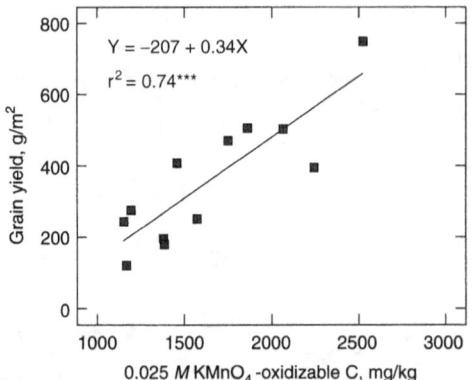

**FIGURE 1.22** Relationship between corn grain yield and active fraction soil organic C (oxidizable by 0.025 m$M$ KMnO$_4$) 4 years after converting a permanent grass pasture to grain production by three tillage systems in Honduras. Open squares: conventional tillage; gray squares: minimum till; black squares: no till. (From Stine, M.A., and R.R. Weil. 2002. *Am. J. Altern. Agric.* 17:2–8. With permission.)

suggested that SOC in the surface layer, especially the active C fraction, markedly affected productivity of the soil through its influence on soil structure. In an experiment in the northern Great Plains of the U.S., Bauer and Black (1994) estimated that over a range of 64 to 142 Mg SOM/ha, every additional Mg of SOM/ha was responsible for approximately an additional 15 kg/ha of wheat grain yield and 35 kg/ha of total aboveground dry matter. They indicated that the primary effect of SOM in their experiment was the contribution of available N. In another rotation experiment (Mitchell and Entry, 1998), corn and cotton yields were correlated with SOM. Organic C in the experimental plots ranged from ca. 30 to more than 150 mg/kg. Although not able to separate out nitrogen or other possible effects, they concluded that "winter legumes and crop rotations result in larger amounts of both C and N in soil which ultimately contribute to higher cotton and corn yields regardless of other practices." Unfortunately, in these types of experiments, cause and effect cannot be distinguished. Where differences in SOM are achieved by imposing treatments such as different rotations, tillage systems, or rates of manure application, the SOM effects on crop yields are usually confounded with other effects of these treatments, such as water and temperature alterations due to surface mulch, effect of N and other essential elements contained in applied legume residues or manure, and rotation effects on plant pathogens and pests. Also, higher levels of SOM and POM are likely to be caused by the more abundant residues from the higher plant productivity in amended treatments rather than vice versa. In contrast to such experiments, Strickling (1975) was successful in isolating the effect of SOM. He imposed various crop rotations for two decades to alter soil C levels and then treated all plots alike for several more years (including adequate mineral fertilization for all plots). Therefore, his experiment assessed the residual effects of the buildup or depletion of SOM. He found that SOM levels accounted for 82 to 84% of the variation in corn yield, regardless of level of N fertilization (Figure 1.23). He attributed the SOM effects on yield mainly to enhancement of water infiltration resulting from improved aggregation.

## CONCLUSIONS

A farmer attempting to optimize soil management on a given farm or field is likely to view SQ as the condition of an existing soil relative to the potential functional capacity of the soil. This view of SQ is essentially concerned with the health of the soil ecosystem. Relatively immutable soil properties such as slope and texture are of limited value in indicating the impact of management

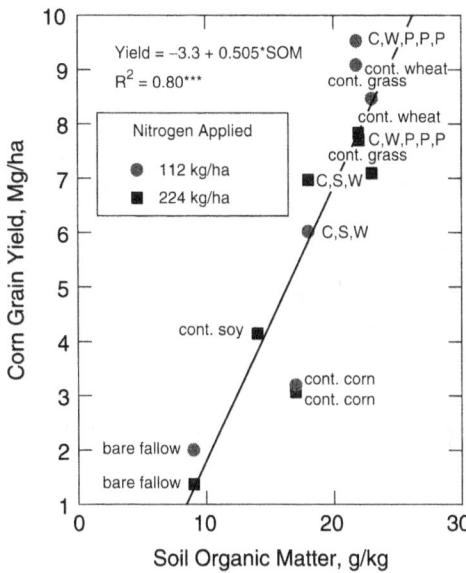

**FIGURE 1.23** Relationship between corn yield and soil organic matter level resulting from various crop rotations during the previous 20 years. The yields were measured in the second year of conventional tillage corn production following the 20-year rotation period on a Beltsville silt loam (Typic Fragiudults) in Maryland. C: corn; P: perennial hay; W: wheat; S: soybean. (Plotted from data in Strickling, E. 1975. In *North East Branch American Society of Agronomy Abstracts*. Agronomy Society of America, Madison, WI, pp. 20–29.)

on SQ. On the other hand, such ephemeral soil properties as water content, temperature, surface soil pH, soluble nutrient levels, and field respiration rates are so easily altered by irrigation, weather, tillage, liming, and fertilization that they normally contribute little to the understanding of long-term management effects on SQ. Between these extremes of stability lie a number of key soil properties that are affected by management practices and can have a critical influence on how soil performs its ecosystem functions. Among these important indicators of SQ are aggregate stability, active SOC, potentially mineralizable N, available water holding capacity, and cation exchange capacity.

Although SOM management might not have much influence on such permanent properties as texture, slope, depth, and mineralogy, there is little else about a soil that cannot be dramatically impacted by changes in SOM. Increased SOM can even mitigate some problems associated with too much clay or too much sand in the soil matrix. Management practices that increase SOM set off a chain of cause-and-effect alterations, most of which improve SQ and health.

The multifarious effects of SOM on SQ and ecological functions described in this chapter highlight the conflict between the need to use SOM and the need to conserve it. The decomposition of SOM is necessary for its use as a source of N, P, and S for plant growth and organic compounds that fuel the food web, promote biological diversity and disease suppression, stabilize aggregates, and chelate metals. In contrast, accumulation of SOM is necessary for these functions in the long term, as well as for the use of SOM as a means of sequestering C and as a source of water-holding capacity and sorptive surface area to hold cations, immobilize pesticides, and detoxify metals. The management of soils to maximize the benefits from SOM will require intelligent compromises in the pursuit of these two contradictory SOM goals.

## REFERENCES

Acton, D.F., and G.A. Padbury. 1993. A conceptual framework for soil quality assessment and monitoring. In D.F. Acton (Ed.), *A Program to Assess and Monitor Soil Quality in Canada: Soil Quality Evaluation Program Summary.* Research Branch, Agriculture and Agri-Food Canada, Ottawa, pp. 2-1–2-10.

Alvarez, R., M. Alconada, and R. Lavado. 1999. Sewage sludge effects on carbon dioxide-carbon production from a desurfaced soil. *Commun. Soil Sci. Plant Anal.* 30:1861–1866.

Anderson, T.H., and K.H. Domsch. 1990. Application of eco-physiological quotients ($qCO_2$ and qD) on microbial biomasses from soils of different cropping histories. *Soil Biol. Biochem.* 22:251–255.

Andrén, O., and T. Kätterer. 1997. ICBM; The introductory carbon balance model for exploration of soil carbon balances. *Ecol. Appl.* 7:1226–1236.

Andrén, O., and T. Kätterer. 2001. Basic principles for soil carbon sequestration and calculating dynamic country-level balances including future scenarios. In R. Lal et al. (Eds.), *Assessment Methods for Soil Carbon.* Lewis Publishers, Boca Raton, FL.

Arrouays, D., W. Deslais, and V. Badeau. 2001. The carbon content of topsoil and its geographical distribution in France. *Soil Use Manage.* 17:7–11.

Arshad, M.A., and G.M. Coen. 1992. Characterization of soil quality: Physical and chemical criteria. *Am. J. Altern. Agric.* 7:25–31.

Arshad, M.A., and J.W.T. Frankenberger. 1998. Plant growth-regulating substances in the rhizosphere: Microbial production and functions. *Adv. Agron.* 62:45–151.

Batjes, N.H. 1996. Total carbon and nitrogen in the soils of the world. *Eur. J. Soil Sci.* 47:157–161.

Batjes, N.H. 1999. *Management Options for Reducing $CO_2$ Concentrations in the Atmosphere by Increasing Carbon Sequestration in the Soil.* International Soil Reference and Information Centre (ISRIC), Wageningen, the Netherlands.

Bauer, A., and A.L. Black. 1994. Quantification of the effect of soil organic matter content on soil productivity. *Soil Sci. Soc. Am. J.* 58:185–193.

Bayer, C., L. Martin-Neto, J. Mielniczuk, C.N. Pillon, and L. Sangoi. 2001. Changes in soil organic matter fractions under subtropical no-till cropping systems. *Soil Sci. Soc. Am. J.* 65:1473–1478.

Blair, N., and G.J. Crocker. 2000. Crop rotation effects on soil carbon and physical fertility of two Australian soils. *Aust. J. Soil Res.* 38:71–84.

Boix-Fayos, C., A. Calvo-Cases, A.C. Imeson, and M.D. Soriano-Soto. 2001. Influence of soil properties on the aggregation of some Mediterranean soils and the use of aggregate size and stability as land degradation indicators. *Catena* 44:47–67.

Bottner, P., M. Pansu, and Z. Sallih. 1999. Modelling the effect of active roots on soil organic matter turnover. *Plant Soil* 216:15–25.

Bowen, G.W., and A.D. Rovira. 1999. The rhizosphere and its management to improve plant growth. *Adv. Agron.* 66:1–102.

Bradford, M.A., T.H. Jones, R.D. Bardgett, H.I.J. Black, B. Boag, M. Bonkowski, R. Cook, T. Eggers, A.C. Gange, S.J. Grayston, E. Kandeler, A.E. McCaig, J.E. Newington, J.I. Prosser, H. Setala, P.L. Staddon, G.M. Tordoff, D. Tscherko, and J.H. Lawton. 2002. Impacts of soil faunal community composition on model grassland ecosystems. *Science* 298:615–618.

Brady, N.C., and R.R. Weil. 2002. *The Nature and Properties of Soils,* 13th ed. Prentice-Hall, Upper Saddle River, NJ.

Brejda, J.J., M.J. Mausbach, J.J. Goebel, D.L. Allan, T.H. Dao, D.L. Karlen, T.B. Moorman, and J.L. Smith. 2001. Estimating surface soil organic carbon content at a regional scale using the National Resource Inventory. *Soil Sci. Soc. Am. J.* 65:842–849.

Bresson, L.M., C. Koch, Y. Le Bissonnais, E. Barriuso, and V. Lecomte. 2001. Soil surface structure stabilization by municipal waste compost application. *Soil Sci. Soc. Am. J.* 65:1804–1811.

Bundy, L.G., T.W. Andraski, and J.M. Powell. 2001. Management practice effects on phosphorus losses in runoff in corn production systems. *J. Environ. Qual.* 30:1822–1828.

Carpenter-Boggs, L., J.P. Reganold, and A.C. Kennedy. 2000. Biodynamic preparations: Short-term effects on crops, soils, and weed populations. *Am. J. of Altern. Agric.* 15:110–118.

Carter, M.R. 2001. Organic matter and sustainability, In R.M. Rees et al. (Eds.), *Sustainable Management of Soil Organic Matter.* CABI Publishing, New York, pp. 9–22.

Carter, M.R., D.A. Angers, and H.T. Kunelius. 1994. Soil structural form and stability, and organic matter under cool-season perennial grasses. *Soil Sci. Soc. Am. J.* 58:1194–1199.

Chan, K.Y., A.M. Bouwman, W. Smith, and R. Ashley. 2001. Restoring soil fertility of degraded hardsetting soils in semi-arid areas with difference pastures. *Aust. J. Exp. Agric.* 41:507–514.

Chen, Y., M. De Nobili, and T. Aviad. 2003. Stimulatory effects of humic substances on plant growth. In F. Magdoff and R. R. Weil (Eds.), *Function and Management of Soil Organic Matter in Agroecosystems.* CRC Press, Boca Raton, FL.

Childers, S.E., S. Ciufo, and D.R. Lovely. 2002. *Geobacter metallireducens* accesses insoluble Fe(III) oxide by chemotaxis. *Nature* 416:767–769.

Corbeels, M., A. Shiferaw, and H. Mitiku. 2000. Farmers' knowledge of soil fertility and local management strategies in Tigray, Ethiopia. Managing Africa's Soils Working Paper 10, NUTNET — Networking on Soil Fertility Mangement, Drylands Programme, IIED, London.

Dalal, R.C. 1998. Soil microbial biomass: What do the numbers really mean? *Aust. J. Exp. Agric.* 38:649–665.

De Bruyn, L., and A. Conacher. 1990. The role of termites and ants in soil modification: A review. *Aust. J. Soil Res.* 28:55–93.

Dekker, L.W., S.H. Doerr, K. Oostinde, A.K. Ziogas, and C.J. Ritsema. 2001. Water repellency and critical water content in a dune sand. *Soil Sci. Soc. Am. J.* 65:1667–1674.

DeLeo, G.A., L. Rizzi, A. Caizzi, and M. Gatto. 2001. Carbon emissions: The economic benefits of the Kyoto Protocol. *Nature* 413:478 – 479.

DeLuca, T.H., and D.R. Keeney. 1993. Soluble organics and extractable nitrogen in paired prairie and cultivated soil of central Iowa. *Soil Sci.* 155:219–228.

Dietrich, A.M., D.L. Gallagher, and K.A. Klawiter. 2001. Inputs of copper-based crop protectants to coastal creeks from plasticulture runoff. *J. Am. Water Resour. Assoc.* 37:281–293.

Doerr, S.H., R.A. Shakesby, and R.P.D. Walsh. 2000. Soil water repellency: Its causes, characteristics and hydro-geomorphological significance. *Earth Sci. Rev.* 51:33–65.

Dumansky, J. 1994. Workshop summary. In J. Dumansky (Ed.), *Proceedings of the International Workshop on Sustainable Land Management for the 21st Century*, Vol. 1. Agricultural Institute of Canada, Ottawa.

Duxbury, J.M., M.S. Smith, and J.W. Doran. 1989. Soil organic matter as a source and a sink of plant nutrients. In D.C. Coleman et al. (Eds.), *Dynamics of Soil Organic Matter in Tropical Ecosystems.* NifTAL Project, Dept. of Agronomy and Soil Science, College of Tropical Agriculture and Human Resources, University of Hawaii, Honolulu, HI, pp. 33–67.

Elliott, J.A., and A.A. Efetha. 1999. Influence of tillage and cropping system on soil organic matter, structure and infiltration in a rolling landscape. *Can. J. Soil Sci.* 79:457–463.

Ferris, H., R. Venette, H. van der Meulen, and S. Lau. 1998. Nitrogen mineralization by bacterial-feeding nematodes: Verification and measurement. *Plant Soil* 203:159–171.

Forge, T.A., E. Hogue, G. Neilsen, and D. Neilsen. 2003. Effects of organic mulches on soil microfauna in the root zone of apple: Implications for nutrient fluxes and functional diversity of the soil food web. *Appl. Soil Ecol.* 22:39–54.

Franzluebbers, A.J., J.A. Stuedemann, and S.R. Wilkinson. 2001. Bermudagrass management in the Southern Piedmont of the USA: I. Soil and surface residue carbon and sulfur. *Soil Sci. Soc. Am. J.* 65:8334–8341.

Gajda, A.M., J. Doran, T. Kettler, J.L. Weinhold, J.L. Pikul, Jr., and C.A. Cambardella. 2001. Soil quality evaluations of alternative and conventional management systems in the Great Plains, In R. Lal et al. (Eds.) *Assessment Methods for Soil Carbon.* Lewis Publishers, Boca Raton, FL, pp. 381–400.

Gewin, V.L., A.C. Kennedy, R. Veseth, and B.C. Miller. 1999. Soil quality changes in eastern Washington with Conservation Reserve Program (CRP) take-out. *J. Soil Water Conserv.* 54:432–438.

Gilbert, G.S., J. Handelsman, and J.L. Parke. 1994. Root camouflage and disease control. *Phytopathology* 84:222–225.

Glover, J.D., J.P. Reganold, and P.K. Andrews. 2000. Systematic method for rating soil quality of conventional, organic and integrated apple orchards in Washington State. *Agric. Ecosyst. Environ.* 80:29–45.

Greenland, D.J., and I. Szabolcs. 1994. *Soil Resilience and Sustainable Land Use.* CAB International, Wallingford, U.K.

Gregorich, E.G., M.R. Carter, D.A. Angers, C.M. Monreal, and B.H. Ellert. 1994. Towards a minimum data set to assess soil organic matter quality in agricultural soils. *Can. J. Soil Sci.* 74:367–385.

Gruver, J.B., and R.R. Weil. 1997. The relationship between farmer perceptions of soil quality and a soil quality index. Presentation to the American Society of Agronomy, Anaheim, CA, October 30, 1997 (Agronomy Abstracts, p. 55).

Guant, J.L., S.P. Sohi, H. Yang, N. Mahieu, and J.R.M. Arah. 2001. A procedure for isolating organic matter fractions suitable for modelling. In R.M. Rees et al. (Eds.), *Sustainable Management of Soil Organic Matter*. CABI Publishers, New York, pp. 90–95.

Gunapala, N., and K.M. Scow. 1998. Dynamics of soil microbial biomass and activity in conventional and organic farming systems. *Soil Biol. Biochem.* 30:805–816.

Hamel, C., Y. Dalpe, V. Furlan, and S. Parent. 1997. Indigenous populations of arbuscular mycorrhizal fungi and soil aggregate stability are major determinants of leek (*Allium porrum* L.) response to inoculation with *Glomus intraradices* Schenck & Smith or *Glomus versiforme* (Karsten) Berch. *Mycorrhiza* 7:187–196.

Hargrove, W.L., and G.W. Thomas. 1981. Effect of organic matter on exchangeable aluminum and plant growth in acid soils. Chemistry on the Soil Environment, Special Publication No. 40, American Society for Agronomy and Soil Science Society of America, Madison, WI.

Hatcher, P.G., K.J. Dria, S. Kim, and S.W. Frazier. 2001. Modern analytical studies of humic substances. *Soil Sci.* 166:770–794.

Haynes, R.J., and M.H. Beare. 1997. Influence of six crop species on aggregate stability and some labile organic matter fractions. *Soil Biol. Biochem.* 29:1647–1653.

Helling, C.S., G. Chesters, and R.B. Corey. 1964. Contribution of organic matter and clay to soil cation exchange capacity as affected by the pH of the saturated solution. *Soil Sci. Soc. Am. J.* 28:517–520.

Hermawan, B., and A.A. Bomke. 1977. Effects of winter cover crops and successive spring tillage on soil aggregation. *Soil Tillage Res.* 44:109–120.

Hodge, A., C.D. Campbell, and A.H. Fitter. 2001. An arbuscular mycorrhizal fungus accelerates decomposition and acquires nitrogen from organic material. *Nature* 413:297–299.

Hoitink, H., and M. Boehm. 1999. Biocontrol within the context of soil microbial communities: A substrate-dependent phenomenon. *Annu. Rev. Phytopathol.* 37:427–446.

Hubbs, M.D., M.L. Norfleet, and D.T. Lightle. 2002. Interpreting the soil conditioning index. In E. van Santen (Ed.) *Making Conservation Tillage Conventional: Building a Future on 25 Years of Research*. Proceedings of the 25th Annual Southern Conservation Tillage Conference for Sustainable Agriculture, 24–26 June 2002, Auburn, AL. Special Report No. 1, Alabama Agricultural Experimental Station and Auburn University, Auburn, AL.

Hudson, B.D. 1994. Soil organic matter and available water capacity. *J. Soil Water Conserv.* 49:189–194.

Hue, N.V., and D.L. Licudine. 1999. Amelioration of subsoil acidity through surface applications of organic manures. *J. Environ. Qual.* 28:623–632.

Hulugalle, N.R., D.L. Larsen, and S. Henggeler. 1996. Organic by-product effects on soil chemical properties and microbial communities. *Compost Sci. Util.* 4:70–80.

Islam, K.R., C.L. Mulchi, and A.A. Ali. 1999. Tropospheric carbon dioxide or ozone enrichments and moisture effects on soil organic carbon quality. *J. Environ. Qual.* 28:1629–1636.

Islam, K.R., and R.R. Weil. 2000. Soil quality indicator properties in mid-Atlantic soils as influenced by conservation management. *J. Soil Water Conserv. (Ankeny)* 55:69–78.

Jannsen, B.H. 1984. A simple method for calculating decomposition and accumulation of "young" soil organic matter. *Plant Soil* 76:297–304.

Jenkinson, D.S. 1991. The turnover of organic carbon and nitrogen in soil. *Phil. Trans. R. Soc. London, Series B* 329:361–368.

Jenny, H. 1941. *Factors of Soil Formation: A System of Quantitative Pedology*. McGraw-Hill, New York.

Juma, N.G. 1993. Interrelationships between soil structure/texture, soil biota/soil organic matter and crop production. *Geoderma* 57:3–30.

Kanchikerimath, M., and D. Singh. 2001. Soil organic matter and biological properties after 26 years of maize-wheat-cowpea cropping as affected by manure and fertilization in a Cambisol in semiarid region of India. *Agric. Ecosyst. Environ.* 86:155–162.

Kapkiya, J.J., N. Karanja, P.L. Woomer, and J.N. Quresh. 1998. Soil organic carbon fractions in a long-term experiment and the potential for their use as a diagnostic assay in highland farming systems of central Kenya. *African Crop Sci. J.* 6:19–28.

Karlen, D.L., S.S. Andrews, and J.W. Doran. 2001. Soil quality: Current concepts and applications. *Adv. Agron.* 74:1–40.

Karlen, D.L., N.S. Eash, and P.W. Unger. 1992. Soil and crop management effects on soil quality indicators. *Am. J. Altern. Agric.* 7:48–55.

Karlen, D.L., M.J. Mausbach, J.W. Doran, R.G. Cline, R.F. Harris, and G.E. Schuman. 1997. Soil quality: A concept, definition, and framework for evaluation. *Soil Sci. Soc. Am. J.* 61:4–10.

Karlen, D.L., M.J. Rosek, J.C. Gardner, D.L. Allan, M.J. Alms, D.F. Bezdicek, M. Flock, D.R. Huggins, B.S. Miller, and M.L. Staben. 1999. Conservation reserve program effects on soil quality indicators. *J. Soil Water Conserv.* 54:439–444.

Kayuki, K.C., and C.S. Wortmann. 2001. Plant materials for soil fertility management in subhumid tropical areas. *Agron. J.* 93:929–935.

Kennedy, A.C., and R.I. Papendick. 1995. Microbial characteristics of soil quality. *J. Soil Water Conserv.* 50:243–248.

Kern, J.S. 1994. Spatial patterns of soil organic carbon in the contiguous United States. *Soil Sci. Soc. Am. J.* 58:439–455.

Kögel-Knabner, I. 2002. The macromolecular organic composition of plant and microbial residues as inputs to soil organic matter. *Soil Biol. Biochem.* 34:139–162.

Krishnaswamy, J., and D.D. Richter. 2002. Properties of advanced weathering-stage soils in tropical forests and pastures. *Soil Sci. Soc. Am. J.* 66:244–253.

Kucey, R.M.N., H.H. Janzen, and M.E. Leggertt. 1989. Microbially mediated increases in plant-available phosphorus. *Adv. Agron.* 42:100–228.

Lal, R. 1997. Effects of N fertilizer treatments on biologically active N pools in soils under plow and no tillage. *Biol. Fertil. Soils* 24:4, 406–412.

Lal, R., J. Kimble, and R.F. Follett. 2001. Methodological challenges toward balancing soil C pool and fluxes. In R. Lal et al. (Eds.), *Assessment Methods for Soil Carbon*. Lewis Publishers, Boca Raton, FL, pp. 659–668.

Larink, O., and S. Schrader. 2000. Rehabilitation of degraded compacted soil by earthworms. In R. Horn et al. (Eds.), *Subsoil Compaction: Distribution, Processes and Consequences. Advances in Geoecology*, Vol. 32. Catena Verlag, Reiskirchen, Germany, pp. 282–294.

Larson, W.E., and F.J. Pierce. 1991. Conservation and enhancement of soil quality evaluation for sustainable land management in the developing world. International Board for Soil Research and Management (IBSRAM), Proceedings No. 12, Vol. 2, Bangkok, Thailand.

Lavahun, M.F.E., R.G. Joergensen, and B. Meyer. 1996. Activity and biomass of soil microorganisms at different depths. *Biol. Fertil. Soils* 23:38–42.

Lavelle, P., D. Bignell, M. Lepage, V. Wolters, P. Roger, P. Ineson, O.W. Heal, and S. Dhillion. 1997. Soil function in a changing world: The role of invertebrate ecosystem engineers. *Eur. J. Soil Biol.* 33:159–193.

Le Bissonnais, Y. 1996. Soil characteristics and aggregate stability. In M. Agassi (Ed.), *Soil Erosion Conservation and Rehabilitation*. Marcel Dekker, New York, pp. 41–60.

Lemaitre, A., R. Chaussod, Y. Tavant, and S. Bruckert. 1995. An attempt to determine a pool of labile organic matter associated with the soil microbial biomass. *Eur. J. Soil Biol.* 31:121–125.

Lewandowski, A. 2002. Organic matter management. Soil Management Series BU–07402, Minnesota Institute for Sustainable Agriculture, University of Minnesota Extension Service and Soil Quality Institute, Natural Resources Conservation Service, United States Department of Agriculture, St. Paul, MN.

Lewandowski, A., M. Zumwinkle, and A. Fish. 1999. Assessing the soil system: A review of soil quality literature. Minnesota Department of Agriculture, Energy and Sustainable Agriculture Program, St. Paul, MN.

Liebig, M., and J. Doran. 1999. Evaluation of farmers' perceptions of soils quality indicators. *Am. J. of Altern. Agric.* 14:11–21.

Liebman, M., and A.S. Davis. 2000. Integration of soil, crop, and weed management in low-input farming systems. *Weed Res.* 40:27–47.

Losi, M.E., C. Amrheim, and W.T. Frankenberger. 1994. Bioremediation of chromatic contaminated groundwater by reduction and precipitation in surface soils. *J. Environ. Qual.* 23:1141–1150.

Loveland, P., and J. Webb. 2003. Is there a critical level of organic matter in the agricultural soils of temperate regions: A review. *Soil Tillage Res.* 70:1–18.

Lu, Y., R. Wassmann, H. Neue, and C. Huang. 2000. Dynamics of dissolved organic carbon and methane emissions in a flooded rice soil. *Soil Sci. Soc. Am. J.* 64:2011–2017.

Lucas, R.E., J.B. Hotman, and L.J. Connor. 1977. Soil carbon dynamics and cropping practices. In W. Lockeretz (Ed.), *Agriculture and Energy.* Academic Press, New York, pp. 333–351.

Macks, S.P., B.W. Murphy, H.P. Cresswell, and T.B. Koen. 1996. Soil friability in relation to management history and suitability for direct drilling. *Aust. J. Soil Res.* 34:343–360.

Magdoff, F., and R.J. Bartlett. 1985. Soil pH buffering revisited. *Soil Sci. Soc. Am. J.* 62:145–148.

Magdoff, F., and R.R. Weil. 2003. Strategies for managing organic matter. In F. Magdoff and R.R. Weil (Eds.), *Functions and Management of Soil Organic Matter in Agroecosystems.* CRC Press, Boca Raton, FL.

Magdoff, F.R. 1996. Soil organic matter fractions and implications for interpreting organic matter tests. In F.R. Magdoff et al. (Eds.), *Soil Organic Matter: Analysis and Interpretations.* SSSA Special Publication No. 46, Soil Science Society of America, Madison, WI, pp. 11–19.

McCarl, B.A., and U.A. Schneider. 2001. Greenhouse gas mitigation in U.S. agriculture and forestry. *Science* 294:2481–2482.

McQuaid, B.F., and G.L. Olson. 1998. Impact of carbon sequestration on functional indicators of soil quality as influenced by management in sustainable agriculture. In *Soil Processes and the Carbon Cycle,* CRC Press, Boca Raton, FL.

Mitchell, C.C., and J.A. Entry. 1998. Soil C, N, and crop yields in Alabama's long-term "old rotation" cotton experiment. *Soil Tillage Res.* 47:331–338.

Molina, J.A.E., and P. Smith. 1998. Modeling carbon and nitrogen processes in soils. *Adv. Agron.* 62: 253–298

Murage, E.W., N.K. Karanja, P.C. Smithson, and P.L. Woomer. 2000. Diagnostic indicators of soil quality in productive and non-productive smallholders' fields of Kenya's Central Highlands. *Agric. Ecosyst. Environ.* 79:1–8.

Myneni, S.C.B. 2002. Formation of stable chlorinated hydrocarbons in weathering plant. *Mater. Sci.* 295:1039–1041.

Nardi, S., D. Pizzeghello, A. Muscolo, and A. Vianello. 2002. Physiological effects of humic substances on higher plants. *Soil Biol. Biochem.* 34:1527–1536.

Neilands, E.B. 1996. A saga of siderophores. In T.R. Swinburne (Ed.), *Iron Siderophores and Plant Diseases.* Plenum, New York, pp. 289–298.

Neue, H.U., R. Wasserman, R.S. Lantin, M.C. Alberto, and J.B. Aduna. 1994. Methane prodcution potential of soils. *Int Rice Res. Notes* 19:37–38.

Nichols, J.D. 1984. Relation of organic carbon to soil properties and climate in the southern Great Plains. *Soil Sci. Soc. Am. J.* 48:1382–1384.

Nichols, K.A., and S.F. Wright. 2003. Contributions of fungi to soil organic matter in agroecosystems. In F. Magdoff and R.R. Weil (Eds.), *Functions and Management of Soil Organic Matter in Agroecosystems.* CRC Press, Boca Raton, FL, pp. 289–298.

Oades, J.M. 1995. Recent advances in organomineral interactions: Implications for carbon cycling and soil structure. In P.M. Huang et al. (Eds.), *Environmental Impact of Soil Component Interactions, Vol. I: Natural and Anthropogenic Organics.* Lewis Publishers, Boca Raton, FL, pp. 119–134.

Olayinka, A., A. Adebayo, and A. Amusan. 1998. Evaluation of organic carbon oxidation efficiencies of a modified wet combustion and Walkley-Black procedures in Nigerian soils. *Commun. Soil Sci. Plant Anal.* 29:2749–2756.

Omay, A., C. Rice, D. Maddux, and W. Gordon. 1998. Corn yield and nitrogen uptake in monoculture and in rotation with soybean. *Soil Sci. Soc. Am. J.* 62:1596–1603.

Parfitt, R.L., G. Yuan, and B.K.G. Theng. 1999. A 13C-NMR study of the interactions of soil organic matter with aluminum and allophane in podzols. *Eur. J. Soil Sci.* 50:695–700.

Parton, W.J., J.W.B. Stewart, and C.V. Cole. 1988. Dynamics of C, N, P and S in grassland soils: A model. *Biogeochemistry* 5:109–131.

Paustian, K., J. Six, E.T. Elliott, and H.W. Hunt. 2000. Management options for reducing $CO_2$ emissions from agricultural soils. *Biogeochemistry* 48:147–163.

Pieri, C.J.M.G. 1992. *Fertility of Soils: A Future for Farming in the West African Savannah.* Springer-Verlag, Berlin.

Pikul, J.L., Jr., and J.K. Aase. 1994. Landscape-scale changes in indicators of soil quality due to cultivation in Saskatchewan, Canada. *Geoderma* 1994. 64:1–2, 1–19.

Ponomarenko, E.V., and D.W. Anderson. 2001. Importance of charred organic matter in Black Chernozem soils of Saskatchewan. *Can. J. Soil Sci.* 81:285–297.

Popp, J., D. Hoag, and J. Ascough II. 2002. Targeting soil conservation policies for sustainability: New empirical evidence. *J. Soil Water Conserv.* 57:67–74.

Reddy, G.B., and K.R. Reddy. 1993. Fate of nitrogen-15 enriched ammonium nitrate applied to corn. *Soil Sci. Soc. Am. J.* 57:111–115.

Redeker, K., N. Wang, J. Low, A. McMillan, S. Tyler, and R. Cicerone. 2000. Emissions of methyl halides and methane from rice paddies. *Science* 290:966–969.

Rees, R.M., B.C. Ball, C.D. Campbell, and C.D. Watson (Eds.). 2001. *Sustainable Management of Soil Organic Matter.* CAB International, Wallingford, U.K.

Reicosky, D.C., J.L. Hatfield, and R.L. Sass. 2000. Agricultural contributions to greenhouse gas emissions. In K.R. Reddy and H.F. Hodges (Eds.), *Climate Change and Global Crop Production.* CABI International, London, pp. 37–55.

Reynolds, C.M., D.C. Wolf, T.J. Gentry, L.B. Perry, C.S. Pidgeon, B.A. Koenen, H.B. Rogers, and C.A. Beyrouty. 1999. Plant enhancement of indigenous soil microorganisms: A low cost treatment of contaminated soils. *Polar Rec.* 35:33–40.

Rice, P.J., L.L. McConnell, L.P. Heighton, A.M. Sedeghi, A.R. Isensee, J.R. Teasdale, A. Abdul-Baki, J.A. Harman-Fetcho, and C.J. Hapeman. 2002. Comparison of copper levels in runoff from fresh-market vegetable production using polyethylene mulch or a vegetative mulch. *Environ. Toxicol. Chem.* 21:24–30.

Rickman, R.W., C.L. Douglas, Jr., S.L. Albrecht, L.G. Bundy, and J.L. Berc. 2001. CQESTR: A model to estimate carbon sequestration in agricultural soils. *J. Soil Water Conserv.* 56:237–242.

Sanchez, J.E., T.C. Willson, K. Kizilkaya, E. Parker, and R.R. Harwood. 2001. Enhancing the mineralizable nitrogen pool through substrate diversity in long term cropping systems. *Soil Sci. Soc. Am. J.* 65:1442–1447.

Sauve, S., M. McBride, and W. Hendershot. 1998. Soil solution speciation of lead (II): Effects of organic matter and pH. *Soil Sci. Soc. Am. J.* 62:618–621.

Schnitzer, M. 1995. Infiltration and soil properties as affected by annual cropping in the northern Great Plains. *Agron. J.* 87:4, 656–662.

Schrader, S., and H. Zhang. 1997. Earthworm casting: Stabilization or destabilization of soil structure. *Soil Biol. Biochem.* 29:469–475.

Sekamatte, M.B., M.W. Ogenga-Latigo, and A. Russell-Smith. 2001. The effect of maize stover used as mulch on the termite damage to maize and activity of predatory ants. *Afr. Crop Sci. J.* 9:411–419.

Seybold, C.A., M.J. Mausbach, D.L. Karlen, and H.H. Rogers. 1996. Quantification of soil quality. In T.S.Q. Institute (Ed.), *The Soil Quality Concept.* USDA/NRCS, Washington, D.C., pp. 53–68.

Six, J., E.T. Elliott, and K. Paustian. 2000a. Soil structure and soil organic matter. II. A normalized stability index and the effect of mineralogy. *Soil Sci. Soc. Am. J.* 64:1042–1049.

Six, J., P. Callewaert, S. Lenders, S. De Gryze, S.J. Morris, E.G. Gregorich, E.A. Paul, and K. Paustian. 2002. Measuring and understanding carbon storage in afforested soils by physical fractionation. *Soil Sci. Soc. Am. J.* 66:1981–1987.

Six, J., K. Paustian, E.T. Elliott, and C. Combrink. 2000b. Soil structure and organic matter. I. Distribution of aggregate-size classes and aggregate-associated carbon. *Soil Sci. Soc. Am. J.* 64:681–689.

Sojka, R.E., and D.R. Upchurch. 1999. Reservations regarding the soil quality concept. *Soil Sci. Soc. Am. J.* 63:1039–1054.

Spaccini, R., A. Zena, C.A. Igwe, J.S.C. Mbagwu, and A. Piccolo. 2001. Carbohydrates in water-stable aggregates and particle size fractions of forested and cultivated soils in two contrasting tropical ecosystems. *Biogeochemistry* 53:1–22.

Sparling, G. 1992. Ratio of microbial biomass carbon to soil organic matter carbon as a sensitive indicator of changes in soil organic matter. *Aust. J. Agric. Res.* 30:195–207.

Stevenson, F.J. 1994. *Humus Chemistry.* John Wiley & Sons, New York.

Stine, M.A., and R.R. Weil. 2002. The relationship between soil quality and crop productivity across three tillage systems in south central Honduras. *Am. J. Altern. Agric.* 17:2–8.

Stone, A.G., and H.M. Darby. 2003. Organic matter-mediated supression of soilborne plant diseases in field agricultural systems. In F. Magdoff and R.R. Weil (Eds.), *Functions and Management of Soil Organic Matter in Agroecosystems.* CRC Press, Boca Raton, FL.

Strek, H.J., and J.B. Weber. 1982. Adsorption and reduction in bioactivity of polychlorinated biphenyl (Aroclor 1254) to redroot pigweed by soil organic matter and montmorillonite clay. *Soil Sci. Soc. Am. J.* 46:318–322.

Strickling, E. 1975. Crop sequences and tillage in efficient crop production. In *North East Branch American Society of Agronomy Abstracts*. Agronomy Society of America, Madison, WI, pp. 20–29.

Strong, D.T., and I.R.P. Fillery. 2002. Denitrification response to nitrate concentrations in sandy soils. *Soil Biol. Biochem.* 34:945–954.

Sui, Y., and M.L. Thompson. 2000. Phosphorus sorption, desorption, and buffering capacity in a bio-solids-amended Mollisol. *Soil Sci. Soc. Am. J.* 64:164–169.

Swift, M.J., and P.L. Woomer. 1993. Organic matter and the sustainability of agricultural systems: Definition and measurement. In K. Mulongoy and R. Merckx (Eds.), *Soil Organic Matter Dynamics and Sustainability of Tropical Agriculture*. John Wiley & Sons, New York, pp. 1–18.

Syers, J.K., and E.T. Craswell. 1995. Soil organic matter management for sustainable agriculture: A workshop. In R.D.B. Lefroy et al. (Eds.), *Role of Soil Organic Matter in Sustainable Agricultural Systems*. ACIAR Proceedings No. 56, Australian Center for International Agricultural Research, Canberra, pp. 7–14 (Workshop held on 24–26 August 1994 in Ubon, Thailand).

Syers, J.K., and J.A. Springett. 1984. Earthworms and soil fertility. *Plant Soil* 76:93–104.

Theng, B.K.G., K.R. Tate, and P. Sollins. 1989. Constituents of organic matter in temperate and tropical soils. In D.C. Coleman et al. (Eds.) *Dynamics of Soil Organic Matter in Tropical Ecosystems*. NifTAL Project, Dept. of Agronomy and Soil Science, College of Tropical Agriculture and Human Resources, University of Hawaii, Honolulu, HI, pp. 5–32.

Tindall, J.A., R.B. Beverly, and D.E. Radcliff. 1991. Mulch effect on soil properties and tomato growth using micro-irrigation. *Agron. J.* 83:1028–1034.

Tisdall, J.M., and J.M. Oades. 1982. Organic matter and water-stable aggregates in soils. *Soil Sci. Soc. Am. J.* 33:141–163.

Torsvik, V., L. Ovreas, and T.F. Thingstad. 2002. Prokaryotic diversity: Magnitude, dynamics, and controlling factors. *Science* 296:1064–1066.

Unger, P.W. 1978. Straw mulch effects on soil temperatures and sorghum germination and growth. *Agron. J.* 70:858–864.

USDA-NRCS. 2001. National Agronomy Manual [Online]. Available by USDA-Natural Resources Conservation Service at ftp.ftw.nrcs.usda.gov/pub/Nat_Agron_Manual, accessed 20 August 2002.

Walkley, A., and I.A. Black. 1947. Determination of organic matter in the soil by chromic acid digestion. *Soil Sci.* 63:251–264.

Wander, M.M., and G.A. Bollero. 1999. Soil quality assessment of tillage impacts in Illinois. *Soil Sci. Soc. Am. J.* 63:961–971.

Weil, R.R. 1984. Morphology and properties of some soils of Sri Lanka's central mountains. *Sri Lankan J. Agric. Sci.* 21:108–127.

Weil, R.R. 1992. Inside the heart of sustainable farming: An intimate look at soil life and how to keep it thriving. *New Farm* 12:45–48.

Weil, R.R. 2000. Soil and plant influences on crop response to two African phosphate rocks. *Agron. J.* 92:1167–1175.

Weil, R.R. 2001. Soil management for sustainable intensification: Some guidelines. In W.A. Payne et al. (Eds.), *Sustainability in Agricultural Systems*. ASA Special Publication No. 64, American Society of Agronomy, Madison, WI, pp. 145–154.

Weil, R.R., and W. Kroontje. 1979. Physical condition of a Davidson clay loam after five years of heavy poultry manure applications. *J. Environ. Qual.* 8:387–392.

Weil, R.R., R.A. Weismiller, and R.S. Turner. 1990. Nitrate contamination of groundwater under irrigated coastal plain soils. *J. Environ. Qual.* 19:441–448.

Weil, R.R., K.R. Islam, and C.L. Mulchi. 2000. Impacts of elevated $CO_2$ and ozone on carbon cycling processes in soil. In *Agronomy Abstracts for the Annual Meetings of the Soil Science Society of America*, 5–9 November 2000, Minneapolis, MN, p. 47.

Weil, R.R., K.R. Islam, M.A. Stine, J.B. Gruver, and S.E. Samson-Liebig. 2003. Estimating active carbon for soil quality assessment: A simplified method for lab and field use. *Am. J. Altern. Agric.* 18:3–17.

Whitbread, A. 1995. Soil organic matter: Its fractionation and role in soil structure. In R. Lefroy et al. (Eds.), *Soil Organic Matter Management for Sustainable Agriculture: A Workshop* ACIAR Proceedings No. 56, Australian Center for International Agricultural Research, Canberra, pp. 124–130 (Workshop held on 24–26 August 1994 in Ubon, Thailand).

Whitbread, A.M., G.J. Blair, and R.D.B. Lefroy. 2000. Managing legume leys, residues and fertilisers to enhance the sustainability of wheat cropping systems in Australia. 2. Soil physical fertility and carbon. *Soil Tillage Res.* 54:77–89.

Wildner, L.D.P. 2000. Soil cover. In *Manual on Integrated Soil Management and Conservation Practices.* FAO Land and Water Bulletin No. 8, Food and Agriculture Organization of the United Nations, Rome, Italy, pp. 87–90.

Wood, S., K. Sebastian, and S.J. Scherr. 2000. Pilot analysis of global ecosystems: agroecosystems [Online]. Available by International Food Policy Research Institute and World Resources Institute at www.wri.org/wr2000, accessed 18 December 2000.

Woomer, P.L., and M.J. Swift. 1994. *The Biological Management of Tropical Soil Fertility.* Tropical Soil Biology and Fertility Programme and Sayce Publishing, John Wiley & Sons, New York.

Yermiyahu, U., R. Keren, and Y. Chen. 1995. Boron sorption by soil in the presence of composted organic matter. *Soil Sci. Soc. Am. J.* 59:405–409.

Yoo, M.S., and B.R. James. 2002. Zinc extractability as a function of pH in organic waste-amended soils. *Soil Sci.* 167:246–259.

# 2 Soil Organic Matter Management Strategies

*Fred Magdoff and Ray R. Weil*

## CONTENTS

Farmers are demonstrating a growing interest in soil health and how to improve the quality of their soils through changes in management. This interest among farmers in soil health, as well as the interest among scientists in the broader concept of soil quality, has produced a focus on soil organic matter (SOM). The profound effects that organic matter has on almost all soil properties — chemical, biological, and physical — places it at the center of soil health and soil quality (Magdoff and van Es, 2000; Weil and Magdoff, Chapter 1, this volume). Although SOM and its management are central to the development of healthy soils, there are important concerns such as compaction and nutrient management that, while impacted by SOM, need to be understood and approached as important separate issues.

Almost all soil and crop management practices have implications for SOM. Practices influencing SOM include tillage and planting techniques, methods of handling crop residue, application of organic amendments, crop rotations, and use of cover crops. SOM in its totality includes soil organisms, simple organic compounds, large and complex humic substances, as well as relatively fresh residue in various stages of decomposition. Soil and crop management practices usually influence all these SOM fractions. An important perspective in a new ecological approach to farming is to manage soils, crops, and animals keeping in mind the critical importance of SOM to healthy soils, crops, and agroecosystems.

Improved SOM management is, in essence, a preventive approach to agroecosystem health. Its purpose is to help build the soil's internal strengths — biological, physical, and chemical — so that it is better able to accept and store water (reducing runoff and erosion as well as providing more water to plants) and roots have a healthy environment in which to fully develop and function.

0-8493-1294-9/04/$0.00+$1.50

Although it might not solve them all, better SOM management can go a long way in alleviating many soil and crop problems and make it easier to deal with such other problems as acidification, nutrient availability, and compaction.

Organic management practices that enhance soil health usually do one or more of the following: (1) increase the addition of organic materials to soils, (2) use a variety of different types of organic materials, and (3) decrease the rate of loss of organic matter. Under most soil conditions, it is possible to build up the total SOM content or maintain an already high level. However, for coarse-textured soils, it takes extraordinary inputs to raise SOM to very high levels. However, it is possible to have a high-quality soil even with a moderate level of SOM as long as sufficient quantities of a variety of residues are routinely present to supply nutrients, aggregating substances, and particulate organic matter (POM; to provide food sources for an active and diverse population of soil organisms).

## PATTERNS OF SOM CHANGES

The total amount of SOM present and the changes in total SOM with management are not the only criteria by which to evaluate practices. Levels of active or labile carbon fractions, such as POM or simple organic chemicals, as well as populations of soil organisms respond more quickly to changes in soil and crop management practices than does measurable change in total SOM. However, a simple model of organic matter changes over time can help visualize how different management practices influence the gain or loss of organic matter from soils as a result of management:

$$\text{Net organic C change} = \text{C gains} - \text{C losses} \qquad (2.1)$$

Four main patterns of SOM change result from crop, soil, and manure management practices (Figure 2.1). Organic matter accumulates in soils to the degree that C inputs exceed C outputs, or, simply put, gains exceed losses. A rapid increase in SOM (Scenario a, Figure 2.1) occurs when large quantities of crop residues or organic amendments are applied and when annually cropped soils are planted to perennial forage crops, which add large quantities of residues and reduce the losses that occur by erosion and tillage. During this buildup of SOM, POM is the major fraction that increases, whereas little new humus is added in the short run. Soil biology also undergoes dramatic shifts during this period as activities of organism increase and populations change in relative abundance (see Chapter 5 and Chapter 10).

When C outputs (losses) exceed inputs, total SOM decreases. The conversion of land from natural vegetation to agriculture generally results in a marked decline in SOM levels, with losses of 40% or more not uncommon. (See Scenario b, Figure 2.1, and discussion in Chapter 1.) Although C additions to soil in agroecosystems might be similar to those of natural systems, they are frequently less. However, the major reason for SOM declines during conversion from natural ecosystems to agricultural production is usually the acceleration of C losses. In general, losses are increased dramatically as soil disturbance by tillage stimulates oxidative loss of organic matter as $CO_2$ and greatly accelerates erosion. Erosion can be a significant pathway for SOM losses because topsoil, the part of the soil usually lost during erosion, is greatly enriched in SOM. When total SOM decreases, POM and microbial numbers and activities decrease whereas the stable humus fraction remains practically unchanged. Simplification of the soil food web because of low levels of available C can allow greater attack of plants by disease and nematodes (see Chapter 5).

When C gains balance losses, SOM remains unchanged (Scenario c, Figure 2.1). This can happen even when intensive tillage is practiced and most of the aboveground residues are removed, as long as sufficient organic materials are introduced from off the field. For example, a 20% loss of SOM over a 5-year period occurred for corn silage on a clay soil under conventional tillage with no added manure, but application of 44 Mg/ha dairy manure (fresh weight) maintained SOM at the original level of 5.2% (Magdoff and Amadon, 1980).

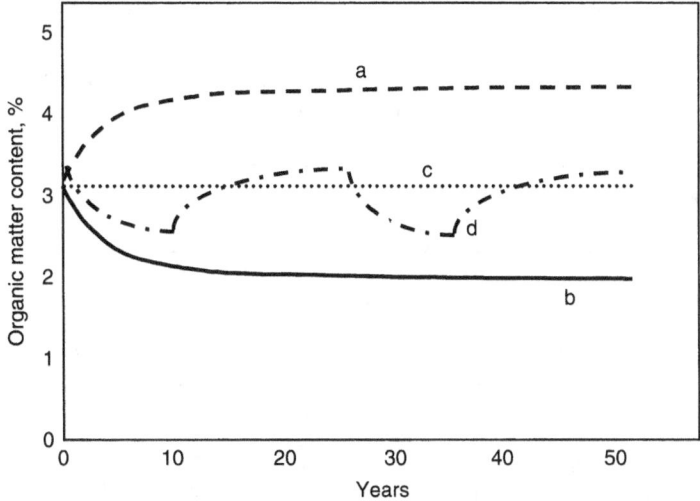

**FIGURE 2.1** Examples of common patterns of change in soil organic matter resulting from crop, tillage, and manure management in (a) gains exceed losses, (b) gains are less than losses, (c) gains equal losses, and (d) cyclical fluctuations.

For fields that are in rotations with low-residue crops for a few years alternating with high-residue crops, or intensively tilled crops alternating with no tillage (including perennial forages), SOM alternates between a decreasing and increasing phase (Scenario d, Figure 2.1).

If the same tillage, cropping, and residue or manure addition pattern is followed for a long time, a steady-state situation develops in which gains and losses are the same and SOM change is 0 (Scenarios a and b after many years, Figure 2.1). The following simple model can be used to calculate the equilibrium situation that would occur after SOM levels have finished changing. Starting from the proposition that the amount of organic matter in soils is a result of the balance between the gains and losses of organic materials, the change in SOM during one year (SOM change) is represented by Equation 2.1.

If gains are more than losses, organic matter accumulates and SOM change is positive; when gains are less than losses, organic matter decreases and SOM change is negative. Gains refer not to the amount of residues added to the soil each year but to the amount of residue remaining at the end of the year. This is the fraction ($f$) of the fresh residues added that is not decomposed during the year multiplied by the amount of fresh residues added ($A$), or:

$$\text{Gains} = (f)(A) \tag{2.2}$$

Losses are the percentage of organic matter that is lost by either mineralization (respiratory loss of $CO_2$) or erosion in a given year ($k$) multiplied by the amount of organic matter (SOM), i.e., losses = $k$(SOM). The amount of organic matter that will remain in a soil under steady-state conditions can then be estimated as follows:

$$\text{SOM change} = 0 = \text{Gains} - k \text{ (SOM)} \tag{2.3}$$

Because in steady-state situations gains equal losses, gains = $k$(SOM), or:

$$\text{SOM} = \text{Gains}/k \tag{2.4}$$

**TABLE 2.1**
**Estimated Levels of Soil Organic Matter (%) for Steady-State Conditions with Various Mineralization Rates (k) and Levels of Annual Residue Additions**

| Residue Additions (kg/ha/year) | k (SOM Loss Rate) | | | | |
|---|---|---|---|---|---|
| | **0.01** | **0.02** | **0.03** | **0.04** | **0.05** |
| 500 | 2.5 | 1.3 | 0.8 | 0.6 | 0.5 |
| 1000 | 5.0 | 2.5 | 1.7 | 1.3 | 1.0 |
| 1500 | 7.5 | 3.8 | 2.5 | 1.9 | 1.5 |
| 2000 | 10.0 | 5.0 | 3.3 | 2.5 | 2.0 |

Under steady-state conditions, the effects of residue addition and the rate of SOM loss as $CO_2$ or in eroded sediments can be calculated by Equation 2.4. Table 2.1 gives the results of these calculations, assuming that the upper 15 cm of soil weighs $2 \times 10^6$ kg/ha.

A conventionally tilled clay soil with moderate drainage can have a SOM loss rate (decomposition plus erosion) of ca. 2% each year ($k = 0.02$). With 2000 kg/ha dry matter residues added annually (a high rate of residue application), this soil will have 5% SOM at equilibrium after the initial rapid decomposition phase of the fresh material. However, for a very well-drained sandy loam soil with a higher rate of SOM loss, $k = 0.04$, the same residue application rate results in only 2.5% organic matter at equilibrium.

It is possible for different conditions to result in the same percentage of SOM. For example, a soil with 2.5% organic matter can result from low annual rates of addition (500 kg/ha) and loss (1%). However, it is also possible to obtain 2.5% organic matter because of high annual rates of both additions (1500 kg/ha) and loss (3%).

## PRACTICES FOR IMPROVED SOM MANAGEMENT

Soil management practices that usually improve soil health through influences on SOM include (1) more complex crop rotations, especially those with high-residue crops, (2) reduced tillage, (3) intensive use of cover crops, and (4) use of a variety of organic amendments. These management practices, in many variations and combinations, usually accomplish one or more of the following goals: increase C inputs, decrease C outputs, harm pests living in the soil, and encourage beneficial organisms (Table 2.2). In addition, improved soil properties that result from these practices, such as more available water, less compaction, better timing of nutrient availability to crop needs, and production of growth-promoting substances, promote the growth of plants that can better defend themselves from pests.

### INCREASING C INPUTS TO SOIL

The accumulation of SOM is greatly influenced by the amount of C inputs. Paustian et al. (1997) showed that the change in SOM over time is linearly related to the level of C inputs in each of seven long-term experiments when the change in C was averaged across the duration of the experiment. In another example, Campbell and Zentner (1993) found a direct relation between the quantity of crop residue and its N content to SOM after 24 years of a crop rotation experiment in Saskatchewan. Reductions in the frequency or duration of bare fallow periods and increases in the frequency of perennial vegetation in rotations are examples of practices that increase C inputs. Both practices increase long-term water and nutrient use efficiency of plants and thereby increase C inputs to the soil (Paustian et al., 2000).

**TABLE 2.2**
**Influence of Soil and Crop Management Practices on SOM**

| Practice | Influence on SOM | | |
|---|---|---|---|
| | **Increased Gains** | **Decreased Losses** | **Increased Beneficials or Decreased Pathogens, Parasites, and Weeds** |
| Rotations | | | Regardless of effect on POM or total SOM levels, soil biology usually more favorable to crops in rotation |
| High-residue crops included | Higher average annual residue | Higher amount of residue leads to higher water infiltration and less runoff and erosion (especially if maintained on surface) | Same as above |
| Perennial forages | Higher average annual residue | Soil continuously covered leads to reduced raindrop impact and physical holding of soil by roots | Same as above, especially because these are usually longer rotations |
| Cover crops | Increase production of biomass when otherwise no primary production<br><br>POM increased or maintained | Same as above | Weeds smothered or suppressed (allelopathy)<br><br>Higher AM inoculation of following crop<br><br>Nematode or diseases suppressed |
| Use of organic amendments | Significant amounts of organic material usually applied along with nutrients (as with compost and dairy or beef manure) | If higher infiltration and drainage lead to less water runoff, less erosion occurs | Diseases sometimes suppressed<br><br>Plants might acquire systemic resistance to diseases<br><br>Insects might find plants less attractive |
| Reduced tillage | Increased water infiltration can increase yields and residues, especially on medium-to-coarse soils | More residue on surface (because of reduced tillage) reduces runoff and erosion | Reduced weed seed survival and emergence |

## Rotations and Crop Residue Management

Cropping practices influence SOM in a number of ways (discussed later). Some rotation crops leave especially large quantities of residue, contributing greatly to the addition side of the gains–losses equation. Some rotation crops, such as legume, grass, or grass–legume forage crops, supply a lot of root biomass, which can contribute to residue, whereas the lack of tillage and continuous soil cover of such crops decreases SOM losses by soil food web respiration or by erosion. The quality of the crop residue also impacts SOM in that higher N content of residue enhances the incorporation of C into stable SOM. In addition, rotations influence soil biology and reduce problems with many plant pests.

Compared with monoculture cropping practices, rotations can result in about a 10% increase in yields (Karlen et al., 1994), and thus more residue usually remains after harvest. In an extensive review of the literature, West and Post (2002) found that transitioning to more complex crop rotations, except for changes such as from continuous corn to alternating between corn and soybean, can increase C in soil by an average of $20 \pm 14$ g $C/m^2/year$. One of the most effective means of increasing the level of SOM and improving soil quality is to include perennial forage crops in the crop rotation. Studdert et al. (1997) reported that inclusion of pastures in a rotation can reverse the soil-degrading effects of conventional cropping and tillage. They studied a long-term crop rotation experiment by using conventional tillage on a prairie-derived soil (fine, mixed, thermic Typic Argiudolls) with 2% slope in Balcarce, Argentina. Treatments were continuous cropping and crop-pasture (50:50 and 75:25) rotations. All soil quality indicators decreased with more cropping and increased with more pasture in the rotations. Soil organic C, for example, decreased by 4.4 g/kg in the 6 to 7 years under cropping and rose to the original level ($37.2$ g/kg$^{-1}$) after 3 to 4 years under pasture. Studdert et al. (1997) concluded that rotations that include up to 7 years of conventional cropping alternating with at least 3 years of pasture will maintain soil quality within acceptable limits and meet the goals of sustainable agriculture under conditions similar to those used in their experiment. Similarly, 6 years under continuous, unharvested grass (tall fescue) nearly doubled the SOC content of the upper 15 cm of a sandy loam soil (Aquic Hapludults) in Maryland from ca. 10 to nearly 20 g/kg (Weil et al., 1993). In contrast to the effect of continuous sod, organic C levels in this soil were little affected by 6 years under four other cropping systems that produced corn, soybean, or wheat with management regimes that ranged from no till with fertilizer and herbicides to an organic system with reduced tillage and clover or grass hay interseeded with the wheat.

In a long-term study in New Zealand (Francis et al., 2001), variations in harvest yield and N uptake were explained by differences in soil N fertility and soil structural conditions. Three years under perennial pasture brought about a degree of improvement in soil structure and N fertility that closely matched the degree of decline that occurred during 3 years of row cropping, suggesting that similar lengths of pasture and cropping periods are needed to sustain soil quality in these weakly structured silt loam soils.

Moderate levels of animal grazing can apparently enhance the soil quality benefits from perennial grass vegetation compared with ungrazed grassland. For example, in an 11-year study on mixed grass prairie in Wyoming (Jordan et al., 1995) found that soils had higher amounts of carbon and nitrogen in the surface 30 cm on the grazed pastures compared with a native rangeland where livestock was excluded.

An important approach to increasing SOM levels from the gains side of the balance is the careful management of crop residues, both above- and belowground. Worldwide, $1.4 \times 10^9$ ha of arable land base is estimated to annually produce $3.44 \times 10^9$ Mg of crop residue containing 45% C or ca. 1.5 Pg/year of total C (Reicosky et al., 2000). Only a fraction of this crop residue C is stabilized in SOM, the majority being returned to the atmosphere as $CO_2$ from microbial respiration within a year or two of its addition to the soil. Nonetheless, within a given set of environmental conditions, the level of SOM sustained is usually related to the amount of C added as residues. After 11 years of cropping on a Typic Hapludoll in Iowa, SOC content was linearly related to the amount of crop residue (corn stover or alfalfa hay) added (Larson et al., 1972). About 5.5 Mg/ha of residues was needed under conventional plowed tillage to maintain the SOC content at its initial level of 1.8% C. A very similar level of wheat residue was required to maintain SOC over a 45-year period of conventional tillage in Oregon (Follett et al., 1987).

One way to increase the amount of crop residue C added to soils is through the use of cover crops to fill seasonal niches when commercial crops are not growing. Some of the effects of cover crops on soil properties have been discussed in Chapter 1 and Chapter 6. Cover crops are usually killed or suppressed when still physiologically young and low in slow-to-decompose structural components. Often, little biomass (perhaps 1 Mg/ha) is produced when cover crops are killed very early, although a production of 4 Mg/ha or more is common when the cover crops are allowed to grow in the spring. Compared with crop residues of physiologically mature crops (e.g., wheat straw, corn stalks), a relatively high percentage of cover crop biomass is lost during the initial rapid stage of decomposition. This rapid decomposition plus the sometimes low biomass production means that cover crops, even when used routinely, might not result in increased total SOM (Allison, 1973). However, even low amounts of biomass of a rapidly decomposing cover crop help maintain pools of POM and also contribute to many of the positive effects of cover crops.

Both crop residue quality and quantity influence levels of SOC. Because of their more recalcitrant composition, their placement in intimate contact with soil aggregates, and their continuous input into the soil, root residues appear to be more effective than shoot residues in building SOM. Puget and Drinkwater (2001) studied the fate of C from shoot and root residues of a hairy vetch cover crop incorporated before planting maize in early spring. They found that at the end of one growing season, only 13% of the shoot-derived C remained in the soil whereas nearly 50% of the root-derived C was still present. The root-derived C was present largely as POM occluded inside aggregates or associated with the clay and silt fraction.

These results are in general agreement with a study (Gregorich et al., 2001) of maize-derived soil organic C after 35 years of cropping in Ontario. From the $^{13}C$ natural abundance data, it was found that ca. 10% of maize-derived C added in the 0- to 20-cm depth was retained as SOM, but 40 to 50% of the maize-derived C added in the 20- to 70-cm layer was retained in SOM (Table 2.3). They suggested that the higher retention of maize-derived C in the subsoil was due to a more recalcitrant composition of root tissue than aboveground plant parts and to slower microbial decay in the subsoil because of the presence of less oxygen and fewer organisms. For each year the maize was grown, maize grown in rotation with legumes added more root residues to the subsoil than did monocrop maize, especially if the systems were unfertilized. This result might have been due to improved subsoil penetration by maize roots following channels made by the alfalfa roots. Although the total amount of residue C returned was about the same for the monocrop and rotation, under the rotation more C accumulated in the soil, especially below the plow layer (Table 2.3). Gregorich et al. (2001) concluded that rotations are likely to result in better preservation of crop residue C than what occurs under monocropping, and that "residue quality plays a key role in increasing the retention of soil C in agroecosystems and that soils under legume-based rotation tend to be more 'preservative' of residue C inputs, particularly from root inputs, than soils under monoculture."

**TABLE 2.3**
**Comparative Addition and Retention of Maize-Derived C as SOM in Surface and Subsoil Layers in a 35-Year Experiment in Ontario, Canada**

| Soil Depth (cm) | Monocrop | | Rotation[a] | |
|---|---|---|---|---|
| | Fertilized | Unfertilized | Fertilized | Unfertilized |
| | Maize-Derived C in Soil (Mg/ha/year) | | | |
| 0–20 | 0.40 | 0.26 | 0.45 | 0.39 |
| 20–70 | 0.26 | 0.14 | 0.48 | 0.39 |
| | Maize Residue C Returned (Mg/ha/year) | | | |
| 0–20 | 4.11 | 2.51 | 5.59 | 4.84 |
| 20–70 | 0.52 | 0.35 | 0.95 | 0.98 |
| | Mg Maize-Derived C Retained in SOM/Mg Maize-Derived C Added | | | |
| 0–20 | 0.10 | 0.11 | 0.08 | 0.08 |
| 20–70 | 0.49 | 0.39 | 0.51 | 0.40 |
| | Total Plant Residue C Returned 1959–1994 (Mg C/ha) | | | |
| 0–70 | 162 | 100 | 113 | 104 |

[a] Rotation of maize–oat–alfalfa.

*Note:* Data are means for years when maize was grown in either monocrop or in rotation with legumes.

*Source:* Data from Gregorich, E.G. et al. 2001. *Can. J. Soil Sci.* 81:21–31. With permission.

For a rotation study consisting of a legume phase followed by three wheat crops on a degraded Ferric Luvisol soil in New South Wales, Whitbread et al. (2000) reported that the concentration of labile soil C significantly increased in the treatments with wheat stubble retention rather than removal. They suggested that the use of legume species is more likely to improve the overall fertility of the farming system when combined with cereal stubble retention. Sanchez et al. (2001) also reported increases in SOC on soils producing a diversity of crops in Michigan. Soil under a diverse rotation with hay, winter legume cover crops, and several grain crops contained 50% more SOC than did soils under continuous monocrop maize with only mineral fertilizer added. However, total C inputs were not reported in their paper; therefore, it is not possible to ascertain how much of the increased SOC was because of altered litter quality and C cycling processes and how much was simply because of additional C inputs for the diverse rotation in the form of composted manure applied to that rotation but not to the monocrop.

## Use of a Variety of Sources of Organic Materials

Crop residues and organic amendments, each with its unique characteristics, have different effects on soil biological, chemical, or physical properties (Chapter 1). Thus, one of the strategies of SOM management is to use a variety of types of organic materials. Growing the same crop in the same field for many years in a row, without organic amendments, exposes the soil flora and fauna to the same type of residues year after year. This constant exposure to the same residue frequently selects for organisms that are harmful to the plant. (Over time, this can be mitigated by developing populations of biological control organisms sufficient to keep disease organisms at low levels.) On the other hand, different types of residue remain on a field when crop rotations and cover crops are used. Also, many different organic amendments can be brought from off the field, including various types of animal manures, crop residues, tree leaves, grass clippings, food processing wastes, and sewage sludges. These materials can be added as such or can be composted alone, combined with one another, or with other materials such as bark or woodchips added as bulking agents.

In addition to the amount of C added, the type of material in which C is added to the soil also influences SOM accumulation. Paustian et al. (2000) reported that when either 250 or 500 g C/m²/year was added to moderately coarse-textured soils in Canada or Sweden, SOM increased in the order alfalfa < straw < manure < peat (manure called peat in Sweden). Manure is often presumed to result in higher increases in SOM because it consists of relatively recalcitrant compounds, the most easily oxidized compounds in the original plant tissue having been broken down by the animal digestive system before excretion of the manure. Therefore, manure additions have been known to impact SOM for many years after additions have ceased (e.g., Jenkinson and Johnson, 1977).

The quantities of organic material additions to soils needed to maintain or increase SOM can be substantial. The use of amendments such as animal manure is especially important when growing crops that leave little residue in the field. In an experiment of growing silage corn, in which almost all the aboveground residue was harvested, with conventional tillage on a clay soil, Magdoff and Amadon (1980) found that approximately 44 Mg of dairy manure (wet weight)/ha/year was needed to maintain SOM at the initial level of 5.2% that resulted from the long-term cropping to a mixed grass–legume hay. The initial SOM level of 5.2% represented a near steady state under previous cropping to a mixed legume–grass hay. The manure plus bedding input per hectare needed to maintain this SOM level was approximately the annual amount expected from 2.2 large (636 kg) Holstein dairy cows. About 2.5 ha are needed to supply all the feed (forage and grain) for the 2.2 lactating cows (Magdoff et al., 1997). Thus, ca. 2.5 ha are necessary to grow the feed for cows that produce enough manure to maintain SOM levels of 1 ha of silage corn.

Amending soil with composted organic wastes is often an effective means of increasing SOC. Because the most labile C fractions are lost during the composting process, much of the C in the final compost as applied to the soil is more recalcitrant than in the uncomposted material. For example, Eghball (2002) reported that after 4 years of amendment, 36% of the C added as composted manure was retained as SOC compared with 14 to 25% of the C added as uncomposted manure. Similarly, in Spain, five annual 24 Mg/ha applications of composted municipal refuse resulted in more SOC than did fresh cow manure or fresh sewage sludge applied at the same rate of dry matter (Albiach et al., 2001). The authors also reported that recommended rates of commercial humic acids (100 L/ha/year) or commercial vermicompost (2.4 Mg/ha/year) did not produce any significant change in SOC or other soil quality indicators. Higher rates of application of the expensive commercial products that might have raised SOC levels were deemed uneconomic.

Compost has also been found to offer advantages over raw organic materials in such environmental applications as bioremediation, slope stabilization, and artificial wetland construction (Alexander, 1999). However, these and similar compost amendment studies failed to make the ecologically relevant comparison of the fate of a given amount of organic C either used to make compost that is then applied to soil or applied directly without composting. In one study in which amendments were added on the basis of the same amount of C in original materials, Chromec and Magdoff (1984) found significantly higher C, N, and CEC levels after 199 d of incubation when uncomposted sludge and hardwood sawdust were added directly to the soil, compared with composting of the sludge–sawdust mixture before addition.

An important factor in the production of plant biomass, for both harvest and soil building, is the availability of soil nitrogen. Many long-term field experiments have demonstrated that the use of N fertilizer can increase SOC when the added N enhances plant productivity (Halvorson et al., 1999). The effect of increasing application rate of N fertilizer on increasing SOC might be most pronounced in the surface layers of soil under no-till management (Rickman et al., 2001). However, increased residue return might not be the only means by which N fertilization increases the accumulation of SOM. Many researchers have reported increased SOM as N levels increase and have argued that higher residue returns do not account for the difference (e.g., see Paustian et al., 1997 and Drinkwater et al., 1998). However, the experiments are far from conclusive in demonstrating this effect. For example, less soil disturbance in the highest N treatment, high levels of inorganic N in the conventional (low N) treatment, and no determination of root residue differences between treatments make it difficult to interpret the experiment of Drinkwater et al. (1998) as demonstrating that higher N levels lead to more efficient conversion of crop residues to stable SOM. Likewise, in their discussion of N effects on SOM accumulation, Paustian et al. (1997) present only four treatments out of a larger experiment reported by Paustian et al. (1992). When all treatments are included, there is a good correlation between residue additions and changes in SOM after 31 years (see Figure 10 of Paustian et al., 1992). In addition, more root growth with added N might have a relatively large influence, because, per unit of C added, root residue might contribute more to building SOM than residues originating aboveground (Campbell et al., 1991; Puget and Drinkwater, 2001). A number of explanations might account for the phenomenon of increased SOM with increasing N, in addition to the effect on quantities of residue returned, if it actually occurs. First, adequate N is needed to build humus by increasing the amount of amino compounds that act as precursors in the formation of recalcitrant humic structures. Second, N additions might have negative effects on decomposition, including repression of lignolytic enzymes by ammonium. Third, the efficiency of microbial C assimilation is reduced with insufficient N (i.e., more is $CO_2$ respired per unit C assimilated).

Most farmers who apply organic amendments such as manures and compost do so to provide nutrients for their crops as well as to increase SOM levels. Because of concerns about the potential for excessive levels of soil N and P to cause water pollution by leaching and runoff losses, the nutrient content of organic amendments must be carefully considered in any SOM management program. Two aspects of organic nutrient management are often overlooked. The first is that many organic materials, especially those of animal origin, such as manure or sewage, contain a higher ratio of P to available N than do crops. Therefore, supplying all the N needs of a crop from manure will oversupply P, eventually leading to excessive buildup of soil P, which will cause runoff from the land to threaten eutrophication of lakes, streams, and estuaries. This problem will be aggravated when manures are from animal feed P sources in addition to the feed grown on the land that receives the manure.

A second aspect of organic amendments often overlooked is their residual release of nutrients. Farm advisors sometimes calculate the rate of amendment application to supply the N needs of a crop in the first year and then recommend this as a routine annual application rate. In fact, the second-year application should be reduced by an amount that takes into consideration the residual release of N from the first year's application, and so on. However, even by using properly declining rates of application to supply only the N needed, manures and many types of compost oversupply P. In the case of fresh manure, which has a relatively high initial decomposition rate, the routine, annual rate of application is soon less than half the initial recommendation (Figure 2.2A). Even though the manure application rate is reduced in subsequent years, soil P continues to accumulate (Figure 2.2B). In the case of compost, which consists of stabilized material with much lower N availability than fresh manure has, there should be a more dramatic decrease in application rates to take into account mineralization resulting from previous applications (Figure 2.2C). These trends can make it unsustainable to use composts with low N availability rates and high P levels as the principal nutrient source for crops (Figure 2.2D).

Ultimately, a sustainable approach to soil improvement will require that inputs of C and associated nutrients be grown in place or be recycled from the unused portions of plant production harvested from that soil for use by animals or people. Application of organic amendments that originated elsewhere can be expected to involve degradation of the soil from which the C and nutrients were originally harvested, making the practice questionable from an overall sustainability viewpoint (Magdoff et al., 1997). Crop rotations, such as those mentioned previously, have long been recognized as an important means of maintaining soil quality in the production of cereal crops. High-value horticultural crops might involve intensive tillage and many produce very little residue mass to return to the soil. We have worked with several farmers who find that a practical approach to building soil quality consists of rotating soil-degrading horticultural crops with soil-building agronomic crops, especially hay crops and cover crops. They report that improved soil quality makes growing low-profit, soil-improving crops on 50 to 75% of their land in rotation with very intensively managed high-value horticultural crops on the remaining land more profitable than growing the high-value, soil-depleting crops on all the land all the time.

## DECREASING SOM LOSSES FROM SOIL

Loss of soil SOM can be reduced by minimizing (1) removal of plant material in harvest, (2) erosion losses by water and wind, and (3) C losses (as $CO_2$) by accelerated microbial respiration. Harvest removal for use or sale is usually the main point of practicing agriculture; therefore, the improvements in the first category are generally limited to careful retention of as much of the plant residues as possible, as discussed in the previous section.

Losses of SOM by erosion are higher than what might be deduced from the magnitude of soil losses reported for cultivated soils (generally 5 to 50 Mg/ha/year), because SOM is typically enriched in the eroded material compared with the bulk soil on the eroded site. That enrichment would occur is logical because erosion takes place at the surface of the soil where the SOM content is highest and because the organic fractions of soil often erode more easily than the mineral fractions. As a result, SOM content is generally higher in soils with less steep slopes and in soils in lower landscape positions, because these soils suffer smaller erosion losses and might acquire SOM through sedimentation from the upslope of steeper sites. For example, in one land resource area in Minnesota (MLRA 105), the average SOC contents for soils on slopes 0–2%, 3–5%, and 6–12% were 22.3, 13.5, and 8.9 g/kg, respectively. Mean annual erosion losses of SOC from these soils ranged from 273 to 758 kg C/ha for conventional tilled fields and 94 to 274 kg C/ha for no-till fields (Follett et al., 1987).

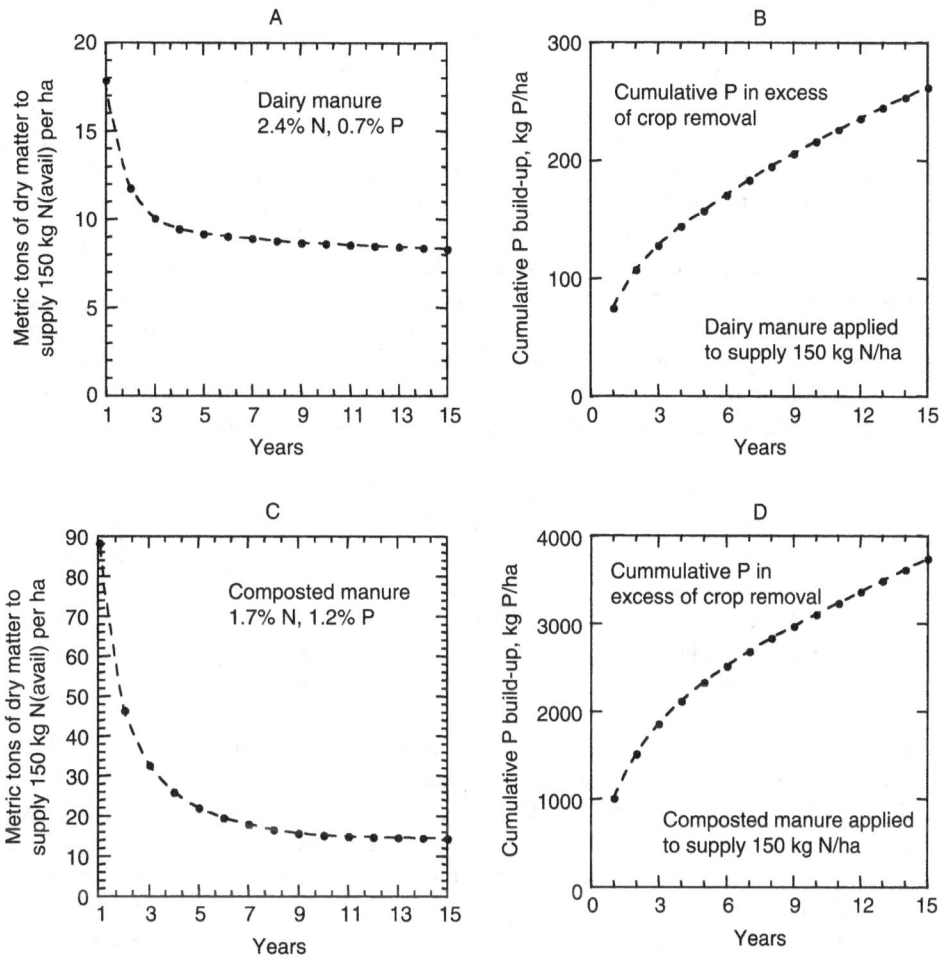

**FIGURE 2.2** Example showing how residual nutrient release from two types of organic amendment results in a declining application rate needed to annually supply 150 kg N/ha for corn silage. The accumulation of phosphorus in excess of that removed by the crop is also shown. This example assumes that raw dairy manure contains 2.4% N and 0.7% P and releases 35% of its N in the first year, with a declining proportion released thereafter. The composted manure is assumed to contain 1.7% N and 1.2% P, because P losses during composting are less than N losses. The compost releases 10% of its N in the first and subsequent years.

In the absence of significantly accelerated erosion, rates of microbial respiration largely govern SOC losses. SOM decomposition by microbial activity is very sensitive to alternating drying and wetting cycles (Birch, 1958) and temperature. The enhanced solubility and availability of SOM when soils are moistened following drying (Bartlett, 1981) is most likely responsible for accelerated microbial respiration and SOM decomposition. Management practices that cause both high soil temperatures and more pronounced alternate wetting and drying cycles include keeping the soil surface bare in summer, forming ridges that create soil surfaces more perpendicular to solar radiation, and subsurface drainage that lowers soil water content and heat capacity. Improved drainage, whether by tiles, ditches, or raised beds, enhances soil aeration, which in turn can accelerate respiration and result in lower levels of SOC accumulation.

Nonetheless, tillage is arguably the management practice that has the greatest influence on the loss side of the SOM balance sheet. Several mechanisms operating together probably explain why tillage accelerates SOM losses. First, tillage on slopes tends to move topsoil enriched with organic matter downhill (Magdoff and van Es, 2000). Second, crop residues decay more rapidly when mixed into the soil, because the soil maintains conditions of moisture, nitrogen availability, and temperature suitable for microbial decomposition (Wilson and Hargrove, 1986). In addition, the diverse population of soil decomposers is in direct contact with the residues tilled into the soil. Reicosky et al. (2000) reported that the use of a moldboard plow to turn wheat stubble caused one third more C to be lost from the soil within 19 d of plowing than was contained on the residues to begin with. The physical disturbance associated with plowing speeds the decomposition of SOM associated with aggregates and allows rapid degassing of $CO_2$ already present in soil pores. The use of no-till planting techniques can greatly reduce SOC losses due to these mechanisms as well as losses due to erosion. As a result, measurements of $^{13}C$ natural abundance indicate that mean residence time of SOM is nearly twice as long under no-till as that under intensive tillage (Paustian et al., 2000). Tilled soils are also usually drier, warmer, and more susceptible to erosion than their untilled counterparts, and these three factors have already been mentioned as promoting the loss of SOM.

In some cropping systems experiments, the effects of tillage intensity might be confounded with the effects of biomass input, as when a wheat–fallow sequence is compared to a wheat-pasture sequence. The fallow period, if the soil is kept bare by repeated tillage, represents a lost opportunity for plant biomass production and a higher average number of tillage operations per year. This is why in five long-term experiments (in the states of South Dakota, North Dakota, Nebraska, and Colorado) alternatives that involved less intense tillage and fewer fallow periods than the conventional wheat–fallow system resulted in 25 to 45% higher SOM levels in the upper 7.5 cm of the soil (Gajda et al., 2001).

Other cropping studies, such as the 13-year study on silt loam Haploborolls in southwestern Saskatchewan (Curtin et al., 2000) have provided unambiguous evidence that less intense tillage, especially no-till management, increases SOM accumulation by reducing organic matter losses in the surface soil layers. This study compared continuous wheat to wheat–fallow under both conventional till (CT) and no-till (NT) management and showed that surface residues accounted for about half of the additional C sequestered in NT compared with that in CT. The authors suggest that switching from CT wheat–fallow to NT continuous wheat in the Canadian prairie could sequester 5 to 6 Mg C/ha in SOM and surface residues during a 13- to 14-year period.

Similar SOM increases over the levels with CT occurred in NT systems at five sites in Germany on soils (Inceptisols and Entisols) ranging from loamy sand to silt loam (Tebrugge and During, 1999). The authors reported that surface residue cover and higher aggregate stability under NT "protected soil fertility by avoiding surface sealing and erosion." They further suggested that the accumulation of organic matter and nutrients near the soil surface under NT resulted from the lack of soil inversion and the maintenance of surface mulch associated with enhanced biological activity. In reviewing the results of NT vs. CT comparisons in five countries, Tebrugge (2001) estimated that NT was associated with ca. 1 Mg/ha/year more SOM accumulation (Table 2.4). Franzluebbers (Chapter 8) reviews 111 published observations in North America comparing inversion tillage or shallow tillage to NT. His analysis shows that NT sequestered more C than conventional tillage in ca. 80% of the cases, but only rarely as much as 1 Mg/ha/year.

The influence of tillage on the conservation of SOC apparently depends on climatic conditions and soil texture, among other factors. When the soil inventory approach of the Intergovernmental Panel on Climate Change is used, the potential for C sequestration as SOM probably varies more than threefold among regions in the U.S. (Eve et al., 2002). For example, converting from conventional tillage cropping with no cover crops to no-till cropping with winter cover crops might initially increase SOC by 0.4 to 1.1 Mg C/ha annually (Table 2.5). Eve et al. (2002) estimated the weighted mean C sequestration (Mg C/ha/year) in various regions, if conventionally tilled cropland is converted to no-till cropland as ranging from 0.20 in the Mountain region to 0.52 in the Delta States. For increasing residues (by use of cover crops) in a no-till system, the values were 0.22–0.57

## TABLE 2.4
## Accumulation of Soil Organic Matter under No-Till (NT) Management Relative to Conventional Tillage (CT) in Several Diverse Locations

| Variable Characterized | Canada | Germany | Italy | Spain | Portugal |
|---|---|---|---|---|---|
| Duration of trial (years) | 18 | 10 | 5 | 12 | 4 |
| SOM in NT (Mg/ha) | 62 | 120 | 108 | 88 | 52 |
| SOM in CT (Mg/ha) | 82 | 105 | 112 | 78 | 48 |
| SOM accumulation (Mg/ha/year) | 1.1 | 0.8 | 0.8 | 0.8 | 1.0 |

*Source:* Modified from Tebrugge, F. 2001. In L. Garcia-Torres et al. (Eds.), *Conservation Agriculture: A Worldwide Challenge.* XUL, Cordoba, Spain. (Proceedings of the 1st World Congress on Conservation Agriculture, 1–5 October 2001, Madrid, Volume I. Keynote Contributions, pp. 303–316.) Original sources of data cited therein.

## TABLE 2.5
## Estimates of Carbon Sequestration Potential (Mg C/ha/year) for Various Management Changes in Two Contrasting Regions of the U.S.

| | Annual Rate of C Accumulation Resulting | |
|---|---|---|
| Change in Land Management | Mountain | Delta States |
| Convert conventional till cropland to no till | 0.20 | 0.52 |
| Add winter cover crops to a no-till system | 0.22 | 0.57 |
| Convert a low residue crop to continuous hay | 0.78 | 1.99 |
| Convert low residue crop to CRP[a] | 0.49 | 1.25 |

[a] CRP: Perennial grass or trees in the Conservation Reserve Program.
*Note:* The Mountain region had the lowest and the Delta States region had the highest potential for C sequestration for most scenarios studied.
*Source:* Data compiled from Eve, M.D. et al. 2002. *J. Soil Water Conserv.* 57:196–204.

for these same regions, and therefore the sum of converting conventionally tilled crop to no till with cover crops might be 0.4–1.09 Mg C/ha annually. Converting a low-residue crop to continuous hay or CRP resulted in even more C sequestration.

## EFFECTS OF MANAGEMENT PRACTICES ON GENERAL SOM LEVELS

Individual SOM management practices can increase C gains, decrease C losses, or accomplish both (Table 2.6). As discussed in Chapter 1, SOM has positive effects on soil physical properties and promotes water infiltration, storage, and drainage. This means that, within reason, any practice that causes large increases in SOM gains above what was previously done will probably result in lower SOM losses over time. However, in the short run, adding a lot more residue than the previous year (i.e., increased gains) might not decrease losses unless used as a surface mulch with reduced tillage.

**TABLE 2.6**
**Effects of Different Management Practices on Gains and Losses of Organic Matter**

| Management Practice | Gains Increased | Losses Decreased |
|---|---|---|
| Add materials from off the field (manures, composts, other organic materials) | Yes | No (unless residue on surface) |
| Better utilize crop residue | Yes | No (unless residue on surface) |
| Include high-residue-producing crops in rotation | Yes | No (unless residue on surface) |
| Include sod crops (grass–legume forages) in rotation | Yes | Yes |
| Grow cover crops | Yes | Yes |
| Reduce tillage intensity | Yes/no[a] | Yes |
| Use conservation practices to reduce erosion | Yes/no[a] | Yes |

[a] Practice might increase crop yields, resulting in more residue.
*Source:* Modified from Magdoff, F.R., and H. van Es. 2000. *Building Soils for Better Crops*, 2nd ed., Sustainable Agriculture Network Handbook No. 4. Burlington, VT.

However, other soil management practices by their very nature can do both. For example, either planting a cover crop or incorporating a perennial forage crop into a rotation provides increased biomass as well as erosion (and therefore SOM loss) reduction. Cover crops or sod-forming perennials promote improved near-surface soil structure, which in turn enhances infiltration and decreases runoff and erosion. In addition, the roots of living plants, including their mycorrhizal fungi, help physically hold onto soil, thus reducing erosion even when runoff occurs. In addition, with perennial forages comes less frequent plowing in the rotation and therefore less respiration loss of SOC.

The main mechanisms by which reduced tillage increase SOM are through reduced erosion and respiration losses. However, on some medium- to coarse-textured soils that have propensities toward doughtiness, surface crusting, and surface runoff, no-till management can result in higher crop yields and larger residue returns than with conventional tillage. Thus, in some situations, reduced tillage increases C additions from crop residues while erosion and respiration losses are being reduced. On the other hand, reduced tillage on soils with drainage problems, especially in cool climates, can result in depressed crop yields and therefore lower crop residue inputs.

## USE OF MULTIPLE MANAGEMENT PRACTICES

Improving crop rotations, using cover crops, reducing tillage, conserving crop residues, using organic amendments such as animal manures and composts, and practicing specific erosion control measures are positive steps to better SOM management. However, the greatest benefits to soil health usually occur from a combination of several improved organic matter management practices. For example, combining no-till management with a diverse rotation that maintains constant soil cover with various high-residue and leguminous cover crops during both warm and cold seasons can have a remarkable impact on soil health.

Management practices have been discussed previously with respect to their impacts on C additions and losses. However, management practices have multiple effects on soils. Cover crops have proven very useful to many farmers because they promote soil health by many different ways. For example, in addition to adding organic matter to soils, cover crops can enhance aggregation and water infiltration, improve water-stable aggregation and increase water infiltration rates (McVay et al., 1989), increase arbuscular mycorrhizae (AM) spore numbers to enhance inoculation of

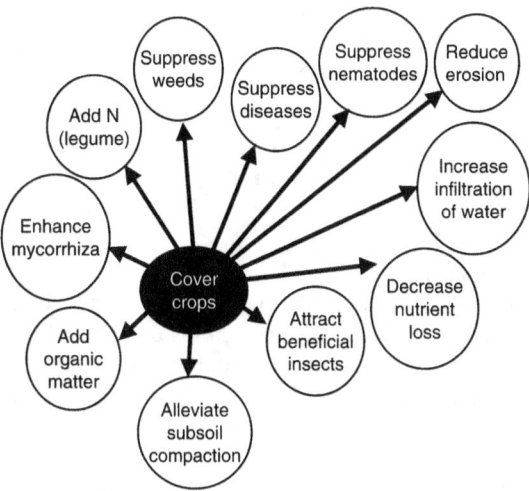

FIGURE 2.3 Some potential beneficial effects of using cover crops.

following commercial crops, decrease plant diseases, increase N availability (through $N_2$ fixation if a legume and through uptake of available soil N, decreasing $NO_3^-$ loss to leaching), decrease soil erosion, suppress weeds, and provide a habitat for beneficial organisms (Figure 2.3).

## COMPENSATION FOR C SEQUESTRATION

If farmers are to be compensated for sequestering C as a measure for mitigating $CO_2$ emissions, there will likely be the need for computer models that can predict, within an appropriate range of conditions, the amount of C sequestered because of improved management practices. Molina and Smith (1998) reviewed 24 such models and concluded that modelers still had to do much work to refine their models, and even the best models need local calibration to be effective in predicting long-term changes in SOC. Rickman et al. (2001) recently developed a computer-based model that predicts tillage and crop rotation effects on SOM trends by using data on soil properties and crop biomass production that are readily available from such sources as the C-factor files in the widely used RUSLE (Revised Universal Soil Loss Equation) program. They calibrated their model by using 60 years of SOM observations at Pendleton, OR, and then successfully validated the model at two sites in Kentucky and Wisconsin. The USDA Natural Resources Conservation Service has developed a soil conditioning index (Hubbs et al., 2002) that attempts to predict long-term trends in soil organic C concentration by using information on climate (mean annual temperature and precipitation), erosion (as predicted by RUSLE), and mean annual organic matter additions (mainly plant residues associated with various cropping systems). This index has had better success in predicting SOC losses and gains in long-term experiments in humid regions than in drier regions of the U.S. (Hubbs et al. 2002).

## SOM MANAGEMENT AS A FOUNDATION FOR AGRICULTURAL PEST MANAGEMENT

The conventional integrated pest management (IPM) approach to controlling agricultural pests has relied on a number of techniques, such as scouting to determine pest numbers and to determine numbers of beneficial organisms, using pesticides that are less damaging to the environment, and releasing beneficial organisms. The IPM approach has been mainly reactive in the sense that inspection determines whether there is a problem, and if one exists, the farmer follows some suggested practices. Lewis et al. (1997) proposed the use of a systems approach to pest management that works to strengthen internal functioning of agroecosystems and aims, to the greatest degree possible, to prevent pest problems from developing. A whole system approach to pest management has the following features:

- Promotes the growth of healthy plants that are able to better defend themselves from or are less attractive to pests
- Uses practices that stress pests
- Encourages the presence of appropriate populations of beneficial organisms, such as ground beetles that eat weed seeds and parasitoids that attack insect pests of plants
- Uses multiple tactics (many little hammers are better than one big hammer)
- If necessary to react to a pest outbreak, uses the least environmentally harmful effective control available

Practices that positively influence SOM frequently lead to the reduction of many plant pests (Abawi and Widmer, 2000; Hoitink et al., 1997; Liebman and Davis, 2000; Phatak, 1998; Phelan et al., 1995; van Bruggen, 1995). Practices that promote soil health commonly contribute to the first two strategies — growing healthy plants and stressing pests — and can also contribute to the enhancing populations of beneficial organisms. By reducing plant stress, inherent plant capabilities to resist pests are better expressed. In addition, healthier soils have more diverse and active populations of soil organisms. Populations of many soilborne pests are maintained at low numbers because of competition for resources or outright antagonism from other organisms. For example, species of beneficial *Trichoderma* can compete for occupation of wound sites, produce antibiotics, and parasitize other fungi (Tronsmo and Hjeljord, 1998). *Sclerotinia* fungal species, which attack many agricultural crops including sunflower, lettuce, peanuts, alfalfa, clover, and common turf-grasses, are attacked by other fungi such as *Trichoderma* and *Penicillium* species as well as collembola and fungus gnats, which can also serve as vectors for parasites of *Sclerotinia* (Zhou and Boland, 1998). Beneficial fluorescent *Pseudomonas* species of bacteria, commonly abundant in the rhizosphere, reduce iron availability to fungal disease organisms such as *Fusarium* wilts by producing siderophores (Sher and Baker, 1982). Fluorescent *Pseudomonas* species also produce a variety of antifungal metabolites (Weller, 1988).

Rotation, a basic agricultural practice that originated thousands of years ago, helps control plant diseases, plant parasitic nematodes, insects, and weeds (Karlen et al., 1994). The importance of rotations in plant diseases, shown through numerous studies and farmer practice, has been summarized by Vilich and Sikora (1998): "Microbial diversity in the rhizosphere is closely related to plant diversity, especially with regards to qualitative and quantitative shifts in populations of pathogens, nonpathogens and antagonists."

Low soil disturbance in reduced till systems and those including perennials — practices that usually lead to SOM increases — can encourage ground beetles and crickets that eat weed seeds and the higher biological activity in soils resulting from large applications of organic materials can promote fungal attack on weed seeds (Liebman and Davis, 2000). During germination and early growth, weeds with small seeds are at a competitive disadvantage with most crops under moderate to low available N levels. This occurs because most crop plants have relatively large seeds, giving them more N reserves for early seedling growth than small-seeded weeds have. In general, N converted to available forms by mineralization of organic matter is timed to the growth of crops. Thus, relying on organic sources of N or sidedress N rather than a large dose of highly available inorganic N fertilizer before planting can provide this early-season low available N advantage. Organic sources of N, such as manures, composts, and legume cover crops, can furnish adequate crop nutrition to full-season crops while maintaining relatively low levels of available N for most of the growing season. On the other hand, N mineralization from organic sources might lag behind in the needs of early short-season crops and might continue in the fall after full-season crops have been harvested. Therefore, when organic sources of fertility are used, additional available N might be needed for early-season crops, and cover crops should be used to prevent excess N leaching following the growing season.

Cover crops can have significant positive effects on soil health and are an important tool for SOM management. The use of cover crops usually leads to better nutrient use, more available N (especially if a legume is grown), better rainfall and irrigation infiltration rates, better soil aggregation, higher AM spore number, etc. Cover crops can also have important pest management implications. Growing a cover crop of corn or barley between annual crops of potatoes was found to essentially eliminate *Verticillium* wilt and yields returned to precontinuous cropping levels, even though inoculum (spore) levels remained unchanged (Davis et al., 2001). In Georgia, planting into strip tilled crimson clover eliminated the need for fungicides on tomatoes, peppers, eggplant, cucumbers, and lima beans, and fungal disease control was achieved for peanuts by no-till planting into killed cereal rye (Phatak, 1998).

Although many cover crops suppress plant parasitic nematodes, if a nematode problem exists it will influence the choice of cover crops. For example, hairy vetch is an excellent cover crop in many respects, but should be avoided if lesion nematode (*Pratylenchus* spp.) is a problem, because it is a good host for the pest (Abawi and Widmer, 2000). On the other hand, ryegrass, sudangrass, and rape help maintain low lesion nematode populations. The root-knot nematode (*Meloidogyne* spp.) has an extremely wide host range, and is thereby a challenge to control by crop rotation. However, this nematode is strongly suppressed by the nonhost grass and grain species such as sudangrass, oats, and rye (Viaene and Abawi, 1998). On-farm research in the state of Washington showed that potato yields were equally good in on-farm trial when root-knot nematodes were controlled by metham sodium fumigation or by use of a preceding white mustard cover crop (McGuire, 2001).

Brassica cover crops can provide practical suppression of many plant parasitic nematodes (in addition to root-knot) as well as pathogenic fungi. Tissues of Brassica cover crops contain variable but generally high levels of glucosinolates (Carlson et al., 1987). When these plants are incorporated into the soil as green manure, the glucosinolates can hydrolyze to produce isothiocyanates (ITCs) that exhibit fungicidal and nematicidal properties (Ettlinger and Kjaer, 1968). Rapeseed cover crops (*Brassica napus* and *B. campetris*) have been implicated in reducing densities of soil nematodes (Mojtahedi et al., 1991). For two consecutive years, planting Jupiter rapeseed in the fall and incorporating it in the spring as green manure limited *M. chitwoodi* damage on potato tubers in field experiments (Mojtahedi et al., 1993). In a California micro plot study (Gardner and Caswell-Chen, 1993), oilseed radish and white mustard were found to be effective trap crops for *Heterodera* (cyst) nematode In a sugar-beet cropping system on loam sand soils in Poland, the population of sugar beet cyst-nematode (*Heterodera schachtii*) was reduced by 40% after growing a white mustard cover crop, and by 41 to 48% after three oilseed radish cultivars (Nowakowski and Szymczak-Nowak, 1998). Preliminary results in Pennsylvania (Halbrendt, 1992) suggest that white mustard, black mustard, and rapeseed can be generally as effective as commercial nematicides in suppressing dagger nematode in fruit orchard and strawberry systems.

Practices that either reduce soil disturbance or maintain growing crops for long periods, such as reduced tillage, rotations with perennial crops, and routine use of cover crops, promote AM presence and spore levels. These practices can help maintain plant health because the association of AM fungi with roots offers some protection against plant diseases (Azcon-Aguilar and Barea, 1996).

Cover crops usually contribute to weed control by using resources that can be used by weeds — water, nutrients, and light. However, certain cover crops, including Brassica species and winter rye, also contain phytotoxic (allelopathic) chemicals that can suppress weeds. Although the exact mechanism is not clear and the effects differ widely among Brassica species and cultivars, numerous studies have shown that dramatic suppression of weeds can occur with Brassica cover crops. For example, incorporation of rape (*Brassica napus*) residue inhibited emergence and biomass of annual grassy weeds (Purvis et al., 1985). Research shows that the concentration and potency of the allelopathic compounds depend on environmental conditions and plant maturity at the time of cover crop kill. Petersen et al. (2001) evaluated the allelopathic potential of ITCs released by turnip-rape mulch and demonstrated strong suppression of germination in numerous weed species, including spiny sowthistle (*Sonchus asper*), smooth pigweed (*Amaranthus hybridus*), barnyardgrass (*Echinochloa crusgalli*), and blackgrass (*Alopecerus myosuroides*). They postulated that ITCs interacted with weed seeds in the soil solution and as vapor in soil pores. In a study in the state of Washington (Boydston and Vaughn, 2002), rapeseed incorporated before flowering and 1 month before potato planting reduced weed density by 73% in the middle of the potato growing season and reduced weed biomass by 50% by the end of the season. The authors stated that the rapeseed provides commercially acceptable weed control without any other control measure in one of the two study years.

Residues of many legume cover crops also provide a degree of weed control in the following crop by allelopathy (Dyck and Liebman, 1994). Undersowing a grain crop with red clover is a practice that can help build organic matter by biomass production and the erosion reduction potential of a sod growing after grain harvest. (Straw is also removed in many cases.) Undersown red clover has been found to reduce not only weeds during the period immediately following grain harvest but also weed populations in the following year (Liebman and Davis, 2000).

## CONCLUSIONS

SOM management is part of a preventive approach to crop production that seeks to maximize the beneficial chemical, physical, and biological characteristics of soils. Individual practices for better SOM management have multiple effects on soils and crops and contribute to SOM dynamics by accomplishing one or more of the following:

1. Increasing the amount of organic carbon added
2. Increasing the diversity of organic materials added
3. Decreasing the rate of organic matter loss

Most SOM management practices also significantly contribute to a whole system approach for pest management by promoting the growth of healthy crops that can better resist pests, stressing crop pests, and enhancing populations of beneficial organisms. Creative SOM management will combine a number of practices that work together on individual farms. For example, better crop rotations, reduced tillage, intensive use of cover crops, and efficient use of animal manures offers a powerful combination of practices to improve soil and agroecosystem health.

## REFERENCES

Abawi, G.S., and T.L. Widmer. 2000. Impact of soil health management practices on soilborne pathogens, nematodes, and root diseases of vegetable crops. *Appl. Soil Ecol.* 15:37–47.
Albiach, R., R. Canet, F. Pomares, and F. Ingelmo. 2001. Organic matter components and aggregate stability after the application of different amendments to a horticultural soil. *Bioresour. Technol.* 76:125–129.
Alexander, R. 1999. Compost markets grow with environmental applications. *Biocycle* 40(4):43–44, 46, 48.
Allison, F.E. 1973. Soil organic matter and its role in crop production. In *Developments in Soil Science,* Vol. 3. Elsevier, Amsterdam.
Azcon-Aguilar, C, and J. M. Barea. 1996. Arbuscular mycorrhizas and biological control of soil-borne plant pathogens: An overview of the mechanisms involved. *Mycorrhiza* 6:457–464.
Bartlett, R.J. 1981. Oxidation–reduction status of aerobic soils. In Dowdy, R.H., J.A. Ryan, V.V. Volk, and D.E. Baker (Eds.), *Chemistry in the Soil Environment.* American Society of Agronomy, Madison, WI, pp. 77–102.

Birch, H.F. 1958. The effect of soil drying on humus decomposition and nitrogen availability. *Plant Soil* 10:9–31.

Boydston, R.A., and S.F. Vaughn. 2002. Alternative weed management systems control weeds in potato (*Solanum tuberosum*). *Weed Technol.* 16:23–28.

Campbell, C.A., G.P. Landford, R.P. Zentner, and V.O. Biederbeck. 1991. Influence of fertilizer and straw baling on soil organic matter in a thin black Chernozem in western Canada. *Soil Biol. Biochem.* 23: 443–446.

Campbell, C.A., and R.P. Zentner. 1993. Soil organic matter as influenced by crop rotations and fertilization. *Soil Sci. Soc. Am. J.* 57:1034–1040.

Carlson, D.G., M.E. Daxenbichler, C.H. Van Etten, W.F. Kwolek, and P.H. Williams. 1987. Glucosinolates in crucifer vegetables: Broccoli, Brussels sprouts, cauliflower, collards, kale, mustard greens, and kohlrabi. *J. Am. Soc. Hortic. Sci.* 112:173–178.

Chromec, F.W. and F.R. Magdoff. 1984. Alternative methods for using organic materials: Composting vs. adding directly to soil. *J. Environ. Sci. Hlth.* 19(6):697–711.

Curtin, D., H. Wang, F. Selles, B.G. McConkey, and C.A. Campbell. 2000. Tillage effects on carbon fluxes in continuous wheat and fallow-wheat rotations. *Soil Sci. Soc. Am. J.* 64:2080–2086.

Davis, J.R., O.C. Huisman, and D.O. Everson. 2001. Cultural management of potato for *Verticillium* wilt control, changes in rhizosphere, increased potato yield, and quality. In *XXXVIth International Horticultural Congress*, August 11–17, 2002, Toronto, Canada, Abstract 218.

Drinkwater, L.E., P. Wagoner, and M. Sarrantonio. 1998. Legume-based cropping systems have reduced carbon and nitrogen loss. *Nature* 396:262–265.

Dyck, E., and M. Liebman. 1994. Soil fertility management as a factor in weed control: The effect of crimson clover residue, synthetic nitrogen fertilizer, and their interactions on emergence and early growth of lambsquarters and sweet corn. *Plant Soil* 167:227–237.

Eghball, B. 2002. Soil properties as influenced by phosphorus- and nitrogen-based manure and compost applications. *Agron. J.* 94:128–135.

Ettlinger, M.G., and K. Kjaer. 1968. Sulfur compounds in plants. In T.J. Mabry (Ed.), *Recent Advancements in Phytochemistry*. Appleton-Century-Crofts, New York, pp. 59–144.

Eve, M.D., M. Sperow, K. Howerton, K. Paustian, and R.F. Follet. 2002. Predicted impact of management changes on soil carbon storage for each cropland region of the coterminous United States. *J. Soil Water Conserv.* 57:196–204.

Follett, R.F., S.C. Gupta, and P.G. Hunt. 1987. Conservation practices: Relation to the management of plant nutrients for crop production. In R.F. Follett et al. (Eds.), *Soil Fertility and Organic Matter as Critical Components of Production Systems*. SSSA Special Publication No. 19, Soil Science Society of America, Madison, WI, pp. 19–51.

Francis, G.S., F.J. Tabley, and K.M. White. 2001. Soil degradation under cropping and its influence on wheat yield on a weakly structured New Zealand silt loam. *Aust. J. Soil Res.* 39:291–305.

Franzluebbers, A.J. 2003. Tillage and residue management effects on soil organic matter. In F. Magdoff and R.R. Weil (Eds.), *Functions and Management of Soil Organic Matter in Agroecosystems*. CRC Press, Boca Raton, FL.

Gajda, A.M., J. Doran, T. Kettler, J.L. Weinhold, J.L. Pikul, Jr., and C.A. Cambardella. 2001. Soil quality evaluations of alternative and conventional management systems in the Great Plains. In R. Lal et al. (Eds.), *Assessment Methods for Soil Carbon*. Lewis Publishers, Boca Raton, FL, pp. 381–400.

Gardner, J., and E.P. Caswell-Chen. 1993. Penetration, development, and reproduction of *Heterodera schactii* on *Fagopyprum esculentum*, *Phacelia tanacetifolia*, *Raphanus sativus*, *Sinapis alba*, and *Brassica oleracea*. *J. Nematol.* 25(4):695–702.

Gregorich, E.G., C.F. Drury, and J.A. Baldock. 2001. Changes in soil carbon under long-term maize in monoculture and legume-based rotation. *Can. J. Soil Sci.* 81:21–31.

Halbrendt, J.M. 1992. Novel rotation crops as alternatives to fumigant nematicide treatment in deciduous tree fruit production. Final Report LNE90-22, USDA/EPA Sustainable Agriculture Research and Education Program and Agriculture in Concert with the Environment, Washington, D.C.

Halvorson, A.D., C.A. Reule, and R.F. Follett. 1999. Nitrogen fertilization effects on soil carbon and nitrogen in a dryland cropping system. *Soil Sci. Soc. Am. J.* 63:912–917.

Hoitink, H.A.J., D.Y. Han, A.G. Stone, M.S. Krause, W. Zhang, and W.A. Dick. 1997. Natural suppression. *Am. Nurseryman* October 1, 1997:90–97.

Hubbs, M.D., M.L. Norfleet, and D.T. Lightle. 2002. Interpreting the soil conditioning index. In E. van Santen (Ed.), *Making Conservation Tillage Conventional: Building a Future on 25 Years of Research*. Proceedings of the 25th Annual Southern Conservation Tillage Conference for Sustainable Agriculture, 24–26 June 2002, Auburn, AL. Special Report No. 1, Alabama Agricultural Experimental Station and Auburn University, Auburn, AL, pp. 192–196.

Jenkinson, D.S., and A.E. Johnson. 1977. Soil organic matter in the Hoosfield barley experiment. In *Annual Report 1976*. Rothamstead Experiment Station, Reading, U.K., pp. 87–102.

Jordan, D., R.J. Kremer, W.A. Bergfield, K.Y. Kim, and V.N. Cacnio. 1995. Rangeland soil carbon and nitrogen responses to grazing. *J. Soil Water Conserv.* 50:294–298.

Karlen, D.L., G.E. Varvel, D.G. Bullock, and R.M. Cruse. 1994. Crop rotations for the 21st century. *Adv. Agron.* 53:1–45.

Kennedy, A.C., T.L. Stubbs, and W.F. Schillinger. 2003. Soil and crop management effects on soil biology. In F. Magdoff and R.R. Weil (Eds.), *Functions and Management of Soil Organic Matter in Agroecosystems*. CRC Press, Boca Raton, FL.

Larson, W.E., C.E. Clapp, W.H. Pierce, and Y.B. Morochan. 1972. Effects of increasing amounts of organic residues on continuous corn. II. Organic carbon, nitrogen, phosphorus and sulfur. *Agron. J.* 64:204–208.

Liebman, M., and A.S. Davis. 2000. Integration of soil, crop, and weed management in low-input farming systems. *Weed Res.* 40:27–47.

Lewis, W.J., J.C. van Lenteren, S.C. Phatak, and J.H. Tumlinson III. 1997. A total system approach to sustainable pest management. *Proc. Nat. Acad. Sci.* 94:12243–12248.

Magdoff, F., L.E. Lanyon, and W. Liebhardt. 1997. Nutrient cycling, transformations, and flows: Implications for a more sustainable agriculture. In D.L. Sparks (Ed.), *Advances in Agronomy*, Vol. 60. Academic Press, Boca Raton, FL, pp. 2–73.

Magdoff, F.R. and J.F. Amadon. 1980. Yield trends and soil chemical changes resulting from N and manure application to continuous corn. *Agron. J.* 72:161–164.

Magdoff, F.R., and H. van Es. 2000. *Building Soils for Better Crops*, 2nd ed. Sustainable Agriculture Network Handbook No. 4. Burlington, VT.

McGuire, A. 2001. On-Farm Research Results 1999-2001: Dale Gies Farm, Mustard Green Manures. [Online]. Available from Washington State University Cooperative Extension, Center for Sustaining Agriculture and Natural Resources, http://grant-adams.wsu.edu/agriculture/covercrops/green_manures/index.htm., accessed October 26, 2002.

McVay, K.A., D.E. Radcliff, and W.L. Hargrove. 1989. Winter legume effects on soil properties and nitrogen fertilizer requirements. *Soil Sci.* 53:1856–1862.

Mojtahedi, H., G.S. Santo, A.N. Hang, and J.H. Wilson. 1991. Suppression of root-knot nematode populations with selected rapeseed cultivars as green manure. *J. Nematol.* 23:176–174.

Mojtahedi, H., G.S. Santo, J.H. Wilson, and A.N. Hang. 1993. Managing *Meloidogyne chitwoodi* on potato with rapeseed as green manure. *Plant Disease* 77:42–46.

Molina, J.A.E., and P. Smith. 1998. Modeling carbon and nitrogen processes in soils. In D.L. Sparks (Ed.), *Advances in Agronomy*, Vol. 62. Academic Press, Boca Raton, FL, pp. 253–298.

Nichols, K.A., and S.F. Wright. 2003. Contributions of fungi to soil organic matter in agroecosystems. In F. Magdoff and R. R. Weil (Eds.), *Functions and Management of Soil Organic Matter in Agroecosystems*. CRC Press, Boca Raton, FL.

Nowakowski, M., and J. Szymczak-Nowak. 1998. Growth dynamics, yielding and antinematode effects of some cultivars of oil radish and white mustard cultivated as a catch crop. [In Polish.] *Rosliny Oleiste* 19:671–678.

Paustian, K., W.J. Parton, and J. Persson. 1992. Modeling soil organic matter in organic-amended and nitrogen-fertilized long-term plots. *Soil Sci. Soc. Am. J.* 56:476–488.

Paustian, K., H.P. Collins, and E.A. Paul. 1997. Management controls on soil carbon. In E.A. Paul et al. (Eds.), *Soil Organic Matter in Temperate Agroecosystems*. CRC Press, Boca Raton, FL, pp. 15–49.

Paustian, K., J. Six, E.T. Elliott, and H.W. Hunt. 2000. Management options for reducing $CO_2$ emissions from agricultural soils. *Biogeochemistry* 48:147–163.

Petersen, J., R. Belz, F. Walker, and K. Hurle. 2001. Weed suppression by release of isothiocyanates from turnip-rape mulch. *Agron. J.* 93:37–43.

Phatak, S.C. 1998. Managing pests with cover crops. In *Managing Cover Crops Profitably*, 2nd ed. Sustainable Agriculture Network, Burlington, VT, pp. 25–33.

Phelan, P.L., J.R. Mason, and R.B. Stinner. 1995. Soil-fertility management and host preference by European corn borer, *Ostrinia nubilalis* (Hubner), on *Zea mays* L.: A comparison of organic and conventional chemical farming. *Agric. Ecosyst. Environ.* 56:1–8.

Puget, P., and L.E. Drinkwater. 2001. Short-term dynamics of root- and shoot-derived carbon from a leguminous green manure. *Soil Sci. Soc. Am. J.* 65:771–779.

Purvis, C.E., R.S. Jessop, and J.V. Lovett. 1985. Selective regulation of germination and growth of annual weeds by crop residues. *Weed Res.* 25:415–421.

Reicosky, D.C., J.L. Hatfield, and R.L. Sass. 2000. Agricultural contributions to greenhouse gas emissions. In K.R. Reddy and H.F. Hodges (Eds.), *Climate Change and Global Crop Production*. CAB International, London, pp. 37–55.

Rickman, R.W., C.L. Douglas, Jr., S.L. Albrecht, L.G. Bundy, and J.L. Berc. 2001. CQESTR: A model to estimate carbon sequestration in agricultural soils. *J. Soil Water Conserv.* 56:237–242.

Sanchez, J.E., T.C. Willson, K. Kizilkaya, E. Parker, and R.R. Harwood. 2001. Enhancing the mineralizable nitrogen pool through substrate diversity in long term cropping systems. *Soil Sci. Soc. Am. J.* 65:1442–1447.

Sher, F.M., and R. Baker. 1982. Effect of *Pseudomonas putida* and a synthetic iron chelator on induction of soil suppressiveness to Fusarium wilt pathogens. *Phytopathology* 72:1567–1573.

Stone, A.G., and H.M. Darby. 2003. Organic matter-mediated suppression of soilborne plant diseases in field agricultural systems. In F. Magdoff and R.R. Weil (Eds.), *Functions and Management of Soil Organic Matter in Agroecosystems*. CRC Press, Boca Raton, FL, pp. 37–55.

Studdert, G.A., H.E. Echeverria, and E.M. Casanovas. 1997. Crop-pasture rotation for sustaining the quality and productivity of a Typic Argiudoll. *Soil Sci. Soc. Am. J.* 61:1466–1472.

Tebrugge, F. 2001. No-tillage vision: Protection of soil, water and climate and influence on management and farm income. In L. Garcia-Torres et al. (Eds.), *Conservation Agriculture: A Worldwide Challenge*. XUL, Cordoba, Spain. (Proceedings of the 1st World Congress on Conservation Agriculture, Vol. I. Keynote Contributions, 1–5 October 2001, Madrid, pp. 303–316.)

Tebrugge, F., and R.A. During. 1999. Reducing tillage intensity: A review of results from a long-term study in Germany. *Soil Tillage Res.* 53:15–28.

Tronsmo, A., and L.G. Hjeljord. 1998. Biological control with *Trichoderma* species. In G.J. Boland and L.D. Kuykendall (Eds.), *Plant–Microbe Interactions and Biological Control*. Marcel Dekker, New York.

van Bruggen, A.H.C. 1995. Plant disease severity in high-input compared to reduced-input and organic farming systems. *Plant Dis.* 79:976–984.

Vilich, V., and R.A. Sikora. 1998. Diversity in soilborne microbial communities: A tool for biological system management of root health. In G.J. Boland and L.D. Kuykendall (Eds.), *Plant-Microbe Interactions and Biological Control*. Marcel Dekker, New York.

Viaene, N.M., and G. Abawi. 1998. Management of *Meloidogyne hapla* on lettuce in organic soil with sudangrass as a cover. *Plant Dis.* 82:45–952.

Weil, R.R., K.A. Lowell, and H.M. Shade. 1993. Effects of intensity of agronomic practices on a soil ecosystem. *Am. J. Altern. Agric.* 8:5–14.

Weller, D.M. 1988. Biological control of soilborne plant pathogens in the rhizosphere with bacteria. *Annu. Rev. Phytopathol.* 26:379–407.

West, T.O., and W.M. Post. 2002. Soil organic carbon sequestration rates by tillage and crop rotation: a global data analysis. *Soil Sci. Soc. Am. J.* 66:1903–1946.

Whitbread, A.M., G.J. Blair, and R.D.B. Lefroy. 2000. Managing legume leys, residues and fertilisers to enhance the sustainability of wheat cropping systems in Australia. 2. Soil physical fertility and carbon. *Soil Tillage Res.* 54:77–89.

Wilson, D.O., and W.L. Hargrove. 1986. Release of nitrogen from crimson clover residue under two tillage systems. *Soil Sci. Soc. Am. J.* 50:1251–1254.

Zhou, T., and G.J. Boland. 1998. Biological control strategies for *Sclerotinia* diseases. In G.J. Boland and L.D. Kuykendall (Eds.), *Plant–Microbe Interactions and Biological Control*. Marcel Dekker, New York.

# 3  Soil Organic Matter Fractions and Their Relevance to Soil Function

*Michelle Wander*

## CONTENTS

Improved management of soil organic matter (SOM) in arable soils is essential to sustain agricultural lands and the urban and natural ecosystems with which they interact. Humus, which has historically been equated with inherent soil fertility, can be efficiently extracted from mineral soils in alkali. The resulting humic and fulvic fractions of SOM continue to be widely studied despite these fractions, which are procedural artifacts existing only in the laboratory that have not proven to be particularly useful guides to adaptive management or contributed notably to our understanding of either SOM dynamics or soil quality. The quest continues to understand organic matter's contributions to soil productive capacity, its ability to transform and store matter and energy, and its capacity to regulate water and air movement. Successful efforts will identify consistently defined and derived SOM fractions that impart fundamental characteristics to soils. This chapter provides an overview of commonly measured SOM fractions and the kinetically or theoretically defined dynamic pools with which they are commonly identified. Organic matter of recent origin is most closely associated with biological activity in soils, whereas materials of recent and intermediate age contribute notably to soil's physical status. Materials with longer residence times typically comprise the largest reservoirs in soils and exert the greatest influence on the physicochemical reactivity of soils. The

0-8493-1294-9/04/$0.00+$1.50

characteristics of individual SOM fractions often vary as a result of the techniques used to isolate them or the experimental context. Amino sugars, glomalin, and particulate organic matter (POM) fractions have multiple identities. In addition to providing information about biologically active SOM, all these fractions provide information about physically active and passive SOM pools. The final section of this chapter is devoted to POM, an increasingly popular measure of labile SOM because it responds readily to soil management, often identifying statistically significant trends when measures of total SOM would not. POM plays important biological and physical roles in soil. Even though POM is most often used as an index of labile SOM, POM fractions include materials that are heterogeneous in age and function. Size, density, and energy can be combined in a variety of ways to recover materials that can be associated with active, slow, and recalcitrant pools. As is true for humic substances, POM's value as an index of SOM will not be proven until the relationships between its characteristics and *in situ* soil processes are clearly demonstrated. The utility of POM will be increased by standardizing the approaches used to subdivide its constituents and through better articulation of criteria used to interpret results.

## HISTORY AND PURPOSE OF ORGANIC MATTER MEASUREMENT

### IMPORTANCE OF SOM AND ITS RELATIONSHIP TO MANAGEMENT

As is true for soil science in general, the study of SOM has emphasized its relationship to soil productivity. Even in well-fertilized soils, soil productivity is reduced by loss of SOM (Johnston, 1991; Aref and Wander, 1997). Accompanying these losses in productive potential are losses in agroecosystem efficiency. Crop response to mineral inputs is increased in soils where organic matter status and biological and physical properties influenced by organic matter are enhanced (Cassman, 1999; Avnimelech, 1986). What exactly "enhanced" means in this context remains a critical question. Several studies have suggested that cropping systems that rely on mixed-crop production and organic sources of fertility are better able to maintain or accumulate organic matter and improve its quality than are mono- or bicropped systems that rely on inorganic nutrient sources (Reganold et al., 1987; Wander et al., 1994; Glendining et al., 1997; Liebig and Doran, 1999). Enhancements of SOM status (based on labile fraction characterization) and crop performance are reported for a variety of management practices, including organic (Wander et al., 1994), compost amended (Stone et al., 2001; Willson et al., 2001), pasture (Sbih et al., 2003), mixed-crop and cover cropped (Drury et al., 1991; Collins et al., 1992; Angers and Mehuys, 1988; Stevenson et al., 1998), and no-till systems (Beare et al., 1994b; Dick, 1997; Frey et al., 1999). Competitive crop yields achieved with fewer external inputs are attributed to cropping systems that enhance organic matter characteristics (Liebhardt et al., 1989; Johnston, 1991; Poudel et al., 2001; Nissen and Wander, 2003).

Despite our long understanding of the relationship between soil building practices and their benefits to SOM (Russell, 1973), and the general appreciation that SOM underpins ecosystem function in terrestrial systems (Odum, 1969), our ability to quantify or manipulate its characteristics remains quite limited. Results from long-term experiments provide critical insights into the influences of management on SOM and its contributions to agricultural sustainability (Rasmussen et al., 1998). Results such as these demonstrate shortfalls in our understanding of SOM's contributions to soil productivity. The general benefits of crop rotation to SOM and soil productivity are suggested by yield trends expressed in Morrow Plots (Wander et al., 2002). In general, differences between the various systems' yields and SOM levels increase with the complexity, or length, of the crop rotation (Figure 3.1A). If maize yield serves as a bioassay, then results in the three-year rotation (corn–oats–hay, COH) suggests that the productive potential of that soil is higher than that of the soil maintained under the two-year corn–soybean (CS) or continuous corn (CC) rotations. Increases in maize yield that result from increased inputs, which include lime, manure, and N, P, K additions and seeding densities adjusted to different rates, are not mirrored by increases in SOM contents (Figure 3.1B). Soils with the highest SOM contents have a history of manure application. The yield

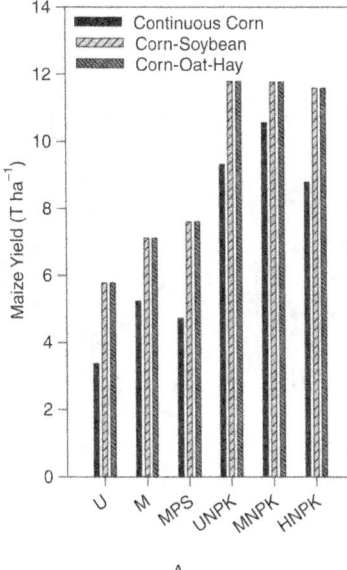
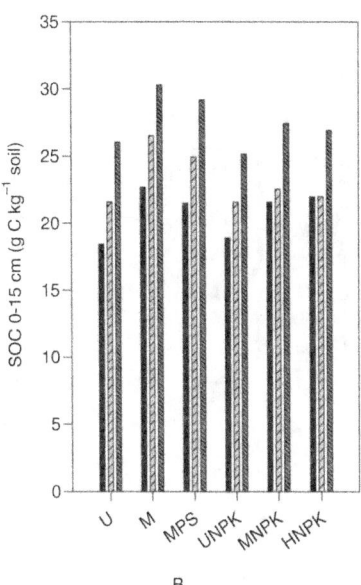

A  B

**FIGURE 3.1** Morrow Plots yield (A) and SOM (B) contents in 1997, when all plots were in corn. This trial, begun in 1876, is the oldest agricultural experiment in the Northern Hemisphere. Since 1967, the plots have included three crop rotations: continuous corn, *Zea mays* L. (CC); corn–soybean, *Glycine max* (CS), and corn–oats–hay, *Avena sativa* and *Melilotus alba* or *Trifolium pratense* (COH). Before that time, the corn–soybean rotation was a corn–oat system. The trial presently compares five fertility regimes, added over the course of the trial: unfertilized controls (U) and combinations of manure (M and MPS, which has a higher seeding density), plots without (UNPK) and with (MNPK) a history of manure amendment that receive inorganic NPK, and plots that had received manure up until 1967 that have subsequently only been amended with the highest P and K rates (HNPK). Since 1967, N has been applied as urea at 200 lb ac$^{-1}$ in NPH and MNPK plots and 300 lb ac$^{-1}$ in HNPK plots. In NPK and MNPK plots, P as triple superphosphate and K as muriate of potash are applied at 49 and 93 lb ac$^{-1}$, respectively, when test values are lower than 45 or 336 lb of available P or K, respectively. The HNPK plots have received 98 and 186 lb ac$^{-1}$ P and K, respectively, when test values fall below 112 and 560 values.

achieved in the higher-input treatments in the CS and COH systems is quite similar even though total SOM levels are not. The SOM contents of the CS and CC rotations are quite similar, but the yields differ markedly. Soil test levels for pH, P, and K (Figure 3.2) do not account for differences in yield achieved in the different rotations or amendment regimes. Phosphorus buildup is apparent in CC plots amended with manure every year. This is a common problem in plots receiving higher manure application rates. A comparatively low seeding rate in the manure-amended plots (M and MPS) likely limits yield in, and associated nutrient removal from, those soils. Nutrients are relatively depleted in fertilizer-amended plots producing the highest system yield. Plots that receive application rates higher than those recommended by the state are an exception and accumulate both P and K. Interestingly, the highest yield is achieved in the COH system even though K test levels are below the reported optimum values. Differences in SOM quality, not quantity, and SOM-dependent microbial and physical properties are thought to explain why these three systems differ in their productive potential and the degree to which crops can exploit the soil resource.

Our understanding of SOM's specific contributions to soil function has not advanced notably in the past 50 years and remains primarily descriptive in nature (Table 3.1). Cation exchange capacity (CEC), a function of SOM, pH, and mineral characteristics, and percentage of surface residue cover are rare examples wherein quantitative relationships between SOM-dependent characteristics and adaptive management practices are established. Soil CEC influences lime and herbicide application rates, whereas residue cover determines eligibility for participation in

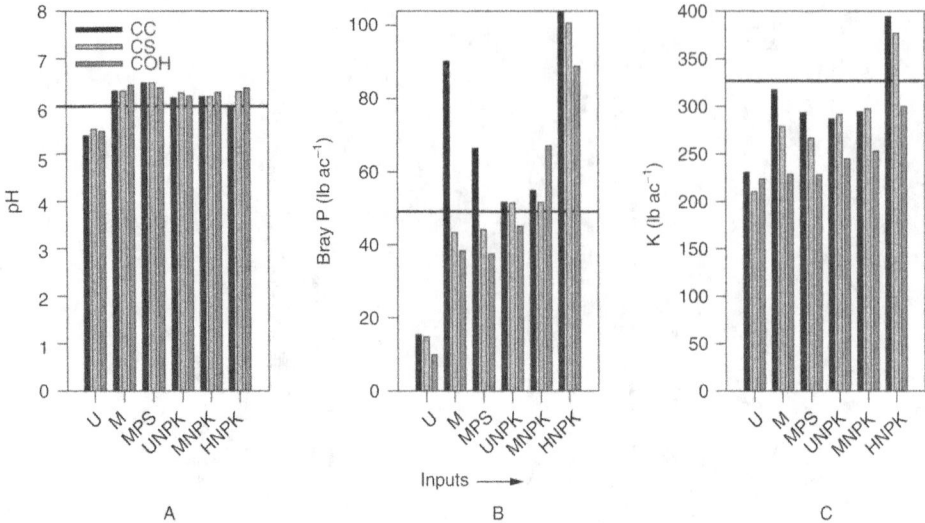

**FIGURE 3.2** Inorganic nutrient status of Morrow Plots, pH in 1:1 water, (A) P via Bray P-1 (B), and K extractable in NaOAc (C). Inputs including fertilizers and seed density increase from left to right and include unfertilized controls (U) and combinations of manure (M and MPS, which has a higher seeding density), plots without (UNPK) and with (MNPK) a history of manure amended that receive lime and inorganic NPK additions, and plots that had received manure up until 1967 and have subsequently been amended with lime plus very high fertility rates (HNPK). See Figure 3.1 legend for additional details about amendments. Solid horizontal lines indicate recognized optimum test values needed to achieve maximum production.

---

**TABLE 3.1**
**Summaries of Physical, Chemical, and Biological Contributions of Organic Matter to Soil Function**

| Summary by Waksman (1938) | Summary by Stevenson (1994) |
|---|---|
| **Physical Functions** | |
| Modifies soil color, texture, structure, moisture-holding capacity, and aeration | Color, water retention, helps prevent shrinking and drying, combines with clay minerals, improves moisture-retaining properties, stabilizes structure, permits gas exchange |
| **Chemical Functions** | |
| Solubility of minerals; formation of compounds with elements such as Fe, making them more available for plant growth; increases the buffer properties of soils | Chelation improves micronutrient availability; buffer action maintains uniform reaction in soil and increases cation exchange |
| **Biological Functions** | |
| Source of energy for microorganisms, making the soil a better medium for the growth of plants; supplies a slow but continuous stream of nutrients for plant growth | Mineralization provides source of nutrients; combines with xenobiotics, influencing bioavailability and pesticide effectiveness |

conservation programs. Despite SOM's importance to food and fiber production, routine methods to quantify its contribution to soil productivity do not exist or are not widely agreed on. The contribution of SOM to soil N supply is still so poorly described that it is estimated by fertilizer equivalency trials, expected yields, and cropping history or by the preplant soil profile $NO_3$ (PPNT) or presidedress $NO_3$ (PSNT) tests that then serve as a basis for N fertilizer application rates (Magdoff et al., 1984; Dahnke and Johnson, 1990). The need to predict organic N supply by using measures that are not as dynamic as nutrient fluxes (Spycher et al., 1983; Mulvaney et al., 2001) explains the research interest in biologically active SOM fractions. However, SOM fractions that are highly labile, varying within a season or a year, might prove to be as difficult to use as indices as are inorganic nutrients. Assessments of labile SOM will improve with separation of fractions most closely associated with fresh inputs, which have annual dynamics tightly coupled to edaphic factors, from constituents that reflect the recent (decadal) influence of management. Fractions of SOM that reflect management deserve particular attention because they predict trends in soil productivity and the efficiency with which the soil cycles matter and energy. The quality and quantity of SOM reserves and the edaphic factors that regulate their dynamics will need to be considered. Measures of SOM that effectively predict nutrient supply, soil–water relations, aeration, pesticide immobilization, and trends in carbon sequestration are likely to differ.

## APPROACHES TO ORGANIC MATTER FRACTIONATION

According to Waksman (1936), the term *humus* dates back to the Romans and was used by the ancients in reference to soil and the "fatness of the land," where fatness connoted fertility. Wallerius in 1761 first defined humus in terms of decomposed organic matter. In 1808, Thaer, cited in Walksman (1936), wrote, "Humus is the product of living matter, and the source of it." Even though Walksman cautioned in 1936 that "any attempt to divide humus on the basis of its practical utilization would prove to be largely artificial," this objective remains a top priority of many wishing to better manage the soil resource. Humus classification schemes probably began with Linneaus's classification of soils in accordance with humus types. Archard (1786) was probably the first to attempt to extract humic substances from soils. De Saussure (1804) equated the Latin term for soil, *humus*, with dark material produced from decayed plants. Wallerius (1761), cited in Walksman (1936), speculated that chalk and likely salts helped dissolve humus to make it available to plants. He advised that alkali be used alternately with dung to satisfy plant demand. The perception that alkali-soluble humic materials contributed to soil's native fertility and the fact that alkali extracts humic materials efficiency, removing typically 20 to 50% and up to 80% in some cases of the organic material from the soil (Stevenson, 1982; Rice, 2001), explain why the study of SOM has focused on humic substances recovered after their dissolution in a dilute base (typically 0.10 *N* NaOH, or, increasingly, $Na_4P_2O_7$). Many extraction methods have been vigorously explored, because separation of organic matter from the mineral matrix facilitates chemical characterization of SOM by HPLC, GC-MS, wet chemistry, and elemental analyses. These techniques would be impossible to apply to intact soils. Separation methods have commonly been judged on their ability to isolate pure, reproducible, and homogenous components (Stevenson, 1994). This quest reflects a historical desire to describe SOM in primarily biochemical terms by using molecular formulae. Berzelius proposed the first chemical formulas for two organic matter fractions: crenic ($C_{24}H_{12}O_{16}$) and apocrenic ($C_{24}H_6O_{12}$) acids (Stevenson, 1994). Crenic acid was isolated from iron- and mud-rich waters by treating them with potassium hydroxide followed by acetic acid and then copper acetate. Apocrenic acid was obtained by treating coal with nitric acid. The often-unstated assumption that legitimate organic matter fractions will be pure in composition with tractable and uniform, or at least consistent, routes of origin has oriented research in only a few directions. The classical method for humic substance fractionation is to acidify the organic colloids obtained after dispersion in dilute sodium hydroxide. Humic acids (HAs) precipitate in acidified solutions whereas fulvic acids (FA) remain in suspension (Swift, 1996). This method continues to be widely used despite criticism that the strategy is archaic and does not produce chemically discrete fractions (Russell, 1973). Humic substances are understood to include a continuum of complex biogenic amorphous heterogeneous molecules that are both chemically reactive and refractory in nature and that are ubiquitously formed through random chemical alteration of diverse precursor molecules. The average properties of FA and HA are distinct and remarkably uniform across soils (Rice and MacCarthy, 1991; Mahieu et al., 1999). The abundance of C in FAs is lower (40–50%) than that in HAs (53–60%), and the abundance of O in FAs higher (40–50%) than that in HAs (32–38%). This is consistent with the higher exchange capacity of FAs, which is 640–1420 cmol (+) kg$^{-1}$ FA, compared with 560–890 cmole (+) kg$^{-1}$ HA (Stevenson, 1994). Reported molecular weight ranges are 3000 Da for HA, 1000 to 3000 Da for FA, and lower (<1000) for dissolved organic constituents (DOC) that are not considered humic substances. Although the macromolecularity of humic substances has long been assumed, Piccolo (2002) points out how difficult it is to accurately measure molecular size in polydisperse systems. Scientific focus on HA and FA has been so dominant that some equate SOM with these fractions, forgetting that these are procedurally defined and do not exist *per se* in nature. Humic and fulvic acids are used as SOM proxies even though half or more of the organic material in mineral soils resides in nonextractable humin (HN). In a review of chemical abstracts, Rice (2001) found that only ca. 3% of the citations addressing humic substances dealt with HN. Interest in humin, which is believed to include the more persistent components of SOM, has increased greatly with the desire to sequester C in soils on a permanent basis.

## TYPES OF ORGANIC MATTER IN MINERAL SOILS AND THEIR PROBABLE FUNCTIONS

### RELATIONSHIP BETWEEN DYNAMICS AND MEASURED FRACTIONS

The quest to define the molecular structure of humic substances has finally been abandoned by most soil chemists (Hayes et al., 1989; MacCarthy, 2001). Models or psuedostructures that portray abundance and proportions of elements and functional groups have been proposed (Stevenson, 1994; MacCarthy, 2001). Current efforts to classify SOM favor techniques that also consider the physical nature of its constituents, which is expressed in scales ranging from the molecular to the macroscopic. Recognition of the need to consider both the quality and location of organic matter in soils has resulted from the failure of classical fractionation schemes to separate SOM components that were kinetically or functionally distinct (Stevenson et al., 1989; Christensen, 1996). Factors that influence the dynamics of SOM constituents include not only recalcitrance but also the interactions between organic and mineral compounds and the accessibility of materials to organisms and enzymes (Sollins et al., 1996). Numerous studies show that organic residues added to temperate soils decompose quickly, with approximately one third of the original C and N persisting as SOM after 1 year, unless edaphic factors, typically physical or chemical extremes, restrict biological activity. Studies of isotopes of C and N added to soils in the organic form indicate that once organic residues enter soil, their turnover rate or half-life slows asymptotically (Stevenson, 1994). Most multicompartment models of SOM assume first-order decay and use three or more pools to describe the diverse timescales of organic C and N turnover (van Veen et al., 1984; Jenkinson et al., 1987; Parton et al., 1987). Abiotic influences on decay are reflected through rate-modifying factors. Multipool models provide greater insight into short- and intermediate-term dynamics (Benbi and Richter, 2002), whereas single-pool (Feng and Li, 2001) and noncompartment models (Ångren and Bosatta, 1987) satisfactorily describe long-term trends in organic C and N equilibrium levels. Although the number, size, and turnover rates of pools used in multicomponent models vary (Benbi and Richter, 2002), common divisions include compartments with time constants or rates of decay $(1/K)$, where $K$ is measured in years for the most active fraction, in decades for the slowly decomposing pool, and in centuries or millennia for the most persistent pool (Elliott et al., 1996; Feng and Li, 2001). Efforts are ongoing to relate such kinetically defined pools to the chemical or biophysical characteristics of measurable organic fractions and associated nutrient and C dynamics (Elliott et al., 1996; Collins et al., 2000; Guggenberger and Haider, 2002; Chapter 1). Organic matter fractions associated with the active and slowly cycled pools are influenced in the near term, and most closely reflect management practices, influence nutrient supply, and determine soil tilth. More recalcitrant SOM fractions that are equated with the slow or passive or resistant pools have greater relevance for long-term C sequestration, sorption, CEC, and soil water-holding capacity.

## COMMONLY DESCRIBED SOM POOLS AND RELATED FRACTIONS

Table 3.2 summarizes the general relationships between kinetically conceived SOM pools and related organic matter fractions. The term *fraction* is used to describe measurable organic matter components. The term *pool* is used to refer to theoretically separated, kinetically delineated components of SOM. The desire to relate procedurally defined SOM fractions to ecosystem processes has prompted the use of the same terms to describe pools and fractions. This interchangeable use of terminology suggests that theoretically defined pools can be equated with SOM fractions (e.g., Stevenson, 1994; Paul and Clark, 1996; Paul and Collins, 1997). Unfortunately, overlapping terminology is often applied to fractions and pools that are not closely related and this has led to confusion. The divisions between active, slow, or passive SOM pools are likely to differ with emphasis on biologically, physically, or chemically regulated dynamics. For example, Motavalli et al. (1994) used soluble-, microbial- and light-fraction C (a measure of POM) values to initialize the C submodel of Century, a leading ecosystem process model, and pointed out that a variety of criteria can be used to determine which fraction is the most suitable proxy for the active fraction. Table 3.2 gives examples of measures used to quantify the biologically active components of SOM that support heterotrophs, and those that are likely to be mineralized are followed by the letter *B* in parentheses. Fractions produced by methods used to separate physically active from protected organic matter or that isolate material associated with physical function are followed by the letter *P* in parentheses. The SOM fractions produced by methods designed to isolate chemically labile from persistent matter or separate matter that usefully describes soil's exchange and sorption characteristics are followed by the letter *C* in parentheses. The functional importance of SOM of different ages varies systematically, with the youngest materials being most biologically active and materials of recent origin and intermediate age contributing notably to the physical status of soils. Materials with longer residence times exert more influence on the physicochemical reactivity of soils.

## FRACTIONS EQUATED WITH THE BIOLOGICALLY ACTIVE POOL

Even though some definitions of SOM exclude fresh plant residues, residues can be important components of the active fraction. The division between plant residues and true SOM is apparent in dynamic models of soil C and N pools in which turnover of fresh residues are characterized separately or treated as distinct pools (Heal et al., 1997). Nonetheless, residues play significant biological and physical roles in soils and represent a principal means by which SOM can be managed. Studies of the factors controlling microbial decay of litter provide the basis for the understanding of how reside quality influences SOM dynamics. Litter quality is equated with the rate at, or ease with, which organic substrates are decomposed (Paustian et al., 1997). Models that effectively describe residue dynamics typically include five or more compartments. Information about plant litter composition and decomposition rates can provide valuable information about the contributions of fresh amendments to nutrient supply (Honeycutt et al., 1993; Vanlauwe et al., 1994; Preston et al., 2000; Cobo et al., 2002; Ruffio and Bollero, 2003). This information can be most relevant for unmanaged or minimally managed systems or for depleted or infertile soils, in which fresh residues contain a notable proportion of available nutrients, or in forest soils, where root activity is concentrated in litter layers. Studies of litter or residues collected from the soil surface

## TABLE 3.2
## Soil Organic Matter Pools and Related Fractions

| Organic Matter Pools, Theorized Kinetics and Function | Procedurally Defined Fractions of Organic Matter[a] |
|---|---|
| Labile or Active SOM | |
| **Half-life days to a few years** | **Microbial biomass** |
| Equated with material of recent origin or embodied living components of SOM | Chloroform-labile SOM (B) |
| | Microwave-irradiation-labile SOM (B) |
| Material of high nutrient or energy value | Amino compounds (B, P) |
| Physical status (not physically protected) makes soil incorporated matter likely to participate in biologically or chemically based reactions | Phospholipids (B) |
| | **Labile substrates** |
| | Mineralizable C or N, estimated by incubation (B) |
| Physical role of materials located at the soil surface and of compounds that promote macroaggregation is transient | Substrate-induced activity (B) |
| | Soluble, extractable by hot water or dilute salts (C, B) |
| | Easily oxidized by permanganate or other oxidants (C, B) |
| | **Residues for which chemical formula can be described, inherited from living organisms** |
| | Litter, vegetative fragments or residues (B,P) |
| | Nonaggregate protected POM (B, P) |
| | Polysaccharides, carbohydrates (C, P) |
| Slow or Intermediate SOM | |
| **Half-life of a few years to decades** | **Partially decomposed residues and decay products** |
| Physical protection, physical status, or location help separate this fraction from the other two fractions | Amino compounds, glycolproteins (B, P) |
| | Aggregate protected POM (B, P) |
| | **Some humic materials** |
| | Acid/base hydrolyzable (B, C) |
| | Mobile humic acids (B, C) |
| Recalcitrant, Passive, Stable, and Inert SOM | |
| **Half-life of decades to centuries** | **Refractory compounds of known origin** |
| Recalcitrance because of biochemical characteristics and/or mineral association | Aliphatic macromolecules (lipids, cutans, algaenans, suberans) (C) |
| | Charcoal (C) |
| | Sporopollenins (C) |
| | Lignins (C) |
| | **Some humic substances** |
| | High molecular weight, condensed SOM (C, P) |
| | Humin (C) |
| | Nonhydrolyzable SOM (C) |
| | Fine-silt, coarse-clay associated SOM (C, P) |

[a] Letters in parentheses that follow fraction labels identify measures commonly used to study biologically active matter (B) associated with nutrient supply or microbial growth, physically active or sequestered matter (P) associated with matter accessibility and soil structure, and chemically active or inactive matter (C) that explains or influences material persistence and its chemical reactivity, including exchange and sorption–desorption properties.

or that consider freshly incorporated materials provide little information about the characteristics of resident SOM.

The physical activity of litter- and plant-derived carbohydrates is important. Surface litter provides protection from erosion. Polysaccharides exuded from roots and microorganisms, which include sugar and nonsugar forms, adsorb strongly to negatively charged soil particles through cation bridging (Chenu, 1995) and contribute notably to aggregate stabilization (Angers and

Mehuys, 1989; Cheshire et al., 1989; Martens and Frankenberger, 1992) and hydraulic conductivity (Robertson et al., 1991). The positive relationship seen between polysaccharides and aggregate stability can be obscured by the presence of living roots (Carter et al., 1994) or fungal hyphae (Miller and Jastrow, 1990). Although beyond the scope of this chapter, the direct contributions of living roots to labile SOM and their ability to influence the dynamics of native SOM should not be overlooked.

Fractions of SOM most often used to estimate the active pool are commonly equated with biological activity. These fractions include measures of the microbial biomass (Paul and van Veen, 1978), which is very often related to chloroform-labile C and N (Brookes et al., 1985) and is one of the few measurable SOM fractions included in several multipool models of SOM dynamics (Jenkinson, 1990; Hansen et al., 1991). Dendooven et al. (2000) attempted to use biomass C:N ratios measured by fumigation extraction to predict C and N dynamics in a simple three-pool model and found that ratios were not related to observed differences. According to Franzluebbers et al. (1999), the microbial biomass, estimated by fumigation extraction, is a good general measure of active SOM if the C recovered from control soils is not subtracted from treatment soils. They found that subtraction of control obscured resolution of differences. Needelman et al. (2001) found that extractant-to-soil ratios used in fumigation extraction techniques markedly influenced the quantity of C extracted from nonfumigated samples and that typical solution-to-soil ratios used were not high enough to ensure complete recovery of C from control soils. Microwave-labile C has also been used to estimate the size of biomass (Islam and Weil, 1998). Phospholipid-P, a more direct measure of the living biomass than are chloroform- or microwave-based estimates (Findlay et al., 1989), has been used effectively to reflect the biomass component of active SOM (Kerek et al., 2002).

The respiratory response of soils to substrate addition is also used to estimate size of the biomass pool (Beare et al., 1991; Stenström, 1998) and to describe the soil's metabolic status (Garland and Mills, 1991). Related measures are very sensitive to the quality of active SOM. For example, microbial substrate utilization characteristics were altered by short-term (18 months) application of organic management practices to soils that had been conventionally cash grain cropped for 20 years even though other measures of labile SOM were unaltered (Bending et al., 2000). Measures of easily oxidized SOM have also been used to estimate the size of the labile C fraction (Weil et al., 2003). Estimates of readily mineralizable organic C or N are used widely to estimate active SOM (Paul, 1984; Woods, 1989; Motavalli et al., 1994; Kelly et al., 1996). Nitrogen mineralized during laboratory incubations are effectively described in simple multifraction models (Cabrera and Kissel, 1988; Benbi and Richter, 2002). Information from incubations and extraction is commonly used to estimate plant-available N (Waring and Bremner, 1964; Michrina et al., 1982; Vanotti et al., 1995).

Amino sugars, which occur in soils as macropolysaccharides including chitin (Stevenson, 1994), have been related to bacterial and fungal biomass and can be used to estimate contributions to the biologically active pool. Newly immobilized N is disproportionately incorporated into the acid-soluble fraction that contains microbially derived amino compounds (Kelly and Stevenson, 1985; He et al., 1988). Amino compounds are sensitive to organic amendments and have been related to plant N acquisition (Appel and Mengel, 1993; Xu et al., 2003). Parveen-Kumar et al. (2002) found that cultivation of legumes increased amino acid and amino sugar fractions during a 4-month cropping season, whereas simultaneous cultivation of pearl millet decreased amino sugar stocks. By using the improved diffusion methods of Mulvaney and Khan (2001), Mulvaney et al. (2001) found that amino sugar N content was predictive of whether maize responded to N fertilization. Collectively, these examples suggest amino compounds, in particular amino sugars, hold great promise as indices of the active N pool. However, the persistence of elevated amino sugar levels in soils historically amended with manure suggests that, at a minimum, this fraction also includes components that might be more appropriately equated with the slow pool. Work by Zhang et al. (1998) suggests that the decay dynamics of individual amino sugars vary and that ratios of individual forms can be used to distinguish decay dynamics. They equated particle-size fractions with stages of organic matter decomposition and found that biomass (amino sugar composition) was most dynamic in the larger-sized particle fractions. This is consistent with findings that microbially derived sugars associated with silt-associated amino sugars dissipate faster than material associated with the fine-sized fraction (Kiem and Kögel-Knabner, 2003). The findings of Amelung et al. (2002) suggest that cultivation-induced shifts from fungal to bacterially derived amino sugar residues are perceptible for up to a century.

Measures of organic matter that is extractable in water or dilute salt solutions have also been used as indices of biologically and chemically labile pools (Wander et al., 1994; Haynes, 2000; Gregorich et al., 2003). The characteristics of the dissolved organic matter fraction (DOM), sometimes operationally defined as SOM that is <0.45 μm in solution, can be of particular importance to fate and transport processes or as a source of energy for microorganisms in subsurface environments that do not receive fresh organic residues as inputs (Herbert and Bertsch, 1995). DOC concentrations, which are typically lower in arable than in grasslands or forested systems, are greater under legumes than grassses, and are affected by additions of lime, organic amendments, and mineral fertilizers, and by tillage practices with species (Chantigny, 2003). Dissolved organic matter has both hydrophilic and hydrophobic components, the latter favoring sorption to natural organic matter and mineral surfaces such as Al and Fe oxyhydroxides (Kaiser and Zech, 1998). The strength of this sorption is related both to specific hydrophilic functional groups and mineral surface properties (Gu et al., 1995). Low-molecular-weight organic anions included in the FA fraction and released by roots in root exudates can disperse clays (Reid et al., 1982; Shanmunananthan and Oades, 1983). The chemical activity of DOC fractions can be particularly important in systems amended with sludge or manure, in which dissolved matter increases the dissolution of sorbed organic and inorganic elements and facilitates transport through the soil profile (Kaschl et al., 2000).

## FRACTIONS ASSOCIATED WITH PHYSICALLY ACTIVE AND SLOW POOLS

Many SOM fractions that are principally equated with the active pool also enhance a soil's physical characteristics (Table 3.2). This is true for amino sugars. Chantigny et al. (1997) used ratios of muramic acid to glucosamine to assess contributions of bacteria and fungi to soil aggregation. Low ratios in well-aggregated soils indicated the greater contributions of fungally derived amino sugars to structure. Glomalin is reputed to be a fungally derived glycoprotein that forms on hyphae of arbuscular mycorrhizal fungi (AMF) in the order Glomoles (Wright, 2000; Chapter 6). It is a procedurally defined fraction that includes methodologically based subfractions obtained by extraction in citrate buffer under varying heat or energy treatments (Wright and Upadhyaya, 1998). Only portions of this fraction are immunoreactive with an antibody raised against spores of an AMF (Wright et al., 1996). This finding and the observation that glomalin C concentrations vary between 27.9 and 43.1% of the organic C in soils (S. Wright, personal communication, cited in Rillig et al., 2003) strongly suggest that glomalin is not a gene product. It is more likely a heterogenous SOM fraction that contains moieties that are immunoreactive against appropriate probes. Accordingly, treatment of the glomalin fraction as a direct measure of the mycorrhizally derived SOM pool is an example of misleading labeling. Glomalin also suffers from the common problem of having multiple identities. As noted for amino sugars, glomalin has been tied to both biological and physical activity. Rapid increases in glomalin contents on growing hyphae (Wright et al., 1996) suggest it is part of the active pool. Correlation with aggregate stability (Wright, 1998) and persistence in incubated soils have been cited as evidence that it contributes to slow or even passive SOM pools (Rillig, 2003). Division of glomalin into subfractions has not yet improved the conceptual or kinetic resolution of the material recovered by citrate extraction. During a study of hyphal decomposition, Steinberg and Rillig (2003) found that, contrary to expectation, easily extractable immunoreactive glomalin content, which is presumably the fraction most enriched in the glycoprotein produced by AMF, increased rather than decreased as decay progressed. Glomalin fractionation schemes need to be improved to allow separation of entities that are kinetically and functionally distinct.

During the past few decades, numerous methods have evolved to isolate and characterize relatively undecomposed particulate or macroorganic matter (referred to as POM). Related measures recover incompletely decomposed residues that were previously removed and typically discarded before humic substances were assessed (Christensen, 1992; Gregorich and Ellert, 1995). Like amino sugars and glomalin, POM fractions too suffer from multiple identities. This results from heterogeneity in materials included in this fraction and from its perceived multi-functionality. Greenland and Ford (1964) and Ladd and Amato (1980) were among the first to suggest that more labile material could be concentrated in low-density solutions, finding that densiometrically obtained POM-N contents had 18 to 23 times more N than N in mineral soil. Tiessen and Stewart (1983) observed that SOM in large-sized particle-size classes mineralized more rapidly than finer components. Measures of POM have been tied to microbial growth and nutrient supply and suggest that it is closely related to biologically mediated C, N, and in some soils P availability (Gregorich et al., 1994 ; Hassink, 1995b; Barrios et al., 1996a; Phiri et al., 2001; Salas et al., 2003). Accordingly, POM is commonly used as an index of the labile SOM pool (Buyanovsky et al., 1994; Carter, 1996; Wander and Bollero, 1999). The enrichment of nutrients, metals, and xenobiotics in POM fractions suggests that this is a site where biological and chemical sorptions are concentrated (Janzen et al., 1992; Barriuso and Koskinen, 1996; Besnard et al., 2001; Eriksson and Skyllberg, 2001; Balabane and van Oort, 2002; Dorado et al., 2003). Even though there is abundant evidence that POM is biologically and chemically active, measures of POM are commonly used to estimate the size of the slowly mineralized pool (Delgado et al., 1996; Elliott et al., 1996; Kelly et al., 1996; Sitompul et al., 2000). As is true for amino sugars and glomalin, POM contributes to aggregate formation and stabilization (Waters and Oades, 1991). Efforts to explain the division between active and slow components of POM and to understand its multiple roles in soils are discussed further in the final section of the chapter.

## Fractions Associated with Recalcitrant Pools

Humic substances are divided into labile and recalcitrant fractions based on the ease with which they can be removed from soil (Table 3.2). Pretreatment of soils with dilute acids greatly increases extraction efficiency by likely increasing dissolution of Fe and Al oxides that act as cementing agents and hydrolysis of clay–humate linkages (Pignatello, 1990). Olk et al. (1995) found that humic acids recovered without acid pretreatment, which they termed *mobile humic acids*, and were less decomposed and more closely related to soil N supply than were humic acids obtained after acid pretreatment. Recovery after acid pretreatment is the standard method for HS extraction. Research on the environmental fate of toxins and pollutants has also increasingly focused on physically based mechanisms to understand differences among labile, slowly available, and persistent SOM fractions. Fractionation procedures, however, emphasize physicochemical rather than biochemical aspects of SOM. Several works suggest that diffusion-based mechanisms account for the cumulative effect of aging on compound recalcitrance in soils and sediments (Wu and Gschwend, 1986; Pignatello, 1989; Bruseau et al., 1991; Scow and Hutson, 1992). Explanations for increased recalcitrance include diffusion-limited sorption and desorption due to movement into nanopores (<100 nm; Nam and Alexander, 1998) or into regions of humified, O-depleted SOM (Huang and Weber, 1997; Xing and Pignatello, 1997). Macroscale heterogeneity occurring within the soil matrix also influences sorption. Increasingly, SOM is described as a dual-mode sorbent, containing both rubbery and glassy fractions, organic chemicals preferentially sorbing in the glassy fraction (Xing and Pignatello, 1996; Huang and Weber, 1997; Leboeuf and Weber, 1997). A range of methods intended to oxidize the rubbery, expanded, and presumably surface-exposed organic matter by removing the carboxylic, aliphatic, and carbohydrate constituents of SOM include subcritical water extraction (Johnson et al., 1999) and persulfate oxidation (Cuypers et al., 2000). In general, glassy SOM is characterized by higher C:H ratios and a greater degree of aromaticity (Li and Werth, 2001; Xing, 1999). Works by Chiou and Kile (1998) and Gustafsson et al. (1997) indicate that only a fraction of the glassy material, characterized by a very high surface area, can control sorption.

The relatively low substrate value and higher recalcitrance of humic substances is a central concept in SOM description, where modifications in structure along with increased mineral affiliation increase the persistence of humic constituents in soils. Questions about the permanence of C sequestered in soils have fueled interest in SOM constituents contributing to the recalcitrant pool. Molecular heterogeneity (MacCarthy, 2001), and chemical composition, principally aromatic and aliphatic constituents (Kiem et al., 2000) remain the primary explanations for the refractory nature of humic materials. Refractory materials include persistent structures of known origin that are naturally resistant as well as molecules that become resistant through condensation and aromatization processes (Derenne and Largeau, 2001). The importance of the physical arrangement of aliphatic and nonaliphatic constituents (principally aromatic and carbohydrate) in recalcitrant SOM and humin fractions is being increasingly recognized (Kiem and Kögel-Knabner, 2003). The arrangement of aliphatic constituents likely influences SOM's recalcitrance and sorptive properties (Preston and Newman, 1992). This change in how humin and stable organic matter are perceived is consistent with the shift to a more physically based understanding of SOM dynamics. Efforts to quantify the passive fraction to initialize SOM models have relied on a variety of methods, including the use of radiocarbon signatures (Hsieh, 1992; Trumbore, 1993; Paul et al., 1997) and measurement of the nonhydrolyzable fraction (Leavitt et al., 1996; Paul et al., 1997). The fact that refractory macromolecules that resist drastic acid or base hydrolysis also resist degradation under natural conditions lends credence to hydrolysis-based separation of resistant SOM. Chemical characteristics and the arrangement of constituent structures only partially explain the recalcitrance of humic substances. Refractory SOM in arable soils is primarily stored in fine-particle-size fractions (Kiem and Kögel-Knabner, 2002). Organic structures that are chemically recalcitrant by nature do not contribute to recalcitrant pools unless they are affiliated with fine-particle-size separates; exceptions include charcoal, which is highly resistant to degradation and recovered in POM fractions (Kiem and Kögel-Knabner, 2003). Measurement of SOM fractions associated with fine-silt, coarse-clay-sized separates (Six et al., 2000b; Christensen 2001; Guggenberger and Haider, 2002) is often used to estimate size of the stable pool. Stable SOM constituents are related primarily to the proportion and characteristics of fine particles in soils (Zinke et al., 1984; Carter et al., 2003). Particle surface area and the abundance of Fe and Al oxides appear to play a key role in SOM stabilization in the fine fraction (Curtin, 2002; Vitorino et al., 2003). The upper limit of carbon content associated with primary particles <20 μm may determine the capacity of soil to protect C and thus establish the size of the stable SOM fraction (Hassink, 1995; Ruhlmann, 1999).

## MEASURES OF POM AND THEIR INTERPRETATION

### POM AS AN INDEX

Labile SOM can be assessed effectively by characterizing POM fractions. POM fractions estimated by measuring low-density (typically 1.4–2.2 g cm$^{-3}$) or coarse-size fractions (>53–100 μm or 53–250 μm) are strongly influenced by soil management (Christensen, 1992; Quiroga et al., 1996). Focus on POM, in lieu of other measures of labile SOM, is warranted largely because this fraction typically has a higher proportional response to management than do other measures of labile SOM (Conteh et al., 1998; Alvarez and Alvarez, 2000; Franzluebbers et al., 2000; Carter, 2002). The material captured in POM fractions is composed primarily of plant-derived remains with recognizable cell structure and typically includes fungal spores, hyphae, and charcoal (Spycher et al., 1983; Molloy and Speir, 1977; Waters and Oades, 1991; Gregorich and Ellert, 1995). The proportion that is charcoal is related to the site's history of burning (Elliott et al., 1991) and geomorphology (Di-Giovanni et al., 1999). Most often, studies comparing measures of labile SOM discover that related measures increase or decrease in parallel (e.g., Janzen et al., 1992; Wander et al., 1994; Carter et al., 1998; Guggenberger et al., 1999; Needelman et al., 1999). Exceptions to this generality exist. Disproportional responses in selected labile fractions might provide insight into resource limitations or surpluses present within the system considered. Collectively, the size of the POM fraction, its relatively distinct nature, and its sensitivity to management, including inputs, support statistical resolution of differences between soils; this, rather than purity or kinetic fidelity, explains the popularity of POM fractions as indices of labile SOM. Ability to sort or classify differences in soils arising from management history does not, however, prove that POM fractions have functional meaning. For example, Letey (1991) and Young et al. (2001) suggest that measures of water-stable aggregation, which are quite sensitive to management (Russell, 1973; Dexter, 1988) and are notably influenced by characteristics of POM (Waters and Oades, 1991; Tisdall, 1996), have had little practical application. They attribute this to the failure of these measures to adequately describe undisturbed soil structure. The development of predictive relationships between measures of POM, or other properties including aggregate and processes of interest, will be proof of their utility.

At present, POM's value as an indicator of early trends in SOM status in managed soils is well recognized (Bremer et al., 1994; Wander et al., 1994, 1998; Gregorich and Carter, 1997; Yakovchenko et al., 1998; Carter, 2002). Care must be taken to control timing, intensity, and pattern of sampling, because POM contents, which are quite sensitive to plant inputs and soil mixing, can vary seasonally (Spycher, 1983; Wander and Traina, 1996b; Willson et al., 2001), spatially (Burke et al., 1999; Bird et al., 2001), according to handling (Yang and Wander, 1999; Rovira et al., 2003), and with soil depth (Guggenberger et al., 1994; Wander et al., 1998; Aoyama et al., 1999). Management's influence on POM fractions appears to interact with texture in various ways. Some works have found that sensitivity to management (Carter et al., 1998; Needelman et al., 1999) and the proportion of SOM in POM (Liang et al., 2003) increase as sand contents increase. According to Hook and Burke (2000), POM is especially important to N retention and availability in sandy soils, because the proportion of total N in POM is higher than in finer-textured soils. In coarser-textured soils, POM contents decline with clay contents if other factors do not limit decay. The inability to conserve POM can limit a sandy soil's ability to respond to management. Malhi et al. (2003b) attributed the failure of N fertilization to increase POM (cited in Noyborg et al., 1999) to its being too sandy, because similar N amendment of a loamy site had increased POM contents.

In addition to texture, soil background, or history of use, also influences the sensitivity of POM fractions to management. Differences in outcomes reflect how close to or far from equilibrium or saturation an individual soil is when subject to new management and whether the regime or condition aggrades or degrades labile SOM. In a study of cotton production on a Vertisol, Conteh et al. (1998) found that the amount of POM obtained after 3 years in a stubble-incorporated soil was almost double that obtained from a soil in which stubble was burnt. This suggests that both management and soil status were conducive to POM, and, presumably, gains in SOM. In contrast, Franzluebbers and Arshad (1996a) found little to no effect of conservation tillage practices on SOM accretion in POM in northern temperate soils, where cold climate minimized decay. In that instance, POM trends indicate that alternative management was not sufficient to prompt SOM aggradation. Carter et al. (2003) suggest that although POM fractions reach saturation later than organic matter affiliated with mineral surfaces, SOM-saturated soils fail to accumulate POM under practices that would typically be considered aggrading. It is important to remember that SOM equilibrium levels are dynamic, varying in individual soils with the pattern, intensity inputs, and disturbance. The quantity and character of organic residues added to any soil, including one in which soils are considered to be SOM saturated or at equilibrium under present management, can be adjusted upward to outpace decomposition rates and thus result in SOM accumulation.

## Approaches to POM Fractionation and Interpretation of Results

Methods used to separate POM from finer-sized, mineral-associated fractions rely on a variety of size- and density-based techniques that are ideally tailored to meet specific objectives (Table 3.3 and Figure 3.3). Material size, shape, and density influence partitioning when separation methods rely on sedimentation (Stevenson et al., 1989; Elliott and Cambardella, 1991). Density of the fluid or size cut-off of the sieves used to separate particulate from organomineral constituents influences the quantity and chemical character of the fractions obtained (Figure 3.4). Procedures should be tailored to suit both the soils and experimental scenarios to which they are applied. The simpler size- or density-based methods listed in Table 3.3 are well suited to study the influence of land use and management practices on SOM characteristics. Measures of coarse fraction (CF) organic matter, typically defined as the material that is sand sized or larger, grew out of methods developed to describe particle-size separates in the 1960s to characterize mineralogical controls over SOM dynamics. The common use of 53 μm, the lower boundary for sand-sized material, as the cut-off for POM is operationally convenient but somewhat arbitrary. For example, Christensen (1992) used 63 μm as the size dimension that, after dispersion in water, separated finer organomineral complexes from the CF. The upper boundary of the CF is also arbitrary and varies notably with sample handling. Often, studies include only fragments smaller than 2 mm, and studies seeking to concentrate plant remains use larger dimensions. Hassink et al. (1993) used materials retained in the 200- to 8000-μm range to characterize macroorganic matter. Willson et al. (2001) found that the average C:N ratio of the 250–2000 μm POM fraction was 17.0 and the ratio of the 53–250 μm POM fraction was 15.5. Densities used to float out the light fraction (LF) of SOM vary, with values between 1.85 and 1.40 g cm$^{-3}$ being common. A variety of liquids are also used for density-based separations, sodium or potassium iodide, sodium polytungstate (NaPT), and silica gels being popular choices. These solutions alter the chemical characteristics of SOM fractions: iodide solutions are strong reducing agents and silica gels have a pH of 8 or more and thus can extract humic substances. Even though it is reported that NaPT is relatively inert, it is difficult — if not impossible — to completely remove from POM. According to Meijboom et al. (1995), silica gels are relatively easy to remove from the sample, and this, plus their lower cost and toxicity, makes them a good choice for density-based separation of the LF. Chemically assisted dispersion of soils before POM isolation is quite common, with hexametaphosphate or calgon being frequently used before both size- and density-based separations. Dispersants influence the chemical properties of SOM (Ahmed and Oades, 1984), but their effect on POM composition has not been investigated in detail.

Soil dispersion also requires physical disruption. Separation by particle size most often employs ultrasonic energy for dispersion, whereas density-based methods typically rely on shaking. According to comparisons of shaking and ultrasonic dispersion summarized in Christensen (1992), long-term shaking can alter SOM properties as much as ultrasonic dispersion can. When sonication is used, energy output and soil solution ratios need to be optimized for POM recovery. Diaz-Zorita et al. (2002) have shown that the size of fragments obtained is inversely related to the mechanical stress applied. Work by Elliott et al. (1996) and Gregorich et al. (1988) suggests that energies lower (300 to 500 J mL$^{-1}$) than the 1500 J mL$^{-1}$ dispersion energy commonly used to obtain complete dispersion of aggregates (cited in Christensen, 1992) should be used to separate POM. Optimum dispersion energies vary among soils. In a study of grassland soils, Amelung and Zech (1999) found that dispersion of macroaggregates (250 to 2000 μm) was achieved at an ultrasonic energy of 1 kJ for most of the sites considered. Soils from wet extremes in the prairie were an exception, for

**TABLE 3.3**
**Approaches to POM-Fractionation-Related Experimental Objectives and Associated Studies**

| Method | Objectives and References |
|---|---|
| **Size-Based Methods** | |
| *Macro organic matter (MOM)*: Typically emphasizes large residues, clearly identifiable as plant residues. Upper boundary of subdivision is variable, e.g., 100–250, 250–2000, 8000-200 µm | *Concentrate recent inputs of plant and organic residues and biologically active SOM*: Hassink et al., 1993; Magid and Kjaergaard, 2001; Willson et al., 2001 |
| *POM or coarse fraction (CF)*: Typically refers to SOM that is sand sized or larger. Common subdivisions include separation of >53-µm material into >53–250 µm and > 250 µm | *Concentrate labile SOM influenced by management*: Cambardella and Elliott, 1992; Angers et al., 1993; Barrios et al., 1996b; Wander et al., 1998; Needelman et al., 1999; Bowman et al., 2000b; Nissen and Wander, 2003 |
| *Sand-sized class as a constituent of particle-size separates*: Methods separate organomineral associations into a range of sand-, silt-, and clay-sized components | *Characterize dynamics of organic matter and the (a) influence of management or amendment*: Christensen, 1986; Quiroga et al., 1996; Lehmann et al., 1998 *and (b) decomposability of SOM or constituents associated with separates*: Christensen, 1987; Cheshire et al., 1990 |
| **Density-Based Methods** | |
| *Light fraction (LF, sometimes referred to as POM)*: Common density ranges 1.6–1.75, 1.8–1.95, 2.0–2.6 g cm$^{-3}$; solutions used to recover LF vary, influence on chemical properties not well characterized; dispersion followed by flotation in liquid, denser fractions not collected; energy used to disperse is source of variability as are methods used to recover suspended matter from heavy fraction. | *Influence of management and relationship to biologically available or unavailable C or N*: Greenland and Ford, 1964; Strickland and Sollins, 1987; Motavalli et al., 1994; Wander et al., 1994; Gregorich et al., 1996; Barrios et al., 1996b; Alvarez et al., 1998; Carter et al., 1998; Curtin and Wen, 1999; Fliessbach and Mader, 2000; Haynes, 2000 |
| **Combined Size and Density Techniques** | |
| *Active POM fraction*: Separate large-sized fraction and then light fraction; size and densities and fraction labels vary as, e.g., >53 µm < 1.6 g cm$^{-3}$, 150–3000 µm <1.13 g cm$^{-3}$; <250 µm and then 1.37 g cm$^{-3}$, >250 µm and then 1.6 g cm$^{-3}$ | *Concentrate the most active fraction, including plant residues*: Cambardella and Elliott, 1992; Hassink, 1995; Meijboom et al., 1995; Barrios et al., 1996a; Magid et al., 1996; Baldock et al., 1997; Magid et al., 1997 |
| *Loose and occluded POM (active and slow)*: Removal of POM or LF predispersion or with gentle shaking by using density followed by complete dispersion and collection of released material using size or density; energy applied before separation of loose and occluded varies — some estimates are based on indirect assessments | *Assess biologically and physically active constituents or mineral-protected POM*: Golchin et al., 1994b; Puget et al., 1996; Jastrow et al., 1996a; Wander and Yang, 2000 |
| *POM or LF isolated in concert with sieving water-stable aggregates*: Methods range from simple, producing a few POM fractions, to highly detailed methods | *Aggregation and C dynamics, interpretation based on SOM characteristics*: Cambardella and Elliott, 1993; Golchin et al., 1994a; Hassink and Dalenberg, 1996; Six et al., 1998; Gale et al., 2000; Puget et al., 2000 |

which 3 kJ was needed for dispersion and ≥5 kJ was required to disperse microaggregates (20 to 250 µm). However, use of energies >5 kJ disrupted POM. Incomplete dispersion can leave air entrapped in microaggregates, which can then contaminate the LF (Turchenek and Oades, 1979; Gregorich et al., 1989). This might explain why Golchin et al. (1994a) found that selected POM

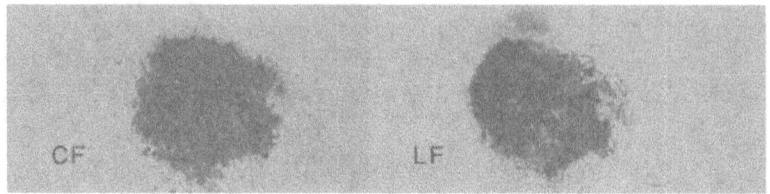

**FIGURE 3.3** Coarse (>53 μm) and light fractions (<1.6 g cm$^{-3}$) isolated from a Typic Fragiudalf supporting the Rodale Institute's Framing Systems Trial. Sand-sized mineral particles were removed from the coarse fraction by sedimentation.

**FIGURE 3.4** Light fraction obtained from the Jornada Experimental Range in New Mexico, where total SOM contents are less than 1%. Increasing density of NaPT from back to front, left to right increase as density of NaPT is increased from 1.2 to 2.2 g cm$^{-3}$.

samples isolated in one study by using a lower density had $^{13}$C-NMR spectral characteristics which indicated that it was more decomposed (had lower $O$-alkyl and higher alkyl C abundance) than did POM obtained at higher densities. Along with litter, they might have recovered microaggregates. Recovery of charcoal in the LF skews fraction characteristics, increasing the abundance of chemical traits attributed to recalcitrant SOM (Roscoe and Buurman, 2003).

LF yields are influenced by density of the solution used, with yield increasing with solution density. Use of lower densities favors recovery of larger POM constituents (Ladd and Amato, 1980). Temperature and actual density of the liquid are difficult to control even though they are important variables and interact with the amount of energy applied. Small differences in these properties can significantly influence the proportion of C recovered in this fraction (Christensen, 1992). Very few systematic studies have considered the energy of the solution. Cleanliness (purity) of the fraction is decreased when excess energy is applied (Kerek et al., 2002). For example, Dalal and Meyer (1986) found that ultrasonic treatment led to greater recovery of total C in the LF, but the average C contents of the material recovered were lower than those in the LF obtained by shaking alone. Ultrasonic treatment caused contamination of the LF with mineral matter. Aspiration and decantation following centrifugation have been used to separate the LF from heavier constituents. Physical entrapment of LF by the heavy fraction and adhesion of the LF to container sides can reduce the efficiency of LF recovery. Maximization of the area rather than the volume of solution to which soil is exposed can reduce entrapment. Efficient decantation can be facilitated by adding fresh solution after shaking to rinse the adhered material from container sides and increase the distance between suspended light and heavy materials that will then be pelleted by centrifugation (Wander et al., 1998).

The combined effects of fragmentation (soil characteristics, dispersion energy, density or size cut-off) are variable enough to make systematic comparison between results obtained from different studies difficult. Consistent, or, at least, stated criteria for methods optimization are needed. Efforts to optimize sonication energy can be tailored to maximize the yield or concentration of selected constituents, including biological activity recovered from soils or retained within selected fractions (e.g., De Cesare et al., 2000). Fraction yield, elemental enrichment, purity, C:N ratios, and chemical properties have been used to assess the validity of fractions obtained by separatory techniques (Golchin et al., 1994a, 1994b; Hassink, 1995b; Kerek et al., 2002). The utility of POM will be increased greatly by standardizing, or at least systematizing, techniques and strategies to interpret related results.

## Methods Yielding a Single POM Fraction

When procedures are designed to isolate POM in its entirety, size- and density-based techniques should provide similar information. According to Cambardella and Elliott (1992), sand-sized organic matter constitutes a major part of the LF. The amount of C and N recovered in the CF (g C or N in per gram fraction of soil), typically <2000 to 53 μm, is often more than that obtained in the LF, and the concentration of these elements (g C or N per gram fraction) and their C:N ratio is lower (Gregorich et al., 1996; Barrios et al., 1996b). Carter (2002) found that the proportion of POM-C in surface soils from Eastern Canada was ca. 20% in the CF (>53 μm) and ≤7% in the LF (<1.7 g cm$^{-3}$ NaI). Magid and Kjaergaard (2001) found that the amounts of C and N, mineralization characteristics, and the appearance under the microscope of CF >400 μm and LF <1.4 g cm$^{-3}$ were similar. The advantages of size-based separatory methods are the relative simplicity and lower input requirements. Coarse-fraction measures are well suited for use in studies requiring large numbers of samples. Simple measures of the CF isolated in various ways have been effectively used to document the influences of land use, including cultivation of forested and grassland soils, tillage practices, and crop rotations on labile SOM (Table 3.3). For example, in a 5-year study of rotations based on continuously cropped wheat and wheat–fallow systems, Bowman et al. (1999) found that increases in the 0 to 5 cm depth in CF-C doubled whereas CF-N and soluble organic C increased by one third. The result was compared with total soil SOC and N, which only increased by ca. 20%. In a separate study, Bowman et al. (2000) found that declines in sunflower yield prompted declines in SOC and proportionally higher losses in CF-C in the surface depth.

Despite the general robustness of size-based separation, CF measures might not be as sensitive to agronomic treatments as are LF materials. Carter et al. (1998) found that the LF was more sensitive to tillage treatments than was the CF fraction. In several studies of fertility sources, LF characteristics differed among treatments but CF characteristics did not (Carter et al., 2003). Sohi et al. (2001) concluded that particle-size fractions confuse POM with matter attached to mineral surfaces. Accordingly, they argued that the distinct chemical properties of the LF provide a better basis for models of SOM turnover. Dalal and Meyer (1987) found density fractionation to be better than size-based methods to document the influence of cultivation on continuous wheat culture on organic N fractions. By using $^{13}$C natural abundance, Gregorich et al. (1995) found that isotopic contents of the LF were more similar to plant residues than were CF contents. In a related study, Gregorich et al. (1996) found that the composition of the LF isolated from soils under forest or maize culture better reflected litter chemistry than did CF composition. They found that CF was more degraded, suffering losses of lignin derivatives, carbohydrate constituents, and aliphatic compounds. The suggestion that the

LF is more closely related to plant residues than is the CF might be misleading. The study and several other works show that the LF also includes microbial residues and humic substances (Baldock et al., 1990; Wander and Traina, 1996b; Kerek et al., 2002). The typically darker color of LF materials (Figure 3.3) is consistent with the presence of humified materials in this fraction. Also noteworthy is the presence of high-surface-area carbonaceous material associated with coal and charcoal present in small amounts in the LF (Kleineidam et al., 1999).

Quality aspects of POM, isolated by size or density, can provide important information about the status of labile SOM. Many studies have investigated the effects of management on POM quality indirectly by relating the quantity of POM to short-term soil C and N mineralization rates or to the size of the microbial biomass, and have found these characteristics to be positively related (Hassink, 1995b; Monaghan and Barraclough, 1995; Fliessbach and Mader, 2000). Typically, POM fractions are more rapidly decayed than heavier or finer-sized fractions. The relationship between POM-C and biomass C has been used as an indicator of C availability (Alvarez et al., 1998). Hassink (1995a) found that decay rates of individual fractions decreased with increasing density and decreasing size, and proposed that these rates and fractions be used to delineate SOM models. Specific mineralization rates of POM-C or POM-N (e.g., mg C mineralized per g C in POM) are sometimes used to assess POM quality. For example, differences in the specific mineralization rates of the LF recovered from soils under organic and conventional management suggest that the LF from manure-amended organic systems was more labile than the LF recovered from organic cash-grain systems (Wander et al., 1994). In a study of soils obtained from six Douglass fir forests, Swanston et al. (2002) noted that the specific C mineralization rates of the LF and heavy fractions estimated during a 300-d incubation did not differ. This led them to conclude that the LF and heavy fractions had similar recalcitrance. Both the long duration of the incubation and fact that the soils studied were forest soils that had 60% of total SOC in the LF contributed to their findings. Plotted results show that considerable site-based differences prevented statistical resolution of differences between mean LF and HF decay rates and that the LF did decay faster during the initial phase of the incubation. If the results had been analyzed by a technique that did not average rates over the entire incubation period, their results would likely have led to a different conclusion.

Treatment effects on SOM are generally more apparent in the POM-C than in the POM-N fraction for various reasons. Management practice influences on soil N reservoirs are manifested primarily in the fine fractions, which constitute a far larger reservoir for N (Cambardella and Elliott, 1993). The proportion of C in POM has a wide range, with reports for mineral soils varying with depth ranging typically from 2 to 30%. The range of the proportion of N in POM is less, varying typically from 1.5 to 10%. Higher proportions of POM-C and -N result when perennial roots are abundant (Garten and Wullschleger, 2000). In addition to varying more, the quantity of C in POM appears to be more dynamic than the N content (Dalal and Meyer, 1986). Even though POM-C and -N contents are highly correlated, their kinetics differ. Observation of decadal retention of $^{15}$N in POM-N led Delgado et al. (1996) to conclude that POM-N should be equated with the slow N pool. Numerous works have shown that N mineralized during incubation studies is not derived exclusively, or in many cases primarily, from POM fractions (Boone, 1994; Gaiser et al., 1998). By using samples collected from agroecological regions of Saskatchewan, Canada, Curtin and Wen (1999) compared potentially mineralizable N with POM, soluble organic matter measured in saturated paste extracts, and $NH_4$-N released by digestion in 2 $M$ KCl or by steam distillation in phosphate-borate buffer. LF-N, which was the largest N fraction measured, mineralized slowly compared with chemically extracted N fractions, but did not account for net N mineralized. Also, POM-N was related with net mineralization potential but not with early mineralization rates. This finding and other results suggest that although POM-N is not a direct measure of labile N, it can be used as an index of this pool. Some constituents of POM might be too dynamic for their effective use as predictors of N mineralization even though they influence N dynamics directly. In a study of LF dynamics in the Harvard forest, Boone (1994) found, through litter removal, that LF mass and mineralization potentials varied monthly and that even though LF mass was 1/3 to 2/3 residue-derived, its direct contributions to net N mineralization was only 13% in pine forests, 2% in maple forests, and 11% maize cropped soil. Willson et al. (2001) also noted a seasonal variability in POM mass and lability and found higher potential N mineralization per unit POM in soils collected in April and November than in those collected in September or October.

The C content and quality of POM may provide more valuable information about N dynamics than can POM-N content. POM's ability to support heterotropic activity, which is solely responsible for ammonification, is likely to be a more significant determinant of N dynamics in most soils than the N it might supply. Janzen et al. (1992) found that soil respiration rates and microbial biomass were strongly correlated with the LF content as well as with N mineralization. The latter relationship, however, was less consistent because high C:N ratios of the LF promoted temporary N immobilization. Even though N incorporation into POM is less important in a quantitative sense than is N incorporation into finer fractions (Balabane and Balesdent, 1992; Schwenke et al., 2002), short-term dynamics of POM-N can provide valuable information about soil N cycling characteristics. The benefits of organic additions and use of diversified rotations to SOM quantity and quality become quite apparent after long-term treatment (Rasmussen et al., 1998). However, associated enhancement of nutrient use efficiency that is expected in SOM-aggraded systems is not reflected in short-term $^{15}$N incorporation into total SOM (Kramer et al., 2002). The short-term dynamics of $^{15}$N cycling through the POM fraction, however, varies with SOM status. Proportional differences in $^{15}$N cycling that are due to residue handling (Schwenke et al., 2002), fertilization (Balabane and Balesdent, 1992), and tillage or crop rotation (Nissen and Wander, (2003) are greater in POM-N than in total N or finer-N fractions. Nissen and Wander (2003) found that assimilation of added $^{15}$N into SOM increased with POM contents, which with other studies (Wander et al., 1996b; Wander and Traina, 1996a) suggests that biological mediation of nutrient cycling is more apparent when soils are able to maintain reserves of labile SOM as indicated by POM characteristics.

C:N ratios of SOM and SOM fractions have often been used as indices of quality; however, interpretation of trends can be difficult. Agricultural use of soils reduces total soil C:N ratios, which typically range from 14 to 8, as labile SOM is lost (Duxbury et al., 1989; Kaffka and Koepf, 1989). Typically POM has a C:N ratio of ca. 20:1, with higher ratios in forested systems that accumulate litter. Reported C:N ratios of POM have a narrower range than the plant residues from which they are derived (Molloy et al., 1977). Figure 3.5 shows the relationship between C:N ratios and H:C ratios for whole soils and POM fractions. The C:N ratios decrease and H:C ratios increase as labile constituents of SOM are lost and SOM aromaticity increases. The wider range of C:N ratios in POM and in H:C of whole soils reveals the differences in the SOM described. Aggradation of SOM in whole soils is reflected in increasing C:N ratios, which are associated in part with accumulation of POM. In a study of SOC accumulation in a chronosequence of restored prairie soils, Jastrow et al. (1996) found that macroaggregate-associated C:N ratios increased with time since cultivation but less than 20% of the C accrued was accounted for in the LF. Degradation of SOM results in carbonization and increased H:C ratios.

In soils where POM constitutes a large proportion of total SOM, there is a direct relationship between POM C:N and whole-soil C:N ratios. However, in arable mineral soils, where POM-C usually accounts for a quarter of the SOC or less, there is no clear relation between POM C:N ratios and whole-soil C:N ratios. Accumulation of POM reservoirs that are partially degraded and thus have lower C:N ratios than fresh residues have been cited as evidence of SOM-aggrading cropping practices (Wander and Traina, 1996a). Higher N contents in POM are associated with enhanced soil N supply potential (Koutika et al., 2001). Several studies have shown that POM C:N ratios are higher in soils where grain crop production relies heavily or exclusively on inorganic N sources of fertility than in systems that include diverse rotations that include legumes or add manures (Gregorich et al., 1998; Kandeler et al., 1999; Aoyama et al., 1999; Nissen and Wander, 2003). In studies of fertilized bromegrass, LF C:N ratios were higher when N application was absent or only made at lower rates. Ratios were calculated by using data reported in Malhi et al.

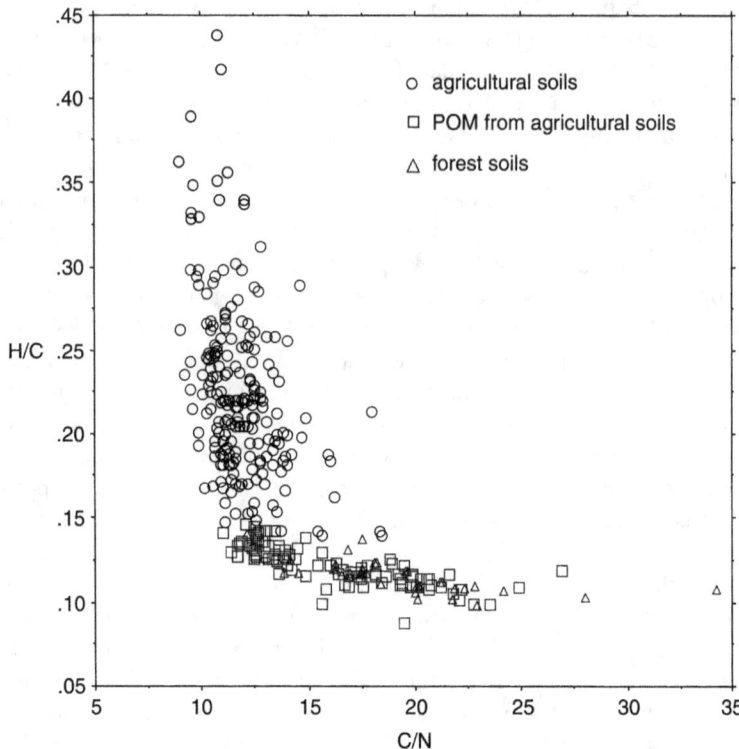

**FIGURE 3.5** Relationship between H/C and C/N of SOM from 228 agricultural soils and 98 POM samples from agricultural soils and 46 POM samples from forest soils; samples from forest soils are whole O or O + A horizon samples. (From F. Magdoff, unpublished data. With permission.)

(2003a, 2003b). Nitrogen limitation likely restricts decay in that highly rooted system. The range of C:N ratios of various SOM fractions obtained from a single location at different times during an annual cycle can vary as much as similarly collected fractions obtained from a wide variety of sites.

Regardless of the methods used to isolate POM, recovered fractions include components of mixed age and origin. Where climate does not restrict decay notably, POM-C is an effective measure of active SOM if contaminants such as charcoal are not present or are accounted for (Gijsman, 1996; Gerzabek et al., 2001). Changes in POM quantity and quality that occur within a season are driven by organic inputs and can influence the mineralization and immobilization of native labile fractions, which vary less in magnitude and decay at a slower rate. Interpretation of fraction composition and stages of decay are more complex in systems where partially decayed organic matter is the input. For example, in a study of sludge-amended minespoil, Fierro et al. (1999) found that residues became denser but remained relatively coarse during decomposition, and, even though sludge addition increased the quantity of POM in soil, inorganic N was best correlated with N in the fine fraction. Residue additions can cause rapid changes in the characteristics of the labile fraction. Amendment with nonplant sources of organic matter can prompt instantaneous changes in the composition of dynamic POM fractions, which would take years to achieve in a system amended exclusively with fresh plant residues. For example, Fortuna et al. (2003) found that addition of compost dramatically increased POM-C contents and changed its chemical characteristics in a manner that reduced its immediate value as an indicator of nutrient dynamics but increased its value as an index of soil tilth. Clearly, it is not safe to assume that POM-C is a measure of the active pool in all environments. It is not safe to assume that POM-C is a measure of the active pool in all environments. Retention of labile organic substrates can serve as evidence of physical limitations to decay (Franzluebbers and Arshad, 1996b; Wander and Bidart, 2000). Gill and Burke (1999) used $^{13}C$ natural abundance to assess the invasion by woody plants of grasslands (Texas and New Mexico) or by grasses of shrublands (Utah) that has occurred during the past 50 years. They found that POM components could be further divided into kinetically and compositionally distinct entities, but in general accounted for the slowly cycled C pool in that environment. Garten and Wullschleger (2000) estimated the turnover times of POM-C (CF) in four switchgrass (*Panicum virgatum* L.) field trials in the southeastern U.S. to be 2.4 to 4.3 years whereas those for mineral-associated organic matter were 26 to 40 years.

## Methods Separating Fresh POM from Resident POM

The material closely associated with residue inputs can be concentrated by focusing on the larger or lighter POM constituents. Changes in the composition of SOM associated with particle-size separates, progressing from large to small, have been related to progressive stages of decay (Baldock et al., 1992; Cambardella and Elliott, 1992). Sitompul et al. (2000) characterized the net monthly decomposition rates of light, intermediate, and heavy fractions of macroorganic matter (size 150 μm to 2 mm) obtained from soils under cultivated sugarcane and used $^{13}C$ natural abundance to estimate their turnover rates. The information was used to estimate the size of the slow pool, and the resistant fraction was equated with the 50- to 150-μm-size fraction. Hassink (1995b) found that the concentration of C, C:N ratios of fractions and chemical indicators of decay varied with density. In a study of a chronosequence of 3 million years of California soils, Baisden et al. (2002) found that the mineral-free POM ($<1.6$ g $cm^{-3}$) contained mostly recognizable plant material, fungal hyphae, and charcoal fractions (1.6 to 2.2 g $cm^{-3}$) were partially or completely humified fine POM, and the dense fraction ($>2.2$ g $cm^{-3}$) consisted of relatively organic-matter-free sand and organic-matter-rich clays. Methodological details no doubt influence findings. In a study of fresh residue recovery, Rovira et al. (1998) found that density-based methods did not efficiently recover freshly added straw residues except in the case of large residues in coarse-textured soils. They suggested that affiliation with light fine particles enhanced residue recovery.

Size and density are often combined in sequence to isolate materials of different age (Table 3.3). Size-based separation of CF materials is often followed with flotation in light solutions to concentrate components that are closely related to fresh residues (e.g., Barrios et al., 1996). In a study, Cambardella and Elliott (1992) first collected the CF, then concentrated the LF ($<1.85$ g $cm^{-3}$) recovered in that fraction, and used sieving to divide the LF into four size-based classes. They found that larger-sized LF was most abundant in native sod, less abundant in arable soils, and least abundant in soils maintained under fallow. Stone et al. (2001) (see Chapter 5) used the reverse sequence, separating the LF into different-sized components. These authors aimed to relate labile constituents of POM derived from compost added to agricultural soils to pythium suppression. Instead of tracking changes in total LF quality, which on average would have appeared more degraded after compost was added, they separated LF fractions by size, assuming that larger components were less decomposed, and used $^{13}C$-NMR to track temporal changes in its composition. Reductions in aromatic and aliphatic structures and losses in alkyl- and *O*-alkyl C in coarse-sized LF were related to loss of pythium suppression. Magid et al. (1996) found the sequence of fractionation by size and then density to be a more powerful approach for separating POM than the use of density alone. They found that density-based separation in a centrifuged container without initial size-based isolation substantially reduced the recovery of freshly added plant material in the LF. This was attributed in part to the loss of air entrapped in the intact tissue during centrifugation and to interactions between small, heavy particles and the large, light plant material. Issues discussed previously regarding entrapment might also have applied.

## Methods Separating Protected from Nonprotected POM

Efforts to separate POM that is biologically active from POM that contributes to aggregation have used a range of approaches including relatively simple strategies and those that rely on multistep procedures that manipulate size, density, and energy to explore the details of aggregate formation (Table 3.3). Golchin et al. (1994a, 1994b) were among the first to attempt to separate free and occluded POM from a series of virgin forest and grassland soils by using a two-step method. Separation of free POM through flotation in 1.6 g cm$^{-3}$ solution occurred after gentle stirring. This was followed by sonication and collection of occluded LF. The proportion of C occluded increased with clay content and, based on characteristics of $^{13}$C-NMR spectra, occluded POM was found to be more degraded, having a lower proportion of $O$-alkyl C and a higher proportion of alkyl C than the free LF. The proportion of C and N in loose or occluded fractions varies notably among studies that have used similar two-step procedures. Studying virgin soils, Golchin et al. (1994b) recovered 19–7% of total C and 22–6% of soil N in loose LF and recovered ≈31–9% of C and ≈14–6% of N in occluded LF fractions. In a study of arable soils, Wander and Bidart (2000) found that the proportion of C and N in the loose LF fraction obtained by flotation in 1.6 g cm$^{-3}$ NaPT after gentle orbital stirring, and in the occluded CF recovered after complete dispersion, was lower than that reported by Golchin et al. (1994b); values were ≈2.5–6.5% of C and ≈1.5–3.5% of N in loose and ≈12–22.5% of C and ≈11–19.5% of N in occluded fractions. This is consistent with findings that POM is lost rapidly when soils are subject to tillage (Tiessen and Stewart, 1983; Cambardella and Elliott, 1992, 1993) and supports the theory that occluded material is less susceptible to decay (Beare et al., 1994a). By using wet sieving to sort loose from occluded POM and $^{13}$C natural abundance to assess C dynamics, Jastrow et al. (1996) also concluded that occlusion protected residues from decay. This is consistent with the findings of Besnard et al. (1996), who found that loose POM derived from maize was rapidly $^{13}$C-depleted in a tilled soil. By using the chemical characteristics of POM, including C:N ratios, abundance of C structures, and isotopic contents as a basis for comparison, results suggest that occluded POM has a lower turnover rate and is typically older in age. For example, Golchin et al. (1994a) found that in occluded POM fractions had high proportions of alkyl C and lower proportions of $O$-alkyl C than loose POM had, and this suggests its greater extent of decay. By using a more detailed separation scheme [where loose POM organic matter <1.6 g cm$^{-3}$ was compared to occluded POM (<1.6 occluded, 1.6–1.8, and 1.8–2.0 g cm$^{-3}$), Golchin et al. (1995) found that the $O$-alkyl C content of occluded POM was inversely related to its stability, which was inferred from $^{13}$C natural abundance. After optimizing their fractionation scheme to maximize recovery in the occluded POM fraction, Sohi et al. (2001) found that $O$-alkyl C to alkyl-C ratios were 1.38 to 2.30 times higher in loose than in occluded POM fractions. Based on the results and accompanying evidence that occluded POM had a greater abundance of aliphatic hydrocarbons, carboxylic anions, and aromatic C, they also concluded that occluded POM was more decomposed. Baldock et al. (1997) found greater carbohydrate abundance in loose POM and concluded that this demonstrated its closer tie to plants. They found that as loose POM aged, its size and quality diminished.

Studies that follow the fate of newly incorporated SOM demonstrate POM dynamics directly. Wander and Yang (2000) followed the $^{13}$C dynamics of newly incorporated maize residues in loose, occluded, and humified fractions for 1.5 years. After maize senescence, $^{13}$C-labeled residues were most concentrated in the loose LF, with $^{13}$C later accumulating in the occluded CF. After 1.5 years, the largest proportion of $^{13}$C resided within the finer and denser fractions. Puget and Drinkwater (2001) used a similar sequence of shaking (1 h, 100 rec$^{-1}$ min) followed by density-based separation of the loose LF, subsequent dispersion, and finally size-based separation of the occluded CF to assess the fate of $^{13}$C-derived from hairy vetch. In that work, and the study by Wander and Yang, occluded POM was less dynamic than loose POM and root-derived residues contributed dispro-portionately more to the occluded fraction in the short term. Wander and Bidert (2000) showed that potential N mineralization was better correlated with total POM than with loose POM recovered after gentle shaking of the sieved soil. This again demonstrated that the concentration of POM most closely associated with residues does not provide valuable information about soil N supply.

Methods that collect POM fractions associated with water-stable aggregates provide the basis for a widely accepted understanding of its contributions to aggregation. Studies show that POM is closely associated with aggregation, particularly water-stable macro- (Angers and Mehuys, 1988), and mesoaggregates (Waters and Oades, 1991). Numerous works suggest that macroaggregates are microaggregates cemented together by residues in progressive states of decay (Tisdall and Oades, 1982; Buyanovsky et al., 1994; Jastrow et al., 1996). Studies exploring the role POM plays in this hierarchy have sorted aggregates according to size and stability in water. Golchin et al. (1995) found $O$-alkyl C in occluded POM was highly correlated with wet aggregate stability. By using $^{13}$C natural abundance, Puget et al. (2000) found that stable macroaggregates were richer in total C and in young C than in unstable macroaggregates. The young SOC in stable macroaggregates, which had a half-life of decades, was only 50% POM. It is well established that physically protected SOM includes POM plus microbial remains (Elliott et al., 1996). Cambardella and Elliott (1994) termed the intermediate density fraction (2.07 to 2.22 g cm$^{-3}$) the enriched light fraction (ELF). They found that material isolated from inside macroaggregates contained the highest percentage of total soil C and N in cultivated soils. That material was largely microbially derived and included POM thought to stabilize microaggregates into macroaggregates. Subsequent studies of the ELF fraction suggest that the origin or function of this fraction can vary among soils or that subtle differences in methods yield fractions with different characteristics (Six et al., 2000b; Rodionov et al., 2001).

A chronology for POM incorporation into different-sized aggregates has been suggested by several authors (Beare et al., 1994b; Six et al., 1998). Fresh residues initially enmesh or cement finer particles into macroaggregates. Thus, the nature of POM occluded into aggregates varies according to aggregate size, with younger POM being enriched in larger aggregates (Angers and Giroux, 1996; Puget et al., 1995). As time progresses decomposition proceeds and POM is altered, reducing both POM size and the size of water-stable aggregates. Baldock (2002) found aggregates isolated at densities <1.6 g cm$^{-3}$ were derived from the most decomposed occluded POM, which included material that cannot support microbial growth. He found that POM isolated in heavier fractions (1.6 to 1.8 cm$^3$ range) was less degraded. Microaggregates then lose stability as the carbohydrates remaining in POM are depleted. Microaggregates not bound into larger units are stabilized by microbial remnants.

Although the general role of POM in aggregates is well supported, the specific dynamics of particular POM fractions derived from highly detailed sequential fractionation schemes can vary from soil to soil. It would be difficult — if not impossible — to optimize methods for a wide range of soils. This and the greater time and resource requirements associated with these methods have restricted their application to comparisons of a relatively small number of samples and treatments. Accordingly, care must be taken when extrapolating findings from limited data sets to generalized theory. For example, as is commonly and logically done, Six et al. (1998) developed their theory of aggregation dynamics by using extremes in C input, basing their work on soils that were in sod or that had been unfertilized while used to produce wheat under no-till-stubble mulch or conventional tillage. The significant influence that fertilization has on POM quantity and chemical characteristics has been noted previously; accordingly, both POM and aggregate characteristics in more typically fertilized arable soils might be unlike those studied. The relationship between POM fractions and their function within aggregates has been inferred from the chemical nature of fractions, including evidence about their turnover dynamics. Ashman et al. (2003) showed that both these aggregate properties depend on the fractionation procedure used. They argued that the observed relationships between aggregate size and biological activity should be interpreted in terms of the disruptive mechanisms used to fractionate soil. Further, they suggested that the aggregate hypothesis might be an artifact of the chosen method of separation. This is most likely true for macroaggregates. In one of the few studies of POM dynamics conducted in the field, Plante and

McGill (2002) demonstrated that the quantity of POM occluded in water-stable aggregates is actually increased by some aggregate turnover. Demonstration of the relationship between fractions obtained in the laboratory and *in situ* processes is required.

## SUMMARY

The association between SOM and soil fertility remains a critical question in need of investigation. During the past century of inquiry, methods of SOM fractionation have proliferated greatly. Appropriate matching of methods, soils, and questions can lead to rapid advances in understanding the biological, physical, and chemical contributions of SOM fractions. Fractions are measurable constituents of organic matter and not the theoretically defined pools of SOM represented in empirically based models. Thus, issues of identity and meaning must be dealt with before information about specific SOM fractions is effectively used to guide management or serve as model inputs. This challenge is complicated by the fact that the ability and desire to sort materials with decomposition rates that vary within months, years, decades, centuries or millennia will vary with the subject of inquiry and resolving power of the fraction. The SOM fractions used to estimate kinetically distinct pools can be divided into three functional classes, with the most dynamic SOM constituents being most closely identified with biological activity, young and intermediate-age materials being associated with physical activity, and materials with the longest half-life contributing most to chemical reactivity. This complexity, along with heterogeneity within SOM fractions, makes interpretation and application of information derived from SOM fractions difficult. This is particularly true for labile SOM fractions, which include components that are dynamic within a year yet have integrative properties that reflect average status of material accrued over several years to a decade. Labile fractions can themselves be subdivided into active or inactive components, and the divisions will likely vary according to whether the subject of interest is biological, physical, or chemical reactivity. Measures of POM are often used as direct estimates of the active C fraction even though this fraction can contain refractory structures, including charcoal; this is because POM kinetics are quite sensitive to resource quality and physical factors impacting decay. Compared to humic substances, POM is more diverse in its route of origin, chemical composition, and function. The composition of POM can effectively reflect the status of labile SOM, providing valuable information about substrate and habitat quality affecting N mineralization or immobilization dynamics N and other microbially regulated processes. Verification of the functional relevance of POM fractions isolated in the laboratory remains a challenge. At present, we know far more about the characteristics of POM isolated by a variety of techniques than we do about how to accurately and consistently interpret this information.

## REFERENCES

Ahmed, M., and J.M. Oades (1984). Distribution of organic matter and adenosine triphosphate after fractionation of soil by physical procedures. *Soil Biol. Biochem.* 16, 465–470.

Alvarez, R., and C.R. Alvarez (2000). Soil organic matter pools and their associations with carbon mineralization kinetics. *Soil Sci. Soc. Am. J.* 64, 184–189.

Alvarez, C.R., R. Alvarez, S. Grigera, and R.S. Lavado (1998). Associations between organic matter fractions and the active soil microbial biomass. *Soil Biol. Biochem.* 30, 767–773.

Amelung, W., and W. Zech (1999). Minimisation of organic matter disruption during particle-size fractionation of grassland epipedons. *Geoderma* 92, 73–85.

Amelung, W., I. Lobe, and C.C. Du Preez (2002). Fate of microbial residues in sandy soils of the South African Highveld as influenced by prolonged arable cropping. *Eur. J. Soil Sci.* 53, 29–35.

Angers, D.A., A. Ndayegamiye, and D. Cote (1993). Tillage-induced differences in organic matter of particle size fractions and microbial biomass. *Soil Sci. Soc. Am. J.* 57, 512.

Angers, D.A., and M. Giroux (1996). Recently deposited organic matter in soil-water-stable aggregates. *Soil Sci. Soc. Am. J.* 60, 1547–1551.

Angers, D.A., and G.R. Mehuys (1988). Effects of cropping on macro-aggregation of marine clay soil. *Can. J. Soil Sci.* 68, 723–732.

Angers, D.A., and G.R. Mehuys (1989). Effects of cropping on carbohydrate content and water stable aggregation of a clay soil. *Can. J. Soil Sci.* 69, 373–380.

Ångren, G.I., and E. Bosatta (1987). Theoretical analysis of the long-term dynamics of carbon and nitrogen in soils. *Ecology* 68, 1181–1189.

Aoyama, M., D.A. Anger, and A. Ndayegamiye (1999). Particulate and mineral-associated organic matter in water-stable aggregates as affected by mineral fertilizer and manure applications. *Can. J. Soil Sci.* 79, 295–302.

Appel, T.K., and K. Mengel (1993). Nitrogen fractions in sandy soils in relation to plant nitrogen uptake and organic matter incorporation. *Soil Biol. Biochem.* 25, 685–691.

Archard, F.W. (1786). Crell's Chemical Analysis 2,391. Cited in Stevenson, F.J. *Humus Chemistry* (1982). John Wiley & Sons, New York, 26 pp.

Aref, S., and M.M. Wander (1997). Long-term trends of corn yield and soil organic matter in different crop sequences and soil fertility treatments. In *Advances in Agronomy*, Vol. 62. Academic Press, San Diego, pp. 153–197, chap. 3.

Ashman, M.R., P.D. Hallert, and P.C. Brookes (2003). Are the links between soil aggregate size class, soil organic matter and respiration rate artifacts of the fractionation procedure? *Soil Biol. Biochem.* 35, 435–444.

Avnimelech, Y. (1986). Organic residues in modern agriculture. In Y. Chen and Y. Avnimelech (Eds.), *The Role of Organic Matter in Modern Agriculture,* Martinus Nijhoff, Dordrecht, pp. 1–10.

Baisden, W., R. Amundson, A.C. Cook, and D.L. Brenner (2002). Turnover and storage of C and N in five density fractions from California annual grassland surface soils. *Glob. Biochem. Cycl.* 16(4), 1117.

Balabane, M., and J. Balesdent (1992). Input of fertilizer derived labelled N to soil organic matter during a growing season of maize in the field. *Soil Biol. Biochem.* 24, 89–96.

Balabane, M., and F. van Oort (2002). Metal enrichment of particulate organic matter in arable soils with low metal contamination. *Soil Biol. Biochem.* 34, 1513–1516.

Baldock, J.A. (2002). Interactions of organic materials and microorganisms with minerals in the stabilization of soil structure. In P.M. Huang, J.-M. Bollag, and S. Senesi (Eds.), John Wiley & Sons, Chichester, U.K., pp. 85–132.

Baldock, J.A., J.M. Oades, A.M. Vassallo, and M.A. Wilson (1990). Solid state CP/MAS $^{13}$C NMR analysis of particle size and density fractions of a soil incubated with uniformly labeled $^{13}$C-glucose. *Aust. J. Soil Res.* 28, 193–212.

Baldock, J.A., J.M. Oades, A.G. Waters, X. Peng, A.M. Vassallo, and M.A. Wilson (1992). Aspects of the chemical structure of soil organic materials as revealed by solid state $^{13}$C NMR spectroscopy. *Biogeochemistry* 16, 1–2.

Baldock, J.A., J.M. Oades, A.G. Waters, X. Peng, A.M. Vassallo, and M.A. Wilson (1997). Assessing the extent of decomposition of natural organic matter using solid state $^{13}$C NMR spectroscopy. *Aust. J. Soil Res.* 35, 1061–1083.

Barrios, E., R.J. Buresh, and J.I. Sprent (1996a). Nitrogen mineralization in density fractions of soil organic matter from maize and legume cropping systems. *Soil Biol. Biochem.* 28, 1459–1465.

Barrios, E., R.J. Buresh, and J.I. Sprent (1996b). Organic matter in soil particle size and density fractions from maize and legume cropping systems. *Soil Biol. Biochem.* 28, 185–193.

Barriuso, E., and W.C. Koskinen (1996). Incorporating nonextractable atrazine residues into soil size fractions as a function of time. *Soil Sci. Soc. Am. J.* 60, 15–157.

Beare, M.H., M.L. Cabrera, P.F. Hendrix, and D.C. Coleman (1994a). Aggregate-protected and unprotected pools of organic matter in conventional and no-tillage ultisols. *Soil Sci. Soc. Am. J.* 58,787–795.

Beare, M.H., P.F. Hendrix, and D.C. Coleman (1994b). Water-stable aggregates and organic matter fractions in conventional and no-tillage ultisols. *Soil Sci. Soc. Am. J.* 58, 777–786.

Beare, M.H., C.L. Neely, D.C. Coleman, and W.L. Hargrove (1991). Characterization of a substrate-induced respiration method for measuring fungal, bacterial and total microbial biomass on plant residues. *Agric. Ecosyst. Environ.* 34, 65–73.

Benbi, D.K., and J. Richter (2002). A critical review of some approaches to modeling nitrogen mineralization. *Biol. Fertil. Soils* 35, 168–183.

Bending, G.D., C., Putland, and F. Rayns (2000). Changes in microbial community metabolism and labile organic matter fractions as early indicators of the impact of management on soil biological quality. *Biol. Fertil. Soils* 31, 78–84.

Besnard, E., C., Chenu, J. Balesdent, P. Puget, and D. Arrouays (1996). Fate of particulate organic matter in soil aggregates during cultivation. *Eur. J. Soil Sci.* 47, 495–503.

Besnard, E., C. Chenu, and M. Robert (2001). Influence of organic amendments on copper distribution among particle-size and density fractions in Champagne vineyard soils. *Environ. Pollut.* 112, 329–337.

Bird, S.B., J.E. Herrick, and M.M. Wander (2001). Exploring heterogeneity of soil organic matter in rangelands: Benefits for carbon sequestration. In R.F. Follett (Ed.), *Potential of U.S. Grazing Lands to Sequester Carbon and Mitigate the Greenhouse Effect*. Lewis Publishers, Boca Raton, FL, pp. 121–138.

Boone, R.D. (1994). Light-fraction soil organic matter: Origin and contribution to net N mineralization. *Soil Biol. Biochem.* 26, 1459–1468.

Bowman, R.A., M.F. Vigil, D.C. Nielsen, and R.L. Anderson (1999). Soil organic matter changes in intensively cropped dryland systems. *Soil Sci. Soc. Am. J.* 63, 186–191.

Bowman, R.A., Nielsen, D.C., Vigil, M.F., and R.M. Aiken (2000). Effects of sunflower on soil quality indicators and subsequent wheat yield. *Soil Sci.* 165, 516–522.

Bremer, E., H.H. Janzen, and A.M. Johnston (1994). Sensitivity of total light fraction and mineralizable organic-matter to management-practices in a Lethbridge soil. *Can. J Soil Sci.* 74, 131–138.

Brookes, P.C., A. Landman, G., Pruden, and D.S. Jenkinson (1985). Chloroform fumigation and the release of soil nitrogen: A rapid direct extraction method to measure biomass nitrogen in soil. *Soil Biol. Biochem.* 17, 837–842.

Bruseau, M.L., R.E. Jessup, and P.S.C. Rao (1991). Nonequilibrium sorption of organic chemicals: Elucidation of rate limiting processes. *Environ. Sci. Technol.* 25, 134–142.

Burke, I., W.K. Lauenroth, R. Riggle, P. Brannen, B. Madigan, and S. Beard (1999). Spatial variability of soil properties in the shortgrass steppe: The relative importance of topography, grazing, microsite, and plant species in controlling spatial patterns. *Ecosystems* 2, 422–438.

Buyanovsky, G.A., M. Aslam, and G.H. Wagner (1994). Carbon turnover in soil physical fractions. *Soil Sci. Soc. Am. J.* 58, 1167–1173.

Cabrera, M.L., and D.E. Kissel (1988). Potentially mineralizable nitrogen in disturbed and undisturbed soil samples. *Soil Sci. Soc. Am. J.* 52, 1010–1015.

Cambardella, C.A., and E.T. Elliott (1992). Particulate soil organic-matter changes across a grassland cultivation sequence. *Soil Sci. Soc. Am. J.* 56, 777–782.

Cambardella, C.A., and E.T. Elliott (1993). Methods for physical separation and characterization of soil organic matter fractions. *Geoderma* 56, 449–457.

Cambardella, C.A., and E.T. Elliott (1994). Carbon and nitrogen dynamics of soil organic matter fractions from cultivated grassland soils. *Soil Sci. Soc. Am. J.* 58, 123–130.

Carter, M., D.A. Angers, E.G. Gregorich, and M.A. Bolinder (2003). Characterizing organic matter retention for surface soils in eastern Canada using density and particle size fractions. *Can. J. Soil Sci.* 83, 11–23.

Carter, M.R. (1996). Analysis of soil organic matter storage in agroecosystems. In M.R. Carter and B.A. Stewart (Eds.), *Structure and Organic Matter Storage in Agricultural Soils*. Lewis Publishers, Boca Raton, FL, pp. 3–11.

Carter, M.R. (2002). Soil quality for sustainable land management: Organic matter and aggregation interactions that maintain soil function. *Agron. J.* 94, 38–47.

Carter, M.R., D.A. Angers, and H.T. Kunelius (1994). Soil structural form and stability, and organic-matter under cool-season perennial grasses. *Soil Sci. Soc. Am. J.* 58, 1194–1199.

Carter, M.R., E.G. Gregorich, D.A. Angers, R.G. Donald, and M.A. Bolinder (1998). Organic C and N storage, and organic C fractions, in adjacent cultivated and forested soils of eastern Canada. *Soil Tillage Res.* 47, 253–261.

Cassman, K.G. (1999). Ecological intensification of cereal production systems: Yield potential, soil quality, and precision agriculture. *Proc. Natl. Acad. Sci.* 11, 5952–5959.

Chantigny, M.H. (2003). Dissolved and water-extractable organic matter in soils: a review on the influence of land use and management practices. *Geoderma* 113, 357–380.

Chantigny, M.H., D.A. Angers, D. Prevost,L.-P. Vezina, and F.-P. Chalifour (1997). Soil aggregation and fungal and bacterial biomass under annual and perennial cropping systems. *Soil Sci. Soc. Am. J.* 61, 262–267.

Chenu, C. (1995). Extracellular polysaccharides: An interface between microorganisms and soil constituents. In P.M. Huang et al. (Ed.), *Environmental Impact of Soil Component Interactions.* CRC Press, Boca Raton, FL, p. 217.

Cheshire, M.V., B.T. Christensen, and L.H. Sorensen (1990). Labeled and native sugars in particle size fractions from soils incubated with $^{14}C$ straw for 6 to 18 years. *J. Soil Sci.* 41, 29–39.

Cheshire, M.V., A. Lomax, and C.M. Mundie (1989). Structure of soil carbohydrates resistant to periodate oxidation. *J. Soil Sci.* 40, 865–872.

Chiou, C.T., and D.E. Kile (1998). Deviations from sorption linearity on soils of polar and nonpolar organic compounds at low relative concentrations. *Environ. Sci. Technol.* 32, 338–343.

Christensen, B. (1996). Matching measurable soil organic matter fractions with conceptual pools in simulation models of carbon turnover: A revision of model structure. In D.S. Powlson, P. Smith, and J.U. Smith (Eds.), *Evaluation of Soil Organic Models Using Long-Term Data Sets.* Springer-Verlag, Berlin, pp. 143–159.

Christensen, B. (2001). Physical fractionation of soil and structural and functional complexity in organic matter turnover. *Eur. J. Soil Sci.* 52, 345–353.

Christensen, B.T. (1986). Straw incorporation and soil organic matter in macro-aggregates and particle size separates. *J. Soil Sci.* 37, 125–135.

Christensen, B.T. (1987). Decomposability of organic matter in particle size fractions from field soils with straw incorporation. *Soil Biol. Biochem.* 19, 429–435.

Christensen, B.T. (1992). Physical fractionation of soil and organic matter in primary particle size and density separates. In *Advances in Soil Science,* Vol. 20, Springer-Verlag, New York, pp. 1–90.

Cobo, J., E. Barrios, D.C.L. Kass, and R. Thomas (2002). Nitrogen mineralization and crop uptake from surface-applied leaves of green manure species on a tropical volcanic-ash soil. *Biol. Fertil. Soils* 36, 87–92.

Collins, H.P., E.T. Elliott, K. Paustian, L.G. Bundy, W.A. Dick, D.R. Huggins, A.J.M. Smucker, and E.A. Paul (2000). Soil carbon pools and fluxes in long-term corn belt agroecosystems. *Soil Biol. Biochem.* 32, 157–168.

Collins, H.P., P.E. Rausmussen, and C.L.J. Douglas (1992). Crop rotation and residue management effects on soil carbon and microbial dynamics. *Soil Sci. Soc. Am. J.* 56, 783–788.

Conteh, A., G.J. Blair, and I.J. Rochester (1998). Soil organic carbon fractions in a Vertisol under irrigated cotton production as affected by burning and incorporating cotton stubble. *Aust. J. Soil Res.* 36, 655–667.

Curtin, D. (2002). Possible role of aluminum in stabilizing organic matter in particle size fractions of Chermozemic and Solonetzic soils. *Can. J. Soil Sci.* 82, 265–268.

Curtin, D., and G. Wen (1999). Organic matter fractions contributing to soil nitrogen mineralization potential. *Soil Sci. Soc. Am. J.* 63, 410–415.

Cuypers, C.G., T.J. Joziasse, and W. Rulkens (2000). Rapid persulfate oxidation predicts PAH bioavailability in soils and sediments. *Environ. Sci. Technol.* 34, 2057–2063.

Dahnke, W.C., and G.V. Johnson (1990). Testing soils for available nitrogen. In R.L. Westerman (Ed.), *Soil Testing and Plant Analysis.* Soil Science Society of America, Madison, WI, pp. 127–139.

Dalal, R.C., and R.J. Mayer (1986). Long-term trends in fertility of soils under continuous cultivation and cereal cropping in southern queensland. IV. Loss of organic carbon from different density fractions. *Aust. J. Soil Res.* 25, 83–93.

Dalal, R.C., and R.J. Mayer (1987). Long-term trends in fertility of soils under continuous cultivation and cereal cropping in southern queensland. VI Loss of total N from different particle size and density fractions. *Aust. J. Soil Res.* 25, 83–93.

De Cesare, F., A.M. Garzillo, V. Buonocore, and L. Badalucco (2000). Use of sonication for measuring acid phosphatase activity in soil. *Soil Biol. Biochem.* 32, 825–832.

de Saussure, T. (1804). *Recherches chimiques sur la végétation,* Paris. Cited in Stevenson, F.J. *Humus Chemistry* (1982). John Wiley & Sons, New York, 26 pp.

Delgado, J.A., A.R. Mosier, D.W. Valentine, D.S. Schimel, and W.J. Parton (1996). Long term N-15 studies in a catena of the shortgrass steppe. *Biogeochemistry* 32, 41–52.

Dendooven, L., E. Murphy, and D.S. Powlson (2000). Failure to simulate C and N mineralization in soil using biomass C to N ratios as measured by the fumigation extraction method. *Soil Biol. Biochem.* 32, 659–668.

Derenne, S., and C. Largeau (2001). A review of some important families of refractory macromolecules: Composition, origin, and fate in soils and sediments. *Soil Sci.* 166, 833–847.

Dexter, A.R. (1988). Advances in the characterization of soil structure. *Soil Tillage Res.* 11, 199–238.

Diaz-Zorita, M., E. Perfect, and J.H. Grove (2002). Disruptive methods for assessing soil structure. *Soil Tillage Res.* 64, 3–22.

Dick, W. (1997). Tillage system impacts on environmental quality and soil biological parameters. *Soil Tillage Res.* 41, 165–167.

Di-Giovanni, C., J.R. Disnar, M. Campy, and J.J. Macaire (1999). Variability of the ancient organic supply in modern humus. *Analysis* 27, 398–402.

Dorado, J., P. Tinoco, and G. Almendros (2003). Soil parameters related with the sorption of 2,4-D and atrazine. *Commun. Soil Sci. Plant Anal.* 34, 1119–1133.

Drury, C.F., J.A. Stone and W.I. Findlay (1991). Microbial biomass and soil structure associated with corn, grasses, and legumes. *Soil Sci. Soc. Am. J.* 55, 805–811.

Duxbury, J.M., M.S. Smith, J. Doran, C. Jordan, L. Szott, and E.Vance (1989). Soil organic matter as a source and sink of plant nutrients. In D.C. Coleman, M. Oades, and G. Urehara (Eds.), *Dynamics of Soil Organic Matter in Tropical Ecosystems* NiFtal Press, Maui, HI, 33–67.

Elliott, E.T., and C.A. Cambardella (1991). Physical separation of soil organic matter. *Agric. Ecosyst. Environ.* 34, 407–419.

Elliott, E.T., K. Paustian, and S. Frey (1996). Modeling the measurable or measuring the modelable: A hierarchical approach to isolating meaningful soil organic matter fractionations. In D.S. Powlson, P. Smith, and J.O. Smith (Eds.), *Evaluation of Soil Organic Matter Models*, Vol. 138. Springer-Verlag, Berlin, pp. 161–179.

Eriksson, J., and U. Skyllberg (2001). Binding of 2,4,6-trinitrotoluene and its degradation products in a soil organic matter two-phase system. *J. Environ. Qual.* 30, 2053–2061.

Feng, Y., and X. Li (2001). An analytical model of soil organic carbon dynamics based on a simple "hockey stick" function. *Soil Sci.* 166, 431–440.

Fierro, A., D.A. Angers, and C.J. Beauchamp (1999). Dynamics of physical organic matter fractions during de-inking sludge decomposition. *Soil Sci. Soc. Am. J.* 63, 1013–1018.

Findlay, R.H., G.M. King, and L. Walting (1989). Efficacy of phospholipid analysis in determining microbial biomass in sediments. *Appl. Environ. Microbiol.* 5, 2888–2892.

Fliessbach, A., and P. Mader (2000). Microbial biomass and size-density fractions differ between soils of organic and conventional agricultural systems. *Soil Biol. Biochem.* 32, 757–768.

Fortuna, A., R.R. Harwood, and E.A. Paul (2003). The effects of compost and crop rotations on carbon turnover and the particulate organic matter fraction. *Soil Sci.* 168(6), 434–444.

Franzluebbers, A.J., and M.A. Arshad (1996a). Soil organic matter pools during early adoption of conservation tillage in northwestern Canada. *Soil Sci. Soc. Am. J.* 60, 1422–1427.

Franzluebbers, A.J., and M.A. Arshad (1996b). Soil organic matter pools with conventional and zero tillage in a cold, semiarid climate. *Soil Tillage Res.* 39, 1–11.

Franzluebbers, A.J., R.L. Haney, F.M. Hons, and D.A. Zuberer (1999). Assessing biological soil quality with chloroform fumigation--incubation: Why subtract a control? *Can. J. Soil Sci.* 79, 521–528.

Franzluebbers, A.J., J.A. Stuedemann, H.H. Schomberg, and S.R. Wilkinson (2000). Soil organic C and N pools under long-term pasture management in the Southern Piedmont USA. *Soil Biol. Biochem.* 32, 469–478.

Frey, S.D., E.T. Elliott, nd K. Paustian (1999). Bacterial and fungal abundance and biomass in conventional and no-tillage agroecosystems along two climatic gradients. *Soil Biol. Biochem.* 31, 573–585.

Gaiser, T., M. Bernard, B.T. Kang, and K. Stahr (1998). Nitrogen mineralization and nitrogen uptake by maize as influenced by light and heavy organic matter fractions. *Z. Pflanzenernaehr. Bodenk.* 161, 555–561.

Gale, W.J., C.A. Cambardella, and T.B. Bailey (2000). Root-derived carbon and the formation and stabilization of aggregates. *Soil Sci. Soc. Am. J.* 64, 201–207.

Garland, J.L., and A.L. Mills (1991). Classification and characterization of heterotrophic microbial communities on the basis of patterns of community-level sole-carbon-source utilization. *Appl. Environ. Microbiol.* 57, 2351–2359.

Garten, C.T., and S.D. Wullschleger (2000). Soil carbon dynamics beneath switchgrass as indicated by stable isotope analysis. *J. Environ. Qual.* 29, 645–653.

Gerzabek, M.H., G. Haberhauer, and H. Kirchmann (2001). Soil organic matter pools and Carbon-13 natural abundance in particle-size fractions of a long-term agricultural field experiment receiving organic amendments. *Soil Sci. Soc. Am. J.* 65, 352–358.

Gijsman, A.J. (1996). Soil aggregate stability and soil organic matter fractions under agropastoral systems established in native savanna. *Aust. J. Soil Res.* 34, 891–907.

Gill, R.A., and I.C. Burke (1999). Ecosystem consequences of plant life form changes at three sites in the semiarid United States. *Oecologia* 121, 551–563.

Glendining, M.J., P.R. Puilton, D.S. Powlson, and D.S. Jenkinson (1997). Fate of $^{15}$N-labeled fertilizer applied to spring barley grown on soils of contrasting nutrient status. *Plant Soil* 195, 83–98.

Golchin, A., J.M. Oades, J.O. Skjemstad, and P. Clarke (1994a). Soil structure and carbon cycling. *Aust. J. Soil Res.* 32, 1043–68.

Golchin, A., J.M. Oades, J.O. Skjemstad, and P. Clarke (1994b). Study of free and occluded particulate organic matter in soils by solid state $^{13}$C CP/MAS NMR spectroscopy and scanning electron microscopy. *Aust. J. Soil Res.* 32, 285–309.

Golchin, A., J.M. Oades, J.O. Skjemstad, and P. Clarke (1995a). Structural and dynamic properties of soil organic-matter as reflected by C-13 natural-abundance, pyrolysis mass- spectrometry and solid-state C-13 NMR-spectroscopy in density fractions of an oxisol under forest and pasture. *Aust. J. Soil Res.* 33, 59–76.

Golchin, A., P. Clarke, J.M. Oades, and J.O. Skjemstad (1995). The effects of cultivation on the composition of organic matter and structural stability of soils. *Aust. J. Soil Res.* 33, 975–993.

Greenland, D.J., and G.W. Ford (1964). Separation of partially humified organic materials from soils by ultrasonic dispersion. *8th International Congress of Soil Science* 3, 137–148.

Gregorich, E., Beare, M.H., Stoklas, U., and P. St-Georges (2003). Biodegradability of soluble organic matter in maize-cropped soils. *Geoderma* 113, 237–252.

Gregorich, E.G., and M.R. Carter (1997). Soil quality for crop production and ecosystem health. In *Developments in Soil Science*, Vol. 25. Elsevier, Amsterdam, pp. 399–428.

Gregorich, E.G., and B.H. Ellert (1995). Light fraction and macroorganic matter in mineral soils. In M.R. Carter (Ed.), *Soil Sampling and Methods of Analysis*. Lewis Publishers, Boca Raton, FL, pp. 397–408.

Gregorich, E.G., G. Kaschanoski, and R.P. Voroney (1988). Ultrasonic dispersion of aggregates: Distribution of organic matter in size fractions. *Can. J. Soil Sci.* 68, 395–403.

Gregorich, E.G., R.G. Kachanoski, and R.P. Voroney (1989). Carbon mineralization in soil size fractions after amounts of aggregate disruption. *J. Soil Sci.* 40, 649–659.

Gregorich, E.G., M.R. Carter, D.A. Angers, C.M. Monreal, and B.H. Ellert, (1994). Towards a minimum data set to assess soil organic matter quality in agricultural soils. *Can. J. Soil Sci.* 74, 367–385.

Gregorich, E.G., B.H. Ellert, and C.M. Monreal (1995). Turnover and storage of corn residue carbon in a Gleysolic soil in eastern Canada. *Can. J. Soil Sci.* 75, 161–167.

Gregorich, E.G., C.M. Monreal, M. Schnitzer, and H.-R. Schulten (1996). Transformation of plant residues into soil organic matter: Chemical characterization of plant tissue, isolated soil fractions and whole soils. *Soil Sci.* 161, 680–693.

Gu, B.H., J. Schmitt, Z. Chen, L.Y. Liang, and J.F. McCarthy (1995). Adsorption and desorption of different organic-matter fractions on iron-oxide. *Geochimica et Cosmochimica Acta* 59, 219–229.

Guggenberger, G., and K.M. Haider (2002). Effect of mineral colloids on biogeochemical cycling of C, N, P and S in soils. In P.M. Huang, J.-M. Bollag, and N. Senesi (Eds.), *Interactions between Soil Particles and Microorganisms*. John Wiley & Sons, Chichester, U.K., pp. 267–322.

Guggenberger, G., B.T. Christensen, and W. Zech (1994). Land-use effects on the composition of organic matter in particle-size separates of soil: I. Lignin and carbohydrate signature. *Eur. J. Soil Sci.* 45, 449–458.

Guggenberger, G., S.D. Frey, J. Six, K. Paustian, and E.T. Elliott (1999). Bacterial and fungal cell-wall residues in conventional and no-tillage agroecosystems. *Soil Sci. Soc. Am. J.* 63, 1188–1198.

Gustafsson, O., F. Haghseta, C. Chan, J. MacFarlane, and P.M. Gschwend (1997). Quantification of the dilute sedimentary soot phase: Implications for PAH speciation and bioavailability. *Environ Sci. Technol.* 31, 203–209.

Hansen, S., H.E. Jensen, N.E. Nielsen, and H. Svendsen (1991). Simulation of nitrogen dynamics and biomass production in winter wheat using the Danish simulation model DIASY. *Fertil. Res.* 27, 245–259.

Hassink, J. (1995a). Decomposition rate constants of size and density fractions of soil organic matter. *Soil Sci. Soc. Am. J.* 59, 1631–1635.

Hassink, J. (1995b). Density fractions of soil macroorganic matter and microbial biomass as predictors of C and N mineralization. *Soil Biol. Biochem.* 27, 1099–1108.

Hassink, J., and J.W. Dalenberg (1996). Decomposition and transfer of plant residue C-14 between size and density fractions in soil. *Plant Soil* 179, 159–169.

Hassink, J., L.A. Bouwman, K.B. Zwart, J. Bloem, and L. Brussaard (1993). Relationships between soil texture, physical protection of organic matter, soil biota, and C and N mineralization in grassland soils. *Geoderma* 57, 105–128.

Hayes, M.H., P. MacCarthy, R.L. Malcom, and R.S. Swift (Eds.) (1989). *Humic Substances II: In Search of Structure*, John Wiley & Sons, Chichester, U.K., pp. 1–30.

Haynes, R.J. (2000). Labile organic matter as an indicator of organic matter quality in arable and pastoral soils in New Zealand. *Soil Biol. Biochem.* 32, 211–219.

He, X.-T., F.J. Stevenson, R.L. Mulvaney, and R.L. Kelly (1988). Incorporation of newly immobilized $^{15}$N into stable organic forms in soil. *Soil Biol. Biochem.* 20, 857–862.

Heal, O.W., J.M. Anderson, and M.J. Swift (1997). Plant litter quality and decomposition: A historical overview. In G. Cadisch, and K.E. Giller (Eds.), *Driven by Nature: Plant Litter Quality and Decomposition*, CAB International, Cambridge, U.K., pp. 3–30.

Herbert, B.E., and P.M. Bertsch (1995). Characterization of dissolved and colloidal organic matter in solutions: A review. In W.W. McFee and J.M. Kelly (Eds.), *Carbon Forms and Functions in Forest Soils*, Soil Science Society of America, Madison, WI, pp. 62–88.

Honeycutt, C.W., L.J. Potaro, K.L. Avila, and W.A. Halterman (1993). Residue quality, loading rate and soil temperature relations with hairy vetch (*Vicia villosa* Roth) residue carbon, nitrogen, and phosphorus mineralization. *Biol. Agric. Hortic.* 91, 181–191.

Hook, P.B., and I.C. Burk (2000). Biogeochemistry in a shortgrass landscape: Control by topography, soil texture, and microclimate. *Ecology* 81, 2686–2703.

Hsieh, Y.P. (1992). Pool size and mean age of stable soil organic carbon in cropland. *Soil Sci. Soc. Am. J.* 56, 460–464.

Huang, W., and W.J. Weber (1997). A distributed reactivity model for sorption by soils and sediments. 10. Relationships between desorption, hysteresis and the chemical characteristics of organic domains. *Environ. Sci. Technol.* 31, 2562–2569.

Islam, K., and R.R. Weil (1998). Microwave irradiation of soil for routine measurement of microbial biomass carbon. *Biol. Fertil. Soils* 27, 408–416.

Janzen, H.H., C.A. Campbell, S.A. Brandt, G.P. Lafond, and L. Townley-Smith (1992). Light fraction organic matter in soils from long-term crop rotations. *Soil Sci. Soc. Am. J.* 56, 1799–1806.

Jastrow, J.D., T.W. Boutton, and R.M. Miller (1996). Carbon dynamics of aggregate associated organic matter estimated by Carbon-13 natural abundance. *Soil Sci. Soc. Am. J.* 60, 801–897.

Jenkinson, D.S. (1990). The turnover of organic carbon and nitrogen in soil. *Philos. Trans. R. Soc. Lond.* 329, 361–368.

Jenkinson, D.S., P.B.S. Hart, J.H. Rayner, and L.C. Parry (1987). Modeling the turnover of organic matter in long-term experiments at Rothamsted. *INTECOL Bull.* 15, 1–8.

Johnson, M., W.L. Huang, Z. Dang, and W.J. Weber (1999). A distributed reactivity model for sorption by soils and sediments. 12. Effects of subcritical water extraction and alterations of soil organic matter on sorption equilibria. *Environ. Sci. Technol.* 33, 1657–1663.

Johnston, A.E. (1991). Fertility and soil organic matter. In W.S. Wilson (Ed.), *Advances in Soil Organic Matter Research: The Impact on Agriculture and the Environment*, The Royal Society of Chemistry, Melksham, Wiltshire, pp. 297–314.

Kaffka, S., and H.H. Koepf (1989). A case study on the nutrient regime in sustainable farming. *Biol. Agric. Hortic.* 6, 89–106.

Kaiser, K., and W. Zech (1998.). Soil dissolved organic matter sorption as influenced by organic and sesquioxide coatings and sorbed sulfate. *Soil Sci. Soc. Am. J.* 62, 129–136.

Kandeler, E., M. Stemmer, and E.-M. Klimanek (1999). Response of soil microbial biomass, urease and xylanase within particle size fractions to long-term soil management. *Soil Biol. Biochem.* 31, 261–273.

Kaschl, A., V. Romheld, and Y. Chen (2000). The influence of soluble organic matter from municipal solid waste compost on trace metal leaching in calcareous soils. *Sci. Tot. Environ.* 291, 45–57.

Kelly, R.H., L.C. Burke, and W.K. Lauenroth (1996). Soil organic matter and nutrient availability responses to reduced plant inputs in shortgrass steppe. *Ecology* 77, 2516–2527.

Kelly, K.R., and F.J. Stevenson (1985). Characterization and extractability of immobilized $^{15}N$ from soil microbial biomass. *Soil Biol. Biochem.* 17, 517–523.

Kerek, M., R.A. Drijber, M.J. Powers, R.C. Shearman, R.E. Gaussoin, and A.M. Streich (2002). Accumulation of microbial biomass within particulate organic matter of aging golf greens. *Agron. J.* 94, 455–461.

Kiem, R., H. Knicker, M. Korschens, and I. Kögel-Knabner (2000). Refractory organic carbon in C-depleted arable soils, as studied by $^{13}C$ NMR spectroscopy and carbohydrate analysis. *Org. Geochem.* 31, 655–668.

Kiem, R., H. Knicker, and I. Kögel-Knabner (2002). Refractory organic carbon in particle-size fractions of arable soils. I: Distribution of refractory carbon between the size fractions. *Org. Geochem.* 33, 1683–1697.

Kiem, R., and I. Kögel-Knabner (2003). Contribution of lignin and polysaccharides to the refractory carbon pool in C-depleted arable soils. *Soil Biol. Biochem.* 35, 101–118.

Kleineidam, S., H. Rugner, and P. Grathwohl (1999). Influence of petrographic composition/organic matter distribution of fluvial aquifer sediments on the sorption of hydrophobic contaminants. *Sediment. Geol.* 129, 311–325.

Koutika, L.S., S. Hauser, and J. Henrot (2001). Soil organic matter assessment in natural regrowth, *Pueraria phaseoloides* and *Mucuna pruriens* fallow. *Soil Biol. Biochem.* 33, 1095–1101.

Kramer, A.W., T.A. Doane, W.R. Horwath, and C. van Kessel (2002). Short-term nitrogen-15 recovery vs. long-term total soil N gains in conventional and alternative cropping systems. *Soil Biol. Biochem.* 34, 43–50.

Ladd, J.N., and M. Amato (1980). Studies of nitrogen immobilization and mineralization in calcareous soils. IV. Changes in the organic nitrogen of light and heavy subfractions of silt- and fine-clay-sized particles during nitrogen turnover. *Soil Biol. Biochem.* 12, 185–189.

Leavitt, S.W., R.F. Follett, and E.A. Paul (1996). Estimation of the slow and fast cycling soil organic pools from 6N HCl hydrolysis. *Radiocarbon* 38, 231–239.

Leboeuf, E.J., and W.J. Weber (1997). A distributed reactivity model for sorption by soils and sediments. 8. Sorbent organic domains: Discovery of humic acid glass transition and an argument for a polymer-based model. *Environ. Sci. Technol.* 31, 1697–1702.

Lehmann, J., N. Poidy, G. Schroth, G., and W. Zech (1998). Short-term effects of soil amendment with tree legume biomass on carbon and nitrogen in particle size separates in Central Togo. *Soil Biol. Biochem.* 30, 1545–1552.

Letey, J. (1991). The study of structure: Science or art. *Aust. J. Soil Sci.* 29, 699–707.

Li, J., and C. J. Werth (2001). Evaluating competitive sorption mechanisms of volatile organic compounds in soils and sediments using polymers and zeolites. *Environ. Sci. Technol.* 35, 568–574.

Liang, B.C., B.G. McConkey, J. Schoenau, D. Curtin. C.A. Campbell, A.P. Moulin, G.P. Lanford, S.A. Brandt, and H. Wang (2003). Effects of tillage and crop rotations on the light fraction organic carbon and carbon mineralization in Chermozemic soils of Saskatchewan. *Can. J. Soil Sci.* 83, 65–72.

Liebhardt, W.C., R.W. Andrews, M.N. Culik, R.R. Harwood, R.R., Janke, J.K. Radke, and S.L. Rieger-Schwartz (1989). Crop production during conversion from conventional to low-input methods. *Agron. J.* 81, 150–159.

Liebig, M., and J.W. Doran (1999). Impact of organic production practices on soil quality indicators. *J. Environ. Qual.* 28, 1601–1609.

MacCarthy, P. (2001). The principles of humic substances. *Soil Sci.* 166, 738–751.

Magdoff, F.R., D. Ross, and J. Amadon (1984). A soil test for nitrogen availability to corn. *Soil Sci. Soc. Am. J.* 48, 1301–1304.

Magid, J., A. Gorissen, and K.E. Giller (1996). In search of the elusive "active" fraction of soil organic matter: Three size-density fractionation methods for tracing the fate of homogeneously $^{14}C$-labelled plant materials. *Soil Biol. Biochem.* 28, 89–99.

Magid, J., L.S. Jensen, T. Mueller, and N.E. Nielsen (1997). Size-density fractionation for *in situ* measurements of rape straw decomposition: An alternative to the litterbag approach? *Soil Biol. Biochem.* 29, 1125–1133.

Magid, J., and C. Kjaergaard (2001). Recovering decomposing plant residues from the particulate soil organic matter fraction: Size versus density separation. *Biol. Fertil. Soils* 33, 252–257.

Mahieu, N., D.S. Powlson, and E.W. Randall (1999). Statistical analysis of published carbon-13 CPMAS NMR spectra of soil organic matter. *Soil Sci. Soc. Am. J.* 63, 307–319.

Malhi, S.S., J.T. Harapiak, M. Nyborg, K.S. Gill, K.S., Monreal, and E.G. Gregorich (2003a). Light fraction organic N, ammonium, nitrate and total N in a think Black Chernozemic soil under bromegrass after 27 annual applications of different N rates. *Nutr. Cycl. Agroecosyst.* 65, 201–210.

Malhi, S.S., J.T. Harapiak, M. Noyborg, K.S. Gill, C.M. Monreal, and E.G. Gregorich (2003b). Total and light fraction organic C in a thick Black Chermozemic grassland soil as affected by 27 annual applications of six rates of fertilizer N. *Nutr. Cycl. Agroecosyst.* 66, 33–41.

Martens, D.A., and W.T. Frankenberger, Jr. (1992). Decomposition of bacterial polymers in soil and their influence on structure. *Biol. Fertil. Soils* 13, 65–73.

Meijboom, F.W., J. Hassink, and M. Van Noordwijk (1995). Density fractionation of soil macroorganic matter using silica suspensions. *Soil Biol. Biochem.* 27, 1109–1111.

Michrina, B.P., R.H. Fox, and P. Piekielek (1982). Chemical characterization of two extracts used in determination of available soil nitrogen. *Plant Soil* 64, 331–341.

Miller, R.M., and J.D. Jastrow (1990). Hierarchy of root and mycorrhizal fungal interactions with soil aggregation. *Soil Biol. Biochem.* 22, 579–584.

Molloy, L.F., and T.W. Speir (1977). Studies on a climosequence of soils in tussock grasslands. 12. Constituents of the light fraction. *N.Z. J. Sci.* 20, 233–255.

Monaghan, R., and D. Barraclough (1995). Contributions to gross N mineralization from [15]N-labelled soil macroorganic matter fractions during laboratory incubation. *Soil Biol. Biochem.* 27, 1623–1628.

Motavalli, P.P., C.A. Palm, W.J. Parton, E.T. Elliott, and S.F. Frey (1994). Comparison of laboratory and modeling simulation methods for estimating soil carbon pools in tropical forest soils. *Soil Biol. Biochem.* 26, 935–944.

Mulvaney, R.L., S.A. Khan, R.G. Hoeft, and H.M. Brown (2001). A soil nitrogen fraction that reduces the need for nitrogen fertilization. *Soil Sci. Soc. Am. J.* 65, 1164–1172.

Mulvaney, R.L., and S.A. Khan (2001). Diffusion methods to determine different forms of organic nitrogen in soil hydrolyzates. *Soil Sci. Soc. Am. J.* 65, 1284–1292.

Nam, K., and M. Alexander (1998). Role of nanoporosity in sequestration and bioavailability: Tests with model solids. *Environ. Sci. Technol.* 32, 71–74.

Needelman, B.A., M.M. Wander, G.A. Bollero, C.W. Boast, G.K. Sims, and D.G. Bullock (1999). Interaction of tillage and soil texture: Biologically active soil organic matter in Illinois. *Soil Sci. Soc. Am. J.* 63, 1326–1334.

Needelman, B., M.M. Wander, and G.S. Shi (2001). Organic carbon extraction efficiency in chloroform fumigated and non-fumigated soils. *Soil Sci. Soc. Am. J.* 65, 1731–1733.

Nissen, T.M., and M.M. Wander (2003). Management and soil-quality effects on fertilizer-use efficiency and leaching. *Soil Sci. Soc. Am. J.* 67,1524–1532.

Noyborg, M., S.S. Malhi, E.D. Solberg, and A.C. Izaurralde (1999). Carbon storage in light fraction C in a grassland Dark Chernozem soil as influenced by N and S fertilization. *Can. J. Soil Sci.* 79, 317–320.

Odum, E.P. (1969). The strategy for ecosystem development. *Science* 164, 262–269.

Olk, D.C., K.G. Cassman, and T.W.M. Fan (1995). Characterization of two chemically extracted humic acid fractions from a calcareous verminculitic soil: Implication for the humification process. *Geoderma* 65, 195–208.

Parton, W. J., D.S. Schimel, C.V. Cole, and D.S. Ojima (1987). Analysis of factors controlling soil organic matter levels in Great Plains grasslands. *Soil Sci. Soc. Am. J.* 51, 1173–1179.

Parveen-Kumar, Tripathi., K.P., and R.K. Aggarwal (2002). Influence of crops, crop residues and manure on amino acid and amino sugar fractions of organic nitrogen in soil. *Biol. Fertil. Soils* 35, 210–213.

Paul, E.A. (1984). Dynamics of organic matter in soils. *Plant Soil* 76, 275–285.

Paul, E.A., and F.E. Clark (Eds.) (1996). *Soil Microbiology and Biochemistry*, 2nd ed., Academic Press, San Diego.

Paul, E.A., and H.P. Collins (1997). The characteristics of soil organic matter relative to nutrient cycling. In *Advances in Soil Science*. CRC Press, Boca Raton, FL, pp. 181–197.

Paul, E.A., and J.A. van Veen (1978). The use of tracers to determine the dynamic nature of organic matter. In *Transactions of the 11th International Congress of Soil Science*, 3, pp. 61–102.

Paul, E.A., R.F. Follett, S.W. Leavitt, A. Halvorson, G.A. Peterson, and D.J. Lyon (1997). Radiocarbon dating for determination of soil organic matter pool size and dynamics. *Soil Sci. Soc. Am. J.* 61, 1058–1067.

Paustian, K., G.I. Ågren, and E. Bosatta (1997). Modelling litter quality effects on decomposition and soil organic matter dynamics. In G. Cadisch and K.E. Giller (Eds.), *Driven by Nature: Plant Litter Quantity and Decomposition.* CAB International, Cambridge, U.K., pp. 313–335.

Phiri, S., E. Barrios, I.M. Rao, and B.R. Singh (2001). Changes in soil organic matter and phosphorus fractions under planted fallows and a crop rotation system on a Colombian volcanic-ash soil. *Plant Soil* 231, 211–223.

Piccolo, A. (2002). The supramolecular structure of humic substances: A novel understanding of humus chemistry and implications in soil science. In *Advances in Agronomy,* Vol. 75. Academic Press, New York, pp. 57–134.

Pignatello, J.J. (1989). Sorption dynamics of organic compounds in soils and sediments. In B.L. Sawhney and K. Brown (Eds.), *Reactions and Movements of Organic Chemicals in Soils,* Vol. 22 (special publ.). Soil Science Society of America, Madison, WI, pp. 45–80.

Pignatello, J.J. (1990). Slowly reversible sorption of aliphatic halocarbons in soils. II. Mechanistic aspects. *Environ. Toxicol. Chem.* 9, 1116–1126.

Plante, A.F., and W.B. McGill (2002). Intraseasonal soil macroaggregate dynamics in two contrasting field soils using labeled tracer spheres. *Soil Sci. Soc. Am. J.* 66, 1285–1295.

Poudel, D.D., W.R. Horwath, J.P. Mitchell, and S.R. Temple (2001). Impacts of cropping systems on soil nitrogen storage and loss. *Agric. Syst.* 68, 253–268.

Preston, C., and J.A. Trofymow (2000). Variability in litter quality and its relationship to litter decay in Canadian forests. *Can. J. Bot. — Revue Canadienne de Botanique* 78, 1269–1287.

Preston, C.M., and R.H. Newman (1992). Demonstration of spatial heterogeneity in the organic matter of de-ashed humin samples by solid state $^{13}$C CPMAS NMR. *Can. J. Soil Sci.* 72, 13–19.

Puget, P., and L.E. Drinkwater (2001). Short-term dynamics of root- and shoot-derived carbon from a leguminous green manure. *Soil Sci. Soc. Am. J.* 65, 771–779.

Puget, P., C. Chenu, and J. Baldesdent (1995). Total and young organic matter distributions in aggregates of silty cultivated soils. *Eur. J. Soil Sci.* 46. 449–459.

Puget, P., E.E. Besnard, and C. Chenu (1996). Particulate organic matter fractionation according to its location in soil aggregate structure. *C.R. Acad. Sci. Ser. II Sci. Terre Planetes* 322, 965–972.

Puget, P., C. Chenu, and J.B. Balesdent (2000). Dynamics of soil organic matter associated with particle-size fractions of water-stable aggregates. *Eur. J. Soil Sci.* 51, 595–605.

Quiroga, A.R., D.E. Buschiazzo, and N. Peinemann (1996). Soil organic matter particle size fractions in soils of the semiarid Argentinian pampas. *Soil Sci.* 161, 104–108.

Rasmussen, P.E., K.W. Goulding, J.R. Brown, P.R. Grace, H.H. Janzen, and M. Korschens (1998). Long-term agroecosystem experiments: Assessing agricultural sustainability and global change. *Science* 282, 893–896.

Reganold, J.P., L.F. Elliot, and Y. Unger (1987). Long-term effects of organic and conventional farming on soil erosion. *Nature* 330, 370–372.

Reid, J.B., M.L. Goss, and P.D. Robertson (1982). Relationship between the decrease in soil stability affected by the growth of maize roots and changes in organically bound iron and aluminum. *J. Soil Sci.* 33, 397.

Rice, J.A. (2001). Humin. *Soil Sci.* 166, 848–857.

Rice, J.A., and P. MacCarthy (1991). Statistical evaluation of the chemical composition of humic substances. *Org. Geochem.* 17, 635–648.

Rillig, M.C., P.W. Ramsy, S. Morris, and E.A. Paul (2003). Glomalin, an arbuscular-mycorrhizal fungal soil protein, responds to land-use change. *Plant Soil* 253, 293–299.

Robertson, E.B., S. Shlomo, and M.K. Firestone (1991). Cover crop management of polysaccharide-mediated aggregation in an orchard soil. *Soil Sci. Soc. Am. J.* 55, 734–739.

Rodionov, A., W. Amelung, I. Urusevskaja, and W. Zech (2001). Origin of the enriched labile fraction (ELF) in Russian Chernozems with different site history. *Geoderma* 102, 299–315.

Roscoe, R., and P. Buurman (2003). Tillage effects on soil organic matter in density fractions of a Cerrado Oxisol. *Soil Tillage Res.* 70, 107–119.

Rovira, P., P. Casals, J. Romanya, P. Bottner, M.M. Couteaux, and V.R. Vallejo (1998). Recovery of fresh debris of different sizes in density fractions of two contrasting soils. *Eur. J. Soil Biol.* 34, 31–37.

Rovira, P., and V.R. Vallejo (2003). Physical protection and biochemical quality of organic matter in Mediterranean calcareous forest soils: A density fractionation approach. *Soil Biol. Biochem.* 35, 245–261.

Ruffio, M.L., and G.A. Bollero (2003). Residue decomposition and prediction of carbon and nitrogen release rates based on biochemical fractions using principal-component regression. *Agron. J.* 95, 1034–1040.

Ruhlmann, J. (1999). A new approach to estimating the pool of stable organic matter in soil using data from long-term field experiments. *Plant Soil* 213, 149–160.

Russell, E.W. (1973). *Soil Conditions and Plant Growth*, 10th ed. Longman, New York.

Salas, A., E.T. Elliott, D.G. Westfall, C.V. Cole, and J. Six (2003). The role of particulate organic matter in phosphorus cycling. *Soil Sci. Soc. Am. J.* 67, 181–189.

Sbih, M., A. Ndayegamie, and A. Karam (2003). Evaluation of carbon and nitrogen mineralization rates in meadow soils from dairy farms under transit to biological cropping systems. *Can. J. Soil Sci.* 83, 25–33.

Schwenke, G., W.L. Felton, D. Herridge, D.F. Khan, and M.B. Peoples (2002). Relating particulate organic matter-nitrogen (POM-N) and non-POM-N with pulse crop residues, residue management and cereal N uptake. *Agronomie* 22, 777–787.

Scow, K.M., and J. Hutson (1992). Effect of diffusion and sorption on the kinetics of biodegradation: Theoretical considerations. *Soil Sci. Soc. Am. J.* 56, 119–127.

Shanmugananthan, R.T., and J.M. Oades (1983). Influence of anions on dispersion and physical properties of the A horizon of a red-brown earth. *Geoderma* 29, 257–277.

Sitompul, S.M., K. Hairiah, G. Cadisch, and M. Van Noordwijk (2000). Dynamics of density fractions of macro-organic matter after forest conversion to sugarcane and woodlots, accounted for in a modified Century model. *Netherl. J. Agric. Sci.* 48, 61–73.

Six, J., E.T. Elliott, K. Paustian, and J.W. Doran (1998). Aggregation and soil organic matter accumulation in cultivated and native grassland soils. *Soil Sci. Soc. Am. J.* 62, 1367–1377.

Six, J., E.T. Elliott, and K. Paustian (2000a). Soil macroaggregate turnover and microaggregate formation: A mechanism for C sequestration under no-tillage agriculture. *Soil Biol. Biochem.* 32, 2099–2103.

Six, J., R. Merckx, K. Kimpe, K., Paustian, and E.T. Elliott (2000b). A re-evaluation of the enriched labile soil organic matter fraction. *Eur. J. Soil Sci.* 51, 283–293.

Sohi, S.P., N. Mahieu, J.R.M. Arah, D.S. Powlson, B. Madari, and J.L. Gaunt (2001). A procedure for isolating soil organic matter fractions suitable for modeling. *Soil Sci. Soc. Am. J.* 65, 1121–1128.

Sollins, P., P. Homann, and B.A. Caldwell (1996). Stabilization and destabilization of soil organic matter: Mechanisms and controls. *Geoderma* 74, 65–105

Spycher, G., P. Sollins, and S. Rose (1983). Carbon and nitrogen in the light fraction of a forest soil: Vertical distribution and seasonal patterns. *Soil Sci.* 135, 79–87.

Steinberg, P.D., and M.C. Rillig (2003). Differential decomposition of arbuscular mycorrhizal fungal hyphae and glomalin. *Soil Biol. Biochem.* 35, 191–194.

Stenström, J., B. Stenberg, and M. Johannson (1998). Kinetics of substrate induced respiration (SIR) theory. *Ambio* 27, 35–39.

Stevenson, F., F.L. Walley, and C. van Kessel (1998). Direct vs. indirect nitrogen-15 approaches to estimate nitrogen contributions from crop residues. *Soil Sci. Soc. Am. J.* 62, 1327–1334.

Stevenson, F.J. (1982). *Humus Chemistry*, John Wiley & Sons, New York.

Stevenson, F.J. (1994). *Humus Chemistry, Genesis, Composition, Reactions*, 2nd ed., John Wiley & Sons, New York.

Stevenson, F.J., E.T. Elliott, C.V. Cole, J. Ingram, J.M. Oades, C. Preston, and P.J. Sollins (1989). Methodologies for assessing the quantity and quality of soil organic matter. In D.C. Coleman, J.M. Oades, and G. Urehara (Eds.), *Dynamics of Soil Organic Matter in Tropical Ecosystems*, NiFTAL, Honolulu, pp. 1–249.

Strickland, T.C., and P. Sollins (1987). Improved method for separating light and heavy-fraction organic material from soil. *Soil Sci. Soc. Am. J.* 15, 1390–1393.

Stone, A.G., S.J. Traina, and H.A.J. Hoitink (2001). Particulate organic matter composition and Pythium damping-off of cucumber. *Soil Sci. Soc. Am. J.* 65, 761–770.

Swanston, C.W., B.A. Caldwell, P.S. Hofmann, P.L. Ganio, and P. Sollins (2002). Carbon dynamics during a long-term incubation of separate and recombined density fractions from seven forest soils. *Soil Biol. Biochem.* 34, 1121–1130.

Swift, R.S. (1996). Organic matter characterization. In D.L. Sparks et al. (Eds.), *Methods of Soil Analysis. Part 3. Chemical Methods. I,* Soil Science Society of America, Madison, WI, pp. 1018–1020.

Tiessen, H., and J.W. Stewart (1983). Particle-size fractions and their use in studies of soil organic matter. II. Cultivation effects on organic matter composition in size fractions. *Soil Sci. Soc. Am. J.* 47, 509–514.

Tisdall, J.M. (1996). *Formation of Soil Aggregates and Accumulation of Soil Organic Matter,* CRC Press, Boca Raton, FL.

Tisdall, J.M., and J.M. Oades (1982). Organic matter and water stable aggregates in soils. *J. Soil Sci.* 33, 141–163.

Trumbore, S.E. (1993). Comparison of carbon dynamics in tropical and temperate soils using radiocarbon measurements. *Glob. Biogeochem. Cycl.* 7, 275–290.

Turchenek, L.W., and J.M. Oades (1979). Fractionation of organo-mineral complexes by sedimentation and density techniques. *Geoderma* 21, 311–343.

van Veen, J.A., and P.J. Kukiman (1990). Soil structural aspects of decomposition of organic matter. *Biogeochemistry* 11, 213–233.

van Veen, J.A., J.N. Ladd, and M.J. Frissel (1984). Modelling C and N turnover through the microbial biomass in soil. *Plant Soil* 76, 257–274.

Vanlauwe, B., L. Dendooven, and R. Merckx (1994). Residue fractionation and decomposition: The significance of the active fraction. *Plant Soil* 158, 263–274.

Vanotti, M.B., S.A. Leclerc, and L.G. Bundy (1995). Short-term effects of nitrogen-fertilization on soil organic nitrogen availability. *Soil Sci. Soc. Am. J.* 59, 1350–1359.

Vitorino, A., M.M. Ferreira, N. Curi, J.M. de Lima, M.l.N. Silva, and P.E.F. da Motta (2003). Mineralogy, chemistry and stability of silt-size aggregates of soils from the southeast region of Brazil. *Pesquisa Agropecuaria Brasileria* 38, 133–141.

Waksman, S.A. (1936). *Humus, Origin, Composition, and Importance in Nature*, Williams & Wilkins, Baltimore.

Wallerius, J.G. (1761). Agriculturae fundamenta chemica spez. Dissertation, Uppsala University, Sweden.

Wander, M.M., and M.G. Bidart (2000). Tillage practice influences on the physical protection, bioavailability and composition of particulate organic matter. *Biol. Fertil. Soils* 32, 360–367.

Wander, M.M., and G.A. Bollero (1999). Soil quality assessment of tillage impacts in Illinois. *Soil Sci. Soc. Am. J.* 63, 961–971.

Wander, M.M., and S.J. Traina (1996a). Organic matter fractions from organically and conventionally managed soils. I. Carbon and nitrogen distribution. *Soil Sci. Soc. Am. J.* 60, 1081–1087.

Wander, M.M., and S.J. Traina (1996b). Organic matter fractions from organically and conventionally managed soils. II. Characterization of composition. *Soil Sci. Soc. Am. J.* 60, 1087–1094.

Wander, M.M., and X.M. Yang (2000). Influence of tillage on the dynamics of loose- and occluded-particulate and humified organic matter fractions. *Soil Biol. Biochem.* 32, 1151–1160.

Wander, M.M., S.J. Traina, B.R. Stinner, and S.E. Peters (1994). Organic and conventional management effects on biologically active soil organic matter pools. *Soil Sci. Soc. Am. J.* 58, 1130–1139.

Wander, M.M., R.B. Dudley, S.J. Traina, D. Kaufman, B.R. Stinner, and G.K. Sims (1996). Acetate fate in organic and conventionally managed soils. *Soil Sci. Soc. Am. J.* 60, 1110–1116.

Wander, M.M., M.G. Bidart, and S. Aref (1998). Tillage impacts on depth distribution of total and particulate organic matter in three Illinois soils. *Soil Sci. Soc. Am. J.* 62, 1704–1711.

Wander, M.M., G. Walter, T.M. Nissen, G.A. Bollero, S.S. Andrews, and D. Cavanaugh-Grant (2002). Soil quality: Science and process. *Agron. J.* 96, 23–33.

Waring, S.A., and J.M. Bremner (1964). Ammonium production in soil under waterlogged conditions as an index of nitrogen availability. *Nature* 201, 951–952.

Waters, A.G., and J.M. Oades (1991). Organic matter in water stable aggregates. In W.S. Wilson (Ed.), *Advances in Soil Organic Matter Research: the Impacts on Agriculture and the Environment*. The Royal Society of Chemistry, Melksham, Wiltshire, U.K., pp. 163–174.

Weil, R.R., K.R. Islam, M.A. Stine, J.B. Gruver, and S.E. Samson-Liebig (2003). Estimating active carbon for soil quality assessment: A simplified method for lab and field use. *Am. J. Altern. Agric.* 18, 3–17.

Willson, T.C., E.A. Paul, and R.R. Harwood (2001). Biologically active soil organic matter fractions in sustainable cropping systems. *Appl. Soil Ecol.* 16, 63–76.

Woods, L.E. (1989). Active organic matter distribution in the surface 15 cm of undisturbed and cultivated soil. *Biol. Fertil. Soils* 8, 271–278.

Wright, S.F. (2000). A fluorescent antibody assay for hyphas and glomalin from arbuscular mycorrhizal fungi. *Plant Soil* 226, 171–177.

Wright, S.F., and A. Upadhyaya (1998). A survey of soils for aggregate stability and glomalin, a glycoprotein produced by hyphae of arbuscular mycorrhizal fungi. *Plant Soil* 198, 97–107.

Wright, S.F., M. Franke-Snyder, M.B. Morton, and A. Upadhyaya (1996). Time-course study and partial characterization of a protein on hyphae of arbuscular mycorrhizal fungi during active colonization of roots. *Plant Soil* 181, 193–203.

Wu, S., and P.M. Gschwend (1986). Sorption kinetics of hydrophobic organic compounds to natural sediments and soils. *Environ. Sci. Technol.* 20, 717–725.

Xing, B., and Z.Q. Chen (1999). Spectroscopic evidence for condensed domains in soil organic matter. *Soil Sci.* 164, 40–47.

Xing, B., and J.J. Pignatello (1996). Time-dependent isotherm shape of organic compounds in soil organic matter: Implications for sorption mechanisms. *Environ. Toxicol. Chem.* 15, 1282–1288.

Xing, B.S., and J.J. Pignatello (1997). Dual-mode sorption of low-polarity compounds in glassy poly(vinyl chloride) and soil organic matter. *Environ. Sci. Technol.* 31, 792–799.

Xu, Y.C., Q.R. Shen, and W. Ran (2003). Content and distribution of organic N in soil and particle size fractions after long-term fertilization. *Chemosphere* 50, 739–745.

Yakovchenko, V.P., L.J. Sikora, and P.D. Millner (1998). Carbon and nitrogen mineralization of added particulate and macroorganic matter. *Soil Biol. Biochem.* 30, 2139–2146.

Yang, X.M., and M.M. Wander (1999). Tillage effects on soil organic carbon distribution and estimation of C storage. *Soil Tillage Res.* 52, 1–9.

Young, I.M., J.W. Crawford, and C. Rappoldt (2001). New methods and models for characterising structural heterogeneity of soil. *Soil Tillage Res.* 61, 33–45.

Zhang, X.D., W. Amelung, Y.A. Ying, and W. Zech (1998). Amino sugar signature of particle-size fractions in soils of the native prairie as affected by climate. *Soil Sci.* 163, 220–229.

Zinke, P.J., A.G. Stangenberger, W.M. Post, W.R. Emanuel, and J.S. Olson (1984). Worldwide organic soil carbon and nitrogen data. Report No. ONRL/TM-8857, Oak Ridge National Laboratory, Oak Ridge, TN.

# 4 Stimulatory Effects of Humic Substances on Plant Growth

*Yona Chen, Maria De Nobili, and Tsila Aviad*

## CONTENTS

## INTRODUCTION

Stimulatory effects of humic substances (HS) on plant growth have been observed and extensively documented in many research and review articles (e.g., Chen and Aviad, 1990; Chen, 1996; Nardi et al., 1996, 2002; Cesco et al., 2002). Beneficial effects of HS on plant growth are usually exhibited by easily measurable parameters, such as leaf chlorophyll concentration, shoot and root fresh and dry weight, the number of root initials, and the number of flower buds (Rautham and Schnitzer, 1981; Chen, 1996). The explanation of these observations, however, has been uncertain, predominantly because of the lack of understanding of the plant growth promotion mechanism.

Studies on the effects of HS on plants were often conducted under conditions of deficient mineral supply. This could have resulted in a positive plant growth response directly related to the addition of macronutrients (N, P, K) with the HS solutions, which were not always purified before the plant growth trial. Because the essential role of major nutritional elements in plant growth is widely recognized, an effort has been made in this review to focus on the synergistic effects of HS and nutritional elements in the presence of complete, and considered to be optimal, nutrient supply.

A number of studies attributed a hormone-like activity to HS. This concept has been thoroughly reviewed by Chen and Aviad (1990), Chen et al. (1994, 2001), Chen (1996), and Cesco et al. (2000, 2002). In these reviews and research articles, the option of plant hormone involvement was investigated and basically negated. An alternative mechanism is discussed in detail in this chapter, along with a critical discussion of the hormone-like effects.

Effects on activity of various enzymes and on membrane permeability have also been suggested (Chen and Aviad, 1990; Pinton et al., 1992). The effects of HS on the activity of plant enzymes were extensively studied during the late 1970s and early 1980s (Vaughan and Malcolm, 1985). Most of these studies showed that HS exhibit an inhibitory effect *in vitro* on a number of enzymatic activities, either measured on homogenated tissues or on purified enzymes. Zaled and Butler (1971) showed that HA inhibited proteolytic enzymes such as carboxypeptidase A, chymotrypsin A, pronase and trypsin, but stimulated papain, ficin, and suptilo peptidase. No effect was observed on phaseolatin and tyrosinase. For proteolytic enzymes, it is obvious that HS can interact not only with the enzyme but also with the substrate, because the substrate itself is a protein, a peptide, or a peptide fragment. The nature of these interactions, however, remains largely unknown.

HS have an overall negative surface charge at all pH values, excluding the very acid pH range (<3) at which acidic functional groups are mostly protonated. Under such conditions, strong electrostatic interactions can be exhibited, provided the pH values are below the protein's isoelectric point and the overall charge of the protein is positive.

Electrostatic interactions are not the only possible mode of interaction between HS and proteins, and are most likely not very important. They actually play a secondary role in stabilizing the tertiary structure of proteins, although proteins possess both positively and negatively charged groups that could, in principle, form intramolecular electrostatic bonds (Tanford, 1961). In solution, protein molecules tend to expose to the solvent as many polar groups as possible in order to form hydrogen bonds with water molecules. Charged groups are therefore preferentially placed on the solvent-exposed protein surface. To form electrostatic bonds with HS, a protein has to break the great number of hydrogen bonds formed with the solvating water molecules, forcing their rearrangement. This process is not altogether an energetic preference.

In the presence of polyvalent cations, electrostatic bonds can be formed through the formation of cationic bridges between negatively charged functional groups of the humic molecule and similarly charged groups on the protein surface. HA can inhibit root surface enzymes such as invertase (Vanylon and Ord, 1978). The particular location of this enzyme facilitates a direct effect of the type and concentration of ions taken up by the roots on the enzyme–HS interaction, through acidification of the rhizosphere. Quite surprisingly, interactions can occur also in diluted solutions of monovalent cations and at pH values higher than the protein's isoelectric point, in spite of the repulsion occurring between two negatively charged molecules. These interactions result in a modification of the protein's tertiary structure. Figure 4.1 shows the modification caused by HS of the solvent perturbation on the UV spectrum of bovine serum albumin (BSA) in the region of absorbance of the aromatic amino acids (240 to 300 nm). The interaction of surface groups of BSA with solvent molecules of lower polarity in a binary solvent mixture (solvent perturbation effect) increases the intensity of adsorption in this region by inducing a rearrangement of the aromatic amino acids, which thereby become more exposed to the surface. Addition of HA to the BSA solution causes a conformational change, which again modifies the exposure of aromatic amino acids, bringing them back inside the folded macromolecule. HS therefore appear to act on the protein's tertiary structure. This is confirmed by the mechanism of activity suppression for most enzymes appearing to be that of noncompetitive inhibition.

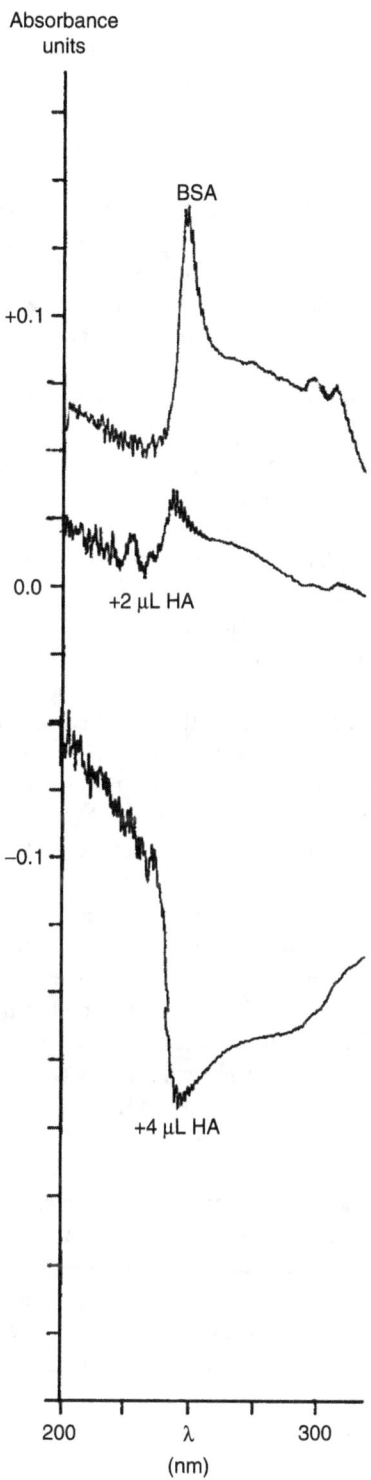

**FIGURE 4.1** Solvent perturbation effect on bovine serum albumin (BSA) caused by changing the solvent from water to an ethylenglycol:water mixture (20:80 w/w) and its modification after adding 2 and 4 mg L$^{-1}$ of humic substance (HS).

Three main mechanisms by which a compound can inhibit enzymatic activity have been identified: (1) the compound reacts like the substrate, competing with it for the active site (competitive inhibition); (2) a reaction with the enzyme that changes the configuration of the active site (incompetitive inhibition); and (3) an interaction with the enzyme that occurs in a part of the molecule other than the active site, which indirectly modifies its exposure or configuration (noncompetitive inhibition).

Kinetic analysis allows one to distinguish between these different mechanisms. Noncompetitive inhibition causes a decrease of the maximum velocity of the reaction, but does not affect the affinity of the enzyme for the substrate. This appears to be the mode of inhibition of HS (Figure 4.2). Besides directly affecting enzyme activity, it should also be remembered that HS can also act indirectly, in some cases, by interacting with the substrate and decreasing the free substrate concentration.

The notion that HS act as hormones in plants cannot be completely ruled out, but evidence provided in the literature for such activity (e.g., Piccolo et al., 1992; Nardi et al., 2002) is not convincing for the following reasons: (1) Hormone-like activity requires uptake into the plant tissue of the HS molecule. Evidence showing such uptake in whole-plant systems is very limited. Studies were usually conducted on cell cultures or excised roots (Nardi et al., 2002), which are not a good simulation of a whole plant. (2) $^{14}$C-uptake studies (Nardi et al., 2002) do not provide the required evidence, because degradation products can be the ones taken up. (3) Appropriate nutrient supply was often not provided to the tested plants. For example, Nardi and coworkers (Piccolo et al., 1992) provided an incomplete nutrient solution (NS) to the studied plants. Instead, they applied a 100 m$M$ KNO$_3$ solution (ca. 10 times more than in an average complete NS) and evaluated NO$_3$ uptake for as little as 4 d on seedlings. (4) Hormone-like activity was determined by this group, whereas the plants were not provided with a complete NS. Because Fe and Zn (and other metals) are of great importance to many physiological activities, it is essential to include these metals in the NS and examine the HS activity in their presence. (5) The extraction and separation procedures used by Nardi and coworkers (e.g., Piccolo et al., 1992; Nardi et al., 2002) vary among publications and even within publications, which creates difficulties in generalizing findings. For example, Piccolo et al. (1992) used 0.5 $M$ NaOH to extract three of the tested samples, but DMSO and acetone for the other two samples. (6) Nardi et al. (2002) used a low-molecular-weight LMS fraction in their study and reviews (e.g., Piccolo et al., 1992; Nardi et al., 2002). This fraction was obtained as the material dialyzed out of a dialysis bag (cut-off <3500 Da) after treatment with 3 to 4 $M$ acetic acid. This fraction could have included a significant quantity of nutritional elements (data not measured) and could have been contaminated with acetic acid, as its removal by vacuum distillation might not have been complete. In addition, many HAs and FAs were shown to have MW in the range of 900 to 3000 Da (Chen et al., 1977). Thus, this procedure is questionable.

The trend to stress the hormone-like effects of HS, which led researchers and commercial enterprises alike to promote HS preparations as alternatives to plant growth hormones rather than as complementary fertilizers, led Chen et al. (1994) to directly examine the presence of auxins, gibberellin, cytokinin, and abscissic acid in HS extracts. Techniques such as bioassays and GC-MS were employed in a search for plant hormones, but failed to provide evidence of their presence. However, a further search for these hormones continues (Nardi et al., 2002).

An alternative hypothesis relates the enhanced growth response of plants in the presence of HS to increased micronutrient availability, Fe and Zn in particular, either via complexing or through binding of colloidal conformations of the metal hydroxide to the HS. The available evidence on beneficial effects to plants of organo-Fe complexes of coal, lignite, and peat strongly supports this hypothesis, and more conclusive evidence is discussed in this chapter.

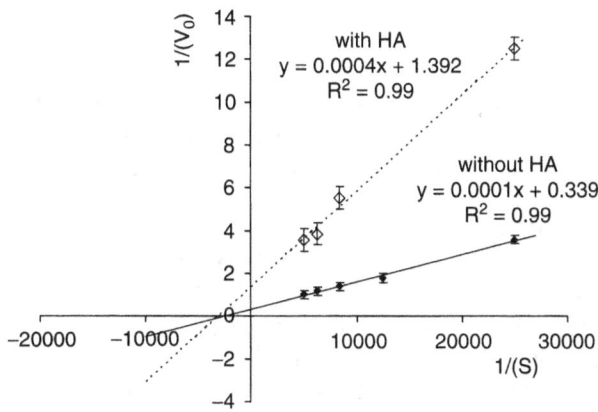

**FIGURE 4.2** Effect of humic substances (HS) on the kinetics of α-glucosidase (De Nobili, unpublished results). The reciprocal plot (reciprocal of the initial velocity $V_0$ of the reaction against the reciprocal of the substrate concentration $S$) demonstrates that HS decreases the $V_{max}$ but not the affinity of the enzyme for the substrate.

## INDIRECT AND DIRECT EFFECTS — AN ESSENTIAL DISTINCTION

### INDIRECT EFFECTS

Soil scientists and plant physiologists have generally accepted that plant growth and yield are largely determined by mineral nutrition, water, and air supply to the roots, and environmental conditions such as light and temperature. However, a number of studies suggest that soil organic matter (SOM) also affects plant growth. Good correlations between SOM content of soils and plant yield are extensively reported in the literature (e.g., Scharpf, 1967; Agboola, 1978; Lykov, 1978; Ojenhiyi and Agbede, 1980; Li et al., 1981; Pilus Zambi et al., 1982; Rebufetti and Labunora, 1982; Olsen, 1986).

It is well established that SOM can affect soil fertility indirectly by the following mechanisms: (1) supply of minerals, mostly N, P, K, and micronutrients to the roots; (2) improved soil structure, thereby improving water–air ratios in the rhizosphere; (3) increase in the soil microbial population, including beneficial microorganisms; (4) increase in the cation exchange capacity (CEC) and the pH buffering capacity of the soil; (5) supply of defined biochemical compounds to plant roots, such as acetamide and nucleic acids (Hutchinson and Miller, 1912); and (6) supply of HS, which serve as carriers of micronutrients or growth factors.

This chapter focuses on the effects of HS on plant growth. There is considerable overlap between the various mechanisms; however, the distinction between them can be difficult to establish.

### DIRECT EFFECTS

Direct effects are those that require uptake of organic macromolecules, such as HS, into the plant tissue, resulting in various biochemical effects at the cell wall, membrane, or in the cytoplasm. A prerequisite for these activities is the decay of the bulk organic matter into humic materials.

Because the soil system is usually too complicated for the direct effects of HS on plant growth to be studied, these effects have usually been studied by using extracted HS supplied to plants grown in NS, with application either to the foliage or to the roots by addition to the NS. Both these methods of application and their effects are discussed.

## SYNERGISTIC EFFECT OF NUTRITIONAL ELEMENTS AND HS
## ON PLANT GROWTH

Thousands of years ago, humans realized that dark-colored soils are more productive than light-colored ones, and that productivity is closely associated with decaying plant and animal residues. This observation later evolved into the belief that prevailed through the earlier decades of the 19th century, namely, that humus is the only or the major soil product supplying nutrients to plants (Thaer, 1808; Grandeau, 1872). This concept was referred to as the humus theory. Liebig, in a number of publications (e.g., 1841, 1856), supported the evidence against the humus theory and provided fundamental information on the role of minerals in plant nutrition.

Lawes and Gilbert (1905) working at Rothamsted, U.K., demonstrated after a long-term field study in the early 20th century that soil fertility could be maintained, for at least several years, by applying mineral fertilizers only (see also Russell, 1921, pp. 1–29). However, these experiments did not end the controversy between the humus and mineral theories. Scientists have realized that more exact experimentation is required to determine the benefit of humus to plant growth and to determine the possible synergistic effects of HS and minerals. From the same long-term experiments started more than 150 years ago at Rothamsted by Lawes and Gilbert, the positive effects of OM have again become evident (Figure 4.3) after the introduction of high-yielding cultivars, which are much more exigent for soil physical conditions and micronutrients supply (Johnston, 1994).

In the early 20th century, Bottomley (1914a, 1914b, 1917, 1920) published a series of papers in which he showed that HS enhanced the growth of various plant species in mineral NS. Bottomley believed that HS acted as plant growth hormones and called them auximones. Other investigators (Olsen, 1930; Burk et al., 1932) attributed the beneficial effects of HS on plant growth to the increased solubilization of some mineral ions such as Fe in either soluble or colloidal forms.

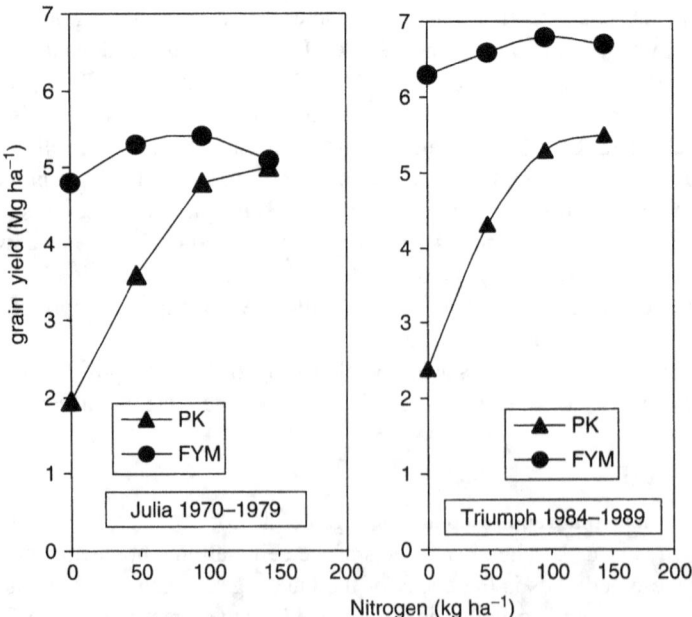

**FIGURE 4.3** Grain yields of two cultivars of spring barley grown on soils with different amounts of organic matter in the Hoosfield classical experiment at Rothamsted, U.K. Soils given farmyard manure (FYM) since 1852 at present contain 4.6% organic matter, and those given mineral fertilizers (NPK) contain 1.8% organic matter. (Modified from Johnston, A.E. 1994. In R.A. Leigh and A.E. Johnston (Eds.), *Long-Term Experiments in Agricultural and Ecological Sciences.* CAB International, Wallingford, U.K., pp. 9–52. With permission.)

## SOURCE AND NATURE OF SOIL AND WATER-BORNE
## ORGANIC MATTER

As organic materials in the soil decay, macromolecules of a mixed aliphatic and aromatic nature are formed. The term *humus* is often synonymous for SOM (e.g., Stevenson, 1994). Some scientists refer to soil humic materials as natural organic matter (NOM or DOM, the latter referring to water-borne or dissolved organic matter, DOM). Others distinguish between the total organic matter in a soil (SOM) and humus. The latter includes the total of NOM in soil and excludes identifiable plant and animal tissues and living biomass. HS, including humic acids (HAs), fulvic acids (FAs) and humin, comprise soil humus. Defined biochemicals are not considered to be a component of the various HS fractions.

The chemical and colloidal properties of NOM can be properly studied only in the free state, i.e., when freed of inorganic components. Thus, a researcher must first separate NOM (and SOM, in particular) from the inorganic matrix. Alkali, usually 0.1 to 0.5 $M$ NaOH, has been a popular extractant of SOM. Various researchers have also used other methods of extraction, but because this subject is beyond the objectives of this chapter, extraction procedures will not be described. A detailed discussion of the extraction procedures can be found elsewhere (Stevenson, 1994; Swift, 1996).

## ORGANIC MATTER COMPLEXES WITH METALS: NATURE,
## PROPERTIES, AND EFFECTS

A complex of a metal ion and organic molecule arises when water molecules surrounding the metal ion are replaced by other molecules or ions, with the formation of a coordination compound. The organic molecule that combines with the metal ion is commonly referred to as the ligand. Examples of groupings in organic compounds that have unshared pairs of electrons and that can form coordinate linkages with metal ions are in the order of decreasing affinity of organic groupings for metal ions as follows (Stevenson, 1994):

| $-O->$ | $-NH_2>$ | $-N=N>$ | $-N>$ | $COO^->$ | $-O->$ | $C=O$ |
|--------|----------|---------|-------|----------|--------|-------|
| enolate | amine | azo | ring N | carboxylate | ether | carbonyl |

If the chelating agent forms two bonds with the metal ion, it is called bidentate; similarly, there are terdentate, tetradentate, pentadentate, and hexadentate complexes. The formation of more than one bond between the metal and the organic molecule usually results in high stability of the complex. The stability of a metal–chelate complex is determined by such factors as the number of atoms that form a bond with the metal ion, the number of rings that are formed, the nature and concentrations of the metal ions, and pH. The stability sequence for some complexes of HS with divalent cations is as follows:

$$Cu^{2+} > Ni^{2+} + Co^{2+} > Zn^{2+} > Fe^{2+} > Mn^{2+}$$

Detailed reviews of various modeling approaches to binding of metals to HS have been published (Stevenson, 1994) and are not described here. Stability constants for various metal ions with HS were presented by Chen and Stevenson (1986), but those for $Fe^{3+}$ were reported in the literature to a very limited extent and only for very low pH (<3.0). The new information on Fe complexes with HS is reported next. The formation of metal–organic complexes has the following effects on the micronutrient cycle in soils (Stevenson and Chen, 1986; Stevenson, 1994):

1. Micronutrient cations that would ordinarily precipitate at the pH values found in most soils are maintained in solution through complexation with soluble organics. Many biochemicals synthesized by microorganisms form water-soluble complexes with trace

elements. Complexes of trace elements with FA are also water soluble. Complexes of HA with trace metals such as Fe are likely to form colloids that will remain in solution or form suspended particles. Either conformation will contribute to the mobilization of the metal to the plant root.

2. Under certain conditions, metal ion concentrations can be reduced to a nontoxic level through complexation with SOM. This is particularly true when the metal–organic complex has low solubility, such as in the case of complexes with HAs and other high-molecular-weight components of humus.
3. Various complexing agents mediate transport of trace elements to plant roots, and, in some cases, to other ecosystems such as lakes and streams.
4. Organic substances can enhance the availability of insoluble phosphates through complexation of Fe and Al in acid soils and Ca in calcareous soils.
5. Chelation plays a major role in the weathering of rocks and minerals. Lichens, for example, enhance the disintegration of rock surfaces to which they are attached by producing chelating agents.

It was recently demonstrated (Leita et al., 2001; De Nobili et al., 2002; Catalano et al., 2002) that HS interact in solution not only with metal ions but also with free ligands and complexes of high stability without causing any ligand exchange. The contribution of HS to the migration of micronutrients and toxic elements might therefore be much more complex than previously considered.

## ORGANICALLY BOUND SPECIES OF MICRONUTRIENT CATIONS

The micronutrient cation pools in soils can be divided into the following (Stevenson, 1991): (1) water soluble (free plus complexed); (2) suspended organo-metal complexes; (3) exchangeable; (4) specifically adsorbed; (5) organically complexed, but water insoluble; (6) insoluble inorganic precipitates; and (7) held in the mineral fraction of soils. The importance of the organically complexed pool arises from findings indicating that organically bound forms of micronutrient cations are more available to plants than are inorganic forms of pools of insoluble inorganic precipitates and those held in primary minerals (McLaren and Crawford, 1973; Murthy, 1982; Mandal and Mandal, 1986). Organically bound micronutrient cations are commonly determined by soil extraction with a complexing agent (e.g., 0.1 $M$ pyrophosphate) or by release to exchangeable forms via OM oxidation (Shuman, 1983, 1985).

In a study conducted by McLaren and Crawford (1973), in 24 soils of diverse types, 20 to 50% of the Cu occurred in organically bound species. They concluded that the amount of Cu available to plants (exchangeable and soluble Cu) was controlled by equilibria involving specifically adsorbed forms (Cu extracted with 2.5% $CH_3COOH$) and the organically bound fraction. The suggested relationship among the three forms was as follows:

Exchangeable and soluble Cu $\Leftrightarrow$ Specifically adsorbed Cu $\Leftrightarrow$ Organically bound Cu

Shuman (1979) obtained the following percentages of three micronutrients that occurred in organically bound species in 10 representative soils of southeastern U.S.: Cu, 1.0–68.6%; Mn, 9.5–82%; and Zn, 0.2–14.3%. Iyengar et al. (1981) obtained a somewhat similar range for Zn (0.1 to 7.4%) for 19 soils from the Appalachian, Coastal Plain, and Piedmont regions of Virginia. Sedberry and Reddy (1976) found that on an average, 2.6% of the Zn in 10 Louisiana soils was organically combined. Considerable variation exists in the distribution of organically bound forms of micronutrients among the various size fractions of soil and among soils differing in pH and other chemical properties (Shuman, 1979).

Boron (B), molybdenum (Mo), and a few others are unique among the micronutrient elements in that they normally occur in anionic forms (e.g., $H_2BO_3^-$ and $MoO_4^{2-}$) and are therefore subject to losses through leaching. However, the main form of B (the only nonmetal in the group) can be combined with OM primarily as borate complexes with compounds that contain the *cis*-hydroxyl group, such as carbohydrates (Yermihayu et al., 1988, 1995).

As SOM is mineralized by microorganisms, B is released to readily available forms. Temporary B deficiency in plants during periods of drought has been attributed to reduced mineralization of B in organically bound forms. Studies by Chen's research group (Yermiyahu et al., 1988, 1995) have shown that the sorption capacity of composted OM for B (on a weight basis) was at least four times that for clay and soils. This was attributed to a chemical association between B and organic molecules, such as those mentioned previously. Our group showed that these binding reactions play a major role in the protection of plants from excessive concentrations of B, which often occur in soils irrigated with reclaimed wastewater. This is one of the prominent and significant detoxifying beneficial effects of HS on plant growth.

Conversion of soluble forms of the micronutrient cations into insoluble organic forms can occur through solid-phase complexation by humates present as coatings on clay surfaces, as well as by the formation of soluble complexes that subsequently become associated with mineral surfaces through adsorption. Some polyvalent cations link humic complexes to clay surfaces; others occupy peripheral sites and are available for exchange with ligands of the soil solution.

## TRANSPORT OF MICRONUTRIENTS AND UPTAKE BY PLANT ROOTS

Plant uptake of elements from the soil solution initially requires positional availability to the plant root. Either the elements must be mobilized or the root must grow toward mineral surfaces containing the element (root interception). The element must then occur in a form that can be transported into the plant via an uptake mechanism. This transfer requires that the element move through a solution phase; thus, water solubility and a variety of complexation, chelation, and other chemical reactions become important. It is likely that micronutrients are predominantly transported to plant roots as soluble chelate complexes. Therefore, chelation in the vicinity of the root plays an important role in their availability.

The transport of micronutrients in general and Fe in particular from the solid phase of the soil to plant roots was studied and elaborated by Lindsay (1974) and O'Connor et al. (1971), who showed that chelated species dominate the transport. Because Fe deficiencies are more widespread than those of Mn, Zn, and Cu, this micronutrient is discussed in greater detail, with an emphasis on both the transport problem and the significance of HS in facilitating the movement of micronutrients to plant roots.

Compared with the macronutrients in general, the Fe content of green plant tissue is low, generally 80 to 150 $\mu g \ g^{-1}$ on a dry-weight basis. Somewhat lower levels are found in cereal grains, tubers, and roots. A soil containing 0.5 $\mu g$ available Fe $g^{-1}$ throughout the plough layer will contain sufficient Fe to meet the requirements of most agricultural plants (Lindsay, 1974). Total Fe levels in soil are substantially higher, usually 2%, or 20,000 $\mu g \ g^{-1}$. Thus, the supply of Fe to plants is dependent on both availability and transport. The solubility of inorganic Fe in soil is highly pH dependent. For sufficient Fe to be transported to roots through mass flow, the total solubility of Fe must be at least $10^{-8} \ M$ (Lindsay and Schwab, 1982; Chen and Barak, 1982), a level that is achieved at pH 3 only. By raising the pH to over 4, only 1% of the Fe demand can be met. At common soil pH levels (6 to 8.5), even with allowance for the contribution of diffusion, the concentration of inorganic Fe in the soil solution is far below that required to meet the Fe requirements of plants. It appears therefore that soluble organic complexes of Fe must play an important role in supplying

Fe to plants. These soluble organic compounds consist of root exudates, HS, metabolites of microorganisms, or applied Fe-chelate fertilizers (Moghimi et al., 1978; Chen, 1996).

Two basic mechanisms — diffusion and mass flow (convection) — can operate in the movement of micronutrients to plant roots or ground water. Seldom has there been a clear division between the two processes; both occur to some degree whenever micronutrients are transported within a soil system. The dominant mechanism depends on the rate and direction of water movement, the micronutrient involved, plant species, and environmental conditions surrounding the plant root, and, in particular, the presence of HS.

Ellis et al. (1983) have reviewed the movement of micronutrients to plant roots through diffusion and mass flow. Therefore, we only briefly discuss some major points related to the two mechanisms.

## DIFFUSION

The activity gradient of ions is the driving force for the net transfer of ions and molecules from regions of high to low activity. This transfer is called diffusion and is described by Fick's law for steady-state diffusion in pure liquid (concentration terms) as follows:

$$\delta Q / \delta t = -DA(\delta C / \delta X)$$

where $Q$ is the quantity of nutrient diffusing across a unit cross-sectional area $A$ in time $t$, D is the diffusion coefficient, $\delta Q / \delta X$ is the activity or concentration gradient, $X$ is the distance in the direction of net movement, and $C$ is the concentration of ions in the bulk solution. As $\delta C / \delta X$ approaches zero, the rate of movement of ions approaches equality in all directions and net diffusion approaches zero. Values of $D$ for the various micronutrients in water were measured or calculated and summarized by Ellis et al. (1983). In general, $D_o$ (the diffusion coefficient in water) for Fe, Mn, Zn, and Cu varies from $10^{-6}$ to $10^{-5}$. Apparent diffusion coefficients ($D_e$) in soils are much lower.

Most work on diffusion rates ($D_e$) of micronutrients in soils has focused on Zn. Several factors affect diffusion, the most important being the moisture content of the soil. An increase in soil moisture content tends to increase $D_e$ by making the paths less tortuous. Bulk density of a soil is related to diffusion via its effect on soil tension and moisture content. Warnacke and Barber (1972) have shown that $D_e$, in general, increases with increasing bulk density. Soil pH also alters the $D_e$ of Zn (Clark and Graham, 1968; O'Connor et al., 1971). Specifically, $D_e$ increases with a decrease in pH below 7.0. The effects of pH might be ion specific and also influenced by soil type. Values of $D_e$ for Zn are also concentration dependent (Nye, 1966; Warnacke and Barber, 1973). Diffusion of cations such as Fe, Mn, Zn, and Cu is also affected by the presence of other ions. To maintain electroneutrality, either co-diffusion (of ions of opposite charge) or counter-diffusion (of ions of the same charge but in the opposite direction) must take place (Low, 1962).

Factors that increase the concentration of a micronutrient in the soil solution should increase its diffusion rate. Accordingly, the presence of chelates and naturally occurring or synthetic organic complexes should increase the $D_e$ of the micronutrient. For example, Lindsay (1974) concluded that diffusion as a chelate complex was a major mechanism for the transport and uptake of Fe by plants.

## MASS FLOW (CONVECTION, MISCIBLE DISPLACEMENT)

The fundamental principle of the movement of micronutrients (and other soluble ions, molecules, and colloidally dispersed particles) through mass flow is simple and obvious: whenever water moves in the soil because of potential gradients, it carries the micronutrients that are in solution with it. Thus, micronutrients as ions or in the form of complexes or chelates can be transported to plant roots by mass flow.

In general, Fe, Mn, Zn, and Cu are considered immobile in soil because they readily form precipitates and are strongly bound to clay surfaces by ion-exchange reactions. Autoradiography (Wilkinson et al., 1968) and thin slicing techniques (Kuchenbuk and Jungk, 1982) were used to differentiate between transport mechanisms of ions to plant roots. If the autoradiograph or thin section shows depletion of a nutrient around the root, absorption is assumed to have exceeded the rate at which the nutrient has been transported through mass flow; thus, diffusion is the major mechanism of ion movement. Accumulation of the micronutrient at the root surface indicates that movement to the root has occurred at a faster rate than absorption; in this case, mass flow is assumed to be the major transport mechanism.

## STIMULATORY EFFECTS OF SOIL-DERIVED HS ON PLANT GROWTH

The beneficial effects of HS on both root and shoot growth under conditions of complete (optimal) NS supply are shown in Table 4.1. The data clearly show growth enhancement compared with pure water by additions of HA (50 mg $L^{-1}$) and further stimulation of growth in Hoagland's solution. The stimulation in the presence of HA exceeded that by Hoagland's solution alone by ca. 25%, which provides evidence for a synergistic effect of combined applications of mineral nutrition and HS.

Along with the actions described in the previous sections, many studies show that HS can stimulate the growth of plant tissues and the total quantity of absorbed nutrients (Vaughan and Malcolm, 1985; Chen and Aviad, 1990). Rauthan and Schnitzer (1981) reported that the yield of cucumber plants grown for 6 weeks in a NS was significantly enhanced as a response to the addition of increasing concentrations of FA (Figure 4.4).

Concomitantly, an increase in the uptake of N, P, K, Ca, Mg, Cu, Fe, and Zn was observed in the shoots. Likewise, Tan and Nopamornbodi (1979) showed that shoots of maize plants grown for 16 d in a NS to which 640 mg $L^{-1}$ of HA was added tripled their dry weight and the uptake of N, P, K, Ca, Mg, Cu, Fe, Zn, and Mn. Other authors reported a much smaller effect or none at all (Mylonas and McCants, 1980; Vaughan and Malcolm, 1985). Differences in the plant organs affected were shown by Dormaar (1975) in *Festuca scabrella* and by Ernst et al. (1987) in *Scrophularia nodosa*. The variability of results can be attributed to the variable sources of HS, because the purification processes were usually overlooked (Lee and Bartlett, 1976) before standard extraction procedures were introduced by the International Humic Substances Society (IHSS).

The effects of HS on growth and nutrition were observed not only in hydroponics but also when plants were grown on sand (Levesque, 1970) or soil, especially those with low OM contents (Lee and Barlett, 1976; Fagbenro and Agboola, 1993), and in callus *in vitro* cultures (Irintoto et al., 1993).

Effects on roots and shoots differ. Root growth was enhanced to a greater extent than shoot growth. Various plant organs were found to respond differently to the presence of HS in the NS (Chen and Aviad, 1990). A summary of the effects of HS on seeds, roots, and shoots of higher plants and relevant concentrations of the HS is presented in Table 4.2.

A distinctive drawback of many published reports on the effects of HS on plant growth is that researchers paid little or no attention to the concentration of the HS in the solution. Most of them employed a single concentration in their studies. A close examination of the data presented by Rauthan and Schnitzer (1981) showed, however, an optimum curve reaching a maximum at a range of concentration of 100 to 300 mg $L^{-1}$ HS in the NS for both roots and shoots (Chen and Aviad, 1990; Figure 4.4). Concentration ranges as well as optimum levels vary between reports and also in relation to the plant organ studied (Table 4.2). Two possible hypotheses can explain the contrasting effects of HS at different concentrations. The first is that by increasing HS concentration, the solution equilibrium shifts toward the formation of HS–metal complexes of higher stability, thus reducing the availability of the micronutrient. This hypothesis, however, is not confirmed by data on stability constants of HS complexes with Fe or Cu, which are much lower than the stability constants of EDTA or EDDHA, the most commonly used chelates in NS cultures. HS can, however,

**TABLE 4.1**
**Effect of 50 mg L⁻¹ Humic Acid (HA) on Growth of Wheat in Water or Hoagland's Nutrient Solution**

| Culture Medium | Plant Organ | Fresh Weight (mg/plant) | Stimulation (%) |
|---|---|---|---|
| Water | Root | 93 | 0 |
| | Shoot | 185 | 0 |
| Water + HA | Root | 146 | 57.5 |
| | Shoot | 252 | 36.2 |
| Hoagland | Root | 182 | 96.3 |
| | Shoot | 342 | 84.9 |
| Hoagland + HA | Root | 203 | 119.0 |
| | Shoot | 390 | 110.8 |

*Source:* Data from Vaughan D., and R.E. Malcolm. 1985. In D. Vaughan et al. (Eds.), *Soil Organic Matter and Biological Activity.* Martinus Nijhoff, Dordrecht, pp. 37–75.

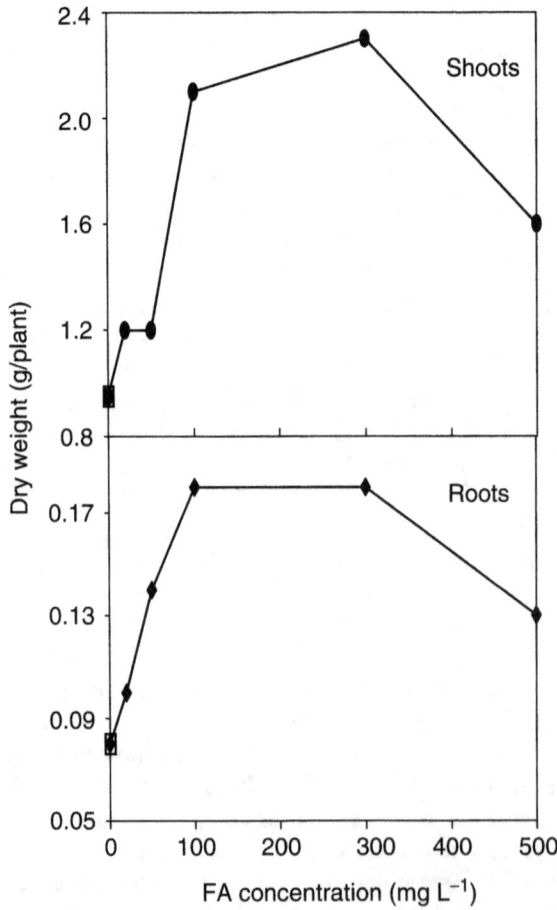

**FIGURE 4.4** Influence of fulvic acid concentration on dry weight of cucumber (*Cucumis sativus* L.) shoots and roots. (From Chen, Y., H. Magen, and J. Riov. 1994. In N. Senesi et al. (Eds.), *Human Substances in the Global Environment and Implication on Human Health.* Elsevier, Amsterdam. With permission.)

**TABLE 4.2**
**Effects of Humic Substances (HS) on Seeds, Roots, and Shoots of Plants**

| Plant Organ Response | Material | Concentration Range (mg $l^{-1}$) | Effect |
|---|---|---|---|
| Seed germination | HA | 0–100 | Enhanced rate; accelerated water uptake |
| Root initiation and elongation | HA, FA | 50–300 | Stimulated root initiation; lateral roots development |
| Excised root elongation | HA | 5–25 | Enhanced growth; cell elongation |
| Intact plant growth | HA, FA | 0–500; optimum 50–300 | Enhanced growth of shoots and roots |

*Note:* Summary from various sources; first shown by Chen and Aviad (1990).

interact with chelates without causing a ligand exchange of the metal ion between the chelate and HS molecules. This interaction is suggested here as a second hypothesis to explain the decreasing effect on high concentrations. It was shown in our laboratories (De Nobili et al., 2002) by UV-Vis absorbance studies that for a number of Fe chelates at concentrations of more than 100 mg $L^{-1}$ (Figure 4.5), the absorbance of the second charge transfer band ($\lambda = 280$ nm for EDDHA) starts to decrease in response to further additions of HS and the absorbance is fully suppressed at ca. 300 mg $L^{-1}$. The differential spectrophotometric titration of EDDHA and other Fe complexes with HS is therefore in accordance with the concentration effects found by Rauthan and Schnitzer (1981), Chen et al. (1994), and Chen (1996). This suggests that HS–chelate interactions are the cause of the lower availability of chelated nutrients to plants at high HS concentrations. Negative effects of this kind can hardly be expected to be observed for plants grown in soil, because the solubility of HS in the soil solution is generally in the range at which positive effects can be observed. Additions of composts rich in DOM could, however, act in this direction as shown for Cd uptake and other heavy metals.

In conclusion, the studies discussed in this section indicate that FAs or HAs, or both, can stimulate shoot growth of various plants when applied either as foliar spray at concentrations of 50 to 300 mg $L^{-1}$ or when applied in the NS at similar concentrations. The stimulatory effect on shoot growth usually correlates to root response, regardless of the mode of application.

## ORGANO-Fe COMPLEXES AND THEIR ROLE IN PLANT GROWTH ENHANCEMENT

### IRON NUTRITION OF PLANTS

The formation of HS complexes with trace elements of importance to plant nutrition, such as Fe, Zn, Mn, and Cu, has been the focus of many publications. The basic concepts are comprehensively described by Stevenson (1994). Reviews on the relationship between these complexes and plant growth have been published by Chen and Stevenson (1986) and Chen (1996).

Solubilization of micronutrients from their inorganic forms can be the major factor in the promotion of plant growth in soils by HS. The same situation can apply to a NS in which solubility of most of the micronutrients is limited. The presence of HS in either a nutrient or soil solution will most likely contribute to improve the availability of metal elements.

Iron concentration in soil solution of most aerated soils is expected to be far below that needed by healthy plants, based on chemical equilibria of Fe minerals (Lindsay, 1974; Marschner et al., 1986a; Kochian, 1991). However, direct measurements show significantly higher levels of Fe because

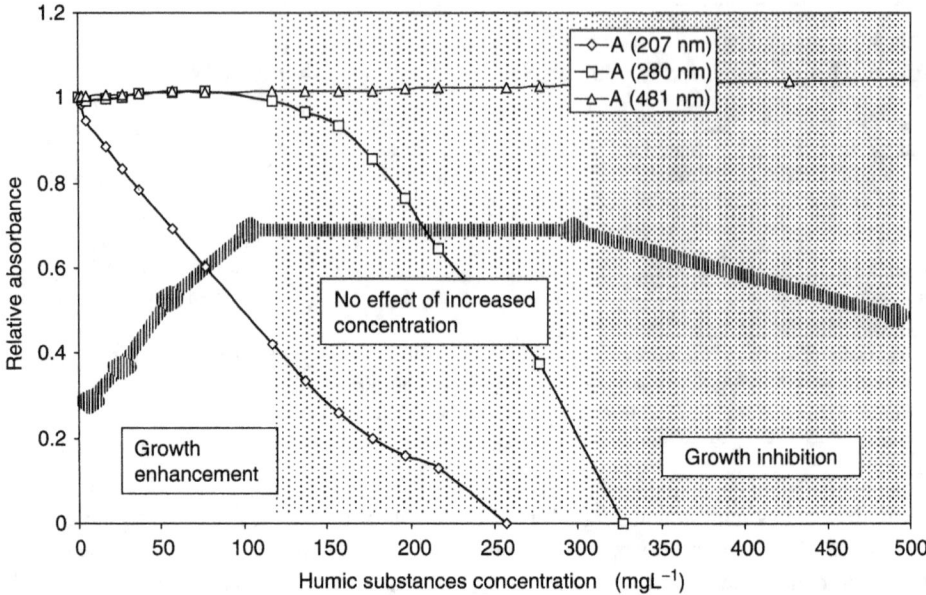

**FIGURE 4.5** Differential spectrophotometric tritration of Fe EDDHA by humic substances (HS). The etched line represents the trend of shoots yield obtained by Rauthan, B.S., and M. Schnitzer. 1981. *Plant Soil* 63:491–495 for cucumber plants grown in nutrient solution (NS) at different fulvic acid (FA) concentrations. (Modified from Catalano, L. et al. 2002. In *Proceedings of the International Humic Substances Society Twentieth Anniversary Conference.* Northeastern University, Boston, MA, pp. 268–271.)

of complexes formed with organic ligands (Lindsay, 1991). These complexing agents include DOM, organic acids, microbial siderophores, and phytosiderophores, as well as the fraction of humified OM (Stevenson, 1991). The involvement of such compounds in the Fe nutrition of plants has been clearly demonstrated (Barak and Chen, 1982; Marschner et al., 1986b; Shenker et al., 1992; Wang et al., 1993; Yehuda et al., 1996). To cope with Fe shortage during growth, the plant kingdom has developed two main strategies (Marschner and Römheld, 1994): (1) Fe-deficient dicotyledonous plants increase their capacity of acidifying the rhizosphere, develop an enhanced level of a reductase enzyme on the root surface, and release reductants to the soil solution. The reduced Fe is then transported through the cell membranes. (2) Monocotyledonous plants of the Gramineae family excrete phytosiderophores to the rhizosphere; these compounds complex soil Fe, and the complex then transfers its Fe to transporters on the root membranes (Marschner and Römheld, 1994).

Based on these mechanisms, soluble complexes between Fe and HS might conceivably contribute to the Fe nutrition of plants by acting as mobilizers and transporters of Fe in the soil or NS.

Because of its scarcity in soil of arid and semiarid areas, Fe has drawn the attention of researchers more than any other microelement has. As early as the mid-1950s, Dekock (1955) concluded that lignite-derived HS enhanced Fe uptake by plants even at high phosphate concentrations. HS not only increased the solubility of Fe in solution but also affected Fe translocation from roots to shoots (Dekock, 1955). Untreated chlorotic plants contained high Fe concentration in their roots, probably because of precipitation as ferric phosphate. Aso and Sakai (1963) found that in NS at pH 7, rice (*Oryza sativa* L.) and barley were chlorotic unless ammonium nitro-HA ($NH_4$-NHA) was added. Addition of Fe–HS complexes significantly reduced chlorosis severity, whereas unferrated HS alone was ineffective. Dyakonova and Maksimova (1967) reported that soluble FeHA complexes occurred in natural peat and prevented chlorosis in plants. Lee and Bartlett (1976) found that 5 mg L⁻¹ Na-humate in a NS enhanced yield of corn plants as well as increased the Fe concentration in roots and shoots. Linehan and Shepherd (1979) compared effects of FA to those of polymaleic acid and other polycarboxylates. Addition of FA to NS at concentrations up to 25 mg L$^{-1}$ enhanced Fe uptake to shoots of wheat seedlings.

In a number of publications, Chen and collaborators (Barak and Chen, 1982; Chen et al., 1982a, 1982b; Chen and Barak, 1983; Bar-Ness and Chen, 1991a, 1991b) have shown that Fe-enriched organic materials such as peat or manure can serve as a remedy for lime-induced chlorosis in soils. The corrective effect was attributed to complexation of Fe by HS in the organic materials. Pinton et al. (1999) investigated the ability of Fe-deficient plants to use Fe bound by the water-extractable (from peat) HS fraction. Studies conducted by our group (Amichai, 2001) have shown that various fractions of DOM extracted, isolated, and purified from compost (see later) bind Fe, maintain it in the NS, and provide this essential element to plants.

In addition to Fe, the effect of HS on the uptake of Zn and Cu was investigated on both intact plants and on plant tissues. Vaughan and McDonald (1976) studied Zn uptake by beet root tissue cut into discs and found that the addition of HA slightly inhibited Zn uptake by aged discs when concentrations exceeded 25 mg L$^{-1}$ HA. Lower concentrations did not affect Zn uptake. Jalali and Takkar (1979) reported that on intact plants, Fe, Cu, and Zn uptake by rice plants was enhanced by increased levels of SOM.

Soil humus can reduce plant uptake of some metal ions by strong adsorption from solution. White and Chaney (1980) followed the uptake of Zn, Cd, and Mn in two soils amended with Zn and Cd. The soils contained 1.2% and 3.8% OM [Sassafras (Typic Hapludult) and Pocomoke (Typic Umbraqualt) soils, respectively]. Toxic effects on plants were reduced in the high-SOM soil because of increased binding of metals to insoluble soil components, possibly fractions of soil SOM. Strickland et al. (1979) reported the reduced uptake of Cd resulting from peat amendments to sand cultures of soybean. The plants were grown in sand mixed with 0.5 to 8% peat (w/w). Up to 20 mg of Cd kg$^{-1}$ soil was added and the yield and Cd uptake recorded. Yields increased with increasing levels of peat. Cd concentrations in roots and shoots increased with low additions of peat to the medium (up to 0.5%), but decreased with high additions in accordance with our explanation discussed previously on the effects of high HS concentrations on metal uptake. Apparently, the solubility of added SOM fractions is an important factor that determines whether uptake enhancement or inhibition will occur.

Chen et al. (2001) designed and conducted two plant growth experiments, aiming to provide evidence for our hypothesis suggesting that plant growth stimulation by HS results from their contribution to trace elements availability, particularly Fe. Since Fe is the main ion of concern because of complex uptake mechanisms, we carried out the tests on two species, one representing monocotyledonous plants and the other dicotyledonous plants, which differ in their Fe uptake strategies (Marschner et al., 1986a, 1986b; Chen et al., 2001) as follows:

1. Strategy I. Mostly present in dicotyledons, represented in this study by melon plants (*Cucumis melo* L., cv. Ein Dor) — Experiment 1.
2. Strategy II. Mostly present in monocotyledons, especially in grasses, represented in this study by ryegrass (*Lolium perenne* L., cv. Omega) — Experiment 2.

## MELON PLANT GROWTH (STRATEGY I PLANTS): EXPERIMENT 1

During the final 6 d of treatment, plants that received Fe in their NS (a nutrient solution that did not contain Fe) underwent a recovery and regreening process. Differences between the treatments became distinct. Chlorophyll concentration in the leaves (young yet mature leaves) differed in accordance with the treatments (Table 4.3).The measured chlorophyll levels were 0.2, 1.3, 1.7, and 2.1 mg g$^{-1}$ (fresh weight) for the NS, NS + Fe(NO$_3$)$_3$, NS + Fe(NO$_3$)$_3$, + HA, and NS + FeEDDHA, respectively. Remedy of the Fe-deficiency-induced chlorosis was only partially achieved by a mineral form of Fe, whereas the presence of either FeHA or FeEDDHA greatly improved chlorophyll synthesis. FeEDDHA seemed to be more effective than FeHA in enhancing

**TABLE 4.3**
**Chlorophyll Concentration in Leaves of Melon Plants Grown in Nutrient Solution after 6 d of Treatment**

| Nutrient Solution (NS) Treatment | Chlorophyll Concentration (mg g$^{-1}$) |
|---|---|
| NS | 0.2 d[a] |
| NS + Fe(NO$_3$)$_3$ | 1.3 c |
| NS + Fe(NO$_3$)$_3$ + HA | 1.7 b |
| NS + FeEDDHA | 2.1 a |

[a] Statistical analysis by Tukey–Krammer test ($p = 0.05$). Letters a through d indicate significant differences.

chlorophyll formation (Table 4.3). This experiment clearly indicates that under neutral to slightly basic pH, HA stimulates plant growth and improves its health by enhancing Fe availability and chlorophyll synthesis. This is in accordance with other reports suggesting that HS are important to the Fe nutrition of dicotyledonous plants (e.g., Chen and Aviad, 1990; Bar-Ness and Chen, 1991a, 1991b).

## Ryegrass Plant Growth (Strategy II Plants): Experiment 2

Ryegrass is a very common plant in grazed pastures and on golf courses. It is a monocotyledonous gramineae and like other members of its family takes up Fe from the soil or from the NS via a ligand exchange mechanism between organo-Fe complexes and mineral Fe compounds or phyto-siderophores (Marschner et al., 1986a, 1986b; Yehuda et al., 1996). As for the dicotyledonous plants, we hypothesized that the formation of organo-Fe complexes in the NS and the resulting maintenance of the Fe in solution would greatly improve the effectiveness of the Fe uptake mechanism via the phytosiderophores. The chlorophyll concentrations for the ryegrass growth study are presented in Table 4.4. Plants grown in a NS devoid of Fe and Zn exhibited very low chlorophyll levels and retarded growth resulting from a lack of proper photosynthesis. Additions of FA or HA to the NS (treatments NS + FA and NS + HA) did not result in a significant improvement in chlorophyll concentration, although a slight improvement seemed to have occurred. This suggests that the HS do not act as growth hormones or growth stimulators when added as purified substances. This observation contradicts the hormone-like activity or plant growth regulators' concept suggested by many researchers (e.g., Nardi et al., 1996, 2002) and supports the results reported by Chen et al. (1994), who tried, but were unable, to extract plant hormones from HS, implying that they were not present. Addition of Fe and Zn salts in the treatments NS + FeSO$_4$ + ZnSO$_4$ resulted in a significant yet insufficient improvement in plant chlorophyll synthesis. The plants were moderately green, but this was insufficient to induce improved plant growth. However, when the Fe and Zn were added to the solutions as organometal complexes, either as an EDTA-chelated form or a HS complex (treatments NS + FeEDTA + ZnEDTA, NS + FA + FeSO$_4$ + ZnSO$_4$, NS + HA + FeSO$_4$ + ZnSO$_4$), the plants synthesized high levels of chlorophyll, reflecting those of healthy plants (almost twice the biomass as that of chlorophyll-deficient plants — data not shown). Results of these experiments clearly show that HS are ineffective in plant growth enhancement when Fe is absent from the NS. Addition of Fe to the NS in the presence of either synthetic chelate or the humic ligands stimulates plant growth.

Because in both Experiments 1 and 2 we did not observe symptoms of Zn deficiency, we concluded that this can be attributed to improved Fe nutrition in solutions containing organo-Fe complexes. However, some contribution of the organo-Zn complexes cannot be completely ruled out at this stage, and therefore further research is being conducted in our laboratory to elaborate on this issue. Because Experiment 2 provides more comprehensive answers (for monocots) than Experiment 1 (dicots), a similar design is employed at present in a study on dicots.

**TABLE 4.4**
**Chlorophyll Concentration of Leaves of Ryegrass Plants Grown in Nutrient Solution and Supplied with Different Nutrient Solutions**

| Nutrient Solution (NS) Treatment | Chlorophyll Concentration (mg g$^{-1}$) |
|---|---|
| NS | 0.66 c[a] |
| NS + FA | 0.91 c |
| NS + HA | 1.04 c |
| NS + FeSO$_4$ + ZnSO$_4$ | 1.76 b |
| NS + FeEDTA + ZnEDTA | 3.28 a |
| NS + FA + FeSO$_4$ + ZnSO$_4$ | 3.39 a |
| NS + HA + FeSO$_4$ + ZnSO$_4$ | 3.47 a |

[a] Statistical analysis by Tukey–Krammer test ($p = 0.05$). Letters a through c indicate significant differences.

The conclusions from this study (Chen et al., 2001) were that growth enhancement of plants in the NS and soil by HS should be attributed mostly to the maintenance of Fe and Zn in solution at sufficient levels. This effect is pH dependent and becomes more prominent at high pH levels.

## COMPOST-DERIVED HS AND PLANT GROWTH

Compost amendment of soils is an attractive way to add organic material to soils for the following reasons: (1) it exhibits a particle-size distribution that favors a uniform application in the field; (2) C:N ratios are optimal; (3) compost is usually devoid of weed seeds (as opposed to fresh manure); (4) it is often suppressive to soil-borne diseases (Hoitink et al., 1993); and (5) it is far richer than manures in HS (Inbar et al., 1990), thereby facilitating plant growth enhancement.

Many publications include discussions on the interaction between composts and plants under growth chamber, greenhouse, and laboratory conditions. Gallardo and Nogales (1987) reviewed many of these publications, specifically those focusing on municipal solid waste (MSW) compost. In general, positive effects on crop yield were reported in a number of crop species. Many of these do not single out Fe nutrition effects, but rather focus on the general beneficial effect of organic matter addition to soils. Compost addition improved water retention and enhanced soil porosity. Also, the pH-buffering capacity of the soil improved, CEC increased, and the level of SOM increased. Nitrogen availability in compost-amended soil depends largely on the maturity of the applied compost, yet the effects of compost application on P, S, Ca, and Mg availability and uptake are inconsistent and vary among reports. K and micronutrient availability improved following compost amendment (Gallardo and Nogales, 1987). Soil amendment with MSW for 4 years or longer improved the physical properties of soils (Giusquianti, 1995) based on the same parameters mentioned previously. These investigators also observed an increase in the HS level, CEC, and microbial activity.

In a long-term experiment conducted in a greenhouse and a field, effects of composted biosolids on the growth of tall fescue (*Festuca arundinacea* Schreb.) and on the physical properties of the soil were studied (Tester, 1989, 1990). Compost was added to the soil at increasing levels, with or without fertilizer addition. Maximum yield was obtained in a combined compost + N + P treatment in accordance with the concepts presented in this review. Maynard (1995) reported on tomato yield and quality enhancement following three consecutive MSW applications of up to 125 Mg ha$^{-1}$.

Experiments conducted with NS enriched with compost-derived DOM (a 1:10 solid:water extract of compost; Chen et al., 1994) showed that growth was significantly enhanced by the DOM, reaching a maximum at 35 to 55 mg L$^{-1}$ (Figure 4.6). This level was found to be optimal for root and shoot weight, leaf chlorophyll concentration, leaf area, and plant height. The patterns of the

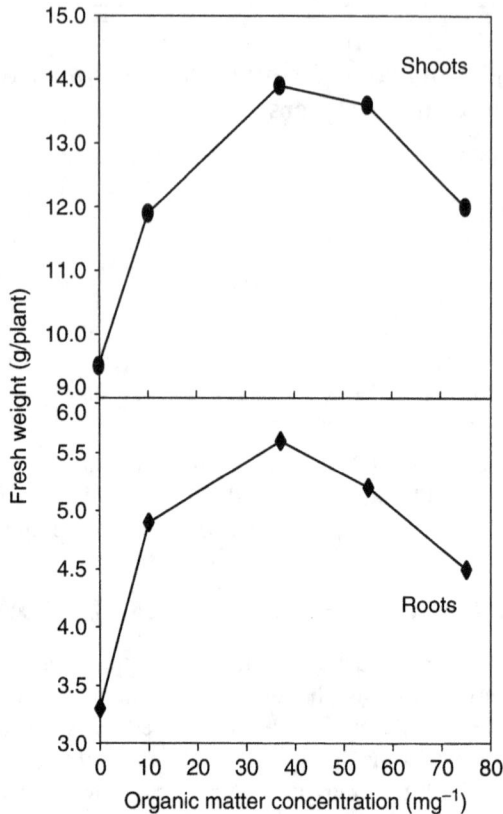

**FIGURE 4.6** Fresh weight of shoots and roots of melon (*Cucumis melo* L.) plants grown in nutrient solutions containing increasing concentrations of humic substances originating from water extract of composted separated manure (CSM). (From Chen, Y. et al. 1994. In N. Senesi et al. (Eds.), *Humic Substances in the Global Environment and Implication on Human Health.* Elsevier, Amsterdam, pp. 427–443. With permission.)

curves presented are similar to those in Figure 4.4 and represent typical optimal response curves. High chlorophyll levels indicate a satisfactory level of Fe supply, possibly due to transport mechanisms of Fe–HS complexes. Higher concentration of the DOM resulted in a growth inhibition when compared with the optimum level. A discussion on the inhibition effect on plant growth at high HS or DOM levels was presented previously in relation to Figure 4.4.

The role of compost-derived DOM in plant growth enhancement was also shown by our group for extracts of composted separated cattle manure (CSM; Figure 4.7; Inbar et al., 1993). Tomato seedlings were inhibited in water extracts obtained from composts less than 40 d old. An increase in inhibition resulting from phytotoxic substances was observed in compost extracts prepared during the thermophilic stages of composting (20 to 40 d). From Day 40 onward, growth in compost extract was significantly better than that in a standard NS. It was concluded that phytotoxic compounds and competition for $O_2$ in the immature compost were responsible for the inhibition in plant growth, whereas growth enhancement was related to the plant growth promotion effects of composts, including those discussed in this chapter. These observations led our research group to focus on extraction and fractionation of compost-derived DOM, aiming to isolate highly effective soluble fractions and to show their involvement in the Fe nutrition of plants.

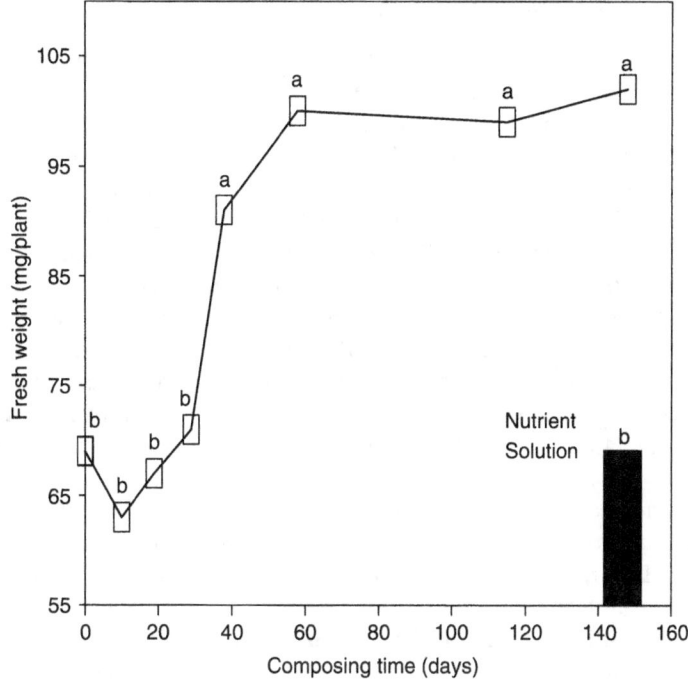

**FIGURE 4.7** Effects of water extract from composted separated manure (CSM) at various stages of decomposition on tomato (*Lycopersicon esculentum*) seedlings' fresh weight. (From Inbar, Y. et al. 1993. *J. Environ. Qual.* 22:857–863. With permission.)

## DOM EXTRACTION FROM COMPOST

Aqueous compost extracts were prepared by our group (Chefetz et al., 1998) by placing 1 kg compost (wet weight) in 10 L of distilled water and shaking (125 rpm) for 2 h at room temperature. The suspension was then centrifuged (10,000 $g$ for 30 min) and the supernatant filtered through a 0.45-µm membrane filter (Supor-450, Gelman Sciences). The concentration of DOM was measured immediately after extraction or fractionation, using a total carbon monitor (TCM 480, Carlo Erba Instruments, Milan, Italy) and transforming the data from DOC to DOM by using the ratio obtained for this material by a volatile solid determination.

## PREPARATIVE FRACTIONATION

Preparative DOM fractionation was performed according to Leenheer (1981), with some modifications. Details of the procedure can be found in Chefetz et al. (1998). The following fractions were obtained:

Hydrophobic fraction — sorbed to XAD-8 (Ho):
- Ho Base (HoB). Desorbed with 0.1 $M$ HCl
- Ho Acid (HoA). Desorbed with 0.1 $M$ NaOH
- Ho Neutrals (HoN). Desorbed with MeOH

Hydrophilic fraction — not sorbed to XAD-8 (Hi) (obtained after sample elution through an XAD-8 column):
- Hi Base (HiB). Sorbed to $H^+$ cation-exchange resin, desorbed with 0.1 $M$ $NH_4OH$
- Hi Acid (HiA). Sorbed to anion-exchange resin, desorbed with 1.0 $M$ NaOH
- Hi Neutrals (HiN). Remains in deionized water solution after all column separations

## Effects of DOM and Its Fractions on Plant Growth

DOM before fractionation as well as HoN and HoA were obtained by our group from a mature MSW compost (Amichai, 2001). The latter fractions were selected because their levels increased with the degree of maturity and correlated with plant growth response. The HoN level in mature compost was higher than in soils because of the enhanced humification in composts. The HoA resembles soil HAs (Guggenberget and Zech, 1994; Chefetz et al., 1998).

The addition of DOM, HoN, or HoA to the NS enhanced their ability to maintain $Fe^{3+}$ in solution (or in a colloidal form) at pH 6 or 7.3 far beyond the levels observed in the control solutions. The option of the formation of colloidal or finely suspended conformations of organo-Fe complexes has also been considered, and further studies on these forms are required. Sticher (1997) and Dolfing et al. (1999) investigated the adsorption of Fe hydroxides on surfaces of organic macromolecules. According to their findings, further crystallization is inhibited by surface adsorption. This hindrance is likely to enhance the availability of the Fe to plants. Determination of molecular sizes of the colloids formed in the NS by size exclusion chromatography (SEC) was conducted by our group and is reported elsewhere.

Our group determined the apparent stability constants ($K_{app}$) for $Fe^{3+}$ with the DOM and its two fractions, using a colored ligand-exchange method. The $K_{app}$ values at pH 5 and ionic strength ($\mu$) of 0.1 $M$ for the DOM, HoA, and HoN were 6.88, 7.91, and 6.76, respectively. The DOM and its two fractions were further tested to evaluate their potential contribution in the Fe nutrition of plants. A distinctive positive response to the Fe addition will substantiate the hypothesis that the growth stimulation of plants by HS is basically a mechanism dependent on Fe nutrition (and possibly additional micronutrients such as Zn).

In earlier experiments, peanuts (*Arachis hypogea*) were shown to provide a good bioassay system for the Fe nutrition status of plants (Barak and Chen, 1982) and were therefore used as test plants for determining the efficacy of the DOM, HoN, and HoA as Fe carriers to plants.

Plant growth trials were conducted in NS buffered to pH 7.3 by excessive $CaCO_3$, thus simulating calcareous soil conditions. Chlorophyll concentration of the leaves, which serves as an indicator for the Fe nutrition status of the plants (green plants = high chlorophyll concentration = sufficient Fe supply; yellow plants = low chlorophyll concentration = deficient Fe supply), were significantly higher in plants grown in the NS containing 50 mg $L^{-1}$ of DOM, HoN, or HoA compared with plants supplied with a mineral form of Fe only. Plants supplied with the FeHS complexes were as healthy and green as those supplied with FeEDDHA, which is the best Fe chelate known to date (for dicotyledonous plants such as peanuts). Plants supplied with only $FeSO_4$ turned yellow during the growth period, and at the end of this period they did not differ from control plants (grown without any Fe supply). Similar results were obtained by Chen et al. (2001), who tested effects of HA or FA on both monocots and dicots. These results are in accordance with a series of studies by our group and other groups, showing the significance of organo-Fe complexes to plant nutrition (Chen and Barak, 1982; Stevenson and Chen, 1986; Chen, 1996). Recent studies by Pinton et al. (1999) also showed the correction of Fe deficiency in cucumber plants (*Cucumis sativus*) by using Fe humates. The experiments summarized in this section clearly show the importance of HS in supplying Fe to plants.

## COMMERCIAL HS AND THEIR EFFECTS ON PLANT GROWTH

Many commercial HS products are available today worldwide. Mostly, although not solely, these products are derived from leonardite or lignite. Another common source is peat. Some examples of reports on trials with commercial products, either via soil or foliar application, are briefly reviewed here. Only a small fraction of laboratory or field trials with commercial products have been conducted in a manner that meets the standards required to report data in scientific articles (using controls and reporting statistical analysis), whereas an even smaller fraction has found its way to reviewed journals.

Brownell et al. (1987) tested the response of various field crops to application of two leonardite extracts in California. No details on the chemical properties of the two extracts were presented. Trials with processing tomatoes (*Lycopersicon esculantum*) produced an average yield increase of 10.5% compared with untreated controls. Trials with cotton (*Gossypium hirsutum* L.) during the same year (Brownell et al., 1987) produced an average yield increase of 11.2%. Unreplicated large-scale field trials with grape vines (*Vitis vinifera*) of various cultivars produced increases in the total yield of 3 to 70%, with an average of 25%. In most of these studies, one of the products that was rich in HS was applied as an early soil treatment, and the other was used as a postemergence foliar spray. Based on their observations, these investigators hypothesized that the tested products, when used singly or in tandem, triggered a flowering response in many crop plant species. Brownell et al. (1987) obtained the most pronounced effect by using a combination of early season soil treatment with a postemergence foliar spray.

Foliar application of FA was tested in field trials by Xudan (1986). Water use, nutrient uptake, and yield of wheat in plants treated with the FA were measured and compared to controls. FA reduced the stomatal conductance of well-watered plants. Drought resistance of the wheat was reduced in FA-treated wheat when the foliar application was conducted during head development, whereas grain yield increased by 7.3 to 18.0%, thus indicating that water was not the growth-limiting factor.

Two HA products originating from peat (CPA) and leonardite (CPB) were bought from companies and their effects on nutrition and growth of tomatoes (*Lycopersicon esculatnum*) tested (Adani et al., 1998). Application of the CPA product resulted in an increase in root weight only. Maximal effect was observed at a HA concentration of 20 mg $L^{-1}$ (22 and 23% increase in dry and fresh weight, respectively). Application of the CPB product resulted in both root and shoot growth enhancement, reaching the highest weight enhancement at 50 mg $L^{-1}$ (fresh and dry weight, respectively, of shoots by 8 and 9% above the control and roots by 16 and 18% above the control). The uptake of N, P, Fe, and Cu was enhanced by CPA treatment, whereas CPA application enhanced the uptake of N, P, and Fe. Both treatments exhibited the strongest effect on the uptake of Fe. In an experiment conducted by Cooper et al. (1998), several products were tested: (1) a solid and granulated HA; and (2) HAs in solution extracted from leonardite, peat, and soil. These products were applied to creeping bentgrass (*Agrostis stolonifera*) grown in NS or sand cultures, or both. Root mass at different depths in the sand cultures, their length, and nutritional elements uptake were tested. The HS granules were mixed with the sand or applied to the NS or foliar-applied by leaf spraying of the HA solutions. Mixing HA granules with the sand resulted in a 45 and 38% increase of root mass at depths of 0–10 cm and 10–20 cm, respectively. The addition of the granules also enhanced root length. Foliar spray, however, did not enhance root mass or length in both NS and sand cultures.

## CONCLUSIONS

Effects of organic amendments, such as enhanced nutrient supply, soil structure improvement, increased cation exchange capacity, increased water retention, and enhanced microorganism populations due to increased C and N sources should be distinguished from the specific effects of HS. This distinction is not always easy to achieve because of the considerable overlap between various sources of HS (i.e., manure, compost, SOM) and their effects on the plant environment.

To assist in evaluating the prospects of finding favorable effects under field conditions, some calculations were presented by Chen and Aviad (1990) and are modified in this chapter. Foliar and soil applications were compared on the basis of concentrations that were frequently reported as those concentrations of HS required to affect plant growth in NS:

Assumptions for foliar spray

1. Required volume of spray: 1500 L ha$^{-1}$
2. Required concentration: 250 mg L$^{-1}$

For 1 ha the following quantity of HS is required:

$$1500 \text{ L} \times 0.25 \text{ g L}^{-1} = 375 \text{ g}$$

Assumptions for soil application:

1. Plow layer weight: 3000 Mg ha$^{-1}$
2. Water content at field capacity (% by weight): 30
3. Increase required in HS concentration: 75 mg L$^{-1}$

For 1 ha the following quantity of HS is required:

$$3000 \text{ Mg} \times 0.3 \text{ m}^3 \text{ Mg}^{-1} \times 0.075 \text{ kg HS m}^{-3} = 67.5 \text{ kg}$$

These calculations, based on extrapolation of numerous laboratory studies, show that the amount of HS required for an effective soil application is ca. 67.5 kg ha$^{-1}$. Foliar spray can be effective at an application rate that is at least 150 times lower than this value. Because of the relatively high cost of commercial preparations of HS, it seems that future prospects for economical use in agriculture of these products are much better for foliar spray application. Soil and foliar applications of complexes of HS with micronutrients such as Fe and Zn require separate evaluation in the field, because of their specific effects under deficiency conditions prevailing in calcareous soils.

In fertile soils of humid climates in which the DOM can reach levels of up to 400 mg L$^{-1}$ (Chen and Schnitzer, 1978), beneficial effects from the application of HS are not likely to occur. Although sandy soils in humid climates might exhibit relatively low DOM levels, Fe deficiency is not likely to occur in these soils and beneficial effects of HS application are not expected. In soils of semiarid or arid zones, however, in which DOM does not exceed 20 to 30 mg L$^{-1}$ (e.g., Chen and Katan, 1980), beneficial effects can be observed at sufficient application rates of HS. HS can be especially effective in soils of arid and semiarid climates when applied in combination with irrigation water along with mineral nutritional elements (fertigation). Applications of very low concentrations of HS in the field, often recommended by extension service advisors and representatives of commercial producers, will probably not lead to positive plant responses. The possibility of obtaining increases of soluble HS by applications of manure or compost should, however, be considered as an economical alternative.

## ACKNOWLEDGMENTS

This research was supported by the von Humboldt Foundation, Germany, The Water Chemistry Institute of the University of Karlsruhe (Prof. Frimmel, Head), and the Compost Management EU Project — The 5th Framework, the European Union.

## REFERENCES

Adani, F., P. Genevini, P. Zaccheo, and G. Zocchi. 1998. The effect of commercial humic acid on tomato plant growth and mineral nutrition. *J. Plant Nutr.* 21:561–575.

Agboola, A.A. 1978. Influence of soil organic matter on cow pea's response to N fertilizer. *Agron. J.* 70:25–28.

Amichai, E. 2001. The effect of dissolved organic matter extracted from municipal solid waste compost on plant growth. M.Sc. Thesis, The Hebrew University of Jerusalem.

Aso, S., and I. Sakai. 1963. Studies on the physiological effects of humic acid. 1. Uptake of humic acid by crop plants and its physiological effects. *Soil Sci. Plant Nutr.* 9:85–91.

Bar-Ness, E., and Y. Chen. 1991a. Manure and peat based Fe-organo complexes. I. Characterization and enrichment. *Plant Soil* 130:35–43.

Bar-Ness, E., and Y. Chen 1991b. Manure and peat based Fe-enriched complexes. II. Transport in soils. *Plant Soil* 130:45–50.

Barak, P., and Y. Chen. 1982. The evaluation of iron deficiency using a bioassay-type test. *Soil Sci. Soc. Am. J.* 46:1019–1022.

Bottomley, W.B. 1914a. Some accessory factors in plant growth and nutrition. *Proc. R. Soc. Lond. B* 88:237–247.

Bottomley, W.B. 1914b. The significance of certain food substances for plant growth. *Ann. Bot.* 28:531–540.

Bottomley, W.B. 1917. Some effects of organic growth-promotion substances (auximones) on the growth of *Lemna minor* in mineral cultural solutions. *Proc. R. Soc. Lond. B* 89:481–505.

Bottomley W.B. 1920. The effect of organic matter on the growth of various plants in culture solutions. *Ann. Bot.* 34:353–365.

Brownell, J.R., G. Nordstrom, J. Marihart, and G. Jorgensen. 1987. Crop responses from two new Leonardite extracts. *Sci. Total Environ.* 62:492–499.

Burk, D., H. Lineweaver, and C.K. Horner. 1932. Iron in relation to the stimulation of growth by humic acid. *Soil Sci.* 33:413–435.

Catalano, L., M. De Nobili, H. Siebner-Freibach, and Y. Chen. 2002. Effect of humic substances on the behaviour of iron siderophores in soil and water. In *Proceedings of the International Humic Substances Society Twentieth Anniversary Conference.* Northeastern University, Boston, MA, pp. 268–271.

Cesco, S., M. Nikolic, V. Römheld, Z. Varanini, and R. Pinton. 2002. Uptake of $^{59}$Fe from soluble $^{59}$Fe-humate complexes by cucumber and barley plants. *Plant Soil* 241:121–128.

Cesco, S., V. Römheld, Z. Varanini, and R. Pinton. 2000. Solubilization of iron by water-extractable humic substances. *J. Plant Nutr. Soil Sci.* 163:285–290.

Chefetz, B., P.G. Hatcher, Y. Hadar, and Y. Chen. 1998. Characterization of dissolved organic matter extracted from composted municipal waste. *Soil Sci. Soc. Am. J.* 62:326–332.

Chen, Y. 1996. Organic matter reactions involving micronutrients in soils and their effect on plants. In A. Piccolo (Ed.), *Humic Substances in Terrestrial Ecosystems.* Elsevier, Oxford, pp. 507–529.

Chen, Y., and T. Aviad. 1990. Effects of humic substances on plant growth. In P. MacCarthy et al. (Eds.), *Humic Substances in Soil and Crop Science: Selected Readings* American Society of Agronomy and Soil Science Society of America, Madison, WI, pp.161–186.

Chen, Y., and P. Barak. 1982. Iron nutrition of plants in calcareous soils. *Adv. Agron.* 35:217–240.

Chen, Y., and P. Barak. 1983. Iron-enriched peat and lignite as iron fertilizer. In K.M. Schallinger (Ed.), *Proceedings of the Second International Symposium on Peat Agriculture.* Publication of the Volcani Center, Bet-Dagan, Israel, pp. 195–202.

Chen, Y., and J. Katan. 1980. Effect of solar heating of soils by transparent polyethylene mulching on their chemical properties. *Soil Sci.* 130:271–277.

Chen, Y., and M. Schnitzer. 1978. The surface tension of aqueous solutions of soil humic substances. *Soil Sci.* 125:7–15.

Chen, Y., and F.J. Stevenson. 1986. Soil organic matter interactions with trace elements. In Y. Chen et al. (Eds.), *The Role of Organic Matter in Modern Agriculture.* Martinus Nijhoff, Dordrecht, pp. 71–116.

Chen, Y., N. Senesi, and M. Schnitzer. 1977. Information provided on humic substances by E4/E6. *Soil Sci. Soc. Am. J.* 41:352–358.

Chen, Y., J. Navrot, and P. Barak. 1982a. Remedy of lime-induced chlorosis with iron-enriched muck. *J. Plant Nutr.* 5:927–940.

Chen, Y., B. Steinitz, A. Cohen, and Y. Elber. 1982b. The effect of various iron-containing fertilizers on growth and propagation of *Gladiolus grandiflorus. Scientia Hortic.* 18:169–175.

Chen, Y., H. Magen, and J. Riov. 1994. Humic substances originating from rapidly decomposing organic matter: properties and effects on plant growth. In N. Senesi et al. (Eds.), *Humic Substances in the Global Environment and Implication on Human Health.* Elsevier, Amsterdam, pp. 427–443.

Chen, Y., H. Magen, and C.E. Clapp. 2001. Plant growth stimulation by humic substances and their complexes with iron. In *Proceedings of the Dalia Greidinger Symposium*. The International Fertiliser Society, Lisbon, Portugal.

Clark, A.L., and E.B. Graham. 1968. Zinc diffusion and distribution coefficients in soil as affected by soil texture, zinc concentration and pH. *Soil Sci.* 105:409–418.

Cooper, R.J., C. Liu, and D.S. Fisher. 1998. Influence of humic substances on rooting and nutrient content of creeping bentgrass. *Crop Sci.* 38:1639–1644.

Dekock, P.C. 1955. Influence of humic acids on plant growth. *Science* 121:473–474.

De Nobili, M., L. Catalano, H. Siebner-Freibach, and Y. Chen. 2002. Sorption of microbial and synthetic ligands and chelates on organic matter and clays: A physico-chemical and iron supply study. In *XI ISINIP Proceedings*. University of Udine Publications, Italy.

Dolfing, J., W.J. Chardon, and J. Japenga. 1999. Association between colloidal iron, aluminum, phosphorus, and humic acids. *Soil Sci.* 164:171–179.

Dormaar, J.F. 1975. Effects of humic substances from chenozemic Ah horizons on nutrient uptake by *Phaseolus vulgaris* and *Festuca scabrella*. *Can. J. Soil Sci.* 55:111–118.

Dyakonova, K.V., and A.E. Maksimova. 1967. Humic substances of the most active part of organic fertilizers and their influence on plants. In *Transactions of the Joint Meeting Committee 2 and 4*, International Society for Soil Science, Dokuchaev Soil Institute, Moscow, pp. 79–85.

Ellis, B.G., B.D. Knezek, and L.W. Jacobs. 1983. The movement of micronutrients in soils. In D.W. Nelson, D. Elrick and K.K. Tanji (Eds.), *Chemical Mobility and Reactivity in Soil Systems*. SSSA Special Publication No. 11, Soil Science Society of America, Madison, WI, pp. 109–122.

Ernst, W.H.O., M.H.S. Kraak, and L. Stoots. 1987. Growth and mineral nutrition of *Scrophularia nodos* with various combinations of fulvic and humic acids. *J. Plant Physiol.* 127:171–175.

Fagbenro, J.A., and A.A. Agboola. 1993. Effect of different levels of humic acid on the growth and nutrient uptake of teak seedlings. *J. Plant Nutr.* 16:1465–1483.

Gallardo, L.F., and R. Nogales. 1987. Effect of the application of town refuse compost on the soil-plant system: A review. *Bio.Wastes* 19:35–62.

Giusquiani, P.L., M. Pagliai, G. Gigliotti, D. Businelli, and A. Benetti. 1995. Urban waste compost: Effects on physical, chemical, and biochemical soil properties. *J. Environ. Qual.* 24:175–182.

Grandeau, L. 1872. Recherches sur le role des matieres organiques du sol dans les phenomenes de la nutrition des vegetaux. Comptes rendus hebdomadaire seances de l'academie des sciences, Paris.

Guggenberger, G., and W. Zech. 1994. Dissolved organic carbon in forest floor leachates: Simple degradation products or humic substances? *Sci. Total Environ.* 152:37–47.

Hoitink, H.A.J., M.J. Boehm, and Y. Hadar. 1993. Mechanisms of suppression of soilborne plant pathogens in compost-amended substrates. In H.A.J. Hoitink et al. (Eds.), *Science and Engineering of Composting*. Renaissance Publications, Worthington, OH, pp. 601–622.

Hutchinson, H.B., and N.H.J. Miller. 1912. The direct assimilation of inorganic and organic forms of nitrogen by higher plants. *J. Agric. Sci.* 4:282–302.

Inbar, Y., Y. Chen, and Y. Hadar. 1990. Humic substances formed during the composting of organic matter. *Soil Sci. Soc. Am. J.* 54:1316–1323.

Inbar Y., Y. Hadar, and Y. Chen 1993. Recycling of cattle mature: The composting process and characterization of maturity. *J. Environ. Qual.* 22:857–863.

Irintoto, B., K.H. Tan, and H.E. Sommer. 1993. Effect of humic acid on callus culture of slash pine (*Pinnus-elliotti engelm*). *J. Plant Nutr.* 16:1109–1118.

Iyengar, S.S., D.C. Martens and W.P. Miller. 1981. Distribution and plant availability of soil zinc fractions. *Soil Sci. Am. J.* 45:735–739.

Jalali, V.K., and P.N. Takkar. 1979. Evaluation of parameters for simultaneous determination of micronutrient cations available to plants from soils. *Indian J. Agric. Sci.* 49:622–626.

Johnston, A.E. 1994. The Rothamsted classical experiments. In R.A. Leigh and A.E. Johnston (Eds.), *Long-Term Experiments in Agricultural and Ecological Sciences*. CAB International, Wallingford, U.K., pp. 9–52.

Kochian, L.V. 1991. Mechanisms of micronutrient uptake and translocation in plants. In J.J. Mortvedt, F.R. Cox, L.M. Shuman, and R.M. Welch (Eds.), *Micronutrients in Agriculture*. Soil Science Society of America, Madison, WI, pp. 229–296.

Kuchenbuk, R., and A. Jungk. 1982. A method for determining concentration profiles at the soil-root interface by thin slicing rhizospheric soil. *Plant Soil* 68:391–396.

Lawes, J.B., and J.H. Gilbert. 1905. Collected papers. In W.H. Hall (Ed.), *The Book of the Rothamsted Experiments*. John Murray, London.

Lee, Y.S., and R.J. Bartlett. 1976. Stimulation of plant growth by humic substances. *Soil Sci. Soc. Am. J.* 40:876–879.

Leenheer, J.A. 1981. Comprehensive approach to preparative isolation and fractionation of dissolved organic carbon from natural waters and wastewaters. *Environ. Sci. Technol.* 15:578–587.

Leita, L., M. De Nobili, L. Catalano, A. Moria, E. Fonda, and G. Vlaic. 2001. Complexation of iron-cyanide by humic substances. In R.S. Swift and K.M. Spark (Eds.), *Understanding and Managing Organic Matter in Soils, Sediments, and Waters,* International Humic Substances Society, St. Paul, MN, pp. 477–482 (IHSS9, 21–25 September 1998, Adelaide, Australia).

Levesque, M. 1970. Fulvic acid and fulvometallic complexes in mineral nutrition of plants. *Can. J. Soil Sci.* 50:385–390.

Li, C.H., G.H. Xu, and Z.W. Feng. 1981. Essential properties of woodland soil in major China fir producing areas and their relationship with growth of China fir. *T'u Jang T'ung Pao* 4:1–6.

Liebig, J.V. 1841. Organic chemistry in its applications to agriculture and physiology. Cambridge, U.K. (translated by J.W. Webster and J. Owens).

Liebig, J.V. 1856. On some points of agricultural chemistry. *J. Royal Agric. Soc.* 17:284–326.

Lindsay, W.L. 1974. Role of chelation in micronutrient availability. In E.W. Carson (Ed.), *The Plant Root and its Environment*. University Press of Virginia, Charlottesville, pp. 507–524.

Lindsay, W.L. 1991. Inorganic equilibria affecting micronutrients in soils. In J.J. Mortvedt, F.R. Cox, L.M. Shuman, and R.M. Welch (Eds.), *Micronutrients in Agriculture*. Soil Science Society of America, Madison, WI, pp. 89–112.

Lindsay, W.L., and A.P. Schwab. 1982. The chemistry of iron in soils and its availability to plants. *J. Plant Nutr.* 5:821–840.

Linehan, D.J., and H. Shepherd. 1979. A comparative study of the effects of natural and synthetic ligands on iron uptake by plants. *Plant Soil* 52:281–289.

Low, P.F. 1962. Effect of quasi-crystalline water on rate processes involved in plant nutrition. *Soil Sci.* 93:6–15.

Lykov, A. 1978. The effect of the organic matter of derno podzolic soil on the yields of field crops. Problemy Zemledeliya. *Referentivnyi Zhurnal Seriya* 5:195–202.

Mandal, L.N., and B. Mandal. 1986. Zinc fractions in soil in relation to zinc nutrition of lowland rice. *Soil Sci.* 132:141–148.

Marschner, H., and V. Römheld. 1994. Strategies of plants for acquisition of iron. *Plant Soil* 165:261–274.

Marschner, H., V. Römheld, W.J. Horst, and P. Martin. 1986a. Root induced changes in the rhizosphere: Importance for the mineral nutrition of plants. *Z. Pflanzenernaer. Bodenk.* 149:441–456.

Marschner, H., V. Römheld, and M. Kissel. 1986b. Different strategies in the higher plants in mobilization and uptake of iron. *J. Plant Nutr.* 9:695–713.

Maynard, A.A. 1995. Cumulative effect of annual addition of MSW compost on yield of field-grown tomatoes. *Compost Sci. Util.* 3:47–54.

McLaren, R.G., and D.V. Crawford. 1973. Studies on soil copper. I. The fractionation of copper in soils. *J. Soil. Sci.* 24:172–181.

Moghimi, A., M.E. Tate, and J.M. Oades. 1978. Characterization of rhizosphere products, especially 2-ketogluconic acid. *Soil Biol. Biochem.* 10:283–287.

Murthy, A.S.P. 1982. Zinc fractions in wetland rice soils and their availability to rice. *Soil Sci.* 133:150–154.

Mylonas, V.A., and C.B. McCants. 1980. Effects of humic and fulvic acids on growth of tobacco. 1. Root initiation and elongation. *Plant Soil* 54:485–490.

Nardi, S., G. Concheri, and Dell'Agnola. 1996. Biological activity of humus. In A. Piccolo (Ed.), *Humic Substances in Terrestrial Ecosystems*. Elsevier Science, Amsterdam, pp. 361–406.

Nardi, S., D. Pizzeghello, A. Muscolo, and A. Vianello. 2002. Physiological effects of humic substances on higher plants. *Soil Biol. Biochem.* 34:1527–1536.

Nye, P.H. 1966. The measurement and mechanism of ion diffusion in soil. I. The relation between self-diffusion and bulk diffusion. *J. Soil Sci.* 17:16–23.

O'Connor, G.A., W.L. Lindsay, and S.R. Olsen. 1971. Diffusion of iron and iron chelates in soil. *Soil. Sci. Soc. Am. Proc.* 35:407–410.

Ojenhiyi, S.O., and O.O. Agbede. 1980. Soil organic matter and yield of forest and tree crops. *Plant Soil* 57:61–67.

Olsen, C. 1930. On the influence of humus substances on the growth of green plants in water culture. *Comptes-rendus du Laboratoire Carlsberg* 18:1–16.

Olsen, S.R. 1986. The role of organic matter and ammonium in producing high corn yields. In Y. Chen and Y. Avnimelech (Eds.), *The Role of Organic Matter in Modern Agriculture*. Martinus Nijhoff, Dordrecht, pp. 29–70.

Piccolo, A., S. Nardi, and G. Concheri. 1992. Structural characteristics of humic substances as related to nitrate uptake and growth regulation in plant systems. *Soil. Biol. Biochem.* 24:373–380.

Pilus Zambi, M., M. Yaacob, A.J.M. Kamal, and S. Paramananthan. 1982. The determination of soil factors on growth of cashew on *bri* soils. Part I. *Pertanika* 5:200–206.

Pinton, R., S. Cesco, S. Santi, F. Agnolon, and Z. Varanini. 1999. Water extractable humic substances enhance iron deficiency responses by Fe-deficient cucumber plants. *Plant Soil* 210:145–157.

Pinton, R., Z. Varanini, V. Vizzota, and A. Maggioni. 1992. Humic substances affect transport properties of tonoplast vesicles isolated from oat roots. *Plant Soil* 142:203–210.

Rauthan, B.S., and M. Schnitzer. 1981. Effects of a soil fulvic acid on the growth and nutrient content of cucumber (*Cucumis sativus*) plants. *Plant Soil* 63:491–495.

Rebufetti, A., and D. Labunora. 1982. Wheat yield in northeastern Uruguay in relation to NPK fertilizers, soil organic matter content and climatic conditions. In C.C. Cerri (Ed.), *Regional Colloquium on Soil Organic Matter Studies*, 18–22 October 1982, University of São Paulo, São Paulo, Brazil, pp. 117–122.

Russell, E.J. 1921. *Soil Conditions and Plant Growth*. Longman, London.

Scharpf, H. 1967. Relationships between the humus content of soil and crop yields in a long term fertilizer trial. *Albrecht-Thaer-Arch.* 11:133–141.

Sedberry, J.E., and C.N. Reddy. 1976. The distribution of Zn in selected soils in Louisiana. *Commun. Soil Sci. Plant Anal.* 7:10–17.

Shenker, M., I. Oliver, M. Helmann, Y. Hadar, and Y. Chen. 1992. Utilization by tomatoes of iron mediated by a siderophore produced by *Rhizopus arrhizus*. *J. Plant Nutr.* 15:2173–2182.

Shuman, L.M. 1979. Zinc, manganese and copper in soil fractions. *Soil Sci.* 127:10–17.

Shuman, L.M. 1983. Sodium hypochlorite methods for extracting microelements associated with soil organic matter. *Soil Sci. Soc. Am. J.* 47:656–660.

Shuman, L.M. 1985. Fractionation method for soil microelements. *Soil. Sci.* 140:11–22.

Stevenson, F.J. 1991. Organic matter micronutrient reactions. In J.J. Morvedt et al. (Eds.), *Micronutrients in Agriculture*. Soil Science Society of America, Madison, WI, pp.145–186.

Stevenson, F.J. 1994. *Humus Chemistry*. John Wiley & Sons, New York.

Sticher, P., M.C.M. Jaspers, K. Stemmler, H. Harms, A.J.B. Zehnder, and J.R. van der Meer. 1997. Development and characterization of a whole-cell bioluminescent sensor for bioavailable middle-chain alkanes in contaminated groundwater samples. *J. Appl. Environ. Microbiol.* 63:4053–4060.

Strickland, R.C., W.R. Chaney, and R.J. Lamoreaux. 1979. Organic matter influences phytoxicity of cadmium to soybeans. *Plant Soil* 52:393–402.

Swift, R.S. 1996. Organic matter characterization. In D.L. Sparks (Ed.), *Methods of Soil Analysis. Part 3. Chemical Methods*. SSSA Book Series No. 5, Soil Science Society of America, Madison, WI, pp. 1011–1069.

Tan, K.H., and V. Nopamornbodi. 1979. Effect of different levels of humic acids on nutrient content and growth of corn (*Zea mays* L.). *Plant Soil* 51:283–287.

Tanford, C. 1961. Molecular structure. In *Physical Chemistry of Macromolecules*, John Wiley & Sons, New York, pp. 15–137.

Tester, C.F. 1989. Tall fescue growth in greenhouse growth chamber, and field plots amended with sewage sludge compost and fertilizer. *Soil Sci.* 148:452–458.

Tester, C.F. 1990. Organic amendment effects on physical and chemical properties of a sandy soil. *Soil Sci. Soc. Am. J.* 54:827–831.

Thaer, A.D. 1808. *Grundriss der Chemie für Landwirte*. Berlin, Germany.

Vaughan, D., and R.E. Malcolm. 1985. Influence of humic substances on growth and physiological processes. In D. Vaughan et al. (Eds.), *Soil Organic Matter and Biological Activity*. Martinus Nijhoff, Dordrecht, pp. 37–75.

Vaughan, D., R.E. Malcolm, and B.G. Ord. 1985. Influence of humic substances on biochemical processes in plants. In D. Vaughan and R.E. Malcolm (Eds.), *Soil Organic Matter and Biological Activity*. Martinus Nijhoff, Dordrecht, pp. 77–108.

Vaughan, D., and I.R. McDonald. 1976. Some effects of humic acid on the cation uptake by parenchyma tissue. *Soil Biol. Biochem.* 8:415–421.

Wang, Y., H.N. Brown, D.E. Crowley, and P.J. Szanislo. 1993. Evidence for direct utilization of a siderophore, ferrioximine B, in axenically grown cucumber. *Plant Cell Environ.* 16:579–585.

Warnacke, D.D., and S.A. Barber. 1972. Diffusion of zinc in soils. I. The influence of soil moisture. *Soil Sci. Soc. Am. Proc.* 36:39–42.

Warnacke, D.D., and S.A. Barber. 1973. Diffusion of zinc in soils. III. Relations to zinc adsorption isotherms. *Soil Sci. Soc. Am. Proc.* 37:355–358.

White, M.C., and R.L. Chaney. 1980. Zinc, cadmium and manganese uptake by soybean from two zinc and cadmium amended coastal plain soils. *Soil Sci. Soc. Am. J.* 44:308–313.

Wilkinson, H.F., J.F. Loneragan, and J.P. Quirk. 1968. The movement of zinc to plant roots. *Soil Sci. Soc. Am. Proc.* 32:831–833.

Xudan, X. 1986. The effect of foliar application of fulvic acid on water use, nutrient uptake and wheat yield. *Aust. J. Agric. Res.* 37:343–350.

Yehuda, Z., M. Shenker, V. Römheld, H. Marschner, Y. Hadar, and Y. Chen. 1996. The role of ligand exchange in the uptake of iron from microbial siderophores by graminaceous plants. *Plant Physiol.* 112:1273–1280.

Yermiyahu, U., R. Keren, and Y. Chen. 1988. Boron sorption on composted organic matter. *Soil Sci. Soc. Am. J.* 52:1309–1313.

Yermiyahu, U., R. Keren, and Y. Chen. 1995. Boron sorption by soil in the presence of composted organic matter. *Soil Sci. Soc. Am. J.* 59:405–409.

# 5 Suppression of Soilborne Diseases in Field Agricultural Systems: Organic Matter Management, Cover Cropping, and Other Cultural Practices

*Alexandra G. Stone, Steven J. Scheuerell, and Heather M. Darby*

## CONTENTS

0-8493-1294-9/04/$0.00+$1.50
© 2004 by CRC Press LLC

## INTRODUCTION

Soil organic matter (SOM) content and quality impact many soil functions related to soil health, such as moisture retention, infiltration, and nutrient retention and release. SOM content and quality also impact an important yet often overlooked soil function: plant health. Soil health is "the capacity of a soil to function as a vital living system…and to promote plant and animal health" (Doran and Zeiss, 2000). However, the impact of SOM management on plant health in field agricultural systems is poorly understood.

Over the past two decades, major advances have been made in understanding how peat and compost quality influence disease suppression in peat- and compost-based container systems. This area has been researched extensively and reviewed recently (Hoitink et al., 1991, 1999). At present, nursery and greenhouse growers successfully use compost-amended potting mixes to suppress soilborne diseases, such as *Pythium* and *Phytophthora* root rots, in container systems (Hoitink et al., 1991). The effect of field-applied organic residues (crop residues, cover crops, and organic wastes) on soilborne pathogens and diseases has also been studied extensively and reviewed previously (Baker and Cook, 1974; Baker, 1991; Cook and Baker, 1983; Forbes, 1974; Huber and Watson, 1970; Lazarovits, 2001; Linderman, 1989; Lumsden et al., 1983b; Palti, 1981; Papavizas and Lumsden, 1980; Patrick and Toussoun, 1965). Organic amendment is an old practice, and examples of organic-amendment-mediated suppression of soilborne diseases were reported as early as the late 19th century. Manures were applied to field soils to reduce the severity of root rot of cotton (causal agent *Phymatotrichum omnivorum*) as early as 1890 (Pammel, 1890). Manure applications were used to control take-all of wheat long before the causal agent was identified (McAlpine, 1904; Tepper, 1892).

Although a well-documented phenomenon in the field, little progress has been made to place organic-residue-mediated disease suppression into a SOM or cropping system perspective. The disjunction between the disciplines of soil science and plant pathology has slowed the incorporation of new views on SOM quality and function into the field of organic matter (OM)-mediated biological control of plant diseases. We attempt to bring together these disparate fields of knowledge to improve our understanding of how OM can be managed to control diseases in field agricultural systems. To this end, we first describe the relationships between OM quality and general suppression of diseases in soilless container mixes and then interpret data from natural and agricultural field systems in the context of the container evidence. We also discuss specific suppression of diseases caused by *Rhizoctonia solani* in both container and field systems and the mechanisms contributing to both specific and general suppression. Finally, we review and discuss a toolbox of cultural strategies and inputs, including SOM management, cover cropping, and rotation, which can be manipulated by growers and scientists to generate disease-suppressive soils and cropping systems.

## DISEASE SUPPRESSION IN FIELD SOILS

### TYPES OF DISEASE SUPPRESSION

#### Suppressive Soils

A suppressive soil is one in which "the pathogen does not establish or persist, establishes but causes little or no damage, or establishes and causes disease for a while but thereafter the disease is less important, although the pathogen may persist in the soil" (Baker and Cook, 1974). Alternatively, a conducive (nonsuppressive) soil is one in which disease occurs and progresses. Suppressive soils have been the subject of considerable research and have been reviewed extensively (Alabouvette, 1986; Alabouvette et al., 1996; Baker and Cook, 1974; Cook and Baker, 1983; Fravel et al., 2003; Hornby, 1983; Schneider, 1982; Shipton, 1981; Weller et al., 2002).

Classic suppressive soils are generally — although not exclusively — either soils (1) consistently suppressive over many years because of stable soil physical, chemical, and biological properties (long-standing suppression, e.g., *Fusarium* wilt suppressive soils, Fravel et al., 2003; Hornby, 1983), or (2) that become suppressive through serial monocropping (e.g., take-all suppressive soils, Fravel et al., 2003; Shipton, 1981; Weller et al., 2002). We discuss in this chapter soil suppressiveness generated through soil or systems management strategies and not serial monocropping or long-standing suppressive soils. However, we refer to the literature on suppressive soils, because many of the mechanisms of suppression in those soils likely work in the suppressive systems we discuss.

## General and Specific Suppression

Historically, suppressiveness to soilborne diseases in field soils has been divided into two major categories: general and specific. General suppression is generated by the sum of the activities of the overall microbial biomass, and specific suppression is generated by the activities of one to a few populations of organisms (Cook and Baker, 1983; Gerlagh, 1968; Hoitink and Boehm, 1999; Weller et al., 2002). According to Cook and Baker (1983):

> *General suppression is related to the total amount of microbiological activity at a time critical to the pathogen. A particularly critical time is during propagule germination and pre-penetration growth in the host rhizosphere. The kinds of active soil microorganisms during this period are probably less important than the total active microbial biomass, which competes for the pathogen for carbon and energy in some cases and for nitrogen in other cases, and possibly causes inhibition through more direct forms of antagonism. In a sense, general suppression is the equivalent of a high degree of soil fungistasis. No one microorganism or specific group of microorganisms is responsible by itself for general suppression.*

In contrast, specific suppression is considered to be generated through the activities of one or several specific populations of organisms. "Specific suppression operates against a background of general suppression but is more qualitative, owing to more specific effects of individual or select groups of microorganisms antagonistic to the pathogen during some stage in its life cycle" (Cook and Baker, 1983).

## OM-MEDIATED GENERAL SUPPRESSION IN CONTAINER MIXES

Our understanding of OM-mediated general suppression is largely derived from work on *Pythium* damping-off (DO) suppression in peat and compost-based soilless container mixes (Hoitink and Boehm, 1999). An understanding of this body of work is fundamental to understanding OM-mediated general suppression in field soils. For this reason, we will first describe this well-documented system.

### Diseases Caused by *Pythium* spp.

OM-mediated biological control of diseases caused by *Pythium* spp. has been widely documented in container systems (Boehm et al., 1997; Chen, 1988a; Erhart and Burian, 1997; Hoitink and Boehm, 1999). Lightly decomposed organic matter colonized by a diverse microflora is typically suppressive to diseases caused by *Pythium* spp. in container systems (Hoitink and Boehm, 1999). This phenomenon is being exploited by nursery growers in compost-amended container mixes. Growers are now using composted materials, including various tree barks, in their container systems to suppress root rots in woody perennials. Growers have observed that different types of organic materials suppress root rots for varying lengths of time. This phenomenon has been documented in the laboratory; composted hardwood barks suppress root rots for ca. 2 years, composted pine barks suppress for up to 9 months, and, in general, peats are not suppressive for more than several weeks to months (described more fully below) (Hoitink, 1980; Hoitink et al., 1991). These observations led to further investigations on the relationship between OM quality and the duration of disease suppression.

The sphagnum peat system has been used as a model system to investigate the impact of OM quality on *Pythium* DO suppression (Boehm and Hoitink, 1992; Boehm et al., 1997). Peats harvested from the top layers of a bog (very slightly decomposed sphagnum moss, or light peat) are suppressive to *Pythium* DO; all other peats (e.g., dark peat) are typically conducive to disease.

As a light peat decomposes, it loses the ability to suppress *Pythium* DO. Suppression is supported for 1 to 7 weeks. The loss of suppressiveness is related to (1) a decline in microbial activity as measured by the rate of hydrolysis of fluorescein diacetate (FDA) activity (Boehm and Hoitink, 1992); (2) a shift in the culturable bacterial community composition from one in which 10% of the isolates have the potential to suppress *Pythium* DO to one in which less than 1% have this potential; and (3) a decline in carbohydrate content, as determined by $^{13}C$ NMR spectroscopy (Boehm et al., 1997).

Functionally, OM-mediated suppression of *Pythium* DO in container experiments is typically characterized by the following phenomena:

1. Many types and sources of organic amendments consistently generate suppression.
2. Suppression is generated immediately after high-rate organic amendment (unless the organic substrate is raw; see section "SOM Quality: Early Stages of Decomposition").
3. Suppression is of fairly short duration (typically weeks to 1 year).
4. Suppression is positively related to microbial activity (specifically FDA activity)

In this chapter, we consider systems that exhibit these phenomena examples of OM-mediated general suppression.

## Diseases Caused by *Phytophthora* spp.

OM-mediated suppression of diseases caused by *Phytophthora* spp. is also considered to be a result of general suppression (Hoitink, 1980; Hoitink and Boehm, 1999), although there is little data on the relationships between OM content or quality and suppression of *Phytophthora* diseases. However, many types of organic materials suppress diseases caused by *Phytophthora* spp., the duration of suppression is similar to that for *Pythium* spp. diseases, and suppression occurs soon after organic amendment (Daft et al., 1979; Hoitink et al., 1975; Hoitink, 1980). However, in contrast to suppression of *Pythium* spp. diseases, in which pathogen populations typically do not decline (Gugino et al., 1973), in most documented systems *Phytophthora* spp. propagules undergo microbial colonization, germination, and lysis (Gray et al., 1968; Hoitink et al., 1977; Nesbitt et al., 1979). However, as is true in many OM-mediated suppressive systems, other mechanisms are also likely at work (Hardy and Sivasithamparan, 1991).

The best-described example of OM-mediated suppression of *Phytophthora* root rot comes from work on root rot of rhododendron. Composted hardwood bark (CHB)-amended container mixes suppress *Phytophthora* root rot of rhododendron under commercial nursery conditions for up to 2 years (Hoitink et al., 1977). In greenhouse bioassays, *Phytophthora* root rot of lupine was suppressed in a fresh CHB–sand medium, whereas a peat–sand mix was conducive to the disease (Hoitink et al., 1977). *Phytophthora* mycelia buried in fresh CHB were colonized by bacteria and protozoans and lysed within 48 h, whereas mycelia buried in the peat–sand mix lysed after 4 d and were not colonized by microorganisms. Zoospores and encysted zoospores, but not chlamydospores, were lysed when exposed to leachates from fresh CHB; zoopores encysted and germinated when exposed to leachates from the peat or 2-year-old CHB mixes (Hoitink et al., 1977). In similar work in North Carolina, CHB was highly suppressive and composted pine bark (CPB) was moderately suppressive to lupine root rot (causal agent *P. cinnamomi*; Spencer and Benson, 1982). Several other studies have reported OM-mediated suppressiveness to *Phytophthora* root rots. Vermicomposted cattle manure suppressed *Phytophthora* root rot (causal agent *P. nicotianae* var. *nicotianae*) of container-grown tomato (Szczech et al., 1993), and an oat straw–chicken manure mulch mixed with sand suppressed *Phytophthora* root rot of *Banksia* (causal agent *P. cinnamomi*; Dixon et al., 1990).

## OM-Mediated General Suppression In Field Soils

OM-mediated general suppression has been documented in container systems and is at present used commercially as a disease control measure. Can this strategy be applied to field soils? We are increasingly looking to natural systems for strategies we can adapt to biological agricultural systems management. Are natural soil systems suppressive to soilborne plant diseases, and is SOM content and quality implicated in suppressiveness? We first describe some examples of general suppression in natural soil systems. In the next section, we describe examples of general suppression in field agricultural soils.

### Natural Soil Systems

In Australia, certain eucalyptus forest soils are suppressive to *Phytophthora* root rot of eucalyptus (causal agent *P. cinnamomi*). These suppressive soils have a thick organic litter layer that supports a high level of microbial activity. The litter overlays a mineral soil of relatively low microbial activity. Introducing *P. cinnamomi* propagules into the litter layer results in their destruction by hyphal lysis and sporangial abortion, whereas this is not observed in the mineral soil. Adding increasing amounts of suppressive litter to mineral soil proportionately increased suppressiveness, as indicated by lysis of hyphae and production of abortive sporangia (Nesbitt et al., 1979). In another experiment in which increasing amounts of suppressive litter was added to a conducive lateritic field soil, hyphal lysis occurred within 24 h in soils containing 50% or more organic matter and reached a maximum level of lysis in 3 to 5 days. In unamended lateritic soil, very little lysis was observed throughout this period (Gray et al., 1968).

Forested soils in the Brazilian Amazon suppress DO caused by *Pythium* spp., and suppressiveness is lost as tillage intensity, and therefore rate of forest litter loss, increases (Lourd and Bouhot, 1987). In a related work, forest soils (clear-cut 2 years previously) in Oregon did not support survival of inoculated *Phytophthora* (*P. drechslera, P. cryptogea, P. megasperma,, P. cactorum,* and an unidentified *Phytophthora* species) and *Pythium* spp., whereas these fungal plant pathogens survived and caused disease in cultivated nursery soils (Hansen et al., 1990; Pratt et al., 1976). Unfortunately, no data were taken on microbial activity or SOM quality to determine whether these factors were related to forest soil suppressiveness (Hansen et al., 1990).

### Field Agricultural Systems

*Orchard Systems*

One of the most notable examples of commercially viable OM-mediated disease suppression in agricultural field soils is organically managed avocado orchards in Australia. Orchards were undersown with *Lablab purpureus* and forage sorghum or corn in the summer and *Lupinus angustifolius* during the winter. All cover crops were slashed and incorporated lightly. Organic amendments such as barley straw, sorghum residues, and native grass hay were also added to soil under the trees as a mulch layer, and poultry litter and dolomite were spread on the surface of the mulches to stimulate rapid decay (Malajczuk, 1979, 1983). After several years, the soil suppressed *Phytophthora* root rot of avocado (causal agent *P. cinnamomi*). Suppressive soils were characterized by high levels of microbial activity, organic matter, and calcium. In a related work, rate of hydrolysis of FDA was positively, and total fungal and actinomycete populations were negatively, related to infectivity of *P. cinnamomi* in oat straw–chicken manure mulch-amended avocado plantation field soils (You and Sivasithamparan, 1994, 1995).

Recent work in California on the use of organic mulches to suppress root rot of avocado has shown that 2 years of annual application of eucalyptus mulch (15 cm deep) prevented *Phytophthora* propagule growth and survival and enhanced root growth in the mulch layer but not in the mineral soil (Downer et al., 2001). Microbial activity (rate of hydrolysis of FDA) was significantly higher in the mulch layers than in mineral soil and was positively associated with lysis of *Phytophthora* propagules (Downer, 2001).

## The Chinampa Agricultural System

The chinampa agricultural system in the Valley of Mexico is ca. 2000 years old (Coe, 1964). The soils in this system are amended each year with large quantities of canal sediments, animal manures, and plant residues (Lumsden et al., 1987). Modern-day plant pathologists noticed that there were fewer soilborne diseases on crops grown in the chinampa systems than in crops grown nearby in conventional fields (Lumsden et al., 1987). Investigations into this phenomenon reported that DO caused by indigenous *Pythium* spp. was reduced in these soils relative to that in conventionally managed soils and suppression was positively correlated to soil dehydrogenase activity (Lumsden et al., 1987). In addition, inoculated *P. aphanidermatum* did not germinate as readily in the chinampa soils even after nutrient addition (Lumsden et al., 1987). The authors concluded that this traditional agricultural system, through its reliance on OM-mediated fertility, generated suppressiveness due in part to biologically mediated fungistasis (Lumsden et al., 1987).

## Field Soils Amended with Paper Mill Residuals

Annual soil amendment with fresh paper mill residuals (PMR; applied at 20 and 30 dry Mg ha$^{-1}$) and composted PMR (applied at 35 and 70 dry Mg ha$^{-1}$) to a sandy loam field soil in Wisconsin suppressed *Pythium* DO of cucumber 1 month after amendment in the first year (with no difference in degree of suppressiveness among treatments) as determined by *in situ* bioassays (Stone et al., 2003). Suppression was lost by 6 months after amendment as determined by growth chamber bioassays (A.G. Stone, unpublished data).

In an adjacent field trial in which snap bean was planted each year for 2 years, treatments included PMR applied to soils both years at 10, 20, and 30 dry Mg ha$^{-1}$; PMR composted without a bulking agent; or composted with bark at 35 and 70 dry Mg ha$^{-1}$ applied both years. All amendments suppressed common root rot of snap bean in the second year (causal agent *Aphanomyces eutiches*; Stone et al., 2003). Root rot severity was too low to evaluate in the first year of the trial. Suppression was generated by both raw and composted PMR amendments in field-grown beans planted 4 weeks after amendment, and suppression was lost by 5 months after amendment as evaluated by greenhouse cone tube bioassays (Cespedes Leon, 2003; Stone et al., 2003).

## Field Soils Amended with Dairy Manure Solids

Most of the previously described examples of OM-mediated general suppression in field soils involve suppression of Oomycete pathogens: *Pythium*, *Phytophthora*, and *Aphanomyces* spp. In this system, we investigated the impact of dairy manure solids (DMS) applications on the root rot disease complexes of sweet corn (causal agents *Drechslera* spp., *Phoma* spp., and *Pythium arrhenomanes*) and snap bean (causal agents *Fusarium solani* and *Pythium* spp.) in the Willamette Valley of Oregon. We then related disease suppression to indicators of SOM quality. Fresh DMS was applied at 16.8 and 33.6 dry Mg ha$^{-1}$ and composted DMS was applied at 28 and 56 dry Mg ha$^{-1}$ each spring for the first 2 years of the trial. Soils were sampled and evaluated with growth chamber cone tube bioassays 2, 6, and 12 months after amendment (Darby, 2003). Root rots of sweet corn and snap bean (as well as cucumber DO) were suppressed 2 months after amendment in all but the low rate of fresh DMS in the first year and in all treatments in the second year (Darby, 2003). Suppression of all diseases was lost between 2 and 6 months after amendment (Darby, 2003). Relationships between soil active OM fractions and disease suppression in this study are described in the section "SOM Quality: Later Stages of Decomposition."

## OM-MEDIATED GENERAL SUPPRESSION AND SOM QUALITY

In systems associated with OM-mediated general suppression, suppression typically occurs as a result of the activation of the indigenous soil microbial community and not of microbial inoculation. Lockwood (1990) stated that his extensive work on the manipulation of soil substrates (energy) for managing plant diseases *involved the exploitation of the indigenous soil microflora, which to me have been much neglected in favor of intensive research on individual antagonistic microorganisms. Possibly, the utilization of the broadly based indigenous soil microbial community could offer greater stability and reliability than are often achieved with single species or strains, since what is sought is the enhancement of natural biological controls already functioning to some extent in soils.*

This sentiment is fundamental to general suppression of plant diseases through the manipulation of SOM. Organisms capable of suppressing a wide range of soilborne diseases through a diversity of mechanisms typically exist in field soils; what is lacking is not biocontrol organisms but the environment that supports high populations and activities related to biological control. The next section addresses this issue: how can farmers manage organic matter in field soils to most efficiently manage plant diseases through general suppression?

### Early Stages of Decomposition

In this section, we review the competitive saprophytic potential of several important genera of fungal plant pathogens, as this impacts the inoculum potential of the pathogen after soil amendment with raw organic residues.

Fresh plant residues or organic wastes support high microbial activity and the activities of biological control organisms in the soil, but they also support the growth and infection potential of saprophytic plant pathogenic fungi. Intrinsic growth rate on a particular substrate (Garrett, 1956), the content and availability of the substrate in the organic material, tolerance to the antagonism or competition of other soil microbes (Rush et al., 1986), and presence of specific antagonists in the soil system (Nelson et al., 1983; Toyota et al., 1996) can play a role in determining the success or failure of a soilborne fungus to colonize fresh organic residues in field soils.

Because some *Pythium* spp. are good primary saprophytes, fresh plant residues incorporated into soil cause an initial increase in *Pythium* spp. populations and the severity of *Pythium* diseases (Grunwald et al., 2000; Hancock, 1977; Rothrock and Hargrove, 1988; Rush et al., 1986; Sawada et al., 1964; Wall, 1984; Watson, 1970). However, suppression is typically generated after several weeks to 1 month of decomposition (Grunwald et al., 2000). The ability of *Pythium* spp. to colonize fresh residues is dependent on rapid spore germination together with very rapid vegetative growth (Stanghellini, 1974). *Pythium ultimum* propagules have been reported to germinate, grow saprophytically on organic matter, and produce new sporangia within 44 h of organic matter incorporation. Populations typically decline slowly thereafter; a half-life of approximately 30 d has been reported in field soils (Hancock, 1981).

*Pythium* spp. are good colonizers of fresh organic residues, but they are not good competitors; prior colonization of organic residues by other microorganisms typically reduces colonization by *Pythium* spp. (Barton, 1961; Hancock, 1977; Rush et al., 1986). For example, wheat chaff collected 1 week after harvest was colonized 90% by inoculated *P. ultimum*, but chaff collected from the field 3 weeks later and then inoculated was only 10% colonized. Autoclaved 4-week-old chaff was colonized ca. 80%, indicating that the biological components of the chaff contributed to suppression of *Pythium* colonization (Rush et al., 1986). Pathogenic species of *Pythium* can also be outcompeted by nonpathogenic species of *Pythium*. *P. nunn*, a highly competitive saprophytic *Pythium* spp., can outcompete pathogenic *P. ultimum* for nutrients and reduces *P. ultimum* numbers even if introduced to a fresh residue after *P. ultimum* colonization (Paulitz and Baker, 1988).

*Phytophthora* and *Aphanomyces* species are typically considered poor saprophytes, but several important exceptions should be taken into account when considering general strategies for controlling these genera. For *Phytophthora* spp., *P. infestans* and *P. megasperma* are considered hemibiotrophs with very little saprophytic potential (Weste, 1983). *P. cinnamomi* and *P. cactorum* can survive either as parasites or saprophytes, depending on environmental conditions (Weste, 1983). *P. parasitica* extensively colonizes papaya residues incubated in field soils within 48 h of inoculation and subsequently produces large numbers of chlamydospores (Trujillo and Hine, 1965). There is little additional evidence for extensive saprophytic colonization of organic matter by other *Phytophthora* spp.

Less evidence of saprophytic activity by *Aphanomyces* has been reported. *Aphanomyces euticles* is considered to have very weak competitive saprophytic potential, because hyphal growth has been observed only in sterilized soil columns and not in natural field soils (Papavizas and Ayers, 1974; Sherwood and Hagedorn, 1961). In contrast, *A. cochloides* increases in crop residues (MacWithey, 1966).

*Fusarium* spp. have good competitive saprophytic abilities and populations can increase after organic amendment. Park (1958) termed *Fusarium oxysporum* a soil inhabitant, because it can persist in soil, is tolerant to antagonism, and can colonize organic substrates. However, similar to *Pythium* spp., many *Fusarium* spp. are poor competitors and cannot colonize organic substrates previously colonized by other organisms (Park, 1958). Precolonization of soils or organic matter with two nonpathogenic *F. oxysporum* isolates reduced *F. solani* f. sp. *pisi* growth and infection of pea (Oyarzun et al., 1994). In studies of soil aggregate colonization, closely related fungal species (other *F. oxysporum formae speciales*) strongly inhibited colonization by *Fusarium oxysporum* f. sp. *raphani*. Other fungal genera moderately, and bacterial species mildly, inhibited colonization. *Burkholderia cepacia*, an antibiotic-producing bacterial species, also strongly inhibited colonization (Toyota et al., 1996).

*Rhizoctonia solani* has high competitive saprophytic ability and degrades cellulose as well as simple sugars and hemicelluloses *in vitro* and in soil systems (Bateman, 1964; Blair, 1943; Papavizas, 1970). *R. solani* populations typically increase during early stages of cover crop or raw residue decomposition and decline as the more labile constituents of the material are exhausted (Croteau and Zibilske, 1998; Grunwald et al., 2000; Papavizas, 1970). This trend is similar to that of *Pythium* spp., but the duration of saprophytic growth is typically longer for *R. solani* than for *Pythium* spp. likely due to its capacity to degrade cellulose, its insensitivity to fungistasis, and a requirement for specific antagonists for suppression (discussed in detail later; Croteau and Zibilske, 1999; Grunwald et al., 2000; Lockwood, 1990).

Metabolic by-products of microbial decomposition of fresh plant residues can also be phytotoxic. The nature, intensity, and duration of phytotoxins released are controlled to a large degree by the type and quantity of amendment and the soil conditions; in general, cold, wet soils enhance production (Toussoun et al., 1968). In addition, phytotoxic reactions can increase plant root permeability and root exudates, factors that predispose plants to increased attack by pathogens (Linderman, 1989). Volatile chemicals released from decomposing plant material can also stimulate dormant pathogen propagules to germinate and grow. A good example is *Sclerotium* spp.; volatiles cause sclerotia to germinate, and extending mycelium can colonize fresh OM or infect susceptible roots (Punja, 1984). For these reasons, planting should be delayed after fresh organic matter is incorporated.

## Later Stages of Decomposition

After the most labile OM constituents (e.g., sugars, proteins, hemicellululoses) have been degraded, considerable energy remains in the organic material, and subsequent decomposition supports OM-mediated general suppression (Grunwald et al., 2000; Stone et al., 2001). As decomposition proceeds, the quality and quantity of the residual substrate dictates the duration of general suppression. This relationship is described for *Pythium* DO and for the root rot disease complexes of snap bean and sweet corn.

### Active OM and Suppression in a Compost-Amended Sand

As a step beyond soilless container mixes, the impact of compost decomposition on suppression of *Pythium* DO of cucumber was investigated in sand amended with composted separated DMS incubated in containers (Stone et al., 2001). DO was suppressed for 1 year after amendment. During the period when suppression was supported, the mass of total particulate organic matter (POM) as well as coarse and mid-sized compost-derived POM declined (Figure 5.1), whereas the composition of the total POM (as determined by $^{13}C$ NMR spectroscopy, Table 5.1) did not change. A change in total POM composition was detected after 1 year, although very little change in mass occurred. Therefore, suppressiveness was sustained by the degradation of the larger-particle-size, less-decomposed POM (Stone et al., 2001). In addition, composition of the suppressive POM was similar to that of unprotected POM (POM not physically protected from microbial attack through association with mineral soil particles) from a variety of soil and forest litter and organic horizons (Stone et al., 2001; Table 5.1).

### Active OM, Microbial Activity, and Suppression in a DMS-Amended Silt Loam

In the DMS-amended snap bean–sweet corn study, microbial biomass, free light fraction (LF) and FDA hydrolytic activity were negatively related to severity of root rot of corn and bean and DO of cucumber (Darby, 2003). β-glucosidase and arylsulfatase activities and soil content of occluded LF were not related to disease suppression. Only FDA hydrolytic activity was always predictive of disease suppression at every sampling date over a 2-year period in both amended and unamended field soils. In contrast, free LF content, when decomposed for a year after a very high rate of amendment, was as high as that of a recently amended suppressive soil but was not suppressive. Microbial biomass was more closely related to free LF content than to FDA activity (Darby, 2003). The lack of suppression in a soil of relatively high free LF content was likely due to the LF being too decomposed to support disease suppression (Darby, 2003). LF quality impacted suppressiveness in this system as reported previously in a compost-amended sand system (Stone et al., 2001; Table 5.1), rendering total LF content a less predictive indicator of disease suppression than FDA activity.

It is not surprising that free POM content is not consistently related to disease suppression. Organic-matter-mediated suppression is of very short duration when considered in organic matter

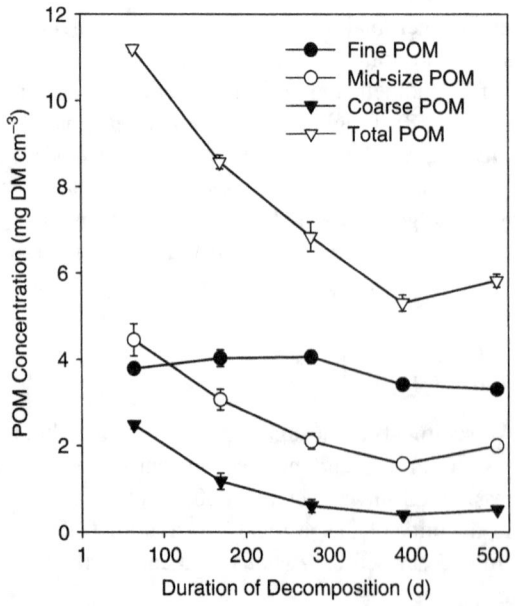

**FIGURE 5.1** Changes in total and size-fractionated POM concentration during decomposition in sand. Suppressiveness to *Pythium* damping-off was sustained from Day 53 to Day 375. (From Stone, A.G. et al., 2001. *Soil Sci. Soc. Am. J.* 65: 761–770. With permission.)

**TABLE 5.1**
**Relative Composition of Compost- and Soil-Derived POM/LF and Forest Soil Organic Horizons as Determined by 13C CPMAS NMR Spectroscopy**

| | 160–200 ppm Carbonyl/ Carboxyl | 110–160 ppm Aromatic | 45–110 ppm O-Alkyl | 10–45 ppm Alkyl | Suppressiveness or Compositional Similarity to Suppressive or Conducive POM[a] |
|---|---|---|---|---|---|
| Compost Day 4 | 4[b] | 12 | 70 | 12 | Less decomposed than POM |
| Compost POM Day 83 | 7 | 20 | 55 | 18 | Suppressive |
| Compost POM Day 391 | 7 | 20 | 54 | 19 | Suppressive |
| Compost POM Day 506 | 10 | 23 | 51 | 16 | Conducive |
| Total soil LF[c] | 6–12 | 16–28 | 39–57 | 16–26 | Similar to suppressive and conducive |
| Free soil LF[d] | 5–7 | 14–18 | 55–63 | 18–25 | Similar to suppressive |
| Occluded soil LF [d] | 7–11 | 15–20 | 33–45 | 28–45 | Similar to conducive |
| L horizon (forest soil)[e] | 5–8 | 14–23 | 54–58 | 16–22 | Similar to suppressive |
| Of horizon (forest soil)[e] | 5–11 | 15–22 | 39–55 | 23–28 | Borderline? |
| Oh horizon (forest soil)[e] | 7–11 | 13–23 | 44–48 | 23–34 | Similar to conducive |
| Aeh horizon (forest soil)[e] | 7–11 | 9–25 | 39–42 | 25–42 | Similar to conducive |

[a] As determined by the relative proportion of 160–200 and 45–110 ppm spectral areas.
[b] Percent contribution of the total ¹³C CPMAS NMR signal intensity for each carbon type.
[c] From Baldock, J.A. et al. 1992. *Biogeochemistry* 16:1–42.
[d] From Golchin, A. et al. 1994. *Aust. J. Soil Res.* 32:285–309.
[e] Forest soil horizons. From Hempfling, R. et al. 1987. *Z. Pflanzenernaehr. Bodenk.* 150:179–186 and Kögel-Knabner, I. et al. 1988. *Z. Pflanzenernaehr. Bodenk.* 151:331–340.
*Source*: Stone, A.G. et al., 2001. *Soil Sci. Soc. Am. J.* 65; 761–770. With permission.

time. Free LF typically resides in soils for 1 to 15 years (Carter, 1996), whereas suppressiveness is supported for several months to a year (Darby, 2003; Stone et al., 2001). It would therefore be expected that free LF content be strongly related to suppression during the first few months of decomposition but not thereafter; this is true in the DMS-amended system (Darby, 2003). Therefore, the rate of hydrolysis of FDA activity remains the best, albeit indirect, measure of OM quality related to OM-mediated general suppression of plant diseases in both soilless container mixes and field soils.

Root rots of snap bean and sweet corn are disease complexes involving multiple pathogens. In this study, we observed many of the phenomena associated with OM-mediated general suppression; to our knowledge, OM-mediated general suppression has not been reported previously for diseases caused by *Phoma* spp., *Drechslera* spp., or *Fusarium solani*, or for disease complexes caused by these pathogens.

### Active OM and Suppression of Pythium DO in Historically Forested Soils

Tillage affects active OM quality and quantity (Cambardella and Elliott, 1992) and should therefore impact general suppression. In historically forested soils in the Brazilian Amazon, suppression of *Pythium* DO was lost as cultivation intensified (Lourd and Bouhot, 1987). Eighty-two percent of undisturbed forested soils, 67% of forest nursery soils, 53% of managed forest soils, 31% of newly cultivated soils, and only 7% of intensively managed annually cropped soils were suppressive to *Pythium* DO. This is further evidence of the active organic matter pool supporting general suppression in field soils.

## ORGANIC MATTER QUALITY: AMENDMENT RATE AND SERIAL AMENDMENT

### High-Rate Organic Amendment

High-rate, single-term amendments can generate disease suppression in the first season after amendment. For example, in the Wisconsin PMR amendment study, PMR applied at 20 and 30 dry Mg ha$^{-1}$ suppressed *Pythium* DO of cucumber 1 month after amendment in the first year (with no difference in degree of suppressiveness), and rates of 10, 20, and 30 dry Mg ha$^{-1}$ suppressed common root rot of snap bean in the second year of amendment (root rot severity too low to detect treatment differences during the first year of the experiment; Stone et al., 2003). In the Oregon DMS amendment study, DMS amended at 33.6 dry Mg ha$^{-1}$ suppressed cucumber DO and root rots of snap bean and sweet corn. The 16.8 dry Mg ha$^{-1}$ rate was not suppressive in the first year (Darby, 2003).

Soil amendment at the rates described previously can suppress certain plant diseases, but what would be the environmental, agronomic, and economic consequences over the long term? It is important to realize that though the use of high-rate compost amendments for disease suppression in container systems might be agronomically, environmentally, and economically responsible, it might not be true for many field systems. In most field agricultural systems, annual high-rate applications of organic wastes such as manures or composts would pose significant problems in the short and long term. The following is a summary of the constraints associated with annual high-rate applications of organic wastes in field agricultural systems.

### Economics

Crop profitability, transportation costs, and level of demand from nonagricultural markets determine the distance that bulk organic materials can be economically hauled and field applied. With the trend toward larger individual and fewer total generators of manure, forest by-products, and green waste, the average hauling distance to cropland, and thus cost, increases (Emerson, 2003; Kellog et al., 2000; McKeever, 2003; Porter and Crockett, 2003; Wright et al., 1998).

### Environmental Considerations

High application rates can result in nutrient pollution of surface and groundwater, leading to eutrophication of aquatic ecosystems and health risks from direct consumption of nitrates or indirectly by increasing human pathogens such as *Pfisteria* (Kellog et al., 2000; Sharpley et al., 1999).

### Agronomic Considerations

For organic amendments of high salt contents, such as some manures and fish or food processing wastes, application rates might need to be limited to avoid salinity-related crop damage (Dickerson, 1996, 1999). Soil and crop nutrient imbalances can result from annual high-rate organic waste amendments; plant tissue nutrient excesses and imbalances can reduce yields and increase pest and disease problems (Cook and Baker, 1983; Graham, 1983; Phelan et al., 1996).

### Efficacy

General suppression generated by annual organic amendment does not suppress all soilborne diseases; some diseases require more sophisticated strategies. Even if a disease can be suppressed, for fields with very high pathogen populations, an agronomically acceptable level of biological control might not be possible in the short term (Johnson, 1994; Paulitz, 2000).

## Low-Rate Organic Amendment

In general, field studies that assess low-rate single-season organic matter amendments report highly variable impacts on disease incidence and yield (Lewis et al., 1992, Lumsden et al., 1986), whereas longer-term studies report more predictable improvements in yield, quality, and disease suppression (Asirifi et al., 1994; Daamen et al., 1989; Darby, 2003; Hannukala and Tapio, 1990; Workneh et al., 1993). In the second year of amendment in the Oregon DMS amendment study, both the low and high rates (16.8 and 33.6 dry Mg ha$^{-1}$, respectively) suppressed all three diseases, with no treatment difference in the degree of suppressiveness (Darby, 2003). In other words, the low rate of DMS amendment was not suppressive in the first year but was as suppressive as the high rate in the second year.

The mechanisms involved in generating disease suppression over several to many years of cover cropping and low-rate organic amendment have not been elucidated. A probable explanation is that single-year, low (agronomic)-rate organic amendments might not significantly increase total or active carbon fractions or microbial biomass, which regulate soil moisture, nutrient mineralization, soil physical properties, and microbial community composition and activities (Darby, 2003; Drinkwater et al., 1995; Wander et al., 1994). Unfortunately, most work on organic amendment disease suppression has been conducted in single-year trials, so little is known about the impact of serial amendment on disease suppression.

## Organic Soil Management, or Long-Term Soil-Building

Comparative studies of organic and conventional cropping systems have been used to study the effects of serial (annual) organic amendment, or soil building, on soil properties and disease incidence. This is because organic soil management is typified by some sort of annual organic amendment, either cover cropping or the application of raw or composted organic materials, and farms must be under organic management for more than 3 years to be considered organic. Microbial biomass, microbial activity, and biologically active carbon are typically higher in soils sampled from organically managed farms than from soils from conventionally managed farms (Andrews et al., 2002; Fraser et al., 1988; Gunapala and Scow, 1998; Reganold et al., 1993; Wander et al., 1994).

A literature review of disease incidence and severity in comparative farming systems trials concluded that root diseases were typically lower on organic and low-input farms than on conventionally managed farms, but that there was no obvious trend for foliar diseases (very little data available on foliar diseases; van Bruggen, 1995). In a comparative study of organic and conventionally managed vineyards, organically managed vineyard soils sustained 9% root necrosis due to *Fusarium oxysporum* and *Cylindrocarpon* spp., whereas conventionally managed soils sustained 31% (Lotter et al., 1999). Drinkwater et al. (1995) investigated the differences between organic and conventionally managed tomato production systems in the Central Valley of California. They reported that corky root on tomatoes grown in organically managed field soils was significantly less severe than on tomatoes grown in conventionally managed soils. Corky root severity on tomatoes grown in soils managed organically for 3 years or less was not different than that on tomatoes grown in conventionally managed soils. Microbial activity (FDA) was significantly higher on organic than on conventional farms, although soil microbial activity on farms in transition (under organic management for 1 to 3 years) was not higher than activity on conventional farms (Workneh et al., 1993).

## OM-Mediated Specific Suppression

All the previous discussions have centered on OM-mediated general suppression. However, not all diseases are reliably suppressed in container mixes or field soils by general suppression alone. For example, in the literature of both suppressive soils (Fravel et al., 2003; Hornby, 1983; Shipton, 1981; Weller et al., 2002) and compost-amended container mixes (Hoitink and Boehm,

1999; Trillas-Gay et al., 1987), suppression of diseases caused by *Rhizoctonia solani* and *Fusarium oxysporum* has been generally considered to be due to specific suppression, or suppression generated through the activities of one or several specific populations of organisms (Cook and Baker, 1983).

## Diseases Caused by *Fusarium oxysporum*

Organic amendments and plant residues suppress diseases caused by *Fusarium oxysporum* in soilless container mixes (Chef et al., 1983; Pera and Calvet, 1989; Pharand et al., 2002; Trillas-Gay et al., 1987), field soils incubated in containers (Oritsejafor and Adeniji, 1990; Pera and Filippi, 1987; Serra-Whitling et al., 1996), and field soils (Lodha, 1995; Sequeira, 1962). General suppression (Serra-Whittling et al., 1986), specific antagonists (Trillas-Gay et al., 1987), propagule lysis (Oritsejafor and Adeniji, 1990; Sequiera, 1962), induced resistance (Pharand et al., 2002), and nonbiotic factors (Kai et al., 1990) have been implicated in OM-mediated suppressiveness of *Fusarium* wilts. Other mechanisms implicated in *Fusarium* wilt suppressive soils, such as competitive colonization of substrate and roots (see section "Mechanisms Involved in Disease Suppression"), can also play a role in OM-mediated disease suppression. However, little is known about the relationships between organic matter quality and suppression of diseases caused by *F. oxysporum* in container systems and field soils.

## Diseases Caused by *Rhizoctonia solani*

The genus *Rhizoctonia* contains a number of important plant pathogens, the 12 currently recognized anastomosis groups of *R. solani* being the most important. *R. solani* causes DO of seedlings, root rots, stem cankers, and aerial blights on a wide range of grain, vegetable, and fruit crops worldwide. In contrast to suppression of *Fusarium* wilts, OM-mediated suppression of *Rhizoctonia* DO in compost-amended container mixes is relatively well described. Decades of research on OM-mediated suppression of *R. solani* in both field soils and soilless container media indicate that diseases caused by *R. solani* can be suppressed by adding SOM or specific microbial antagonists, or both. For *R. solani,* suppression is viewed as specific because suppression is not related to microbial activity (Chung et al., 1988a; Grunwald et al., 2000; Scheuerell, 2002), suppression can be transferred from soil to soil (Cook and Baker, 1983), augmentation of compost or peat with antagonists is often required to generate suppression in soilless media (Krause et al., 2001), and OM-mediated suppression is associated with dramatic population increases of antagonists known to inhibit *R. solani* (Huang and Kuhlman, 1991a, 1991b). In addition, in two broad surveys of compost-amended soilless container mixes, only 20% and 18% of compost-amended container media suppressed *R. solani* DO, whereas 80% and 68% suppressed *P. ultimum* DO (Krause et al., 1997; Scheuerell, 2002).

Aggressive isolates of *R. solani* are difficult to control because of a number of intrinsic properties: a wide host range, large sclerotia insensitive to fungistasis and resistant to decomposition, rapid colonization of fresh organic matter, extensive mycelial growth, mycelium of high biphenolic content that is relatively resistant to degradation, hyperparasitic potential, and capacity to escape soil competition under humid conditions by growing on surface organic matter or aerial plant-to-plant spread (reviewed in Papavizas, 1970). Although organic amendments in some cases suppress diseases caused by *R. solani,* some amendments enhance the saprophytic and pathogenic capacity of *R. solani.* In a survey of compost products blended with peat, more compost samples significantly enhanced *R. solani* DO than suppressed the disease (Scheuerell, 2002).

Whether diseases caused by *R. solani* are suppressed, unaffected, or enhanced by organic matter amendment is modulated by a complex interaction of biotic and abiotic factors. The literature describing OM-mediated management of diseases caused by *R. solani* is extensive and replete with variable and apparently contradictory results (Cook and Baker, 1983; Lewis et al., 1992; Manning and Crossan, 1969; Papavizas, 1970; Papavizas et al., 1975). Considering the great amount of inherent variability across *Rhizoctonia* spp. and isolates, plant susceptibility, soil characteristics, and other environmental factors that influence disease suppression, the lack of uniform, concise management recommendations should not be surprising (Baker et al., 1967). In addition, care must be taken when attempting to develop on-farm management strategies based on research results generated in soilless container media or field soils incubated in containers. The type and rates of organic matter used in potting media are not typically realistic for field application because of logistical, economic, and environmental reasons, and results from disease assays performed in containers often do not correlate well with results from field trials (Manning and Crossan, 1969; Papavizas et al., 1975). For these reasons, we do not offer prescriptive solutions for OM-mediated suppression of diseases caused by *R. solani* in field soils, but instead summarize some general trends that emerge from the literature in the hope of generating research hypotheses for future work in this important area.

In discussing management of *R. solani* from a specific suppression viewpoint, we focus on two key factors: (1) organic matter quality, and (2) activity of specific microbial antagonists. Concepts developed from research on soilless media are presented first and then expanded to more complex field soils.

## Soilless Container Media

The two key factors listed previously have been thoroughly studied in peat- and compost-amended container media (Hoitink and Fahey, 1986; Hoitink et al., 1993; Hoitink and Boehm, 1999; Quarles and Grossman, 1995; Tahvonen, 1982). Organic matter in the initial stages of decomposition is not suppressive to *R. solani* seedling DO (as described previously in the section "Organic Matter Quality: Early Stages of Decomposition"). Lack of suppression is attributed to high levels of easily decomposable OM that support saprophytic growth of both the pathogen and the antagonists, and downregulate the induction of parasitism genes in specific antagonists (Chung et al., 1988a; Cohen et al., 1998; Kuter et al., 1988; Nelson and Hoitink, 1983; Nelson et al., 1983). Stabilization (composting) of OM reduces the potential for saprophytic growth of *R. solani*, but pathogenicity is not reduced until the compost has been sufficiently recolonized by specific microbial antagonists. Recolonization is strongly influenced by the moisture content of curing compost. Moisture contents of 15 to 34% permit fungal growth (including that of *R. solani*) and prevent regrowth of bacterial biological control agents and are therefore more conducive to disease; in contrast, moisture contents of 45 to 55% permit colonization by a full spectrum of competitive saprophytes and antagonists, which increases the likelihood of colonization by specific antagonists (Hoitink et al., 1998).

Adequate stabilization of compost is relatively easy to achieve, but natural recolonization by specific antagonists of *R. solani* is random, often resulting in inconsistent or insufficient suppressive properties (Kuter et al., 1983; Ringer et al., 1997; Scheuerell, 2002; Schuler et al., 1989; Stephens et al., 1981). Only 3 out of 5000 bacterial strains isolated from suppressive soilless media consistently suppressed DO caused by *R. solani in vivo* (Harris et al., 1994). In comparison, a similar study revealed that 10% of bacterial isolates suppressed DO caused by *P. ultimum* (Boehm et al., 1997). The low frequency of *R. solani* suppression observed with compost products is not commercially viable; recent work has demonstrated that augmentation of composts with specific antagonists improves the consistency of suppression (Krause et al., 2001; Kwok et al., 1987; Nakasaki et al., 1998; Ryckeboer et al., 1999; Weindling and Fawcett, 1936).

Added antagonists are effective only when operating against a background of general suppression. For example, suppressiveness of compost was not affected by amendment with small quantities of cellulose, but suppression was destroyed by amendment with 20% cellulose (Chung et al., 1988a). With excess cellulose, saprophytic increase of both *R. solani* and antagonistic *Trichoderma* spp. was observed; however, the *Trichoderma* spp. did not parasitize *R. solani,* most likely due to high levels of free glucose that are known to suppress antibiotic production and parasitic activity in *Trichoderma* (Chung et al., 1998a). Suppression can also be lost as the organic amendment or residue decomposes and the substrates that support the activity of specific antagonists are depleted (Krause et al., 2001). Therefore, successful inhibition of *R. solani* in soilless media relies on maintaining environmental conditions that support (1) general suppression, (2) colonization by specific antagonists, and (3) the activity of specific antagonists.

## Field Soils

The key concepts observed in soilless media can serve as a foundation for the interpretation of data on diseases caused by *R. solani* in field soils. High cellulose content negated suppression of DO caused by *R. solani* in soilless media (Chung et al., 1988a); Chung et al. (1988b) related this observation to the complexities of management of diseases caused by *R. solani* in field soils. Large volumes of fresh crop residues left on the soil surface in arid agricultural cropping systems increase the incidence and severity of *Rhizoctonia* root rot, called bare patch, of wheat in the Pacific Northwest of the U.S. and in Australia (Rovira, 1986; Weller et al., 1986). In these arid regions, the standing residues are of very low moisture content and not readily colonized by other saprophytic organisms; as a result, a very large volume of undecomposed plant residue is available to support saprophytic growth of *R. solani*. In contrast, buried residues decay much more rapidly (three to four times faster) than surface residues, reducing the window for saprophytic growth of *R. solani*. Colonization of fresh residue by *R. solani* peaks 2 to 4 days after incorporation; therefore, soil conditions at the time of incorporation, especially soil moisture content, are critical for increasing competition for added substrate (Papavizas, 1970). Cultivating wheat fields several weeks before planting physically breaks down residue, mixes it with the soil and its associated microbial community, and disrupts hyphal networks of the pathogen. This strategy suppresses bare patch, although the volume of plant residues applied to the soil is equivalent in the tilled and no-till systems (Rovira, 1986).

Suppressiveness can generally be generated by long-term curing of compost or by applying the material to a field soil several months before planting a susceptible crop (Tuitert et al., 1998; Lumsden et al., 1983a). However, simply ensuring that OM is thoroughly precolonized to avoid saprophytic increase of *R. solani* is not necessarily sufficient to make OM suppressive in field soil; for example, stabilizing dairy manure by composting did not increase the suppressiveness of dairy manure (Voland and Epstein, 1994). In addition, black scurf of potato (causal agent *R. solani* AG-3) was suppressed to significantly different degrees in field soil after amendment with two different composted dairy manure sources, although produced by similar methods (Tsror et al., 2001). Other strategies that can increase the consistency of suppression include manipulation of the soil environment to increase the population of specific indigenous antagonists and soil amendment with OM fortified with antagonists (Chet and Baker, 1980; Huang and Kuhlman, 1991a; Nelson et al., 1994).

Suppression of diseases caused by *R. solani* in field soils, as in soilless media, relies on reducing the saprophytic potential of *R. solani* throughout the bulk soil and rhizosphere while protecting root tip infection sites from pathogen ingress. It is thought that reduction of saprophytic activity through competition for nitrogen (by amendment with organic residues of high C:N ratio) effectively limits *R. solani* infection to the inoculum contacting the rhizoplane where specific antagonists can act through antibiosis and direct parasitism (Davey and Papavizas, 1963).

Soil populations of specific antagonists sufficient to sustain suppression can occur in field soils in one of three ways. Some field soils are naturally suppressive because of robust indigenous antagonistic populations. However, very few naturally suppressive soils have been identified. As a result, research has focused on enhancing low levels of indigenous antagonists or introducing biocontrol agents. Enhancing indigenous antagonists has most readily occurred by repeatedly cultivating specific crop, cover crop, or rotation crop species and cultivars that support growth of effective antagonists in the rhizosphere (Mazzola and Gu, 2002; Weller et al., 2002). Although this can be effective, plant selection relies on trial and error and researchers lack phenotypic or molecular markers for identifying plant genotypes that selectively increase specific antagonists (Weller et al.,

2002). In addition, indigenous antagonists of *R. solani* can be selectively enriched by amending soil with organic residues such as chitin, but the exact mechanism of disease suppression is not known (Henis et al., 1967).

Introduction of biocontrol agents to field soils has received considerable attention, although commercially viable disease control has been difficult to achieve. Successful colonization requires sufficient bioavailable food resources and manipulation of environmental conditions such as soil pH to optimize antagonist growth (Chet and Baker, 1980; Katznelson, 1940; Papavizas, 1970; Weindling and Fawcett, 1936). Although inoculation with biocontrol agents incurs an additional production cost, it can be necessary when the indigenous microbial community has been radically altered through the application of biocides or multiple selective forces such as cultivation, modified pH and conductivity, high soil nutrient contents, and irrigation (Deacon and Berry, 1993; Quarles, 1997).

In summary, generating soils suppressive to diseases caused by *R. solani* will require a cropping-system-specific and site-specific suite of management strategies. Successful inhibition of *R. solani* relies on maintaining soil environmental conditions that support both general competition for OM colonization and specific activities of antagonists. The following is a summary of factors that should be considered for cropping systems management of diseases caused by *R. solani*:

1. Isolates of *R. solani* vary in their saprophytic, competitive, and parasitic abilities (Papavizas, 1970; Papavizas et al., 1975).
2. Disease is favored by minimum-tillage systems (Bockus and Shroyer, 1998; Cook and Haglund, 1991), surface residues of low moisture content (Keinath et al., 2003; Rickerl et al., 1992; Stephens et al., 1994), amendment with OM not previously colonized by microbes (Bailey et al., 2000), neutral pH (Chet and Baker, 1980), soils of low moisture content (Gill et al., 2001), high connectivity of soil pore spaces (Otten et al., 1999), low soil bulk density (Gill et al., 2001; Harris et al., 2003; Otten et al., 2001), residual herbicides (Altman and Campbell, 1977), and excess available nitrogen at amendment incorporation (Kundu and Nandi, 1984; Papavizas, 1970).
3. Disease may be suppressed by surface tillage (Lucas et al., 1993; Rovira, 1986) or deep tillage (Tan and Tu, 1995), delayed planting after organic amendment (Dabney et al., 1996; Kundu and Nandi, 1985; Lumsden et al., 1983a; Papavizas and Davey, 1960), rotation with nonhosts (Rovira, 1986; Secor and Gudmestad, 1999), burning field stubble (Mazzola et al., 1997), soil pH below 5.8 (Huang and Kuhlman, 1991b), low available nitrogen (Croteau and Zibilske, 1998; Davey and Papavizas, 1963) or high ammonia concentrations (Tavoularis, 1995), high soil populations of mycophagous soil mesofauna (Scholte and Lootsma, 1998), high earthworm populations (Stephens et al., 1994), high soil $CO_2$ concentration (Croteau and Zibilske, 1998; Durbin, 1959), and high soil water content (Gill et al., 2001).

The development of systems strategies for managing diseases caused by *R. solani* will require an understanding of OM-mediated general suppression, but that alone is insufficient. Much research is required to improve our understanding of the impact of organic matter quality and other soil and system properties on the populations and activities of both the pathogen and its specific antagonists.

## MECHANISMS INVOLVED IN DISEASE SUPPRESSION

In the following section we describe specific mechanisms involved in biologically and OM-mediated disease suppression. For clarity, mechanisms are described individually, but note that OM-mediated suppression is typically supported by multiple mechanisms.

## Microbiostasis

The soil microbial community exists under strong competition for energy-yielding nutrients, and the soil community rapidly utilizes any readily available nutrients entering the soil system (Gray and Williams, 1971). Typically, energy stress results in the repression of microbial spore germination and growth; this phenomenon, called microbiostasis, or fungistasis for repression of fungal spores, has been extensively investigated and reviewed (Lockwood, 1977, 1990). Microbiostasis is an adaptive feature, because it protects the propagule from the energy losses or even death that might occur if germination occurred in the absence of a host. Microbiostasis can be overcome by inputs of external energy-rich nutrients, such as root and seed exudates, or organic amendments, such as plant residues or manures (Lockwood, 1990).

Smaller fungal propagules residing in soil, e.g., conidia and the chlamydospores of *Fusarium* spp., require an external source of energy for germination *in vitro*; germination of these propagules appears to be restricted because of an insufficiency of energy-yielding nutrients (Lockwood, 1977). However, although large conidia and sclerotia germinate *in vitro* without an external energy source, these might also experience fungistasis in field soils (Lockwood, 1977). Germination repression for large propagules can also be generated by competition for energy substrates. Fungal propagules release exudates, and these are competed for as would any nutrient released into the soil system. The competition for energy sources by the microbial community is a strong energy sink; exudation from $^{14}C$-labeled fungal propagules increases in response to energy stress in soil (Bristow and Lockwood, 1975). However, propagules also lose energy and viability because of respiration (Hyakumachi et al., 1989). Losses in propagule energy can lead to a reduction in biological function. Nutrient independence for germination of sclerotia was lost after 20% of sclerotial $^{14}C$ was lost and sclerotial death occurred at 40% loss; virulence declined between 20 and 40% (Filonow and Lockwood, 1983).

New energy sources entering the soil system can initially destroy fungistasis, but fungistasis resumes (and typically at a higher fungistatic level) after the sources have been slightly degraded (Lockwood, 1990). The germination of chlamydospores and conidia of *Thielaviopsis basicola* in soil after the incorporation of 1% alfalfa hay increased for the 4 d immediately following the amendment, but germination was suppressed thereafter. The germination of chlamydospores in the spermosphere of bean was reduced from 38% to 0% by alfalfa hay added at least 7 d before beans were planted (Adams and Papavizas, 1969).

Addition of sucrose and asparagine, or seed exudates, to compost-amended suppressive potting mixes reduces the level of suppressiveness in a dose-dependent, linear relationship (Chen et al., 1988b). In addition, compost harvested from the center, high-temperature region of a hardwood bark compost pile was conducive and of lower microbial activity and biomass and higher reducing sugars than the suppressive, lower-temperature outer region of the same pile. However, within days, the conducive material (incubated at room temperature) became suppressive; during the same period, the microbial activity increased and the reducing sugar content declined to levels comparable to those in the suppressive, outer-region compost (Chen et al., 1988b).

Preemptive metabolism of seed exudates that initiate germination of pathogen propagules can induce microbiostasis and prevent disease; this is an indirect form of biological control because the pathogen is not directly antagonized. This has been most elegantly described for bacterial biocontrol agent (BCA)- and compost-mediated suppression of cotton DO (causal agent *Pythium ultimum*; McKellar and Nelson, 2003; van Dijk and Nelson, 1998, 2000). The antagonistic bacterium *E. cloacae* metabolizes plant exudates required by *P. ultimum* for germination and infection. *P. ultimum* oospores and sporangia germinate, grow, and infect cotton seeds in response to long-chain fatty acids (e.g., linoleic acid) released by the seeds as they germinate. *E. cloacae* inoculated onto cotton seeds competitively metabolizes the fatty acids and prevents *P. ultimum* germination, thereby suppressing the disease. Fatty acid uptake and oxidation mutants of *E. cloacae* do not prevent germination. In addition, there is no evidence that *E. cloacae* produces compounds inhibitory to the *Pythium* propagules (e.g., antibiotics) or is directly engaged in parasitism (van Dijk and Nelson, 1998, 2000). This evidence indicates that DO suppression might be slightly more specific than in the theory put forward by Cook and Baker (1983). However, *Pseudomonas* spp. cultured from cotton spermospheres also inactivate seed exudates, but there is no strong relationship between suppressiveness and exudate inactivation, indicating that other mechanisms are involved in suppressiveness by *Pseudomonas* species (van Dijk and Nelson, 1998).

In subsequent work investigating these mechanisms in a suppressive leaf compost, suppressiveness to *Pythium* DO of cotton was related to reduced *P. ultimum* sporangium germination and subsequent seed colonization and not to parasitism or hyphal or sporangial lysis. Suppression was generated immediately after planting. Only microbial consortia isolated from cotton seeds sown in suppressive compost suppressed DO and metabolized linoleic acid. In addition, populations of linoleic-acid-metabolizing bacteria and actinobacteria were higher in the seed-colonizing microbial consortium from the suppressive compost than from the consortium isolated from the conducive compost. Individual isolates were not as suppressive as the suppressive microbial consortium, and linoleic acid metabolism varied greatly among isolates. The authors concluded that competition for linoleic acid was a strong determinant of DO suppression and that suppression was generated not by single isolates but by the combined activities of the linoleic-acid-degrading microbial consortium supported by the suppressive compost substrate (McKellar and Nelson, 2003).

## Microbial Colonization of Pathogen Propagules

Pathogen propagules incubated in compost-amended potting mixes and organic-residue-amended field soils are typically colonized by higher densities of bacterial and fungal propagules, and in some cases protozoa, than in conducive or nonamended soils (Hoitink et al., 1977; Lumsden et al., 1987; Malajczuk, 1983; Toyota and Kimura, 1993). Colonized fungal spores germinate less readily and lyse and die more rapidly than noncolonized spores (Fradkin and Patrick, 1985; Lockwood, 1990). Bacterial colonization increased the rate of lysis, reduced the germination potential, and decreased the virulence of spores of various *Cochliobolus* spp. (causal agents of root rots of grasses, Filonow et al., 1983; Fradkin and Patrick, 1985). Adherence might be an important component of biological control in and of itself; bacterial–fungal, fungal–fungal, and fungal–nematode interactions might be mediated by specific adherence mechanisms (Barak et al., 1985; Nelson et al., 1986; Nordbring-Hertz and Mattiasson, 1979).

## Destruction of Pathogen Propagules

Pathogen propagules can be destroyed after incubation in suppressive organic substrates, the mechanism of which is poorly understood. Microbial antagonists generate hyphal lysis and degradation of chlamydospores, oospores, conidia, sporangia, and zoospores. Lysis of mycelium is typically associated with high levels of bacterial colonization and breakdown of the hyphal contents (Malajczuk and Theodorou, 1979). Bacterial colonization increased the rate of lysis, reduced the germination potential, and decreased the virulence of spores of various *Cochliobolus* spp. (causal agents of root rots of grasses; Filonow et al., 1983; Fradkin and Patrick, 1985).

Forest floor conifer litter induced germination and subsequent lysis of chlamydospores and macroconidia of *Fusarium oxysporum* (Toussoun et al., 1969). *Phytophthora* spp. propagules were destroyed when introduced into forest floor eucalyptus litter (Malajczuk, 1983) and hardwood-bark-amended container media (Hoitink et al., 1977). Sporangia of *Phytophthora* spp. were destroyed after bacterial colonization of the sporangial surface (Broadbent and Baker, 1974). Sporangia nearing maturity release substances attractive to both microorganisms and microfauna. In soils suppressive to *Phytophthora* root rot of avocado, colonization typically destroys the sporangium without zoospore formation and release. The outer layer of the sporangial cell wall is degraded and the cytoplasm withdraws from the cell wall in the area of bacterial attachment (Broadbent and Baker, 1974).

Many bacterial species have been cultured from hyphae, including *Pseudomonas, Bacillus*, and *Streptomyces* spp. *Trichoderma* spp. and chytrids actively parasitize hyphae (Sneh et al., 1977). Protozoa and fungal mites attack hyphae and chlamydospores (Sneh et al., 1977). Small amoebas ingest and lyse zoospores (Malajczuk, 1983). *Trichoderma* spp. can stimulate oospore formation, hyphal lysis, and chlamydospore formation in *Phytophthora* (Malajczuk, 1983). At least 22 fungal species as well as soil microfaunal species (vampyrellid and testate amoebae and ciliated protozoa) have the potential to antagonize resting structures (Malajczuk, 1983; Old and Darbyshire, 1978; Old and Oros, 1980; Palzer, 1976). *Pseudomonas stutzeri* and *Pimelobacter* spp. isolated from chlamydospores of *Fusarium oxysporum* f. sp. *raphani* (incubated in a manure-amended field soil) prevented chlamydospore formation or reduced chlamydospore germination (Toyota and Kimura, 1993).

## Antibiosis

Antibiosis is "antagonism mediated by specific or nonspecific metabolites of microbial origin, by lytic agents, volatile compounds, or other toxic substances" (Fravel, 1988). The evidence for the role of antibiotics in biocontrol of plant diseases has been extensively reviewed (Fravel, 1988). *Pseudomonas* spp. that produce the antibiotic 2,4-diacetylphloroglucinol have been implicated in suppression of take-all of wheat, *Fusarium* wilt of pea, cyst nematode and soft rot of potato, and *Thielaviopsis* root rot of tobacco (Weller et al., 2002). Antibiotic production has also been implicated in the suppression of DO (causal agent *Pythium ultimum*) by *Gliocladium virens* (Howell and Stipanovic, 1983).

## Competition for Substrate Colonization

Most plant pathogens are weak saprophytes, and competition in the soil environment for organic substrates is strong. Pathogens that grow saprophytically on plant residues can be managed by precolonizing plant residues with nonpathogens, termed the *possession principle* (Bruehl, 1975; Cook and Baker, 1983). Leach (1938) was the first pathologist to use this principle knowingly in the field, and left tea prunings on the soil surface to permit their colonization by saprophytes before burial; this practice reduced colonization of the prunings by the devastating pathogen *Armillaria mellea* when buried. This practice is also used to reduce inoculum increase of rubber root rot pathogens when replanting rubber plantations (Fox, 1965).

In studies of competitive interactions in soil aggregate colonization, closely related fungal species (other *F. oxysporum formae speciales*) strongly inhibited colonization by *Fusarium oxysporum* f. sp. *raphani*. Other fungal genera moderately inhibited colonization, and bacterial species mildly inhibited colonization. *Burkholderia cepacia*, an antibiotic-producing bacterial species, also strongly inhibited colonization (Toyota et al., 1996).

*Pythium nunn*, a saprophytic species of *Pythium*, outcompetes *Pythium ultimum* for colonization of added organic substrates, resulting in nutrient deprivation and production of survival structures by *Pythium ultimum*. In many cases, these structures are of lower inoculum potential, resulting in a reduction in the disease potential of *P. ultimum* (Paulitz and Baker, 1988).

## Competition for Root Infection Sites

In a study on high biomass cover cropping for suppression of *Verticillium* wilt of potato, potato root colonization by the nonpathogenic fungal species *Fusarium equiseti* was positively related to suppression of *Verticillium* wilt. Root colonization by *V. dahliae* was positively related to wilt incidence and negatively related to root colonization by *F. equiseti*. Potatoes grown in soils previously cover cropped for 2 or 3 years had more *F. equiseti* root infections and fewer *Verticillium dahliae* root infections than potatoes grown in the fallow. Sudangrass-cropped fields had the highest soil and root populations of *F. equiseti* and had the lowest wilt incidence. However, it is not clear whether the increased *F. equiseti* colonization directly impacts *V. dahliae* colonization and disease incidence (Davis et al., 1994, 1996). Similarly, broccoli residues amended to field soils (or rotation with broccoli crops) suppressed *Verticillium* wilt of cauliflower (causal agent *V. dahliae*); suppressiveness was due in part to reduced viability of microsclerotia and in part to a reduction in *V. dahliae* root colonization (Shetty et al., 2000). Nonpathogenic strains of *Fusarium oxysporum* compete with pathogenic strains for colonization of the root (Benhamou and Garand, 2001; Olivain and Alabouvette, 1999) and other plant tissues (Postma and Luttikholt, 1996) and might thereby contribute to suppression of *Fusarium* wilt. The presence of mycorrhizal fungi might also decrease plant pathogen and nematode infection of crops by mechanisms that include competition for infection sites (Chapter 6).

## Induced Systemic Resistance

Induced systemic resistance (ISR; or systemic acquired resistance, SAR) is "a state of enhanced defensive capacity developed by a plant when appropriately stimulated" (Bakker et al., 2003; van Loon et al., 1998). ISR can provide protection against viral, fungal, and bacterial plant pathogens and root, vascular, and foliar diseases of plants. A variety of soil and rhizosphere bacteria and fungal isolates have been reported to turn on ISR in plants (van Loon et al., 1998). Microbial metabolites such as salicylic acid, siderophores, antibiotics, and lipopolysaccharides have been implicated in microbially mediated ISR (Bakker et al., 2003). Induced resistance has recently been implicated in some suppressive soil systems. Nonpathogenic *Fusarium oxysporum* soil isolates induced systemic resistance in watermelon to *Fusarium* wilt (Larkin et al., 1996). Paper mill residuals compost induced resistance to *Fusarium* wilt of tomato, resulting in a reduction in fungal colonization of root tissues. Suppression was associated with reduced fungal colonization of the tomato roots due to an increase in physical barriers (callose-enriched, multilayered wall appositions and osmiophilic deposits) to fungal penetration (Pharand et al., 2002). Tomato plants grown in compost-amended peat without inoculation with *Fusarium oxysporum* did not exhibit increased physical barriers. An increased level of suppression and physical protection occurred when suppressive compost was inoculated with *Pythium oligandrum*, a species of *Pythium* known to induce resistance in tomato (Benhamou et al., 1999; Pharand et al., 2002).

Composted pine bark container media were suppressive to *Pythium* root rot and foliar anthracnose of cucumber (Zhang et al., 1996), whereas dark peat container media were not suppressive to either disease. Cucumber and *Arabidopsis* plants grown in the composted pine bark expressed higher levels of β-1,3-glucanase (Zhang et al., 1998) and peroxidase (Zhang et al., 1996) than those grown in peat. Split root experiments suggested that the resistance mechanism in cucumber was systemic (Zhang et al., 1996). Compost-amended container mixes suppress bacterial spot of radish (causal agent *Xanthemonas campestris* pv. *armoraciae*; Miller et al., 1997).

Long-term no-till soils induced suppression of bacterial leaf spot (causal agent *Xanthemonas campestris* pv. *armoraciae*) in radish under field conditions, whereas long-term tilled soils did not (Zhang, 1997). Composted paper mill residuals applied to a sandy field soil suppressed bacterial spot of field-grown snap bean (causal agent *Pseudomonas syringae* pv. *syringae*), angular leaf spot of field-grown cucumber (*P. syringae* pv. *lachrymans*), and anthracnose (causal agent *Colletotrichum lindemuthianum*) in greenhouse-grown snap bean (Stone et al., 2003); suppression of foliar diseases in the compost-amended soils was likely due to induced resistance responses (Vallad et al., 2000). *Pythium irregulare* infection of *Banksia grandis* and *Casuarina fraserian* protected these Western Australian forest species from subsequent infection by *Phytophthora cinnamomi*, putatively through ISR (Borstel, 1979).

## Soil Chemical and Physical Properties

SOM management affects not only soil biological properties but also soil chemical and physical properties and plant nutrient status, all of which might also affect plant health.

### Soil and Plant Nutrient Status

SOM quantity and quality impact soil and plant nutrient status. SOM can impact not only total soil nutrient contents but also nutrient availability through the activities of soil microorganisms (de Brito Alvarez et al., 1995). Nutrients impact disease incidence by increasing plant resistance, improving plant growth (permitting disease escape), and influencing the pathogen's environment (Huber and Wilhelm, 1988). Changes in soil and plant nutrient contents may in some cropping systems dramatically alter plant susceptibility to disease.

Plant pathologists have devised integrated management systems to control plant diseases (e.g., *Fusarium* wilt; Woltz and Jones, 1973) through nutrient management in combination with other management strategies. A thorough review of this literature is beyond the scope of this chapter; several excellent reviews of this subject have been published previously (Goss, 1968; Graham, 1983; Huber and Watson, 1974; Huber and Wilhelm, 1988).

*Macronutrients*

High N supply tends to increase disease incidence and contribute to micronutrient deficiencies (Graham, 1983). It is thought that high plant N removes C from plant defense pathways (e.g., those generating phenolics, alkaloids, and phytoalexins) to support growth pathways (those generating carbohydrates; Horsfall and Cowling, 1980). Excess N increases fungal disease incidence, particularly if P and K are deficient (Mengel and Kirkby, 1978). The form of N can also impact disease incidence. Root diseases caused by pathogenic species of *Fusarium*, *Rhizoctonia*, and *Aphanomyces* are typically reduced by $NO_3$-N and increased by $NH_4$-N, whereas the reverse is true for diseases caused by pathogenic species of *Pythium* and *Ophiobolus* (Huber and Watson, 1974). Severity of root rot of bean (causal agents *Fusarium solani* f. sp. *phaseoli*, *Rhizoctonia solani*, and *Thielaviopsis basicola*) is reduced by application of $NO_3$-N and increased by application of $NH_4$-N (Huber and Watson, 1974).

Moderate P levels tend to decrease disease incidence (in particular fungal diseases such as powdery mildew and *Pythium* root rot), whereas very high or low levels tend to increase disease incidence (Graham, 1983). Potassium fertilizers reduce the severity of a variety of fungal root rots caused by *Fusarium* spp., *Pythium* spp., and *Phytophthora* spp. (Graham, 1983). Potassium fertilizers also reduce the severity of late blight of potato (causal agent *Phytophthora infestans*; Goss, 1968). Calcium fertilization suppresses fungal diseases caused by *Pythium* spp. (Ko and Kao, 1989). Calcium fertilization also reduces the severity of postharvest diseases of potato (Conway et al., 1994). There are few reports on the impact of Mg on disease incidence. Intermediate levels of N, P, K, and Ca reduce the severity of *Pythium* root rot of sugar cane (Heck, 1934). Rice panicle N, P, and Mg contents were positively correlated with panicle blast (causal agent *Pyricularia grisea*) severity, whereas Zn, K, and Ca were negatively related (Filippi and Prabhu, 1998).

*Micronutrients*

Manganese fertilization reduces the incidence of fungal diseases (from *Phytophthora* root rot of avocado to stem rust of cereals; Huber and Wilhelm, 1988). Mn is thought to improve host resistance either by alteration of metabolic status or by the production of toxic metabolites (Huber and Wilhelm, 1988).

High N/low Cu plants are highly disease susceptible (Graham, 1983). Boron-deficient plants are susceptible to a wide range of diseases such as ergot, *Fusarium* wilt, powdery mildew, and rust (Graham, 1983). Foliar applications of Fe have increased plant resistance to smut, *Fusarium* patch, and rust (Graham, 1983). Graham (1983) summarized that nutrition is typically only one of several mechanisms contributing to disease expression, amelioration of nutrient deficiencies typically reduces disease incidence, supraoptimal micronutrient levels can also in some cases reduce disease incidence, and nutrient additions can increase plant disease incidence if the addition creates a nutrient imbalance in the host.

Although SOM quantity and quality can have dramatic impacts on soil and plant nutrient contents, few studies on soil properties and disease incidence have seriously investigated the contribution of soil or tissue nutrient contents to disease suppressive effects. Drinkwater et al. (1995) investigated the relationships between soil chemical, physical, and biological properties and incidence of tomato corky root (causal agent *Pyrenocaeta lycopersici*) in a comparative farming systems trial in the Central Valley of California. Farms with a history of annual organic amendment were characterized by soils of higher microbial activity and K contents and lower $NO_3$ contents. Corky root incidence was positively associated with soil $NO_3$ and tomato tissue N and negatively associated with soil N mineralization potential, microbial activity, total soil N, and soil pH.

Composted biosolids amended to a nutrient-deficient subsoil improved perennial ryegrass establishment and growth and suppressed leaf rust severity (causal agent *Puccinia* spp.), putatively due to enhanced nitrogen nutrition in the amended soil (Loschinkohl and Boehm, 2001).

## Soil Physical Properties

SOM content and quality also impact soil physical properties such as aggregation (Carter, 1996, 2002; Soane, 1990) and thereby affect soil functions such as water-holding capacity, resistance to compaction, workability, infiltration and aeration, and resistance to erosion (Carter, 2002). High-rate organic amendments can dramatically improve soil physical properties in a single season, and these effects persist over several years (Chantigny et al., 1999; Gagnon et al., 2001; Grandy et al., 2002; Ndayegamie and Angers, 1993). Long-term lower-rate amendments and cover cropping and rotation with perennial forage crops also improve physical properties (Angers et al., 1999; Grandy et al., 2002; Perfect et al., 1990; Sommerfeldt et al., 1988).

Poor soil physical properties exacerbate a wide variety of root diseases (Allmaras et al., 1988; Cook and Papendick, 1972). Soil compaction was the factor most strongly related to black root rot of strawberry in a New York survey of cultural and physical factors associated with this syndrome (Wing et al., 1995). Similarly, compaction is a strong determinant of *Aphanomyces* root rot of pea (Allmaras et al., 2003), root rots and wilt of chickpea (Bhatti and Kraft, 1992), and root rot of white bean (Tu and Tan, 1991).

Improvements in soil physical properties typically enhance root growth and health (reviewed by Allmaras et al., 1988; Russell, 1975). Poorly aerated or physically constrained soils reduce the rate of root growth by up to 75%, induce the formation of lateral roots, and increase root exudation (Russell, 1975); all these factors increase the likelihood of a root becoming infected (Allmaras et al., 1988). Most soilborne pathogens survive in soils as resting structures and germinate and infect plant roots when stimulated by root or seed exudates. These stimulants are produced in the highest quantity near the root tip and zone of elongation, and this rhizosphere effect extends to at most 2 mm from the root or seed (Huisman, 1982). Root tips typically move at 0.4 mm $h^{-1}$ (Huisman, 1982). Fungal propagules typically detect the stimulant several hours before the root tip arrives, whereas most fungal propagules require 6 to 10 h to germinate and grow in response to a stimulant. Therefore, rapidly moving root tips are less likely than slow-moving root tips to become infected, as the portion of the root of greatest exudation and susceptibility moves past the fungal propagule by the time it has germinated (Huisman, 1982).

## DESIGNING SUPPRESSIVE SOILS AND CROPPING SYSTEMS

Generating disease suppressive cropping systems requires managing the chemical, physical, and biological properties of soil, as well as other cropping system components, to promote plant health.

Many cultural practices and inputs directly and indirectly affect the soil chemical, physical, and biological properties, as well as other cropping systems factors regulating plant disease suppression. The many cultural practices and inputs available to farmers can be viewed as tools; when used intelligently and in the right combination, these tools can generate disease-suppressive cropping systems. In the following section, we discuss a toolbox of cultural practices and inputs that could be used by farmers and scientists to generate disease-suppressive soils and cropping systems.

## CULTURAL PRACTICES

### Crop Rotation

Crop rotation is the practice of growing a sequence of different crops on the same piece of land. The idea that crop rotation improves overall agricultural productivity is not new; crop rotation was practiced in China during the Han dynasty (ca. 206 B.C. to A.D. 220) to improve productivity (MacRae and Mehuys, 1985). Long-term (more than 100 years) rotation studies indicate that crop rotation, in conjunction with other fertility management practices, is fundamental to long-term agricultural productivity and sustainability (Aref and Wander, 1998; Mitchell et al., 1991). The impact of crop rotation on soil quality and plant health and productivity has been reviewed previously (Bullock, 1992; Curl, 1963; Francis and Clegg, 1990; Glynne, 1965; Hall and Nasser, 1996; Karlen et al., 1994; Leighty, 1938; Palti, 1981; Sumner, 1982; Thurston, 1992).

Recent reviews have discounted the importance of crop rotation for disease management (Cook et al., 1995; Karlen et al., 1994; Weller et al., 2002). However, crop rotation remains one of the most important disease management strategies available in many cropping systems (Hall and Nasser, 1996). The most straightforward principle underlying rotation as a disease control strategy is that plant pathogen propagules have a lifetime in soils, and rotation with nonhost crops starves them out (Curl, 1963; Palti, 1981). This principle is fundamental to rotational strategies in many historic and modern cropping systems (Hall and Nasser, 1996; Lawes and Gilbert, 1894; Thurston, 1980). A good example is the effect of crop rotation on the ca. 50 diseases of common bean (*Phaseolis vulgaris*); the pathogens include 29 fungi, 4 bacteria, 14 viruses, 2 groups of nematodes, and 1 mycoplasma-like organism (MLO; Hall and Nasser, 1996). Crop rotation is considered "the most powerful and most frequently recommended practice for controlling bean diseases. It is moderately to highly effective for 33 bean diseases, including most caused by fungi, all caused by nematodes and bacteria, and 'machismo' caused by an MLO" (Hall and Nasser, 1996).

Although starving out pathogens is an important mechanism contributing to the disease-suppressive effect of rotation, Curl (1963) aptly remarked 40 years ago that "the simple act of planting infested land with nonsusceptible crops is only the beginning of a long and complicated story....Crop rotation and biological control of plant diseases are in many respects closely related." Biological control through crop rotation and cover cropping is at the cutting edge of sustainable plant disease management, as described in the next section.

### Cover and Rotation Crops

Cover and rotation crops can alter soil chemical, physical and biological properties, including the composition of the soil microbial community, and thereby reduce (Abadie et al., 1998; Elmer and LaMondia, 1999; Lyle et al., 1948) or increase (Hansen et al., 1990; Keinath et al., 2003) the severity of plant diseases by a variety of mechanisms. These effects can be frustratingly variable among plant species (Abawi and Widmer, 2000) and cultivar to cultivar (Mazzola and Gu, 2002; Sturtz and Christie, 1998). Cover cropping can generate soil properties that favor the pathogen or disease. For example, planting immediately after cover crop incorporation typically increases diseases caused by saprophytic plant pathogenic species of *Pythium* and *Rhizoctonia* (see section "Organic Matter Quality: Early Stages of Decomposition"; Dabney et al., 1996; Grunwald et al., 2000; Wall, 1984). Cover crops can also serve as alternative hosts to pathogens, which might (or might not; Darby, 2003; Lyle et al., 1948) increase disease incidence or severity in subsequent host crops (Dhingra and Netto, 2001; Dillard and Grogan, 1985; Koike et al., 1996). Certain species or varieties of cover crops can suppress plant diseases by directly destroying pathogen propagules (Candole and Rothrock, 1997; Chan and Close, 1987; Kazmar, 1995; Maizel et al., 1963; Malajczuk, 1979; Muehlchen et al., 1990; Reddy and Patrick, 1989; Sequeira, 1962; Shetty et al., 2000).

## Cover and Rotation Crops and General Suppression

Cover crops can increase the content of active OM in soil, microbial biomass, and microbial activity (Bandick and Dick, 1999; Kuo et al., 1997; Mendes et al., 1999) and thereby contribute to general suppression (Davis et al., 1994; Workneh et al., 1993). Typically, however, these changes in soil quality take several years to generate through cover cropping alone, because of the high lability and low biomass of cover crops relative to manure or compost amendments (Workneh et al., 1993).

## Cover and Rotation Crops and Specific Suppression

In addition to improving overall soil quality, plant species and cultivars can have dramatic impacts on rhizosphere and soil microbial community composition (Marschner et al., 2001; Mazzola and Gu, 2002; Miller et al., 1989). In some cases, these plant-induced changes in microbial community composition affect plant health. Nonpathogenic fungal and bacterial isolates cultured from oilseed rape rhizospheres are antagonistic to propagules of *Verticillium dahliae,* a pathogen of oilseed rape (Alstrom, 2000, 2001; von Berg, 1996). In other cases, the shift in the microbial community persists in the soil and alters the rhizosphere and even the endophytic microbial community of subsequent crops. For example, specific potato and red clover cultivar rotations appear to be mutually beneficial, whereas others are detrimental. Red clover and potatoes in a crop rotation can share specific associations of bacterial endophytes (Sturz and Christie, 1998). Some strains promote growth and development in both potato and clover and enhance root nodulation in clover (Sturz and Christie, 1998). Plantlets of two potato cultivars, Russet Burbank and Shepody, were inoculated individually with seven endophytic bacterial isolates cultured from four red clover cultivars (AC Charle, Altaswede, Marino, and Tempus), and impacts on plant growth were evaluated. The potato cultivar Russet Burbank did best when inoculated with endophytes from Marino red clover, whereas the potato cultivar Shepody performed best with bacteria from Altaswede red clover (Sturz and Christie, 1998). These endophytes can also play a role in suppressing potato tuber diseases (Sturz et al., 1999).

Cover and rotation crops can shift the composition of the nonpathogenic microbial community to one suppressive to the disease (Abadie et al., 1998; Davis et al., 1994, 1996; Mazzola and Gu, 2002; Reddy and Patrick, 1989). A soil suppressive to *Rhizoctonia* root rot of apple (apple replant disease) was recently identified (Mazzola, 1999; Mazzola et al., 2002). This soil had been in continuous wheat production until planted to apple, and suppressiveness to apple replant disease [causal agent, *R. solani* anastomosis group (AG)-5 and AG-8] was maintained for 3 years after orchard establishment. As the soil lost suppressiveness, the populations of culturable actinomycetes declined and the dominant fluorescent pseudomonad species shifted from *Pseudomonas putida* to *P. syringae* and *P. fluorescens* bv. III. Some of the isolates of *P. putida* were antagonistic to *R. solani* (Mazzola, 1999). In a subsequent work, wheat cultivars were identified that generated suppressiveness in infested soils over several cover cropping cycles. Roots of suppressive wheat cultivars were colonized by a fluorescent pseudomonad population that was more likely to inhibit *R. solani* AG-5 in *in vitro* studies than the fluorescent pseudomonads cultured from wheat roots grown in fallow soils or soils cropped to nonsuppressive wheat cultivars (Mazzola and Gu, 2002).

Rotation of wheat with a grass ley controls take-all of wheat (causal agent *Gaeumannomyces graminis* var. *tritici*; *Ggt*). Grasses support the growth of *Phialophora graminicola* (*Pg*), a nonpathogenic fungus very difficult to distinguish from *Ggt* when growing on wheat roots. Precolonization of wheat roots with *Pg* protects the roots from subsequent colonization by *Ggt*. Other common root-colonizing nonpathogenic fungi such as *Fusarium oxysporum, G. graminis* var. *graminis,* and other *Phialophora* spp. also suppress take-all (Sivasithamparan, 1975; Wong and

Southwell, 1979; Wong, 1981). While inoculating wheat or soils with *Pg* suppressed take-all, rotation with a short grass ley was as or more effective, effectively distributing high populations of *Pg* throughout the wheat root zone (Wong, 1981). Wong (1981) summarizes:

> *Starting with a cereal crop with severe take-all, a break crop should be grown to reduce the inoculum of* Ggt. *This is followed by a grass ley of about a year's duration to increase rapidly the resident population of the avirulent fungus, which otherwise would have to be introduced. With a high population of the antagonist and a low population of the pathogen, two or three crops of wheat or barley may be grown relatively free of disease.*

The grass ley would also increase active organic matter content and induce some level of general suppression, another component of biologically mediated suppression of take-all (Rovira and Wildermuth, 1981; Pankhurst et al., 2002).

Cotton root rot (causal agent *Phymatotrichum omnivorum*) is a devastating disease of cotton in the south. As early as 1888, manures were applied to control root rot in continuous cotton. In the 1940s, Rogers reported that cotton grown after hubam sweetclover (*Melilotus alba* var. *annua*) was significantly more productive than continuous cotton or cotton grown after other nonhost crops (Rogers, 1942). Subsequent work demonstrated that hubam grown as a green manure and plowed down and planted to cotton the same year suppressed cotton root rot. However, 2 to 3 years of suppression was generated when hubam was grown to maturity the previous year (Lyle et al., 1948). Suppressiveness was generated even when the mature hubam incurred *Phymatotrichum* root rot (Lyle et al., 1948). *Phymatotrichum* mycelia and sclerotia were destroyed when incubated in soils amended with manure or cover crops (Mitchell et al. 1941).

Root rots of pea (caused by one or combinations of *Fusarium oxysporum* f. sp. *pisi*, *F. solani* f. sp. *pisi*, *Pythium ultimum*, and *Aphanomyces euteches*) cause losses on pea crops worldwide. Recent work on this disease complex in clay soils in Ontario, Canada, demonstrated that interseason green manure crops (planted after pea harvest) of corn, sudangrass, sorghum, or oats significantly reduced the severity of pea root rot in both heavily and moderately infested fields planted to pea each spring; in contrast, peas or beans grown as green manures increased root severity in subsequent pea crops. Oats were the most suppressive green manure crop, and disease severity declined in each of 3 years (Tu and Findlay, 1986).

*Fusarium* wilt of banana (causal agent *F. oxysporum* f. sp. *cubanse*) is a serious disease in replanted banana plantations. *Fusarium* wilt incidence was reduced by 42% in banana stands planted into infested field soils cover cropped with sugarcane (179.2 Mg ha$^{-1}$ fresh weight) in the previous year. Cover crops were allowed to decompose for 3 months before planting to banana. Sweet sorghum (179.2 Mg ha$^{-1}$) and velvet bean (*Stizolobium deeringianum*; 112 Mg ha$^{-1}$) cover crops reduced disease incidence, but the difference was not significant (Sequeira, 1962). Cover-cropped suppressive soils enhanced chlamydospore germination, prevented new chlamydospore formation, and lysed hyphae and thin-walled spores (Sequeira, 1962). Serial cover cropping over a several-year period with high biomass legumes such as *Crotolaria* spp. and velvet bean, accompanied by liming to raise the pH to 7.5, reduced the severity of *Fusarium* wilt of banana (*F. oxysporum* f. sp. *cubanse*) for several years after replanting with banana (Scarseth, 1945). Suppressiveness could have been due to both increased microbial activity and the toxic effects of ammonia and nitrous acid on fungal propagules (Lazarovits, 2001).

There are many reports of *Brassica* spp. cover crops or residue amendments generating suppressiveness to soilborne diseases, particularly in container experiments (Brown and Morra, 1997). Typically, suppressiveness is attributed to inhibition of the growth or survival of the fungus by glucosinolates, which are converted to isothiocyanates or other related toxic compounds during residue decomposition (Brown and Morra, 1997). However, it is likely that plants of the mustard family suppress *Verticillium* wilt through multiple mechanisms. Broccoli residues amended to field soils infested with *Verticillium dahliae* suppressed *Verticillium* wilt of cauliflower (Subbarao et al., 1999). However, broccoli residues did not suppress the growth of *V. dahliae* in *in vitro* studies.

Suppressiveness was associated with a reduction in the soil population of microsclerotia, and also with increased microbial activity and inhibition of cauliflower root colonization by *V. dahliae*; the rate of root colonization per unit inoculum was 70% lower in broccoli-amended soils than in unamended soils (Shetty et al., 2000). *V. dahliae* and species of saprophytic soil fungi colonize cauliflower roots in a similar fashion and can compete for colonization sites (Huisman, 1988). Similarly, suppression of *Verticillium* wilt of potato by cover cropping with sudangrass was associated with increased microbial activity (hydrolysis of FDA), increased root colonization by a nonpathogenic *Fusarium* sp., and reduced root colonization by *V. dahliae* (Davis et al., 1994, 1996).

*Pueraria javanica*, a tropical leguminous cover crop, is commonly planted in oil palm orchards to reduce soil erosion and fix nitrogen. Farmers noticed that palm trees undersown with *P. javanica* exhibited reduced incidence of *Fusarium* wilt (causal agent *Fusarium oxysporum* f. sp. *elaeidis*). Subsequent research demonstrated that *Calapagonium coeruleum*, another tropical leguminous cover crop species, also suppresses *Fusarium* wilt of oil palm (Abadie et al., 1998). Soils collected from pots planted for 49 and 230 d to *P. javanica* were suppressive to *Fusarium* wilt; 230-d soils were more consistently suppressive than 49-d soils. *P. javanica* plants and roots were removed from the pots and not incorporated. Suppressiveness generated by the growth of *P. javanica* was associated with an increase in total soil populations of fungi, as well as *Fusarium* spp. and *F. oxysporum*, but no change in the relative populations of various *Fusarium oxysporum* isolates or other *Fusarium* spp. (Abadie et al., 1998).

Multiple-season/serial or permanent (living mulch) cover cropping is typically more effective in reducing soilborne disease incidence and severity than is single-season cover cropping (Abadie et al., 1998; Davis et al., 1994; Mazzola and Gu, 2002; Muehlchen et al., 1990; Scarseth, 1945; Tu and Findlay, 1986). Gradual changes in suppressiveness over time might be related to changes in a variety of soil factors, including indicators of general suppression such as active organic matter fractions, microbial biomass, and activity (Davis et al., 1994); increases in populations of specific antagonists (Davis et al., 1994; Mazzola and Gu, 2002); or reductions in pathogen propagule densities (Muehlchen et al., 1990).

## Tillage

Reduced tillage is one means of increasing SOM content in field agricultural systems. Reduced tillage conserves SOM, reduces erosion, and reduces energy consumption and production costs (Carter, 1994). However, reduced tillage changes the soil environment, and these changes can result in an increase (Bockus and Shroyer, 1998; Jackson et al., 2002), a decrease (Johnson et al., 2001; Ristaino et al., 1997), or no change (Bailey et al., 2001; Johnson et al., 2001) in disease incidence or severity, depending on the cropping system and disease. This topic is beyond the scope of this chapter and has been reviewed previously (Bockus and Shroyer, 1998; Paulitz et al., 2002; Rothrock, 1992; Rovira et al., 1990; Sumner et al., 1981); however, some fundamental principles are addressed here.

Minimum tillage concentrates residues at the soil surface and therefore concentrates pathogen propagule numbers at the soil surface; this might or might not impact disease incidence or severity (Jackson et al., 2002; Sumner et al., 1981). Most plant pathogens are weak saprophytes (as described previously), and therefore previously colonized plant residues are poor substrates for their growth. Residues incorporated immediately are colonized by a diverse community of soil microorganisms that rapidly decompose the residue. Leaving crop residues on the surface of the soil favors pathogens that can survive and grow on surface crop residues. These organisms grow on the living plant tissues during the cropping season and then grow on or survive in the dead plant residues as they decompose on the soil surface. Standing residues or residues lying on the soil surface are colonized by soil organisms much more slowly; pathogen survival and growth in the undisturbed residue is favored in these systems. Residue-colonizing pathogens are therefore favored over time in reduced-tillage systems and can generate significant yield reductions (Bockus and Shroyer, 1998).

Continuous wheat is one of the most extensively studied cropping systems in terms of the impact of reduced tillage on disease incidence and severity. Examples of pathogens favored by reduced tillage in continuous wheat production systems include *Cephalosporium gramineum, Gaeumannomyces graminis, Pythium* spp., and *Rhizoctonia solani* (Bockus and Shroyer, 1998; Cook and Haglund, 1991). *Cephalosporium graminearum* infests the wheat roots and then grows throughout the plant by colonizing the vascular system. The fungus sporulates during the winter on the infected standing straw, and the spores percolate through the soil and infect the roots of the newly planted wheat. *Pythium* seed and root rots increase in reduced-tillage wheat, because *Pythium* spp. can colonize the wheat residues and are favored by the high moisture content of the soils protected by the surface residues (Bockus and Shroyer, 1998; Cook and Haglund, 1991). *Gaeumannomyces graminearum* grows in the root and basal stem tissues of the wheat plant. Under conventional tillage, these plant parts are chopped into small pieces and mixed with soil and the tissues decompose relatively quickly. In reduced-tillage systems, there is little disturbance of the root and basal stem tissues and therefore they take longer to decompose; *G. graminearum* can survive longer and has more opportunity to infect the following wheat crop (Bockus and Shroyer, 1998). *Rhizoctonia* root rot (causal agent *R. solani*) survives in infected wheat stem bases and roots and infects new wheat plants from this food base. Tillage disturbs the mycelial webs that *R. solani* produces to colonize living tissue and thereby reduces disease incidence; reduced tillage does not disturb the mycelial webs and therefore enhances infection and disease (Bockus and Shroyer, 1998).

Many of the constraints associated with reduced tillage in continuous wheat and other monoculture cropping systems can be managed through crop rotation. Tillage can be minimized throughout the rotation sequence but targeted to the requirements of the crops and soils in the system at any point in the rotation (Carter, 1994). Recent work has demonstrated that leaf spots and root pathogens of wheat were not increased in reduced tillage wheat under rotation (Bailey et al., 2001; Elen, 2002).

In untilled soils, root growth will typically be limited to the same interaggregate spaces and biopores year after year. Disease inoculum will occur at highest density in decaying roots, and therefore each new season of roots will encounter the inoculum from infested roots of previous growing seasons; this process effectively clusters both inoculum and new root growth, enhancing the likelihood of root infection (Gilligan, 1985).

Reduced tillage can reduce disease severity in some crop or disease systems. *Phytophthora* blight (causal agent *P. capsici*) of bell pepper no-till planted into rye cover crop residues was dramatically reduced relative to pepper grown in black plastic-mulched raised beds because of improved soil physical properties (Ristaino et al., 1997). A novel reduced-tillage wheat/grain sorghum/fallow rotation called ecofallow was developed at the University of Nebraska (Doupnik and Boosalis, 1980). In this system, no foliar diseases were observed, and incidence of stalk rot of sorghum (causal agent *Fusarium moniliforme*) was reduced relative to incidence in the conventional tillage wheat/grain sorghum/fallow rotation. Wheat and grain sorghum yields increased by 8–10% and 40–50%, respectively, in the reduced-tillage relative to the conventional-tillage rotation. The benefits of reduced tillage were attributed to improved soil physical properties, reduced soil temperatures, and higher soil moisture retention in the wheat stubble, all of which reduced plant stress (Doupnik and Boosalis, 1980).

Conservation-tillage systems concentrate plant residues in the surface soil layer. As a result, microbial biomass and activity are typically much higher in that layer (Dick, 1984). Microbial community composition can also be altered; in typical corn belt systems, fungal biomass increases in the surface layer under reduced tillage as the pH is lower (Doran, 1980). No-till surface soil layers had soil enzyme activities (e.g., fluorescein diacetate and dehydrogenase) two to seven times those of ploughed soils, whereas lower soil layers had enzyme activities equivalent to or lower than those in ploughed soil (Dick, 1984; Zhang, 1997). Little information is available on the impact of short-term vs. long-term reduced tillage on disease incidence and severity and the relationships between soil properties and disease severity over time. However, it is possible that in long-term reduced-tillage soils, the improvement in soil biological quality (microbial activity and biomass) in the surface soil layers generates biological disease suppression. Zhang (1997) recently demonstrated that long-term no-till soils induced suppression of bacterial leaf spot (causal agent *Xanthemonas campestris* pv. *armoraciae* 704B) in radish under field conditions, whereas long-term tilled soils did not. In a 6-year continuous wheat experiment in Australia, the direct-drilled/stubble-retained treatment suppressed take-all (causal agent *Gaeumannomyces graminis* var. *tritici*) and bare patch (causal agent *Rhizoctonia solani*) compared to a conventional till/stubble-burned treatment. Suppressiveness was related to a gradual increase in microbial biomass and organic C in the surface soil (Pankhurst et al., 2002).

## INPUTS

### Plant Genetics

There is considerable evidence that host genetic variability can strongly influence the composition of plant-associated microbial communities involved in disease suppression (Alstrom, 2000, 2001; Cook et al., 1995; Marschner et al., 2001; Mazzola and Gu, 2002; Miller et al., 1989; Simon et al., 2001; von Berg, 1996). Plant genes involved in regulating BCA–plant interactions are being identified; in tomato, several quantitative trait loci have been identified that are related to population growth of the introduced BCA *Bacillus cereus* UW85 after seed inoculation (Smith et al., 1999). Breeding of plant species to improve disease resistance based on the plant's capacity to support plant-associated biocontrol agents has been suggested (Smith and Goodman, 1999). This could include improving the overall performance and consistency of introduced (Simon et al., 2001) and indigenous (Mazzola and Gu, 2002) biocontrol agents in crop as well as cover or rotation crop species. As the soil environment might affect the composition of plant-associated microbial communities, including those involved in plant resistance (Malajczuk et al., 1977), breeding efforts should take place within the cropping or soil system.

### Organic Amendments

Composted and uncomposted animal manures, organic industrial by-products (e.g., biosolids and food wastes), yard trimmings, and other organic residues are used worldwide as organic soil amendments. Amendment characteristics such as availability, cost, ease of application, and chemical and biological properties (e.g., biological stability, pH, nutrient and heavy metals content, weed seed and human and plant pathogen loads) affect its utility in a specific cropping system. In addition, the problems and opportunities in the cropping system that might be ameliorated or exacerbated by amendment must be considered. Long-term low-rate annual amendment might be a more economically, agronomically, and environmentally desirable alternative to single-year high-rate applications.

### Formulated Amendments

A formulated soil amendment (S–H mixture: 4.4% bagasse, 8.4% rice husks, 4.25% oyster shell powder, 8.25% urea, 1.04% potassium nitrate, 13.16% calcium superphosphate, and 60.5% mineral ash) raised soil pH and microbial activity and suppressed bacterial wilt of tomato (causal agent *P. solanacearum*), *Fusarium* wilt of watermelon (casual agent *Fusarium oxysporum* f. sp. *niveum*), *Fusarium* yellows of radish (causal agent *F. oxysporum* f. sp. *raphani*), clubroot of Chinese cabbage (causal agent *Plasmodiophora brassicae*), and cucumber blight (causal agent *Phytophthora melonis*) (Sun and Huang, 1985). The purpose of this formulated amendment is to manipulate multiple soil factors known to suppress plant diseases: pH, microbial activity, calcium content, and volatiles such as ammonia. The mixture inhibited spore germination of both *formae speciales* of *F. oxysporum* and reduced soil populations of *Fusarium* spp. as well as *Pseudomonas solanacearum*. Mineral ash enhanced germ tube lysis of *Fusarium* spores via the direct action of calcium and ferrous oxides (components of the ash). Oyster shell powder, urea, and rice husk reduced the germination of

*Fusarium* spores. Some of the suppressiveness was biologically mediated, because addition of antibiotics to amended soils increased pathogen spore germination (Sun and Huang, 1985).

Another formulated soil amendment, SF-21, suppressed DO of loblolly pine (causal agents *P. aphanidermatum, R. solani,* and *F. moniliforme* var. *subglutinans*) when incorporated into soil seedbeds at 2400 kg ha$^{-1}$. SF-21 is composed of 750 g milled pine bark, 750 mL 10% glycerine, 150 g $Al_2(SO_4)_3$, 35 g $(NH_4)_2SO_4$, 30 g $CaCl_2$, 25 g KCl, and 10 g triple super phosphate (Huang and Kulman, 1991a). Pathogen inhibition by SF-21 is due to multiple indirect and direct inhibitory mechanisms. Suppression of *Pythium* DO was positively related to microbial activity and chemical components of the mixture. Suppression of *R. solani* DO was negatively related to pH; soils of lower pH contained higher populations of the antagonists *Trichoderma harzianum* and *Penicillium oxalicum* (Huang and Kuhlman, 1991b).

## High N-Content Amendments

Organic amendments of high N content (e.g., poultry manures and plant and meat meals) have the potential to suppress soilborne diseases through the toxic effects of ammonia, nitrous acid, or volatile fatty acids on plant pathogen growth and survival (Lazarovits, 2001). This phenomenon has been reported in *Macrophomina phaseolina* (Papavizas, 1976), *Pythium* spp. (Chun and Lockwood, 1985; Craft and Nelson, 1996), *Phytophthora* spp. (Gilpatrick, 1969; Tsao and Oster, 1981), *Fusarium* spp. (Smiley et al. 1970), *Verticillium* spp. (Lazarovits, 2001; Wilhelm, 1951), *Ralstonia* spp. (Michel and Mew, 1998), *Thielaviopsis basicola* (Candole and Rothrock, 1997), and *Sclerotium* spp. (Henis and Chet, 1968). Suppression of *Verticillium* wilt of tomato was generated by soil amendment with the equivalent of 0.1 to 0.13% N to the soil on a dry weight basis (Wilhelm, 1951). Hairy vetch incorporation to a field soil (at 0.25 and 0.75%, w/w with soil) reduced *Thielaviopsis basicola* (causal agent for cotton root rot) chlamydospore viability by 29% in a container study and 16% in a field study. In the field, 0.11 and 0.14 ppm of ammonia were detected in soil atmospheres 3 and 7 d after incorporation, respectively, and these levels are sufficient to reduce chlamydospore viability (Candole and Rothrock, 1997). Ammonia and nitrous acid (but not ammonium, nitrite, or nitrate) are toxic to microsclerotia of *Verticillium dahliae* (Tenuta and Lazarovits, 2002a). It should be recognized that manipulation of this phenomenon without applying N at environmentally and agronomically irresponsible levels requires a thorough understanding of soil factors regulating the release of these compounds in soil systems, such as moisture content, pH, SOM content and quality, soil texture and buffering capacity, and nitrification rate (Lazarovits, 2001; Tenuta and Lazarovits, 2002b).

## Inorganic Amendments

Inorganic fertilizers, liming and acidifying materials, gypsum, and other amendments adjust soil pH and nutrient contents and ratios and can thereby impact plant growth, saprophytic microbial activity, and organic matter quality. For example, liming of acid soils can reduce severity of diseases caused by *Fusarium oxysporum*, possibly because of reduced micronutrient nutrition of the fungus (Woltz and Jones, 1981). Amendment with lime and gypsum increased the Ca saturation of the cation-exchange complex and reduced inoculum levels of *Fusarium solani* f. sp. *pisi* in a wheat–pea rotation (Allmaras et al., 1987). These soil factors also impacted the populations and activities of specific microbial antagonists such as *Trichoderma* spp. (Chet and Baker, 1980; Huang and Kulman, 1991a). Soil and plant nutrient status can affect plant disease severity in many ways, as described in the section "Soil and Plant Nutrient Status."

### Microbial Inoculants

Specific microbial antagonists, often referred to as biological control agents (BCAs), can be applied to field soils to improve suppression of hard-to-control plant diseases. Although this has been the subject of intense research for decades, few examples of commercial implementation exist because of high costs, logistical problems, and low consistency of efficacy (Becker and Schwinn, 1993). To increase efficacy through improved colonization and persistence, recent research has focused on applying BCAs in formulations with organic amendments. BCAs can be blended with various organic matter carriers to buffer the soil environment and provide nutrients for initial soil exploration (Fravel et al., 1995; Maplestone et al., 1991). Timely inoculation of bulk organic residues with BCAs, as in the case of inoculation after peak heating during controlled composting, permits BCAs to colonize the material at very high populations and thereby increase consistency of disease suppression (Hoitink et al., 1993). Generating reliable suppression can also require manipulating soil pH, moisture content, and temperature. These types of strategies have been most successful in environmentally controlled greenhouses (Paulitz and Bélanger, 2001).

One strategy to overcome the limitations inherent in soil applications of a single BCA strain has been to inoculate with mixtures of BCAs that are active over a diverse range of environmental conditions and that generate suppression through multiple mechanisms (de Boer et al., 2003). The application of compost teas for disease suppression is similar in concept; compost teas inundate the growing environment with a highly diverse microbial community of uncertain composition and potentially multiple biocontrol activities. Microorganisms and soluble nutrients are extracted from compost in water; this watery extract is then incubated (fermented) to increase microbial populations before application as a foliar spray or soil drench. Compost tea has the potential to control a variety of soilborne and foliar diseases; current research is focused on identifying compost tea production practices that result in consistent disease suppression (reviewed in Scheuerell and Mahaffee, 2002). Much research is required to determine whether compost tea applications, in combination with other types of soil building such as cover cropping and crop residue management, could aid in generating general or specific suppression in field agricultural systems and reduce the need for high-rate organic amendments.

Other microbial inoculants, particularly symbiotic mycorrhizae and nitrogen-fixing bacteria, have been successfully used to suppress plant diseases through improvements in plant nutrient status and water relations and by altering the microbial community of the rhizosphere (Linderman, 1992; Smith and Read, 1997). Although relatively little is known about the interactions of symbionts and organic amendments, it is known that excess available nutrients can sometimes reduce colonization efficiency. In greenhouse production, root colonization by obligately symbiotic vesicular arbuscular mycorrhizae was negatively affected by high rates of compost amendment; the negative effect was proportional to the quantity of soluble phosphate in the compost (Miller, 1997). This could affect field performance of transplants that are subsequently transplanted into biocide-treated fields that lack indigenous mycorrhizae. Although not always directly related to managing organic matter, augmenting with microbial plant symbionts is a potentially useful tool when designing suppressive soils and cropping systems.

## EXAMPLES OF DISEASE-SUPPRESSIVE SYSTEMS

Cook et al. (1995) stated:

> *Growth and reproduction of the same plant species at the same sites year after year is the norm in natural plant communities....Selection pressure imposed by soil borne pathogens may favor...plants with the ability to support and respond to populations of rhizosphere microorganisms antagonistic to their pathogens.*

We go a few steps further. Natural plant communities not only contain plant species that have undergone strong selection to form associations with suppressive microorganisms (and not in the rhizosphere only), these species reside within ecosystems composed of multiple components that contribute to suppressiveness.

The best-described example of this is the suppressiveness of the eucalyptus forest in western Australia to *Phytophthora* root rot. Suppressiveness of this system is generated by a litter layer (described previously) as well as the resistance (and suppressiveness) of the native understory community that includes legumes resistant to root rot. *Acacia pulchella*, an indigenous understory legume, is both resistant and antagonistic to root rot; organisms growing in the rhizosphere of *A. pulchella* lyse *P. cinnamomi* propagules (Malajczuk, 1979). In addition, the suppressive forest soil also aids in generating the resistance of the eucalyptus plants; soil microorganisms colonize the rhizosphere of *E. calophylla* (considered resistant to *Phytophthora* root rot infection), and these colonized plants are more resistant than noncolonized plants (Malajczuk et al., 1977; Malajczuk, 1979). Malajczuk (1983) also suggested that numerous nonpathogenic *Pythium* spp. isolated from diseased roots could protect eucalyptus seedlings from infection by *P. cinnamomi*. Routine burning of the forests, which destroys the litter layer and shifts the understory plant community to one that is not suppressive, destroys the suppressiveness of the system (Shea, 1975; Shea et al., 1979).

Can we design suppressive agricultural systems based on natural systems? An Australian avocado grower, Guy Ashburner, noted that there were obvious differences between his avocado orchard system (conducive to *Phytophthora* root rot) and the suppressive native forest system described previously. Absent from his orchard, but present in the forest, were a litter layer and a continuous understory. In addition, the forest soils had higher OM and Ca contents and higher pH. He adopted agricultural strategies to generate these components in his orchard system. His orchard was undersown with *Lablab purpureus* and forage sorghum or corn in the summer and *Lupinus angustifolius* during the winter. All cover crops were slashed and incorporated lightly. Organic amendments such as barley straw, sorghum residues, and native grass hay were also added to soil under the trees as a mulch layer, and poultry litter and dolomite were spread on the surface of the mulches to stimulate rapid decay and raise the pH and Ca contents (Malajczuk, 1979, 1983). After several years, the soil suppressed *Phytophthora* root rot of avocado (Malajczuk, 1979, 1983).

Another very elegant and well-documented example of a systems approach to disease management comes from a unique rubber production system in Kuala Lampur. Suppression of root rots (causal agents *Fomes lignosus* and *Ganoderma pseudoferreum*) when replanting rubber plantations is generated through cultural practices such as saprophytic colonization of old rubber trees, rogueing of diseased plants, and planting mixed understory legumes (*Pueraria phaseoloides*, *Centrosema pubescens*, and *Calapogonium mucunoides*). Rubber is typically replanted immediately after old trees are destroyed, because land is limited. Old trees are poisoned to accelerate colonization by saprophytes before felling (Fox, 1965). Creeping legumes are planted between the rows of rubber seedlings to cover 70% of the soil surface (they are not planted directly within the row of rubber trees) at planting. Over a 60-year period, as the young rubber trees grow, the covers are shaded out and replaced by shade-tolerant indigenous species. This creeping cover system creates a form of crop rotation for what is otherwise continuous rubber production. Rubber seedlings planted into infested soil planted to individual or mixed stands of creeping legumes survive better than seedlings planted into clean cultivated fields or planted to grass species (Cronshey and Barclay, 1939; Fox, 1965; Napper, 1932). The creeping covers provide a decoy substrate (physically separated from the young seedlings) to prevent pathogenic colonization of the rubber roots, and they enhance the decay of diseased roots from previous rubber plantings by fixing N and reducing the C:N ratio in the decaying wood, thereby reducing substrate energy availability to the pathogens (Watson et al., 1964). In addition, the covers generate a litter layer that supports antagonists of the pathogen propagules. This system is termed *decoy, decay, and deter* (Fox, 1965).

## CONCLUSION AND FUTURE RESEARCH DIRECTIONS

Generating disease-suppressive cropping systems requires a systems approach. Cultural practices and inputs affecting soil and cropping system properties regulating plant disease suppression must be identified and then articulated into a site- and cropping-system-specific management strategy.

### OM-MEDIATED GENERAL SUPPRESSION

A fundamental step toward generating disease-suppressive agricultural systems is to maintain general suppression through SOM management. Lightly decomposed, or active, OM drives general suppression of root rot caused by a variety of fungal pathogens in peat- and compost-amended container mixes (Boehm et al., 1997; Hoitink and Boehm, 1999; Stone et al., 2001). These same processes are likely at work in field systems; lightly decomposed OM (derived from plant residues or organic wastes) likely drives general suppression in field soils (Darby, 2003). Soil building, or improving soil properties through a diversity of practices including organic amendment, reduced tillage, cover cropping, and rotation into sod, has long been known to improve agricultural productivity (Mitchell et al., 1991; Aref and Wander, 1998). More recently, these general practices have been shown to improve plant health (Darby, 2003; Drinkwater et al., 1995; Lotter et al., 1999; Pankhurst et al., 2002). An improved understanding of SOM pools and their functions related to disease suppression, and increased cooperation between soil scientists and plant pathologists, should improve our ability to manipulate SOM and other soil properties to induce general suppression in field soils. At the same time, more work is required to determine the effectiveness of OM-mediated general suppression for managing a wide range of soilborne diseases. This work should not rely on high-rate organic amendment, but should investigate the use of low-rate organic amendment, cover cropping, reduced tillage, rotation, and combinations of these practices. In addition, this work should not be conducted in short-term field experiments but in longer-term rotational or comparative cropping systems experiments.

### BEYOND OM-MEDIATED GENERAL SUPPRESSION

Inducing general suppression might not be sufficient to achieve commercially viable disease control in many disease and cropping systems. Furthermore, generating OM-mediated general suppression will not be possible in some cropping systems. In these cases, other strategies or combinations of strategies (as described in the section "Designing Suppressive Soils and Cropping Systems") will be necessary. Important research areas include (1) the effect of serial amendment and long-term soil building on soil properties and general suppression; (2) the impact of serial amendment and long-term soil building on soil properties and specific suppression; (3) the impact of reduced tillage on soil properties and disease severity in the short and long term; (4) screening of cover and rotation crops, plant residues, and other organic wastes for specific suppressive qualities; (5) breeding of crops, cover crops, and rotation crops for enhanced support of plant-associated beneficial microbes; (6) the relationship between plant nutrient status and plant health; (7) the impact of faunal predators of plant pathogens on disease suppression; and (8) biocontrol agents (single and consortia) with consistent field efficacy.

In summary, plant health management should be approached from a biological and cropping systems perspective in which the manipulation of SOM quality and other soil properties are used in combination with rotation, resistant varieties, cover cropping, intercropping (especially in perennial cropping systems), residue management, and other cultural practices. SOM-mediated suppression of soilborne diseases is one — but only one — potentially powerful tool in an overall systems approach to plant health management in field agricultural systems.

# REFERENCES

Abadie, C., V. Edel, and C. Alabouvette. 1998. Soil suppressiveness to *Fusarium* wilt: Influence of a cover-plant on density and diversity of *Fusarium* populations. *Soil Biol. Biochem.* 30:643–649.

Abawi, G.S., and T.L. Widmer. 2000. Impact of soil health management practices on soilborne pathogens, nematodes and root diseases of vegetable crops. *Appl. Soil Ecol.* 15:37–47.

Adams, P.B., and G.C. Papavizas. 1969. Survival of root-infecting fungi in soil. X. Sensitivity of propagules of *Thielaviopsis basicola* to soil fungistasis in natural and alfalfa meal amended soil. *Phytopathology* 59:135–138.

Alabouvette, C. 1986. *Fusarium* wilt-suppressive soils from the Chateaurenard region: Review of a 10-year study. *Agronomie* 6:273–284.

Alabouvette, C., H. Hoeper, P. Lemanceau, and C. Steinberg. 1996. Soil suppressiveness to diseases induced by soilborne plant pathogens. In G. Stotzky and J.M. Bollag (Eds.), *Soil Biochemistry*, Vol. 9. Marcel Dekker, New York, pp. 371–413.

Allmaras, R.R., V.A. Fritz, F.L. Pfleger, and S.M. Copeland. 2003. Impaired internal drainage and *Aphanomyces euiches* root rot of pea caused by soil compaction in a fine-textured soil. *Soil Tillage Res.* 70:41–52.

Allmaras, R.R., J.M. Kraft, and J.L. Pikul. 1987. Lime and gypsum effects on pea-root-pathogen inoculum and related factors in a wheat-peas rotation. *Agron. J.* 79:1987.

Allmaras, R.R., J.M. Kraft, and D.E. Miller. 1988. Effects of soil compaction and incorporated crop residue on root health. *Annu. Rev. Phytopathol.* 26:219–243.

Alstrom, S. 2000. Root colonising fungi from oilseed rape and their inhibition of *Verticillium dahliae*. *J. Phytopathol.* 148:417–423.

Alstrom, S. 2001. Characteristics of bacteria from oilseed rape in relation to their biocontrol activity against *Verticillium dahliae*. *J. Phytopathol.* 149:57–64.

Altman, J., and C.L. Campbell. 1977. Pesticide-plant disease interactions: effect of cycloate on sugar beet damping-off induced by *Rhizoctonia solani*. *Phytopathology* 67:1163–1165.

Andrews, S.S., J.P. Mitchell, R. Mancinelli, D.L. Karlen, T.K. Hartz, W.R. Horwath, G.S. Pettygrove, K.M. Scow, and D.S. Munk. 2002. On-farm assessment of soil quality in California's Central Valley. *Agron. J.* 94:12–23.

Angers, D.A., L.M. Edwards, J.B. Sanderson, and N. Bissonnette. 1999. Soil organic matter quality and aggregate stability under eight potato cropping sequences in a fine sandy loam of Prince Edward Island. *Can. J. Soil Sci.* 79:411–417.

Aref, S., and M.M. Wander. 1998. Long-term trends of corn yield and soil organic matter in different crop sequences and soil fertility treatments on the Morrow plots. *Adv. Agron.* 62:153–197.

Asirifi, K.N., W.C. Morgan, and D.G. Parbery. 1994. Suppression of sclerotinia soft rot of lettuce with organic soil amendments. *Aust. J. Exp. Agric.* 34:131–136.

Bailey, K.L., B.D. Gossen, G.P. Lafond, P.R. Watson, and D.A. Derksen. 2001. Effect of tillage and crop rotation on root and foliar diseases of wheat and pea in Saskatchewan from 1991 to 1998: Univariate and multivariate analyses. *Can. J. Plant Sci.* 81:789–803.

Baker, K.F., and R.J. Cook. 1974. *Biological Control of Plant Pathogens*. Freeman, San Francisco.

Baker, K.F., N.T. Flentje, C.M. Olsen, and H.M. Stretton. 1967. Effect of antagonists on growth and survival of *Rhizoctonia* in soil. *Phytopathology* 57:591–597.

Baker, R. 1991. Diversity in biological control. *Crop Prot.* 10:85–94.

Bakker, P.A.H.M., L.X. Ran, C.M.J. Pieterse, and L.C. van Loon. 2003. Understanding the involvement of rhizobacteria-mediated induction of systemic resistance in biocontrol of plant diseases. *Can. J. Plant Pathol.* 25:5–9.

Baldock, J.A., J.M. Oades, A.G. Waters, X. Peng, A.M. Vassallo, and M.A. Wilson. 1992. Aspects of the chemical structure of soil organic materials as revealed by solid-state $^{13}$C NMR spectroscopy. *Biogeochemistry* 16:1–42.

Bandick, A.K., and R.P. Dick. 1999. Field management effects on soil enzyme activities. *Soil Biol. Biochem.* 31:1471–1479.

Barak, R., Y. Elad, D. Mirelman, and I. Chet. 1985. Lectins: A possible basis for specific recognition in the interaction of *Trichoderma* and *Sclerotium rolfsii*. *Phytopathology* 75:458–462.

Barton, R. 1961. Saprophytic activity of *Pythium mamillatum* in soils II. Factors restricting *P. mamilatum* to pioneer colonization of substrates. *Trans. Br. Mycol. Soc.* 44:105–118.

Bateman, D.F. 1964. Cellulase and the *Rhizoctonia* disease of bean. *Phytopathology* 54:1372–1377.

Becker, J.O., and F.J. Schwinn. 1993. Control of soilborne pathogens with living bacteria and fungi: status and outlook. *Pest. Sci.* 37:355–363.

Benhamou, N., P. Rey, K. Picard, and Y. Tirilly. 1999. Ultrastructural and cytochemical aspects of the interaction between the mycoparasite *Pythium oligandrum* and soilborne plant pathogens. *Phytopathology* 89:506–517.

Benhamou, N., and C. Garand. 2001. Cytological analysis of defense-related mechanisms induced in pea root tissues in response to colonization by nonpathogenic *Fusarium oxysporum*. *Phytopathology* 91:730–740.

Bhatti, M.A., and J.M. Kraft. 1992. Influence of soil bulk density on root rot and wilt of chickpea. *Plant Dis.* 76:960–963.

Blair, I.D. 1943. Behaviour of the fungus *Rhizoctonia solani* Kuhn in the soil. *Ann. Appl. Biol.* 30:118–127.

Bockus, W.W., and J.P. Shroyer. 1998. The impact of reduced tillage on soilborne plant pathogens. *Annu. Rev. Phytopathol.* 36:485–500.

Boehm, M.J., and H.A.J. Hoitink. 1992. Sustenance of microbial activity in potting mixes and its impact on severity of *Pythium* root rot of poinsettia. *Phytopathology* 82:259–264.

Boehm, M.J., T. Wu, A.G. Stone, B. Kraakman, D.A. Iannotti, G.E. Wilson, L.V. Madden, and H.A.J. Hoitink. 1997. Cross-polarized magic-angle spinning ¹³C nuclear magnetic resonance spectroscopic characterization of soil organic matter relative to culturable species composition and sustained biological control of *Pythium* root rot. *Appl. Environ. Microbiol.* 63:162–168.

Borstel, P. J. 1979. Cross protecting several West Australian native plants against *Phytophthora cinnamomi* Rands using *Pythium irregulare* Buism. Master's Thesis, University of Western Australia, 130 pp.

Bristow, P.R., and J.L. Lockwood. 1975. Soil fungistasis: Role of spore exudates in the inhibition of nutrient-independent propagules. *J. Gen. Microbiol.* 90:147–156.

Broadbent, P., and K.F. Baker. 1974. Association of bacteria with sporangium formation and breakdown of sporangia in *Phytophthora* spp. *Aust. J. Agric. Res.* 25:139–145.

Brown, P.D., and M.J. Morra. 1997. Control of soilborne plant pests using glucosinolate-containing plants. In D.L. Sparks (Ed.), *Advances in Agronomy.* Academic Press, San Diego, CA, pp. 167–231.

Bruehl, G.W. 1975. *Biology and Control of Soilborne Plant Pathogens.* APS Press, St. Paul, MN.

Bullock, D.G. 1992. Crop rotation. *Crit. Rev. Plant Sci.* 11(4):309–326.

Cambardella, C.A., and E.T. Elliott. 1992. Particulate soil organic matter changes across a grassland cultivation sequence. *Soil Sci. Soc. Am. J.* 56:777–783.

Candole, B.L., and C.S. Rothrock. 1997. Characterization of the suppressiveness of hairy vetch-amended soils to *Thielaviopsis basicola*. *Phytopathology* 87:197–202.

Carter, M.R. 1994. Strategies to overcome impediments to adoption of conservation tillage. In M.R. Carter (Ed.), *Conservation Tillage in Temperate Agroecosystems.* Lewis Publishers, Boca Raton, FL, pp. 1–22.

Carter, M.R. 2002. Soil quality for sustainable land management: Organic matter and aggregation interactions that maintain soil functions. *Agron. J.* 94:38–47.

Cespedes Leon, M.C. 2002. Organic soil amendments: Impacts on snap bean common root rot and soil quality. Master's Thesis, Oregon State University, Corvallis, OR.

Chan, M.K.Y., and R.C. Close. 1987. Aphanomyces root rot of peas. 3. Control by the use of cruciferous amendments. *N. Z. J. Agric. Res.* 30:225–233.

Chantigny, M.H., D.A. Angers, and C.J. Beauchamp. 1999. Aggregation and organic matter decomposition in soils amended with de-inking paper sludge. *Soil Sci. Soc. Am. J.* 63:1214–1221.

Chef, D.G., H.A.J. Hoitink, and L.V. Madden. 1983. Effect of organic components in container media on suppression of *Fusarium* wilt of chrysanthemum and flax. *Phytopathology* 73:279–281.

Chen, W., H.A.J. Hoitink, and L.V. Madden. 1988a. Microbial activity and biomass in container media for predicting suppressiveness to damping-off caused by *Pythium ultimum*. *Phytopathology* 78:1447–1450.

Chen, W., H.A.J. Hoitink, A.F. Schmitthenner, and O.H. Tuovinen. 1988b. The role of microbial activity in suppression of damping-off caused by *Pythium ultimum*. *Phytopathology* 78:314–322.

Chet, I., and R. Baker. 1980. Induction of suppressiveness to *Rhizoctonia solani* in soil. *Phytopathology* 70:994–998.

Chun, D., and J.L. Lockwood. 1985. Reductions of *Pythium ultimum, Thielaviopsis basicola,* and *Macrophomina phaseolina* populations in soil associated with ammonia generated from urea. *Plant Dis.* 69:154–158.

Chung, Y.R., H.A.J. Hoitink, W.A. Dick, and L.J. Herr. 1988a. Effects of organic matter decomposition level and cellulose amendment on the inoculum potential of *Rhizoctonia solani* in hardwood bark media. *Phytopathology* 78:836–840.

Chung, Y.R., H.A.J. Hoitink, and P.E. Lipps. 1988b. Interactions between organic matter decomposition level and soilborne disease severity. *Agric. Ecosyst. Environ.* 24:183–193.

Coe, M.D. 1964. The chinampas of Mexico. *Sci. Am.* 211:90–98.

Cohen, R., B. Chefetz, and Y. Hadar. 1998. Suppression of soilborne pathogens by composted municipal solid waste. In S. Brown et al. (Eds.), *Beneficial Co-Utilization of Agricultural, Municipal and Industrial By-Products.* Kluwer, Dordrecht, pp. 113–130.

Conway, W.S., C.E. Sams, and A. Kelman. 1994. Enhancing the natural resistance of plant tissues to postharvest diseases through calcium applications. *HortScience* 29:751–754.

Cook, R.J., and K.F. Baker. 1983. *The Nature and Practice of Biological Control of Plant Pathogens.* APS Press, St. Paul.

Cook, R.J., and W.A. Haglund. 1991. Wheat yield depression associated with conservation tillage caused by root pathogens in the soil not phytotoxins from the straw. *Soil Biol. Biochem.* 23:1125–1132.

Cook, R.J., and R.I. Papendick. 1972. Influence of water potential of soils and plants on root disease. *Annu. Rev. Phytopathol.* 10:349–374.

Cook, R.J., L.S. Thomashow, D.M. Weller, D. Fujimoto, and M. Mazzola. 1995. Molecular mechanisms of defense by rhizobacteria against root disease. *Proc. Natl. Acad. Sci. USA* 92:4197–4201.

Craft, C.M., and E.B. Nelson. 1996. Microbial properties of composts that suppress damping-off and root rot of creeping bentgrass caused by *Pythium graminicola. Appl. Environ. Microbiol.* 62:1550–1557.

Cronshey, J.F.H., and C. Barclay. 1939. Replanting in areas infested with root disease. *Arch. Rubbercult. Ned.-Ind.* 23:163–169.

Croteau, G.A., and L.M. Zibilskie. 1998. Influence of papermill processing residuals on saprophytic growth and disease caused by *Rhizoctonia solani. Appl. Soil Ecol.* 10:103–115.

Curl, E.A. 1963. Control of plant diseases by crop rotation. *Bot. Rev.* 29:413–479.

Daamen, R.A., F.G. Wijnands, and G. van der Vliet. 1989. Epidemics of diseases and pests of winter wheat at different levels of agrochemical input. *J. Phytopathol.* 125:305–319.

Dabney, S.M., J.D. Schreiber, C.S. Rothrock, and J.R. Johnson. 1996. Cover crops affect sorghum seedling growth. *Agron. J.* 88:961–970.

Daft, G.C., H.A. Poole, and H.A.J. Hoitink. 1979. Composted hardwood bark: A substitute for steam sterilization and fungicide drenches for control of poinsettia crown and root rot. *HortScience* 14:185–187.

Darby, H.M. 2003. Soil organic matter management and root health. Dissertation, Oregon State University, Corvallis, OR.

Davey, C.B., and G.C. Papavizas. 1963. Saprophytic activity of *Rhizoctonia* as affected by the carbon-nitrogen balance of certain organic soil amendments. *Proc. Soil Sci. Soc. Am.* 27:164–167.

Davis, J.R., O.C. Huisman, D.T. Westermann, S.L. Hafez, D.O. Everson, L.H. Sorenson, and A.T. Schneider. 1996. Effects of green manures on *Verticillium* wilt of potato. *Phytopathology* 86:444–453.

Davis, J.R., O.C. Huisman, D.T. Westermann, L.H. Sorenson, A.T. Schneider, and J.C. Stark. 1994. The influence of cover crops on the suppression of *Verticillium* wilt of potato. In G.W. Zehnder et al. (Eds.), *Advances in Potato Pest Biology and Management.* American Phytopathological Society, St. Paul, pp. 332–341.

Deacon, J.W., and L.A. Berry. 1993. Biocontrol of soilborne plant pathogens: concepts and their applications. *Pest. Sci.* 37:417–426.

de Boer, M., P. Bom, F. Kindt, J.J.B. Keurentjes, I. van der Sluis, L.C. van Loon, and P.A.H.M. Bakker. 2003. Control of *Fusarium* wilt of radish by combining *Pseudomonas putida* strains that have different disease-suppressive mechanisms. *Phytopathology* 93:626–632.

de Brito Alvarez, M.A., S. Gagne, and H. Antoun. 1995. Effect of compost on rhizosphere microflora of the tomato and on the incidence of plant growth-promoting rhizobacteria. *Appl. Environ. Microbiol.* 6:194–199.

Dhingra, O.D., and R.A.C. Netto. 2001. Reservoir and non-reservoir hosts of bean-wilt pathogen, *Fusarium oxysporum* f. sp. *phaseoli. J. Phytopathol.* 149:463–467.

Dick, W.A. 1984. Influence of long-term tillage and rotation combinations on soil enzyme activities. *Soil Sci. Soc. Am. J.* 48:569–574.

Dickerson, G.W. 1996. Compost dressing helps chile peppers. *BioCycle* March:80–82.

Dickerson, G.W. 1999. Damping off and root rot. *BioCycle* August:62–63.

Dillard, H.R., and R.G. Grogan. 1985. Influence of green manure crops and lettuce on sclerotial populations of *Sclerotinia minor*. *Plant Dis.* 69:579–582.

Dixon, K.W., K. Frost, and K. Sivasithamparan. 1990. The effect of amendment of soil with organic matter, a herbicide, and a fungicide on the mortality of seedlings of two species of *Banksia* inoculated with *Phytophthora cinnamomi*. *Acta Horticulturae* 264:123–131.

Doran, J.W. 1980. Soil microbial and biochemical changes associated with reduced tillage. *Soil Sci. Soc. Am. J.* 44:765–771.

Doran, J.W., and M.R. Zeiss. 2000. Soil health and sustainablility: managing the biotic component of soil quality. *Appl. Soil Ecol.* 15:3–11.

Doupnik, B., and M.G. Boosalis. 1980. Ecofallow — a reduced tillage system — and plant diseases. *Plant Dis.* 64:31–35.

Downer, A.J., J.A. Menge, and E. Pond. 2001. Association of cellulytic enzyme activities in eucalyptus mulches with biological control of *Phytophthora cinnamomi*. *Phytopathology* 91:847–855.

Drinkwater, L.E., D.K. Letourneau, F. Workneh, A.H.C. Van Bruggen, and C. Shennan. 1995. Fundamental differences between conventional and organic tomato agroecosystems in California. *Ecol. Applic.* 5:1098–1112.

Durbin, R.D. 1959. Factors effecting the vertical distribution of *Rhizoctonia solani* with special reference to $CO_2$ concentration. *Am. J. Bot.* 46:22–25.

Elen, O. 2002. Plant protection in spring cereal production with reduced tillage. III. Cereal diseases. *Crop Prot.* 21:195–201.

Elmer, W.H., and J.A. LaMondia. 1999. Influence of ammonium sulfate and rotation crops on strawberry black root rot. *Plant Dis.* 83:119–123.

Emerson, D. 2003. Building strong markets for mulch and compost products. *BioCycle* 44:36–40.

Erhart, E., and K. Burian. 1997. Evaluating quality and suppressiveness of Austrian biowaste composts. *Compost Sci. Util.* 5:15–24.

Filippi, M.C., and A.S. Prabhu. 1998. Relationship between panicle blast severity and mineral nutrient content of plant tissue in upland rice. *J. Plant Nutr.* 21:1577–1578.

Filonow, A.B., C.O. Akueshi, and J.L. Lockwood. 1983. Decreased virulence of *Cochliobolus victoriae* conidia after incubation on soils or on leached sand. *Phytopathology* 73:1632–1636.

Filonow, A.B., and J.L. Lockwood. 1983. Loss of nutrient independence for germination by fungal propagules incubated on soils or on a model system imposing diffusive stress. *Soil Biol. Biochem.* 15:567–563.

Forbes, R.S. 1974. Decomposition of agricultural crop debris. In C.H. Dickinson and G.J.F. Pugh (Eds.), *Biology of Plant Litter Decomposition*, Vol. 2. Academic Press, New York.

Fox, R.A. 1965. The role of biological eradication in root-disease control in replantings of *Hevea brasiliensis*. In K.F. Baker, and W.C. Snyder (Eds.), *Ecology of Soilborne Plant Pathogens*. University of California Press, Berkeley, pp. 348–362.

Fradkin, A., and Z.A. Patrick. 1985. Effect of matric potential, pH, temperature, and clay minerals on bacterial colonization of conidia of *Cochliobolus sativus* and on their survival in soils. *Can. J. Plant Pathol.* 7:19–27.

Francis, C.A., and M.D. Clegg. 1990. Crop rotations in sustainable production systems. In C.A. Edwards et al. (Eds.), *Sustainable Agricultural Systems*. St. Lucie Press, Delray Beach, FL, pp. 107–122.

Fraser, D.G., J.W. Doran, W.W. Sahs, and G.W. Lesoing. 1988. Soil microbial populations and activities under conventional and organic management. *J. Environ. Qual.* 17:585–590.

Fravel, D.R. 1988. Role of antibiosis in the biocontrol of plant diseases. *Annu. Rev. Phytopathol.* 26:75–91.

Fravel, D.R., J.A. Lewis, and J.L. Chittams. 1995. Alginate prill formulations of *Talaromyces flavus* with organic carriers for biocontrol of *Verticillium dahliae*. *Phytopathology* 85:165–168.

Fravel, D.R., C. Olivain, and C. Alabouvette. 2003. *Fusarium oxysporum* and its biocontrol. *New Phytol.* 157:493–502.

Gagnon, B., R. Lalande, and S.H. Fahmy. 2001. Organic matter and aggregation in a degraded potato soil as affected by raw and composted pulp residue. *Biol. Fertil. Soils* 34:441–447.

Garrett, S.D. 1956. *Biology of Root-Infecting Fungi*. University Press, London.

Gerlagh, M. 1968. Introduction of *Ophiobolus graminis* into new polders and its decline. *Netherl. J. Plant Pathol.* 74:1–81.

Gill, J.S., K. Sivasithamparan, and K.R.J. Smettem. 2001. Effect of soil moisture at different temperatures on *Rhizoctonia* root rot of wheat seedlings. *Plant Soil* 231:91–96.

Gilligan, C.A. 1985. Construction of temporal models. III. Disease progress of soilborne plant pathogens. *Adv. Plant Pathol.* 3:67–105.

Gilpatrick, J.D. 1969. Effect of soil amendments upon inoculum survival and function in *Phytophthora* root rot of avocado. *Phytopathology* 59:979–985.

Glynne, M.D. 1965. Crop sequence in relation to soilborne pathogens. In K.F. Baker, and W.C. Snyder (Eds.), *Ecology of Soilborne Plant Pathogens*. University of California Press, Berkeley, pp. 423–435.

Golchin, A., J.M. Oades, J.O. Skjemstad, and P. Clarke. 1994. Study of free and occluded particulate organic matter in soils by solid state $^{13}$C CPMAS NMR spectroscopy and scanning electron microscopy. *Aust. J. Soil Res.* 32:285–309.

Goss, R.L. 1968. The effects of potassium on disease resistance. In V.J. Kilmer et al. (Eds.), *The Role of Potassium in Agriculture*. American Society of Agronomy, Madison, WI, pp. 221–242.

Graham, R.D. 1983. Effects of nutrient stress on susceptibility of plants to disease with particular reference to the trace elements. *Adv. Bot. Res.* 10:221–276.

Grandy, A.S., G.A. Porter, and M.S. Erich. 2002. Organic amendment and rotation crop effects on the recovery of soil organic matter and aggregation in potato cropping systems. *Soil Sci. Soc. Am. J.* 66:1311–1319.

Gray, T.R.G., P. Baxby, I.R. Hill, and M. Goodfellow. 1968. Direct observation of bacteria in soil. In T.R.G. Gray, and D. Parkinson (Eds.), *The Ecology of Soil Bacteria*. Liverpool University Press, Liverpool.

Gray, T.R.G., and S.T. Williams. 1971. Microbial productivity in soil. In D.E. Hughes, and A.H. Rose (Eds.), *Microbes and Biological Productivity*. Cambridge University Press, London.

Grunwald, N.J., S. Hu, and A.H.C. Van Bruggen. 2000. Short-term cover crop decomposition in organic and conventional soils: soil microbial and nutrient cycling indicator variables associated with different levels of soil suppressiveness to *Pythium aphanidermatum*. *Eur. J. Plant Pathol.* 106:51–65.

Gugino, J.L., F.A. Pokorny, and F.F. Hendrix. 1973. Population dynamics of *Pythium irregulare* Buis. in container-plant production as influenced by physical structure of the media. *Plant Soil* 39:591–602.

Gunapala, N., and K. Scow. 1998. Dynamics of soil microbial biomass and activity in conventional and organic farming systems. *Soil Biol. Biochem.* 30:805–816.

Hall, R., and L.C.B. Nasser. 1996. Practice and precept in cultural management of bean diseases. *Can. J. Plant Pathol.* 18:176–185.

Hancock, J.G., 1977. Factors affecting soil populations of *Pythium ultimum* in the San Joaquin Valley of California. *Hilgardia* 45:107–122.

Hancock, J.G., 1981. Longevity of *Pythium ultimum* in moist soils. *Phytopathology* 71:1033–1037.

Hannukkala, A.O., and E. Tapio. 1990. Conventional and organic cropping systems at Suitia. V. Cereal diseases. *J. Agric. Sci. Finl.* 62:339–347.

Hansen, E.M., D.D. Myrold, and P.B. Hamm. 1990. Effects of soil fumigation and cover crops on potential pathogens, microbial activity, nitrogen availability, and seedling quality in conifer nurseries. *Phytopathology* 80:698–704.

Hardy, G.E.S.J., and K. Sivasithamparan. 1991. Sporangial responses do not reflect microbial suppression of *Phytophthora drechsleri* in composted eucalyptus bark. *Soil Biol. Biochem.* 23:757–765.

Harris, A.R., D.A. Schisler, R.L. Correll, and M.H. Ryder. 1994. Soil bacteria selected for suppression of *Rhizoctonia solani*, and growth promotion, in bedding plants. *Soil Biol. Biochem.* 26:1249–1255.

Harris, K., I.M. Young, C.A. Gilligan, W. Otten, and K. Ritz. 2003. Effect of bulk density on the spatial organisation of the fungus *Rhizoctonia solani* in soil. *FEMS Microbiol. Ecol.* 44:45–56.

Heck, A.F. 1934. Some indications of a relation of soil fertility and plant nutrition to cane diseases in Hawaii. *J. Am. Soc. Agron.* 26:381–389.

Hempfling, R., F. Ziegler, W. Zech, and H.-R. Schulten. 1987. Litter decomposition and humification in acidic forest soils studied by chemical degradation, IR and NMR spectroscopy and pyrolysis field ionization mass spectrometry. *Z. Pflanzenernaehr. Bodenk.* 150:179–186.

Henis, Y., and I. Chet. 1968. The effect of nitrogenous amendments on the germinability of sclerotia of *Sclerotium rolfsii* and on their accompanying microflora. *Phytopathology* 58:209–211.

Henis, Y., B. Sneh, and J. Katan. 1967. Effect of organic amendments on *Rhizoctonia* and accompanying microflora in soil. *Can. J. Microbiol.* 13:643–650.

Hoitink, H.A.J. 1980. Composted bark, a lightweight growth medium with fungicidal properties. *Plant Dis.* 64:142–147.

Hoitink, H.A.J., and M.J. Boehm. 1999. Biocontrol within the context of soil microbial communities: a substrate-dependent phenomenon. *Annu. Rev. Phytopathol.* 37:427–446.

Hoitink, H.A.J., M.J. Boehm, and Y. Hadar. 1993. Mechanisms of suppression of soilborne plant pathogens in compost-amended substrates. In H.A.J. Hoitink, and H.M. Keener (Eds.), *Science and Engineering of Composting: Design, Environmental, Microbiological and Utilization Aspects.* Renaissance Publications, Worthington, OH, pp. 601–621.

Hoitink, H.A.J., and P.C. Fahy. 1986. Basis for the control of soilborne plant pathogens with composts. *Annu. Rev. Phytopathol.* 24:93–114.

Hoitink, H.A.J., Y. Inbar, and M.J. Boehm. 1991. Status of compost-amended potting mixes naturally suppressive to soilborne diseases of floricultural crops. *Plant Dis.* 75:869–873.

Hoitink, H.A.J., M.S. Krause, D.Y. Han, and T.J.J. de Ceuster. 1998. Soil organic matter quality and induced resistance of plants to root rots and foliar diseases. *The International Plant Propagators Society Proceedings*, Vol. 48, pp. 361–367.

Hoitink, H.A.J., A.F. Schmitthenner, and L.J. Herr. 1975. Composted bark for control of root rot in ornamentals. *Ohio Rep.* 60:25–26.

Hoitink, H.A.J., D.M. Van Doren, and A.F. Schmitthenner. 1977. Suppression of *Phytophthora cinnamomi* in a composted hardwood bark potting medium. *Phytopathology* 67:561–565.

Hornby, D. 1983. Suppressive soils. *Annu. Rev. Phytopathol.* 21:65–85.

Horsfall, J.G., and E.B. Cowling. 1980. *How Plants Defend Themselves.* Academic Press, New York.

Howell, C.R., and R.D. Stipanovic. 1983. Gliovirin, a new antibiotic from *Gliocladium virens*, and its role in the biological control of *Pythium ultimum*. *Can. J. Microbiol.* 29:321–324.

Huang, J.W., and E.G. Kuhlman. 1991a. Formulation of a soil amendment to control damping-off of slash pine seedlings. *Phytopathology* 81:163–170.

Huang, J.W., and E.G. Kuhlman. 1991b. Mechanisms inhibiting damping-off pathogens of slash pine seedlings with a formulated soil amendment. *Phytopathology* 81:171–177.

Huber, D.M., and R.D. Watson. 1970. Effect of organic amendment on soilborne plant pathogens. *Phytopathology* 60:22–26.

Huber, D.M., and D.R. Watson. 1974. Nitrogen form and plant disease. *Annu. Rev. Phytopathol.* 12:139–165.

Huber, D.M., and N.S. Wilhelm. 1988. The role of manganese in resistance to plant diseases. In R.D. Graham (Ed.), *Manganese in Soils and Plants.* Kluwer, Dordrecht, pp.155–173.

Huisman, O.C. 1982. Interrelations of root growth dynamics to epidemiology of root-invading fungi. *Annu. Rev. Phytopathol.* 20:303–327.

Huisman, D.T. 1988. Colonization of field-grown cotton roots by pathogenic and saprophytic soilborne fungi. *Phytopathology* 78:716–722.

Hyakumachi, M., H.J.M. Loffler, and J.L. Lockwood. 1989. Methods for the determination of carbon loss from fungal propagules incubated in soil. *Soil Biol. Biochem.* 21:567–571.

Jackson, L.E., I.R. Ramirez, I. Morales, and S.T. Koike. 2002. Minimum tillage practices affect disease and yield of lettuce. *Calif. Agric.* 56:35–40.

Johnson, K.B. 1994. Dose-response relationships and inundative biological control. *Phytopathology* 84:780–784.

Johnson, W.C., T.B. Brenneman, S.H. Baker, A.W. Johnson, D.R. Sumner, and B.G. Mullinix. 2001. Tillage and pest management considerations in a peanut-cotton rotation in the southeastern coastal plain. *Agron. J.* 93:570–576.

Kai, H., T. Ueda, and M. Sakaguchi. 1990. Antimicrobial activity of bark–compost extracts. *Soil Biol. Biochem.* 22:983–986.

Karlen, D.L., G.E. Varvel, D.G. Bullock, and R.M. Cruse. 1994. Crop rotations for the 21st century. *Adv. Agron.* 53:1–45.

Katznelson, H. 1940. Survival of microorganisms introduced into the soil. *Soil Sci.* 49:283–293.

Kazmar, E.R. 1995. The effect of intercropped oats on *Aphanomyces* root rot of pea. Master's Thesis, University of Wisconsin, Madison, WI.

Keinath, A.P., H.F. Harrison, P.C. Marino, D.M. Jackson, and T.C. Pullaro. 2003. Increase in populations of *Rhizoctonia solani* and wiresttem of collard with velvet bean cover crop mulch. *Plant Dis.* 87:719–725.

Kellogg, R.L., C.H. Lander, D.C. Moffitt, and N. Gollehon. 2000. Manure nutrients relative to the capacity of cropland and pastureland to assimilate nutrients. U.S. Department of Agriculture, Natural Resources Conservation Service and Agricultural Research Service.

Ko, W.H., and C.W. Kao. 1989. Evidence for the role of calcium in reducing root disease incited by *Pythium* spp. In A.W. Engelhard (Ed.), *Soilborne Plant Pathogens: Management of Diseases with Macro- and Microelements*. APS Press, St. Paul, MN, pp. 205–217.

Kögel-Knabner, I., W. Zech, and P.G. Hatcher. 1988. Chemical composition of the organic matter in forest soils. III. The humus layer. *Z. Pflanzenernaehr. Bodenkd.* 151:331–340.

Koike, S.T., R.F. Smith, L.E. Jackson, L.J. Wyland, J.I. Inman, and W.E. Chaney. 1996. Phacelia, lana woollypod vetch, and Austrian winter pea: three new cover crop hosts of *Sclerotinia minor* in California. *Plant Dis.* 80:1409–1412.

Krause, M.S., L.V. Madden, and H.A.J. Hoitink. 2001. Effect of potting mix carrying capacity on biological control of *Rhizoctonia* damping-off of radish and *Rhizoctonia* crown and root rot of poinsettia. *Phytopathology* 91:1116–1123.

Krause, M.S., C.A. Musselman, and H.A.J. Hoitink. 1997. Impact of sphagnum peat decomposition level on biological control of *Rhizoctonia* damping-off of radish induced by *Flavobacterium balustinum* 299 and *Trichoderma hamatum* 382. *Phytopathology* 87:S49 (abstract).

Kundu, P.K., and B. Nandi. 1985. Control of *Rhizoctonia* disease of cauliflower by competitive inhibition of the pathogen using organic amendments in soil. *Plant Soil* 83:357–362.

Kuo, S., U.M. Sainju, and E.J. Jellum. 1997. Winter cover crop effects on soil organic carbon and carbohydrate in soil. *Soil Sci. Soc. Am. J.* 61:145–152.

Kuter, G.A., H.A.J. Hoitink, and W. Chen. 1988. Effects of municipal sludge compost curing time on suppression of *Pythium* and *Rhizoctonia* diseases of ornamental plants. *Plant Dis.* 72:751–756.

Kuter, G.A., E.B. Nelson, H.A.J. Hoitink, and L.V. Madden. 1983. Fungal populations in container media amended with composted hardwood bark suppressive and conducive to *Rhizoctonia* damping-off. *Phytopathology* 73:1450–1456.

Kwok, O.C.H., P.C. Fahy, H.A.J. Hoitink, and G.A. Kuter. 1987. Interactions between bacteria and *Trichoderma hamatum* in suppression of *Rhizoctonia* damping-off in bark compost media. *Phytopathology* 77:1206–1212.

Larkin, R.P., D.L. Hopkins, and F.N. Martin. 1996. Suppression of *Fusarium* wilt of watermelon by nonpathogenic *Fusarium oxysporum* and other microorganisms recovered from a disease-suppressive soil. *Phytopathology* 86:812–819.

Lawes, J.B., and J.H. Gilbert. 1894. Rotation of crops. *J. R. Agric. Soc. Engl.* 5:585–646.

Lazarovits, G. 2001. Management of soilborne plant pathogens with organic soil amendments: A disease control strategy salvaged from the past. *Can. J. Plant Pathol.* 23:1–7.

Leach, R. 1939. Biological control and ecology of *Armillaria mellea*. *Trans. Br. Mycol. Soc.* 23:320–329.

Leighty, C.E. 1938. Crop rotation. *U.S. Department of Agriculture Yearbook,* Vol. 1938. United States Department of Agriculture, Washington, D.C., pp. 406–430.

Lewis, J.A., R.D. Lumsden, P.D. Milner, and A.P. Keinath. 1992. Suppression of damping-off of peas and cotton in the field with composted sewage sludge. *Crop Prot.* 11:260–266.

Lifshitz, R., B. Sneh, and R. Baker. 1984. Soil suppressiveness to a plant pathogenic *Pythium* species. *Phytopathology* 74:1054–1061.

Linderman, R.G. 1989. Organic amendments and soilborne diseases. *Can. J. Plant Pathol.* 11:180–183.

Linderman, R.G. 1992. Vesicular-arbuscular mycorrhizae and soil microbial interactions. In G. Bethlenfalvay and R.G. Linderman (Eds.), *Mycorrhizae in Sustainable Agriculture*. American Society of Agronomy, Madison, WI, pp. 45–70

Lockwood, J.L. 1977. Fungistasis in soils. *Biol. Rev.* 52:1–43.

Lockwood, J.L. 1990. Relation of energy stress to behaviour of soilborne plant pathogens and to disease development. In D. Hornby (Ed.), *Biological Control of Soilborne Plant Pathogens*. CAB International, Wallingford, U.K., pp. 197–214.

Lodha, S. 1995. Soil solarization, summer irrigation, and amendments for the control of *Fusarium oxysporum* f. sp. *cumini* and *Macrophomina phaseolina* in arid soils. *Crop Prot.* 14:215–219.

Loschinkohl, C., and M.J. Boehm. 2001. Composted biosolids incorporation improves turfgrass establishment on disturbed urban soil and reduces leaf rust severity. *HortScience* 36:790–794.

Lotter, D.W., J. Granett, and A.D. Omer. 1999. Differences in grape *Phylloxera*-related grapevine root damage in organically and conventionally managed vineyards in California. *HortScience* 34:1108–1111.

Lourd, M., and D. Bouhot. 1987. Research on and characterization of *Pythium* suppressive soils in Brazilian Amazonia. *Bull. OEPP* 17:569–575.

Lucas, P., R.W. Smiley, and H.P. Collins. 1993. Decline of *Rhizoctonia* root rot on wheat soils in infested with *Rhizoctonia solani* AG-8. *Phytopathology* 83:260–265.

Lumsden, R.D., J.A. Lewis, and P.D. Millner. 1983a. Effect of composted sewage sludge on several soilborne pathogens and diseases. *Phytopathology* 73:1543–1548.

Lumsden, R.D., J.A. Lewis, and G.C. Papavizas. 1983b. Effect of organic amendments on soilborne plant diseases and pathogen antagonists. In W. Lockeretz (Ed.), *Environmentally Sound Agriculture*. Praeger Publishers, New York, pp. 51–70.

Lumsden, R.D., P.D. Milner, and J.A. Lewis. 1986. Suppression of lettuce drop caused by *Sclerotinia minor* with composted sewage sludge. *Plant Dis.* 70:197–201.

Lumsden, R.D., R. Garcia-E, J.A. Lewis, and G.A. Frias-T. 1987. Suppression of damping-off caused by *Pythium* spp. in soil from the indigenous Chinampa agricultural system. *Soil Biol. Biochem.* 19:501–508.

Lyle, E.W., A.A. Dunlap, H.O. Hill, and B.D. Hargrove. 1948. Control of cotton root rot by sweetclover in rotation. Bulletin 699, Texas Agricultural Experiment Station, College Station, TX.

MacRae, R.J., and G.R. Mehuys. 1985. The effect of green manuring on the physical properties of temperate-area soils. *Adv. Soil Sci.* 3:71–94.

MacWithey, H.S. 1966. Relationship between oospore production by *Aphanomyces cochliodes* in crop residue and disease incidence. *Phytopathology* 58:887.

Maizel, J.V., H.J. Burkhardt, and H.K. Mitchell. 1963. Avenacin, an antimicrobial substance isolated from *Avena sativa*. I. Isolation and antimicrobial activity. *Biochemistry* 3:424–426.

Malajczuk, N. 1979. Biological suppression of *Phytophthora cinnamomi* in eucalyptus and avocados in Australia. In B. Schippers and W. Gams (Eds.), *Soilborne Plant Pathogens*. Academic Press, London, pp. 635–652.

Malajczuk, N. 1983. Microbial antagonism to *Phytophthora*. In D.C. Erwin et al. (Eds.), *Phytophthora: Its Biology, Taxonomy, Ecology and Pathology*. American Pathological Society, St. Paul, MN, pp. 197–218.

Malajczuk, N., A.J. McComb, and C.A. Parker. 1977. Infection by *Phytophthora cinnamomi* Rands of roots of *Eucalyptus calophylla* R.Br. and *Eucalyptus marginata* Donn. ex Sm. *Aust. J. Bot.* 25:483–500.

Malajczuk, N., and C. Theodorou. 1979. Influence of water potential on growth and cultural characteristics of *Phytophthora cinnamomi*. *Trans. Br. Mycol. Soc.* 72:15–18.

Manning, W.J., and D.F. Crossan. 1969. Field and greenhouse studies on the effects of plant amendments on *Rhizoctonia* hypocotyl rot of snapbean. *Plant Dis. Rep.* 53:227–231.

Maplestone, P.A., J.M. Whipps, and J.M. Lynch. 1991. Effect of peat-bran inoculum of *Trichoderma* species on biological control of *Rhizoctonia solani* in lettuce. *Plant Soil* 136:257–263.

Marschner, P., C.H. Yang, R. Lieberei, and D.E. Crowley. 2001. Soil and plant specific effects on bacterial community composition in the rhizosphere. *Soil Biol. Biochem.* 33:1437–1445.

Mazzola, M., T.E. Johnson, and R.J. Cook. 1997. Influence of field burning and soil treatments on growth of wheat after Kentucky bluegrass, and effect of *Rhizoctonia cerealis* on bluegrass emergence and growth. *Plant Pathol.* 46:708–715.

Mazzola, M. 1999. Transformation of soil microbial community structure and *Rhizoctonia*-suppressive potential in response to apple roots. *Phytopathology* 89:920–927.

Mazzola, M., D.M. Granatstein, D.C. Elfving, K. Mullinix, and Y.-H. Gu. 2002. Cultural management of microbial community structure to enhance growth of apple in replant soils. *Phytopathology* 92:1363–1366.

Mazzola, M., and Y. H. Gu. 2002. Wheat genotype-specific induction of soil microbial communities suppressive to disease incited by *Rhizoctonia solani* anastomosis group (AG)-5 and AG-8. *Phytopathology* 92:1300–1307.

McAlpine, D. 1904. Take-all and white-heads in wheat. *Vict. Dep. Agric. Bull.* 2:410–426.

McKeever, D. 2003. Taking inventory of woody residuals. *BioCycle* 44:31–35.

McKellar, M.E., and E.B. Nelson. 2003. Compost-induced suppression of *Pythium* damping-off is mediated by fatty-acid-metabolizing seed-colonizing microbial communities. *Appl. Environ. Microbiol.* 69:452–460.

Mendes, I.C., A.K. Bandick, R.P. Dick, and P.J. Bottomley. 1999. Microbial biomass and activities in soil aggregates affected by winter cover crops. *Soil Sci. Soc. Am. J.* 63:873–881.

Mengel, K., and E.A. Kirkby. 1978. *Principles of Plant Nutrition.* International Potash Institute, Berne, Switzerland.

Michel, V.V., and T.W. Mew. 1998. Effect of a soil amendment on the survival of *Ralstonia solanacearum* in different soils. *Phytopathology* 88:300–305.

Miller, H.J., G. Henken, and J.A. van Veen. 1989. Variation and composition of bacterial populations in the rhizospheres of maize, wheat, and grass cultivars. *Can. J. Microbiol.* 35:656–660.

Miller, M.L. 1997. The effect of four composts on the establishment of vesicular-arcuscular mycorrhizae in soilless media. Master's Thesis, Oregon State University, Corvallis, OR.

Miller, S.A., F. Sahin, M.S. Krause, J. Al-Dahmani, A.G. Stone, and H.A.J. Hoitink. 1997. Control of bacterial leaf spot of radish in compost-amended planting mixes. *Phytopathology* 87:66 (abstract).

Mitchell, C.C., R.L. Westerman, J.R. Brown, and T.R. Peck. 1991. Overview of long-term agronomic research. *Agron. J.* 83:24–29.

Mitchell, R.B., D.R. Hooton, and F.E. Clark. 1941. Soil bacteriological studies on the control of Phymatotrichum root rot of cotton. *J. Agric. Res.* 63:535–547.

Muehlchen, A.M., R.E. Rand, and J.L. Parke. 1990. Evaluation of crucifer green manures for controlling *Aphanomyces* root rot of peas. *Plant Dis.* 74:651–654.

Nakasaki, K., S. Hiraoka, and H. Nagata. 1998. A new operation for producing disease-suppressive compost from grass clippings. *Appl. Environ. Microbiol.* 64:4015–4020.

Napper, R.P.N. 1932. Observations on the root disease of rubber trees caused by *Fomes lignosus*. *J. Rubber Res. Inst. Malaya* 4:5–33.

Ndayegamie, A., and D.A. Angers. 1993. Organic matter characteristics and water-stable aggregation of a sandy loam soil after 9 years of wood-residue applications. *Can. J. Soil Sci.* 73:115–122.

Nelson, E.B., L.L. Burpee, and M.B. Lawton. 1994. Biological control of turfgrass diseases. In A.R. Leslie (Ed.), *Integrated Pest Management for Turf and Ornamentals.* Lewis Publishers, Boca Raton, FL, pp. 409–427.

Nelson, E.B., W.L. Chao, J.M. Norton, G.T. Nash, and G.E. Harman. 1986. Attachment of *Enterobacter cloacae* to hyphae of *Pythium ultimum*: possible role in the biological control of *Pythium* preemergence damping-off. *Phytopathology* 76:327–335.

Nelson, E.B., and H.A.J. Hoitink. 1983. The role of microorganisms in the suppression of *Rhizoctonia solani* in container media amended with composted hardwood bark. *Phytopathology* 73:274–278.

Nelson, E.B., G.A. Kuter, and H.A.J. Hoitink. 1983. Effects of fungal antagonists and compost age on suppression of *Rhizoctonia* damping-off in container media amended with composted hardwood bark. *Phytopathology* 73:1457–1462.

Nesbitt, H.J., N. Malajcuk, and A.R. Glenn. 1979. Effect of organic matter on the survival of *Phytophthora cinnamomi* Rands in soil. *Soil Biol. Biochem.* 11:133–136.

Nordbring-Hertz, B., and B. Mattiasson. 1979. Action of a nematode trapping fungus shows lectin mediated host-microorganism interaction. *Nature* 281:477–479.

Old, K.M., and J.F. Darbyshire. 1978. Soil fungi as food for giant amoebae. *Soil Biol. Biochem.* 10:93–100.

Old, K.M., and J.M. Oros. 1980. Mycophagous amoebae in Australian forest soils. *Soil Biol. Biochem.* 12:169–175.

Olivain, C., and C. Alabouvette. 1999. Process of tomato root colonization by a pathogenic strain of *Fusarium oxysporum* f. sp. *lycopersici* in comparison to a pathogenic strain. *New Phytol.* 141:497–510.

Oritsejafor, J.J., and M.O. Adeniji. 1990. Influence of host and non-host rhizospheres and organic amendments on survival of *Fusarium oxysporum* f. sp. *elaeidis*. *Mycol. Res.* 94:57–63.

Otten, W., C.A. Gilligan, C.W. Watts, A.R. Dexter, and D. Hall. 1999. Continuity of air-filled pores and invasion thresholds for a soilborne fungal plant pathogen, *Rhizoctonia solani*. *Soil Biol. Biochem.* 31:1803–1810.

Otten, W., D. Hall, K. Harris, K. Ritz, I.M. Young, and C.A. Gilligan. 2001. Soil physics, fungal epidemiology, and the spread of *Rhizoctonia solani*. *New Phytol.* 151:459–468.

Oyarzun, P.J., J. Postma, A.J.G. Luttikholt, and A.E. Hoogland. 1994. Biological control of foot and root rot in pea caused by *Fusarium solani* with nonpathogenic *Fusarium oxysporum* isolates. *Can. J. Bot.* 72:843–852.

Palti, J. 1981. *Cultural Practices and Infectious Crop Diseases.* Springer-Verlag, New York.

Palzer, C. 1976. Zoospore inoculum potential of *Phytophthora cinnamomi.* Ph.D. Thesis, University of Western Australia, 222 pp.

Pammel, L.H. 1890. Cotton root-rot. *Texas Agric. Exp. Stat. Bull.* 7:1–30.

Pankhurst, C.E., H.J. McDonald, B.G. Hawke, and C.A. Kirkby. 2002. Effect of tillage and stubble management on chemical and microbiological properties and the development of suppression towards cereal root disease in soil from two sites in NSW, Australia. *Soil Biol. Biochem.* 34:833–840.

Papavizas, G.C. 1970. Colonization and growth of *Rhizoctonia solani* in soil. In J.R. Parmeter (Ed.), *Rhizoctonia solani, Biology and Pathology.* University of California Press, Berkeley, pp. 108–122.

Papavizas, G.C., P.B. Adams, R.D. Lumsden, J.A. Lewis, R.L. Dow, W.M. Ayers, and J.G. Kantzes. 1975. Ecology and epidemiology of *Rhizoctonia solani* in field soil. *Phytopathology* 65:871–877.

Papavizas, G.C., and W.A. Ayers. 1974. *Aphanomyces* species and their root diseases in pea and sugarbeet. Technical Bulletin 1485, U.S. Department of Agriculture, Agricultural Research Service, 158 pp.

Papavizas, G.C., and C.B. Davey. 1960. *Rhizoctonia* disease of bean as affected by decomposing green plant materials and associated microflora. *Phytopathology* 50:515–522.

Papavizas, G.C., and R.D. Lumsden. 1980. Biological control of soilborne fungal propagules. *Annu. Rev. Phytopathol.* 18:389–413.

Park, D. 1958. The saprophytic status of *Fusarium oxysporum* Schl. causing vascular wilt of oil palm. *Ann. Bot.* 22:19–35.

Patrick, Z.A., and T.A. Toussoun. 1965. Plant residues and organic amendments in relation to biological control. In K.F. Baker, and G.H. Snyder (Eds.), *Ecology of Soilborne Pathogens.* University of California Press, Berkeley, pp. 440–457.

Paulitz, T. 2000. Population dynamics of biocontrol agents and pathogens in soils and rhizospheres. *Eur. J. Plant Pathol.* 106:401–413.

Paulitz, T., and R. Baker. 1988. The formation of secondary sporangia by *Pythium ultimum*: The influence of organic amendments and *Pythium nunn. Soil Biol. Biochem.* 20:151–156.

Paulitz, T., and R.R. Belanger. 2001. Biological control in greenhouse systems. *Annu. Rev. Phytopathol.* 39:103–133.

Paulitz, T., R.W. Smiley, and R.J. Cook. 2002. Insights into the prevalence and management of soilborne cereal pathogens under direct seeding in the Pacific Northwest, USA. *Can. J. Plant Pathol.* 24:416–428.

Pera, A., and C. Filippi. 1987. Controlling of *Fusarium* wilt in carnation with bark compost. *Biol. Wastes* 22:219–228.

Pera, J., and C. Calvet. 1989. Suppression of *Fusarium* wilt of carnation in a composted pine bark and a composted olive pumice. *Plant Dis.* 73:699–700.

Perfect, E., B.D. Kay, W.K.P. van Loon, R.W. Sheard, and T. Pojasok. 1990. Rates of change in soil structural stability under forages and corn. *Soil Sci. Soc. Am. J.* 54:179–186.

Pharand, B., O. Carisse, and N. Benhamou. 2002. Cytological aspects of compost-mediated induced resistance against *Fusarium* crown and root rot in tomato. *Phytopathology* 92:424–438.

Phelan, P.L., K.H. Norris, and J.F. Mason. 1996. Soil-management history and host preference by *Ostrinia nubilalis*: Evidence for plant mineral balance mediating insect–plant interactions. *Environ. Entomol.* 25:1329–1336.

Porter, J.F., and J. Crockett. 2003. Calculating food residuals generation quantities. *BioCycle* 44:35–39.

Postma, J., and A.J.G. Luttikholt. 1996. Colonization of carnation stems by a nonpathogenic isolate of *Fusarium oxysporum* and its effect on *F. oxysporum* f. sp. *Dianthi. Can. J. Bot.* 74:1199–1205.

Pratt, R.G., L.F. Roth, E.M. Hansen, and W.D. Ostrofsky. 1976. Identity and pathogenicity of species of *Phytophthora* causing root rot of Douglas-fir in the Pacific Northwest. *Phytopathology* 66:710–714.

Punja, Z.K., S.F. Jenkins, and R.G. Grogan. 1984. Effect of volatile compounds, nutrients, and source of sclerotia on eruptive sclerotial germination of *Sclerotium rolfsii. Phytopathology* 74:1290–1295.

Quarles, W. 1997. Alternatives to methyl bromide in forest nurseries. *IPM Practit.* 19:1–14.

Quarles, W., and J. Grossman. 1995. Alternatives for methyl bromide in nurseries: Disease suppressive media. *IPM Practit.* 17:1–13.

Reddy, M.S., and Z.A. Patrick. 1989. Effect of host, nonhost, and fallow soil on populations of *Thielaviopsis basicola* and severity of black root rot. *Can. J. Plant Pathol.* 11:68–74.

Reganold, J.P., A.S. Palmer, J.C. Lockhart, and A.N. Macgregor. 1993. Soil quality and financial performance of biodynamic and conventional farms in New Zealand. *Science* 260:344–349.

Rickerl, D.H., E.A. Curl, J.T. Touchton, and W.B. Gordon. 1992. Crop mulch effects on *Rhizoctonia* soil infestation and disease severity in conservation-tilled cotton. *Soil Biol. Biochem.* 24:553–557.

Ringer, C.E., P.D. Millner, L.M. Teerlinck, and B.W. Lyman. 1997. Suppression of seedling damping-off disease in potting mix containing animal manure composts. *Compost Sci. Util.* 5:6–14.

Ristaino, J.B., G. Parra, and C.L. Campbell. 1997. Suppression of *Phytophthora* blight in bell pepper by a no-till wheat cover crop. *Phytopathology* 87:242–249.

Rogers, C.H. 1942. Cotton root rot studies with special reference to sclerotia, cover crops, rotations, tillage, seeding rates, soil fungicides, and effects on seed quality. Bulletin 614, Texas Agricultural Experiment Station, College Station, TX.

Rothrock, C.S. 1992. Tillage systems and plant disease. *Soil Sci.* 154:308–315.

Rothrock, C.S., and B.D. Hargrove. 1988. Influence of legume cover crops and conservation tillage on soil populations of selected fungal genera. *Can. J. Microbiol.* 34:201–206.

Rovira, A.D. 1986. Influence of crop rotation and tillage on *Rhizoctonia* bare patch of wheat. *Phytopathology* 76:669–673.

Rovira, A.D., L.F. Elliott, and R.J. Cook. 1990. The impact of cropping systems on rhizosphere organisms affecting plant health. In J.M. Lynch (Ed.), *The Rhizosphere*. Wiley & Sons, London, pp. 389–436.

Rovira, A.D., and G.B. Wildermuth. 1981. The nature and mechanisms of suppression. In M.J.C. Asher and P.J. Shipton (Eds.), *Biology and Control of Take-All*. Academic Press, London, pp. 385–415.

Rush, C.M., R.E. Ramig, and J.M. Kraft. 1986. Effects of wheat chaff and tillage on inoculum density of *Pythium ultimum* in the Pacific Northwest. *Phytopathology* 76:1330–1332.

Russell, R.S. 1975. *Plant Root Systems: Their Function and Interaction with Soil*. McGraw-Hill, London.

Ryckeboer, J., K. Deprins, and J. Coosemans. 1999. The influence of biowaste and garden waste composts on diseases caused by *Pythium ultimum* and *Rhizoctonia solani* related to the antagonists *Trichoderma hamatum* and *Flavobacterium balustinum*. ORBIT 99.

Sawada, Y., K. Nitta, and T. Igarashi. 1964. Injury of young plants caused by the decomposition of green manure. I. Causal agent of injury. *Soil Sci. Plant Nutr.* 10:163–170.

Scarseth, G.D. 1945. Reported control of Panama disease. *Rev. Appl. Mycol.* 24:198.

Scheuerell, S. 2002. Compost teas and compost amended container media for plant disease suppression. Dissertation, Oregon State University, Corvallis, OR.

Scheuerell, S., and W. Mahaffee. 2002. Compost tea: Principles and prospects for plant disease control. *Compost Sci. Util.* 10(4):313–338.

Schneider, R.W. 1982. *Suppressive Soils and Plant Disease*. American Phytopathological Society Press, St. Paul, MN.

Scholte, K., and M. Lootsma. 1998. Effect of farmyard manure and green manure crops on populations of mycophagous soil fauna and *Rhizoctonia* stem canker of potato. *Pedobiologia* 42:223–231.

Schüler, C., J. Biala, C. Bruns, R. Gottschall, S. Ahlers, and H. Vogtmann. 1989. Suppression of root rot on peas, beans, and beetroots caused by *Pythium ultimum* and *Rhizoctonia solani* through the amendment of growing media with composted organic household waste. *J. Phytopathol.* 127:227–238.

Secor, G.A., and N.C. Gudmestad. 1999. Managing fungal diseases of potato. *Can. J. Plant Pathol.* 21:213–221.

Sequeira, L. 1962. Influence of organic amendments on survival of *Fusarium oxysporum* f. *cubense* in the soil. *Phytopathology* 52:976–982.

Serra-Wittling, C., S. Houot, and C. Alabouvette. 1996. Increased soil suppressiveness to *Fusarium* wilt of flax after addition of municipal solid waste compost. *Soil Biol. Biochem.* 28:1207–1214.

Sharpley, A.N., T. Daniel, T. Sims, J. Lemunyon, R. Stevens, and R. Parry. 1999. Agricultural phosphorus and eutropphication. ARS 149, U.S. Department of Agriculture, Agricultural Research Service, Washington, D.C.

Shea, S.R. 1975. Environmental factors of the northern jarrah forest in relation to pathogenicity and survival of *Phytophthora cinnamomi*. West Aust. For. Dept. Bull. 85, 83 pp.

Shea, S.R., J. McCormick, and C.C. Portlock. 1979. The effect of fires on regeneration of leguminous species in the northern jarrah forest of Western Australia. *Aust. J. Ecol.* 4:195–205.

Sherwood, R.T., and D.J. Hagedorn. 1961. Studies on the biology of *Aphanomyces eutiches*. *Phytopathology* 52:150–154.

Shetty, K.G., K.V. Subbarao, O.C. Huisman, and J.C. Hubbard. 2000. Mechanism of broccoli-mediated *Verticillium* wilt reduction in cauliflower. *Phytopathology* 90:305–310.

Shipton, P.J. 1977. Monoculture and soilborne plant pathogens. *Annu. Rev. Phytopathol.* 15:387–407.

Simon, H.M., K.P. Smith, J.A. Dodsworth, B. Guenthner, J. Handelsman, and R.M. Goodman. 2001. Influence of tomato genotype on growth of inoculated and indigenous bacteria in the spermosphere. *Appl. Environ. Microbiol.* 67:514–520.

Sivasithamparan, K. 1975. *Phialophora* and *Phialophora*-like fungi occurring in the root region of wheat. *Aust. J. Bot.* 23:193–212.

Smiley, R.W., R.J. Cook, and R.I. Papendick. 1970. Anhydrous ammonia as a soil fungicide against *Fusarium* and fungicidal activity in the ammonia retention zone. *Phytopathology* 60:1227–1232.

Smith, K.P., J. Handelsman, and R.M. Goodman. 1999. Genetic basis in plants for interactions with disease-suppressive bacteria. *Proc. Natl. Acad. Sci. USA* 96:4786–4790.

Smith, K.P., and R.M. Goodman. 1999. Host variation for interactions with beneficial plant-associated microbes. *Annu. Rev. Phytopathol.* 37:473–491.

Smith, S.E., and D.J. Read. 1997. *Mycorrhizal Symbiosis*. 2nd ed. Academic Press, San Diego, CA.

Sneh, B., S.J. Humble, and J.L. Lockwood. 1977. Parasitism of oospores of *Phytophthora megasperma* var. *sojae*, *P. cactorum*, *Pythium* sp., and *Aphanomyces euteiches* in soil by oomycetes, chytridiomyctes, hyphomycetes, actinomycetes, and bacteria. *Phytopathology* 67:622–628.

Soane, B.D. 1990. The role of organic matter in soil compactability: a review of some practical aspects. *Soil Tillage Res.* 16:179–201.

Sommerfeldt, T.G., C. Chang, and T. Entz. 1988. Long-term annual manure applications increase soil organic matter and nitrogen, and decrease carbon to nitrogen ratio. *Soil Sci. Soc. Am. J.* 52:1668–1672.

Spencer, S., and D.M. Benson. 1982. Pine bark, hardwood bark compost, and peat amendment effects on development of *Phytophthora* spp. and lupine root rot. *Phytopathology* 72:346–351.

Stanghellini, M. E. 1974. Spore germination, growth, and survival of *Pythium* in soil. *Proc. Am. Phytopath. Soc.* 1:211–214.

Stephens, C.T., L.J. Herr, H.A.J. Hoitink, and A.F. Schmitthenner. 1981. Control of *Rhizoctonia* damping-off by the use of composted hardwood bark. *Plant Dis.* 65:796–797.

Stephens, C.T., and T.C. Stebbins. 1985. Control of damping-off pathogens in soilless container media. *Plant Dis.* 69:494–496.

Stephens, P.M., C.W. Davoren, M.H. Ryder, and B.M. Doube. 1994. Influence of the earthworms *Aporrectodea rosea* and *Aporrectodea trapezoides* on *Rhizoctonia solani* disease of wheat seedlings and the interaction with a surface mulch of cereal-pea straw. *Soil Biol. Biochem.* 26:1285–1287.

Stone, A.G., S.J. Traina, and H.A.J. Hoitink. 2001. Particulate organic matter composition and *Pythium* damping-off of cucumber. *Soil Sci. Soc. Am. J.* 65:761–770.

Stone, A.G., G.E. Vallad, L.R. Cooperband, D. Rotenberg, H.M. Darby, R.V. James, W.R. Stevenson, and R.M. Goodman. 2003. The effect of organic amendments on soilborne and foliar diseases in field-grown snap bean and cucumber. *Plant Dis.* 87:1037–1042.

Sturz, A.V., and B.R. Christie. 1998. The potential benefits from cultivar specific red clover–potato crop rotations. *Ann. Appl. Biol.* 133:365–373.

Sturz, A.V., B.R. Christie, B.G. Matheson, W.J. Arsenault, and N.A. Buchanan. 1999. Endophytic bacterial communities in the periderm of potato tubers and their potential to improve resistance to soilborne plant pathogens. *Plant Pathol.* 48:360–369.

Subbarao, K.V., J.C. Hubbard, and S.T. Koike. 1999. Evaluation of broccoli residue incorporation into field soil for *Verticillium* wilt control in cauliflower. *Plant Dis.* 83:124–129.

Sumner, D.R. 1982. Crop rotation and plant productivity. In M. Rechcigl (Ed.), *CRC Handbook of Agricultural Productivity*. CRC Press, Boca Raton, FL, pp. 273–313.

Sumner, D.R., B. Doupnik, and M.G. Boosalis. 1981. Effects of reduced tillage and multiple cropping on plant diseases. *Annu. Rev. Phytopathol.* 19:167–187.

Sun, S.-K., and J.-W. Huang. 1985. Formulated soil amendment for controlling *Fusarium* wilt and other soilborne diseases. *Plant Dis.* 69:917–920.

Szczech, M.M., W. Rondomanski, M.W. Brzeski, U. Smolinska, and J.F. Kotowski. 1993. Suppressive effect of a commercial earthworm compost on some root infecting pathogens of cabbage and tomato. *Biol. Agric. Hortic.* 10:47–52.

Tahvonen, R. 1982. The suppressiveness of Finnish light coloured sphagnum peat. *J. Sci. Agric. Soc. Finl.* 54:345–356.

Tan, C.S., and J.C. Tu. 1995. Tillage effect on root rot severity, growth and yield of beans. *Can. J. Plant Sci.* 75:183–186.

Tavoularis, K., A. Papdaki, and V. Manios. 1995. Effect of volatile substances released from olive tree leaf compost on the vegetative growth of *Rhizoctonia solani* and *Fusarium oxysporum* f. sp. *lycopersici*. *Acta Horticulturae* 382:183–185.

Tenuta, M., and G. Lazarovits. 2002a. Ammonia and nitrous acid from nitrogenous amendments kill the microsclerotia of *Verticillium dahliae*. *Phytopathology* 92:255–264.

Tenuta, M., and G. Lazarovits. 2002b. Identification of specific soil properties that affect the accumulation and toxicity of ammonia to *Verticillium dahliae*. *Can. J. Plant Pathol.* 24:219–229.

Tepper, J.G.O. 1892. "Take-all" and its remedies. *Agric. Gaz. New South Wales* 3:69–72.

Thurston, H.D. 1980. International potato disease research for developing countries. *Plant Dis.* 64:252–257.

Thurston, H.D. 1992. *Sustainable Practices for Plant Disease Management in Traditional Farming Systems*. Westview Press, Boulder, CO.

Toussoun, T.A., A.R. Weinhold, R.G. Linderman, and Z.A. Patrick. 1968. Nature of phytotoxic substances produced during plant residue decomposition in soil. *Phytopathology* 58:41–45.

Toussoun, T.A., W. Menzinger, and R.S. Smith, Jr. 1969. Role of conifer litter in ecology of *Fusarium*: stimulation of germination in soil. *Phytopathology* 59:1369–1399.

Toyota, K., and M. Kimura. 1993. Colonization of chlamydospores of *Fusarium oxysporum* f. sp. *raphani* by soil bacteria and their effects on germination. *Soil Biol. Biochem.* 25:193–197.

Toyota, K., K. Ritz, and I.M. Young. 1996. Microbiological factors affecting the colonisation of soil aggregates by *Fusarium oxysporum* f. sp. *raphani*. *Soil Biol. Biochem.* 28:1513–1521.

Trillas-Gay, M.I., H.A.J. Hoitink, and L.V. Madden. 1987. Nature of suppression of *Fusarium* wilt of radish in a container medium amended with composted hardwood bark. *Plant Dis.* 70:1023–1027.

Trujillo, E.E., and R.B. Hine. 1965. The role of papaya residues in papaya root rot caused by *Pythium aphanidermatum* and *Phytophthora parasitica*. *Phytopathology* 55:1293–1298.

Tsao, P.H., and J.J. Oster. 1981. Relation of ammonia and nitrous acid to suppression of *Phytophthora* in soils amended with nitrogenous organic substrates. *Phytopathology* 71:53–59.

Tsror, L., R. Barak, and B. Sneh. 2001. Biological control of black scurf on potato under organic management. *Crop Prot.* 20:145–150.

Tu, J.C., and W.I. Findlay. 1986. The effects of different green manure crops and tillage practices on pea root rots. *Proc. Br. Crop Prot. Conf.* 3:229–236.

Tu, J.C., and C.S. Tan. 1991. Effect of soil compaction on growth, yield and root rots of white beans in clay loam and sandy loam soil. *Soil Biol. Biochem.* 23:233–238.

Tuitert, G., M.M. Szczech, and G.J. Bollen. 1998. Suppression of *Rhizoctonia solani* in potting mixes amended with compost made from organic household waste. *Phytopathology* 88:764–773.

Vallad, G.E., A.G. Stone, R.M. Goodman, and L.R. Cooperband. 2000. Assessment of foliar disease suppression generated by paper mill sludge amendments. *Phytopathology* 90:S79 (abstract).

van Bruggen, A.H.J.C. 1995. Plant disease severity in high input compared to reduced input and organic farming systems. *Plant Dis.* 79:976–984.

van Dijk, K., and E.B. Nelson. 1998. Inactivation of seed exudate simulants of *Pythium ultimum* sporangium germination by biocontrol strains of *Enterobacter cloacae* and other seed-associated bacteria. *Soil Biol. Biochem.* 30:183–192.

van Dijk, K., and E.B. Nelson. 2000. Fatty acid competition as a mechanism by which *Enterobacter cloacae* suppresses *Pythium ultimum* sporangium germination and damping-off. *Appl. Environ. Microbiol.* 66:5340–5347.

van Loon, L.C., P.A.H.M. Bakker, and C.M.J. Pieterse. 1998. Systemic resistance induced by rhizosphere bacteria. *Annu. Rev. Phytopathol.* 36:453–483.

Voland, R.P., and A.H. Epstein. 1994. Development of suppressiveness to diseases caused by *Rhizoctonia solani* in soils amended with composted and noncomposted manure. *Plant Dis.* 78:461–466.

von Berg, G. 1996. Rhizobacteria of oilseed rape antagonistic to *Verticillium dahliae*. *J. Plant Dis. Prot.* 103:20–30.

Wall, R.E. 1984. Effects of recently incorporated organic amendments on damping-off of conifer seedlings. *Plant Dis.* 68:59–60.

Wander, M.M., S.J. Traina, R.B. Stinner, and S.E. Peters. 1994. The effects of organic and conventional management on biologically active soil organic matter pools. *Soil Sci. Soc. Am. J.* 62:1704–1711.

Watson, G.A., P.W. Wong, and R. Narayanan. 1964. Effect of cover plants on the growth of *Hevea*. IV. Leguminous creepers compared with grasses, *Mikania cordata* and mixed indigenous covers on four soil types. *J. Rubber Res. Inst. Malaya* 18:123–145.

Watson, A.G. 1970. The effect of cover crops incorporated into field soil on *Pythium ultimum* populations and inoculum potentials. *Phytopathology* 60:1537 (abstract).

Weindling, R., and H.S. Fawcett. 1936. Experiments in the control of *Rhizoctonia* damping-off of citrus seedlings. *Hilgardia* 10:1–16.

Weller, D.M., R.J. Cook, G. MacNish, E.N. Bassett, R.L. Powelson, and R.R. Petersen. 1986. *Rhizoctonia* root rot of small grains favored by reduced tillage in the Pacific Northwest. *Plant Dis.* 70:70–73.

Weller, D.M., J.M. Raaijmakers, B.B. MacSpadden Gardener, and L.S. Thomashow. 2002. Microbial populations responsible for specific soil suppressiveness to plant pathogens. *Annu. Rev. Phytopathol.* 40:309–348.

Weste, G. 1983. Population dynamics and survival of *Phytophthora*. In D.C. Erwin et al. (Eds.), *Phytophthora: Its Biology, Taxonomy, Ecology and Pathology*. American Phytopathological Society, St. Paul, MN, pp. 237–257.

Wilhelm, S. 1951. Effect of various soil amendments on the inoculum potential of the *Verticillium* wilt fungus. *Phytopathology* 41:684–690.

Wing, K.B., M.P. Pritts, and W.F. Wilcox. 1995. Biotic, edaphic, and cultural factors associated with strawberry black root rot in New York. *HortScience* 30:86–90.

Woltz, S.S., and J.P. Jones. 1973. Tomato *Fusarium* wilt control by adjustments in soil fertility. *Proc. Fla. State Hort. Soc.* 86:157–159.

Wong, P.T.W. 1981. Biological control by cross-protection. In M.J.C. Asher and P.J. Shipton (Eds.), *Biology and Control of Take-All*. Academic Press, London, pp. 417–431.

Wong, P.T.W., and R.J. Southwell. 1979. Biological control of wheat take-all in the field using *Gaeumannomyces graminis* var. *graminis* and related fungi. In B. Schippers, and W. Gams (Eds.), *Soilborne Plant Pathogens*. Academic Press, London, pp. 597–602.

Workneh, F., A.H.C. Van Bruggen, L.E. Drinkwater, and C. Shennan. 1993. Variables associated with corky root and *Phytopthora* root rot of tomatoes in organic and conventional farms. *Phytopathology* 83:581–589.

Wright, R.J., W.D. Kemper, P.D. Millner, and R.F. Korcak. 1998. Agricultural uses of municipal, animal, and industrial byproducts. Conservation Research Report No. 44, U.S. Department of Agriculture, Agricultural Research Service, Washington, D.C.

You, M.P., and K. Sivasithamparan. 1994. Hydrolysis of fluorescein diacetate in an avocado plantation mulch suppressive to *Phytopthora cinnamomi* and its relationship with certain biotic and abiotic factors. *Soil Biol. Biochem.* 26:1355–1361.

You, M.P., and K. Sivasithamparan. 1995. Changes in microbial populations of an avocado plantation mulch suppressive of *Phytophthora cinnamomi*. *Appl. Soil Ecol.* 2:33–43.

Zhang, W. 1997. Disease suppression and systemic acquired resistance induced in plants by compost-amended potting mixes, compost water extracts and no-tillage soil. Dissertation, Ohio State University, OH.

Zhang, W., D.Y. Han, W.A. Dick, K.R. Davis, and H.A.J. Hoitink. 1998. Compost and compost water extract-induced systemic acquired resistance in cucumber and *Arabidopsis*. *Phytopathology* 88:450-455.

Zhang, W., H.A.J. Hoitink, and W.A. Dick. 1996. Compost-induced systemic acquired resistance in cucumber to *Pythium* root rot and anthracnose. *Phytopathology* 86:1066–1070.

# 6 Contributions of Fungi to Soil Organic Matter in Agroecosystems

*Kristine A. Nichols and Sara F. Wright*

## CONTENTS

Soil fungi are important agents of decomposition, pathogenicity, and plant and soil health (i.e., nutrient cycling, soil fertility, aggregate stability, and soil organic matter turnover). In agricultural soils, there are at least 25,000 fungal species (Carlile and Watkinson, 1996a), accounting for ca. 70% of the microbial biomass (Paul and Clark, 1996). Fungal growth is a function of carbon availability. Hyphal lengths often range from 3 to 300 m/g soil (Frey et al., 1999; Miller et al., 1995; Olsson et al., 1999; Rillig et al., 1999). Most fungal organisms are found in the rhizosphere, which is enriched in organic carbon from proteins, amino acids, organic acids, and sugars released by roots; the mucopolysaccharide mucigel on the root; and sloughed root cap cells.

Fungal contributions to agroecosystem function are difficult to quantify because of the lack of accurate methods to measure fungal biomass and activity. The benefits and limitations of some typical quantification methods are discussed in this chapter. Three major groups of soil fungi are important in agroecosystems: (1) saprophytes, (2) pathogens, and (3) mutualists. The mutualistic arbuscular mycorrhizal (AM) fungi account for the majority of fungal biomass (Olsson et al., 1999; Vieira et al., 2000) and are examined in some detail along with glomalin, a glycoproteinaceous substance that coats AM hyphae. Glomalin might act as a hydrophobin, which are a class of biomolecules that protect hyphae from nutrient loss (Wessels, 1997), form and stabilize soil aggregates (Wright and Upadhyaya, 1998), and store soil carbon (Rillig et al., 2001b).

## ESTIMATING FUNGAL BIOMASS

Microbial biomass is often used as an indicator of the microbial contribution to soil organic matter, plant health, and understanding nutrient fluxes (i.e., the transport and storage of nutrients). Estimates of fungal biomass are often included with bacterial biomass numbers in techniques based on the amount of carbon released after chloroform fumigation followed by incubation or extraction (i.e., total microbial biomass). Chloroform fumigation methods are inherently more accurate for bacteria than for fungi (Olsson et al., 1999; Paul and Clark, 1996; Vieira et al., 2000). Even with variations in the incubation procedure, fungal cell walls and spores do not completely lyse (Horwath and Paul, 1994; Paul and Clark, 1996).

Microscopic counts of hyphae and other fungal structures by the grid-line intersect method are tedious and have many technical problems. Hyphal diameters range from 2 to 20 μm and can be used to estimate biovolume or to classify different groups of fungi (Bottomley, 1994; Carlile and Watkinson, 1996a; Miller et al., 1995). Various stains are applied to help visualize hyphae or to determine viability, or both. However, viability stains, such as 4′6 diamidino-2-phenyl indole (DAPI; reacts with active DNA) or fluorescein diacetate (an indicator of cytoplasmic constituents, i.e., esterase), might be ineffective in determining the length of aseptate hyphae when nuclei and cytoplasmic contents are not distributed evenly (Bottomley, 1994; Carlile and Watkinson, 1996a; Paul and Clark, 1996). Nonspecific stains make hyphae more visible but do not correct for errors due to (1) exclusion of spores or yeasts; (2) large differences in counts between individuals or laboratories; (3) heterogeneous distribution of hyphae in the soil; and (4) differences in extraction techniques, such as grinding soil in a mixer compared to shaking free hyphae from soil (Millner and Wright, 2002; Rillig et al., 1999; Stahl et al., 1995). In extracting fungal hyphae from soil, a balance must be achieved between homogenizing soil to effectively release hyphae and excessively fragmenting hyphae (Bottomley, 1994). Inherent variability makes it difficult to determine differences between treatments unless numerous replicate samples are examined (Stahl et al., 1995).

Other methods for measuring fungal biomass quantify a specific substance, such as chitin or ergosterol. The major limitations in these assays are that these substances (1) are not found in all fungi, (2) might be present in other soil organisms, (3) vary in concentration in different fungal species or physiological states, and (4) are not calibrated with fungal biomass (Bottomley, 1994; Paul and Clark, 1996; Vieira et al., 2000). Chitin is found in the cell walls of most fungi, but is missing in Oomycetes and is present in insects and mites. (Although Oomycetes have been reclassified into the Kingdom Chromista in the eight-kingdom system, in this chapter they are still considered as part of Kingdom Fungi.) Ergosterol is also found in other organisms, such as algae and protozoa, and can only be measured in living mycelium.

Seasonal fluctuations and substrate (i.e., carbon) availability influence fungal biomass (Carlile and Watkinson, 1996a; Bottomley, 1994). For example, following an increase in soil moisture from precipitation or irrigation, the germination and proliferation of fungi can increase as soluble carbon compounds are released by plants, but this rapid growth declines when substrates become limited (Carlile and Watkinson, 1996a; Klein et al., 1995). Therefore, it is important not only to take a number of samples from a site but also to note the time of sampling, to sample a number of times a year, or to do repeated sampling over a number of years at the same time. Sampling times should be dictated not by the calendar but by climatic conditions and management events, such as sampling at the same time in reference to precipitation, frost, planting, harvesting, or application of fertilizer, herbicide, or pesticide (Bottomley, 1994).

## SAPROPHYTIC FUNGI

Fungal saprophytes are the primary degraders of plant debris, whereas bacteria and select highly specific fungi decompose animal material and microbes (Bird et al., 2002; Carlile and Watkinson, 1996a; Frey et al., 1999; Stevenson, 1994; Vieira et al., 2000). Because of their relatively benign role as decomposers, saprophytic fungi are often overlooked by agricultural scientists, but life on this planet could not be maintained without these fungi recycling basic nutrients such as C, N, P, and K, and we would have been buried hundreds of times over by undecomposed leaves, roots, and other plant material (Carlile and Watkinson, 1996a; Klein et al., 1995). Although these fungi play a vital role in nutrient cycling, they are mostly on surface residues and account for less than 1% of the total microbial biomass to a depth of 20 cm (Frey et al., 1999).

For the most part, saprophytic fungi are not plant species specific but rather substrate specific. Substrates can be divided into several groups: (1) soluble, simple sugars, (2) insoluble sugars, and (3) lignin and cellulose (Carlile and Watkinson, 1996a). Soluble-sugar-utilizing fungi are mostly Zygomycetes, with a short lifespan consisting of rapid growth and sporulation. Insoluble sugars are degraded primarily by Ascomycetes, which are ubiquitous in soil and often produce or tolerate antibiotics to help them compete successfully for substrates. Lignin and cellulose degraders are mostly slow-growing Basidiomycetes that usually use other substances as carbon energy sources but contain enzymes that break lignin or cellulose down into substrates that are further processed by other microorganisms (Carlile and Watkinson, 1996a).

## FUNGAL PLANT PATHOGENS

Plant pathogens are important to agroecosystems because of economic losses resulting from fungal infection. Fungal pathogens break down plant tissue, decrease yields, or produce animal toxins. Both aboveground tissue (leaves, stems, and fruiting bodies) and belowground roots might become infected by pathogens (Carlile and Watkinson, 1996b). Infection aboveground often causes wide-spread destruction, because spores can be dispersed by wind over long distances. Belowground pathogen spread is slower, because propagules are disseminated in soil solution or by small animals. These propagules exist as fungal spores or infected roots and can remain dormant for long periods of time until a susceptible plant releases the organic C compounds that trigger germination (Carlile and Watkinson, 1996b). Pathogens often enter the plant tissue through the younger parts such as root hairs or through wounds. Some typical examples of root pathogenic fungi are *Fusarium*, *Phytophthora*, *Pythium*, and *Rhizoctonia*.

Plants have several mechanisms to defend against fungal infection. Physical barriers such as the mucigel on plant roots and the plant cell wall are the first lines of defense. Other defense mechanisms include (1) the hypersensitivity response (the death of host tissue around the point of infection to stop spread), (2) lignification of the cell wall, (3) synthesis of cellulose or callose, (4) phytoalexin accumulation, (5) release of hydrolytic enzymes, (6) synthesis of proteinase inhibitors, and (7) accumulation of hydroxyproline-rich glycoproteins (Carlile and Watkinson, 1996b).

Despite these defenses, conventional agricultural practices might help promote disease spread through the use of monocultures or only a few crops in a rotation, introduction of nonnative crop species, or the use of plants with gene-for-gene resistance instead of multiple-gene resistance. Crop varieties with gene-for-gene resistance are not effective over the long term, especially when planted across the whole field instead of being mixed with nonresistant varieties. In gene-for-gene resistance, only a single gene in the plant is active in defense, for which pathogens might evolve mechanisms to overcome. When multiple genes are employed in disease resistance, the pathogens are less likely to compensate and become infective (Carlile and Watkinson, 1996b). An example of the devastating results of a monoculture system with a nonnative crop species was the potato famine in Ireland during the 1840s caused by the fungal pathogen *Phytophthora infestans*, which led to over one million deaths from starvation and to mass emigration to the U.S. (Carlile and Watkinson, 1996b).

Agricultural practices, especially sustainable agricultural practices, can control fungal pathogens by (1) using cultivation to bury propagules away from new roots; (2) eliminating monoculture systems by increasing the number of crops in a rotation, or using buffer strips, shelterbelts, or interrow crops; (3) using resistant cover crops or fallow periods to limit propagule survival; (4) using fungicides or biocontrol methods [such as composting, mycoparasites (i.e., fungi parasitic to other fungi) or microbial competitors]; or (5) growing crops with multiple-gene pathogen resistance or limiting the number of gene-for-gene-resistant plants in a field (Carlile and Watkinson, 1996b). Increasing plant diversity through additional crops in a rotation system or using cover crops, buffer strips, or shelterbelts reduces or eliminates pathogens, because unlike saprophytes and most mutu-alists, fungal plant pathogens are usually host specific.

## BIOTROPHIC MUTUALISTIC FUNGI

In mutualistic relationships, both plant host and fungal invader obtain benefits that outweigh the inherent costs of the symbiosis. The fungi are carbon-limited and form associations with plants to acquire photosynthetic carbon. Some of these fungi might be saprophytic (e.g., many ectomycor-rhizal species or endomycorrhizal species after first germinating) or pathogenic under some con-ditions, but for the most part the mutualistic relationship is the norm. Plant host biomass increases because of low-cost acquisition of nutrients, especially highly immobile nutrients such as P and Zn. Better nutrition can enhance drought tolerance and disease resistance (Bolan, 1991; Hooker and Black, 1995; Paul and Clark, 1996).

The fungal symbiont causes physiological changes in the plant host. Disease resistance increases when the fungus triggers changes in plant cell wall chemistry or a hypersensitivity response to slow or eliminate infection. Plants stimulate colonization by the mutualistic fungus through increased root exudation, which stimulates spore germination and germ tube growth; increased root branching, which provides a greater surface area for colonization; and changes in the permeability of the cell membrane to promote colonization (Carlile and Watkinson, 1996b).

### Arbuscular Mycorrhizal Fungi

Of the four major types of mycorrhizal fungi [orchidaceous, ericoid (etcoendo-), ectomycorrhizal, and endomycorrhizal], the endomycorrhizal (AM) fungi are the most abundant and ubiquitous in agroecosystems (Millner and Wright, 2002; Olsson et al., 1999). AM fungi account for 5 to 50% of the total microbial biomass (Olsson et al., 1999) and are associated with ca. 70% of the vascular plant species (Trappe, 1987), including almost all crop plants. Exceptions are some members of the Brassicaceae (formerly the Cruciferae), namely broccoli, cauliflower, crambe, and canola. Brassicaceae is traditionally regarded as a nonmycorrhizal family. However, AM fungal colonization has been reported in ca. 33% of the plant species examined in this family (Harley and Harley, 1987).

Endomycorrhizal hyphae might colonize up to 80% of plant host root length (Millner and Wright, 2002), penetrating the plant cell wall and forming branched structures, called arbuscules, where nutrients and carbon are exchanged. Intraradical colonization includes hyphae, spores, arbuscules, and vesicles (storage structures). Colonization can be easily measured and used to indicate fungal activity (Giovannetti and Mosse, 1980), but accounts for a small amount of AM biomass (Olsson et al., 1999). Extraradical hyphae and spores account for 80 to 90% of the AM fungal biomass (Olsson et al., 1999). However, it requires some expertise to correctly differentiate AM hyphae from other fungal hyphae when measuring extraradical hyphal length (Steinberg and Rillig, 2003).

About 12 to 30% of plant photosynthetic carbon is translocated belowground in the form of sugars that support fungal growth and development (Paul and Clark, 1996; Tinker et al., 1994). These sugars are rapidly converted into sugar alcohols to maintain C flow to the fungus (Tinker et al., 1994). Carbon cost to the plant is balanced by access to a greater volume of soil through fungal hyphae. Hyphae have a much larger surface area to volume ratio than do root hairs and fan out up to 8 cm beyond nutrient depletion zones around roots (Douds and Millner, 1999; Millner and Wright, 2002; Figure 6.1). This allows AM fungi to scavenge even highly immobile nutrients such as phosphate. Also, the fungal cell membrane is capable of concentrating solutes against a gradient (Bolan et al., 1991; George et al., 1992). The high carbon cost of P uptake is compensated for by

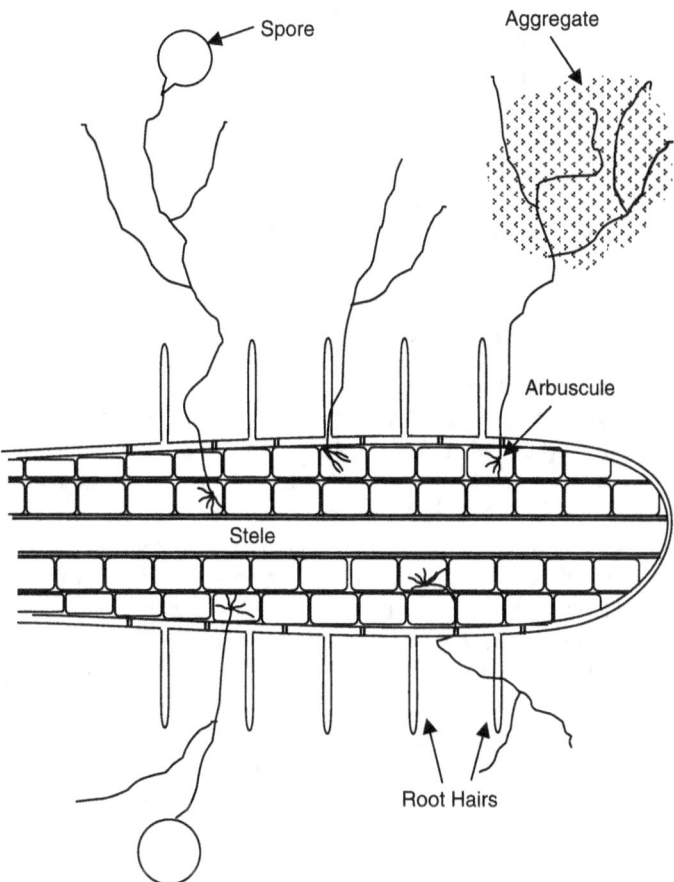

**FIGURE 6.1** Hyphae of arbuscular mycorrhizal fungi can access much more of the soil than can roots and root hairs and form a framework on which aggregates can form.

an increase in photosynthetic capability of the host through increased leaf surface area and photosynthetic efficiency (Bolan et al., 1991; George et al., 1992). Mycorrhiza is the most efficient mechanism for P acquisition, especially under stress conditions.

To varying degrees, mycorrhizal fungi can also provide other benefits, such as more efficient uptake of N, the micronutrients Fe, Cu, and Zn (Clark and Zeto, 1996; Pawlowska et al., 2000), and water; disease suppression; protection from heavy metal toxicity; and improved soil structure. The mycorrhizal relationship reduces the growth of plant pathogens, especially fungal pathogens, by increasing host resistance (triggering a defense response), altering root exudations to stimulate the growth of microbes antagonistic to pathogens, competing for photosynthetic carbon, and reducing the number of infection sites (Borowicz, 2001). The type of pathogen (nematode or fungal), pathogen species, mode of action (necrotrophic or wilt for fungal pathogens and migratory or sedentary for nematodes), and pathogen density help determine the severity of disease (Borowicz, 2001). As with other benefits in the mycorrhizal relationship, the magnitude and direction effects of AM fungi on disease resistance depend on host genotype, AM species and isolate, timing of AM colonization, other soil organisms, and abiotic factors.

Mycorrhizal host plants have been found at many heavy-metal contaminated sites, but the fungi typically are not examined (Pawlowska et al., 2000). In pot culture experiments, it has been shown that mycorrhizal fungi can take up toxic heavy metals, such as Cd and Pb, in addition to micronutrients (Gonzalez-Chavez et al., 2002; Diaz et al., 1996). Metal uptake depends on soil fertility, metal concentration, pH, the host plant, and AM species, and might interfere with P nutrition in the host plant (Gonzalez-Chavez et al., 2002; Diaz et al., 1996).

In addition to improving plant health, fungal hyphae improve soil structure by helping form water-stable soil aggregates (Miller and Jastrow, 1990; Rillig and Steinberg, 2002; Tisdall et al., 1997). Mycorrhizal fungi also improve rhizosphere health by stimulating root exudation, which promotes the growth of other soil microbes (Borowicz, 2001; Paul and Clark, 1996). Many excellent books and review articles have been published on AM fungi and agroecosystems (Bolan et al., 1991; Douds and Millner, 1999; George et al., 1992; Zak and McMichael, 2001).

## GLOMALIN, A GLYCOPROTEIN PRODUCED BY AM FUNGI

The identification of glomalin, a glycoprotein produced by AM fungi, has led to a reevaluation of fungal contributions to SOM and aggregate stability. Glomalin was identified at the United States Department of Agriculture (USDA) in 1993 during work to produce monoclonal antibodies reactive with AM fungi. One of these antibodies reacted with a substance on the hyphae of a number of AM species (Wright et al., 1996). This substance was named glomalin after Glomales, the order to which AM fungi belong. Several other typical soil fungi, such as *Rhizoctonia, Gaeumannomyces, Endogone, Mucor,* and *Phytophthora*, were tested for cross-reactivity with the antibody against glomalin, but were not immunoreactive (Wright et al., 1996). The glomalin fraction is operationally defined by its extraction procedure, but is further characterized by total and immunoreactive protein assays (Wright et al., 1996). Glomalin is found in abundance in both native and agricultural soils (2–14 mg/g soil and 2–5 mg/g soil, respectively; Wright and Upadhyaya, 1998; Wright et al., 1999) and appears to be as ubiquitous as AM fungi themselves (Carlile and Watkinson, 1996b; Olsson et al., 1999; Wright and Upadhyaya, 1998; Wright, unpublished data).

Glomalin was revealed on AM fungal hyphae by using an indirect immunofluorescence procedure that employs the antibody against glomalin and a second antibody tagged with a fluorescein isothiocyanate (FITC) molecule (Wright, 2000). Evidence that glomalin is produced by AM fungi and not plant roots was obtained early in the investigation of the reaction of the monoclonal antibody against glomalin. Colonized and uncolonized roots were submitted for evaluation of the technique by J.B. Morton (West Virginia University) in a blind experiment. Colonization was correctly identified by immunofluorescence only on the roots that were later described as having been inoculated. Immunofluorescence was absent on the roots later described as uninoculated controls (Wright, unpublished data). In more recent work with an axenic culture of transformed carrot roots, glomalin was extracted from hyphae in a root-free zone (Rillig and Steinberg, 2002). Glomalin is also routinely extracted from hyphae up to 7 cm away from roots in pot cultures wherein hyphae is separated from roots by a 38-μm nylon mesh bag (Wright and Upadhyaya, 1999; Figure 6.2). Immunofluorescence assays show that glomalin coats AM fungal hyphae (Figure 6.3A to C); sloughs from hyphae onto colonized roots, organic matter, soil particles, horticultural or nylon mesh (Figure 6.3D), and glass beads (Figure 3E); and is found on arbuscules (green) within autofluorescing (yellow) root cells (Figure 6.3F; Wright et al., 1996; Wright and Upadhyaya, 1999; Wright, 2000).

## POOLS OF GLOMALIN

Glomalin consists of four major pools: (1) easily extractable glomalin (EEG), (2) total glomalin (TG), (3) recalcitrant glomalin (RG), and (4) scum. The EEG pool is extracted with 20 m$M$ citrate, pH 7.0, for 0.5 h (Wright and Upadhyaya, 1998). Total glomalin is extracted with 50 m$M$ citrate, pH 8.0, in 1-h intervals (Wright and Upadhyaya, 1998), and recalcitrant glomalin is soluble only in 50 m$M$ citrate, pH 8.0, at 121°C after harsh treatment of the soil (i.e., treatment with dilute acid for 1 h followed by three 16- to 18-h extractions in alkaline solutions; Nichols, 2003). When mature

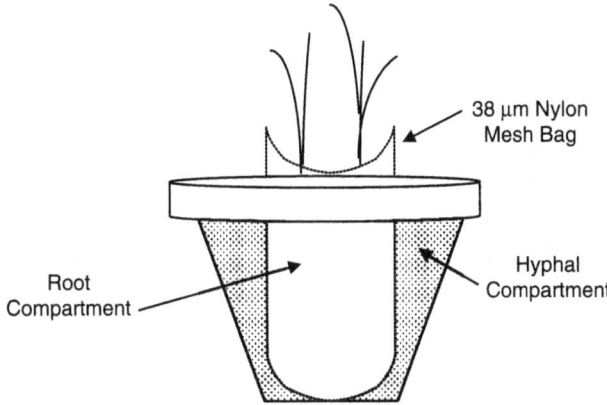

**FIGURE 6.2** Single-species arbuscular mycorrhizal fungal cultures can be grown with different plant hosts to examine glomalin accumulation in sterile sand, or, as in this case, sand and crushed coal medium where the roots are contained in the root compartment within the nylon mesh bag, and the fungal hyphae, which grows through the mesh and into the surrounding media in the hyphal compartment, can be examined under single-species conditions.

sand-based pot cultures are submerged in water, an unattached fraction of glomalin forms tan-colored foam on the surface of water. This scum is apparently a sloughed component of glomalin and is very hydrophobic. We speculate that scum floats on soil water until it attaches to soil or organic matter particles, but the chemistry of this interaction is not currently defined. Our lab postulates that hydrophobic or cationic interactions, or both, might be the mechanisms by which glomalin becomes deposited on soil or organic particles and mesh or glass beads (Wright and Upadhyaya, 1996; Nichols and Wright, unpublished data). Glomalin contains high concentrations of iron (2 to 12%), and recently it has been speculated that Al- and Fe-hydroxyls are involved in aggregate formation by bridging organic matter to clay particles (Bird et al., 2002; Chenu et al., 2000). It appears that glomalin can move in and out of these operationally defined pools (i.e., EEG becomes scum and scum becomes TG). Steinberg and Rillig (2003) found that during an incubation experiment EEG increased while TG decreased. They speculated that partial degradation decreases sorption of glomalin to soil particles, which might increase the solubility and amount in the EEG pool.

   Glomalin concentration in these pools is measured by a Bradford total protein assay (i.e., TG and EEG), immunoreactive protein (i.e., IRTG and IREEG) assays (Wright et al., 1996), or as gravimetric or carbon weight. The Bradford protein assay is nonspecific and detects any proteinaceous material. Bradford concentrations are based on comparison with a bovine serum albumin (BSA) standard curve. The immunoreactive protein assay (ELISA) uses the monoclonal antibody specific for glomalin, but certain artificial conditions might reduce immunoreactivity. The ELISA values are determined by comparison to 100% immunoreactive glomalin extracted from hyphae or soil (Wright et al., 1996). The total protein assay measures concentrations from 1.25 to 5.0 μg, whereas ELISA measures concentrations from 0.005 to 0.04 μg (Wright and Upadhyaya, 1999). Because the range of Bradford values is ~100 times higher than that for ELISA, it can support values of more than 100%. Both gravimetric and carbon weight have been used to quantify glomalin partially purified by acid precipitation and dialysis against water (Nichols, 2003; Wright et al., 1996). These weights are not based on structural components of glomalin but are rather direct measurements on lyophilized material.

   Comparisons of the total and immunoreactive pools of glomalin extracted from soil or pot culture show that not all the extracted material is immunoreactive. Reduction in immunoreactivity can be due to exposure to conditions that affect the site of binding of the antibody. The reactive site for a monoclonal antibody is very specific (Goding, 1986), and some reactivity is lost probably

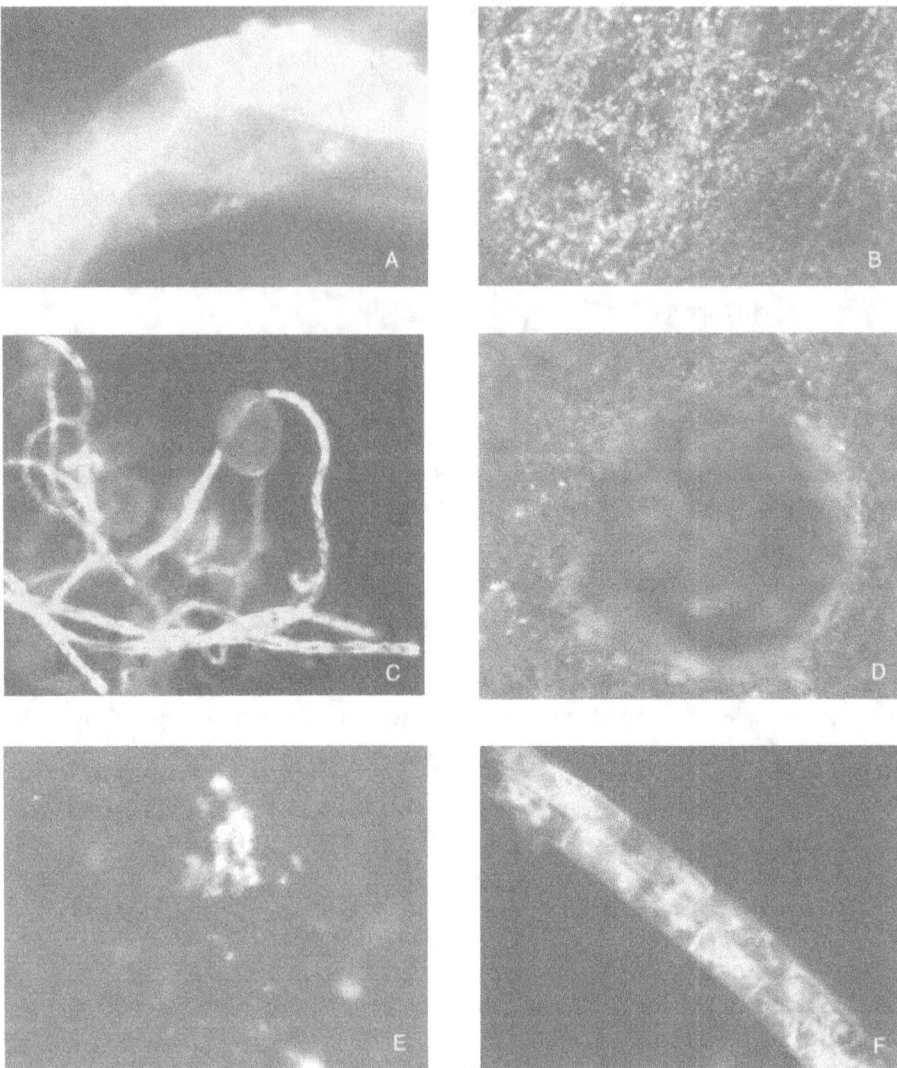

**FIGURE 6.3** Arbuscular mycorrhizal fungi can be cultured in hydroponic pot cultures with a mycorrhizal host plant, and glomalin can be examined by an immunofluorescence assay with a monoclonal antibody against glomalin (seen as bright spots). Glomalin has been found coating and sloughing from hyphae of *Acaulospora morrowiae* (CL551) (A), on a *Gigaspora rosea* (FL224) hyphal mat adhering to a horticultural mesh (B), on *Glomus intraradices* hyphae grown in liquid cultures media by Dr. Yair Shachar-Hill at the New Mexico State University (C), deposited on and around a hole in a horticultural mesh by *Gi. rosea* (FL224) (D), on a glass bead by *A. morrowiae* (CL551) (E), and on arbuscules of *G. etunicatum* (BR220) in a corn root (F).

because of conformational changes by exposure to high heat (121°C) for a long time period (at least 0.5 to 1.0 h) during extraction (Wright and Upadhyaya, 1999; Wright, unpublished data). In the soil, organic matter, metals (such as iron), clays, and other substances might bind to glomalin, causing conformational changes or masking the reactive site and thereby interfering with immunoreactivity. Also, conformational changes can occur in the molecule because of hydrophobic interactions when it sloughed from the hyphae and is in the scum pool. Degradation is a factor in soil extracts and might result in a decline in immunoreactivity (Wright and Upadhyaya, 1999). Differences in immunoreactivity and extraction techniques are used to further describe some of the glomalin pools, such as the highly immunoreactive EEG (IREEG), lower immunoreactive TG (IRTG; Wright and Upadhyaya, 1996), and very low immunoreactive RG (IRRG; Nichols, 2003).

## CHARACTERIZATION OF GLOMALIN

Glomalin extracted from soil is very similar to that extracted from single-species pot cultures. Samples have been examined by SDS-PAGE (Nichols, 2003; Rillig et al., 2001b; Wright et al., 1996; Wright and Upadhyaya, 1996); nuclear magnetic resonance (NMR) (Nichols, 2003; Rillig et al., 2001b; Nichols and Wright, unpublished data); carbohydrate analyses by a colorimetric assay; gas chromatography–mass spectroscopy (GC-MS) and capillary electrophoresis (CE; Wright et al., 1998; Nichols and Wright, unpublished data); and C, H, N analysis by combustion (Nichols, 2003; Rillig et al., 2001b). There are minor variations in elemental constituents of glomalin among samples, but the structural group assays (NMR, GC-MS, and CE) and SDS-PAGE demonstrate that glomalin extracted from soil is similar to that from hyphae.

Rillig et al. (2003) and Steinberg and Rillig (2003) examined decomposition of glomalin following soil incubation. One of the incubation studies (Steinberg and Rillig, 2003) showed that hyphal length declined by 60% after 150 d of incubation, whereas the TG of glomalin declined by 25%, the IRTG disappeared almost completely, the EEG did not change, but the IREEG increased fivefold. In the other study (Rillig et al., 2003), the TG declined by 48–81% and the EEG declined by 51–88% after 413 d of incubation. By $^{14}$C data, Rillig et al. (1999) calculated a turnover time for glomalin of 6 to 42 years. These recent incubation studies suggest that a long-lived, recalcitrant glomalin fraction exists with a much longer turnover time.

Experiments to identify structural units of glomalin are currently underway. Information obtained to date shows that glomalin is composed of proteinaceous, carbohydrate, and aliphatic (potentially polymerized) components and binds multivalent cations (i.e., Fe and Al; Nichols, 2003; Wright and Anderson, 2000; Nichols and Wright, unpublished data). The protein component appears to be 30 to 40% of the molecular structure, measured by comparisons of gravimetric and protein weight and preliminary amino acid measurements. The carbohydrate component is 3 to 6% according to a colorimetric assay, which measures oligosaccharide concentration. Aliphatic groups comprise 20 to 70% according to mass balance and NMR spectroscopy. Glomalin has 2 to 12% iron based on acid hydrolysis and atomic adsorption measurements.

## GLOMALIN, A MAJOR COMPONENT OF SOIL ORGANIC MATTER

A study was conducted to compare concentrations of glomalin to humic acid (HA), fulvic acid (FA), and particulate organic matter (POM) in eight undisturbed soils in the U.S. All the fractions have been operationally defined by extraction techniques. The appropriate extraction method was used to remove each fraction: (1) alkaline extraction of HA and FA followed by acidic separation, (2) citrate extraction of glomalin, and (3) density separation of POM. Quantities were measured by using gravimetric and carbon weights and comparing total and immunoreactive protein concentration. The protein values also were used to correct for coextraction of glomalin in HA. The study showed that glomalin represents a major fraction of soil organic carbon (SOC; 22 to 27%) and the extractable part of the material previously identified as HA and humin contains glomalin (Figure 6.4; Nichols, 2003; Nichols and Wright, unpublished data).

## QUANTITIES OF GLOMALIN

Soils from a variety of ecosystems throughout the U.S. and the world have been extracted for glomalin, with TG concentrations ranging from 2 to 14 mg/g in most soils (Wright and Upadhyaya, 1998). High glomalin (TG) amounts have been found in undisturbed, volcanic soils from Japan and Hawaii (19 mg/g soil and >60 mg/g soil, respectively; Rillig et al., 2001b; Wright, unpublished data) and a humoferric podzol oak forest soil in Ireland (69 mg/g soil; Nichols and Wright,

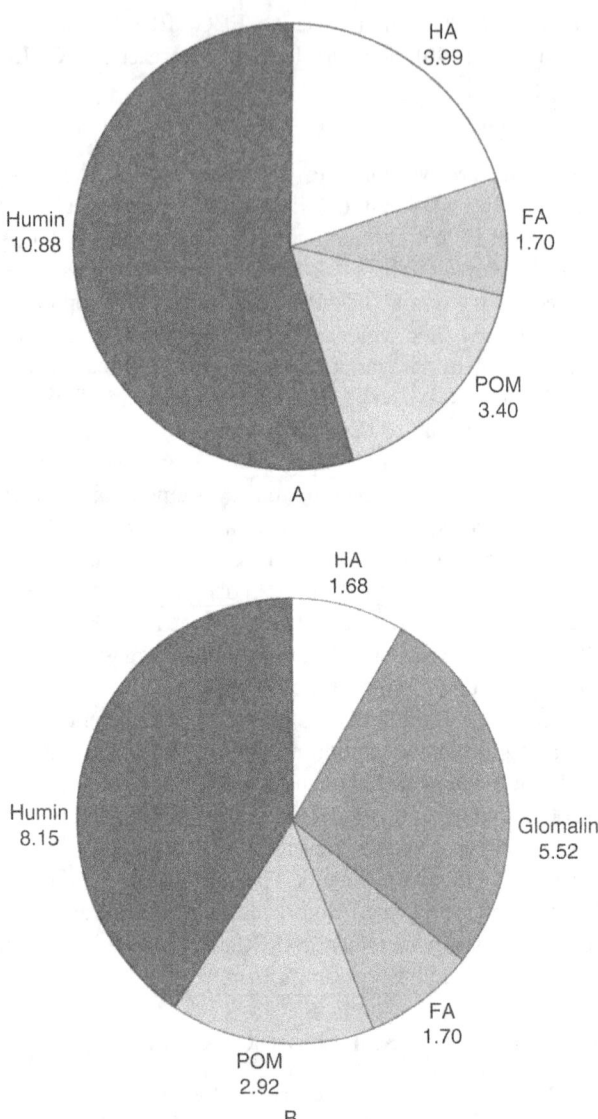

**FIGURE 6.4** The major fractions of soil organic carbon (SOC) have historically been (A) humic acid (HA), fulvic acid (FA), humin (or humus), and particulate organic matter (POM), but with the identification of glomalin and its separation from humic components (B), a sizable amount of SOC has been found in this fraction. Units are mg C in the fraction per g soil extracted. (Adapted from Nichols, K.A. 2003. Ph.D. thesis, University of Maryland, College Park. With permission.)

unpublished data). Typically, acidic soils have higher glomalin concentrations than do calcareous soils (Wright and Upadhyaya, 1996; M. Haddad, personal communication), as do undisturbed soils compared with agricultural soils (Wright and Upadhyaya, 1998; Nichols, 2003). Acidic soils have lower decomposition rates and more soluble metals (such as Fe and Al), which might increase glomalin concentrations by interactions with the molecules that inhibit degradation. Undisturbed soils have lower decomposition rates than agricultural soils and a greater presence of AM fungi because undisturbed soils have no P inputs from fertilizers and tillage has not disrupted hyphal networks.

Glomalin concentrations have been measured in a number of agricultural soils with different tillage treatments; crop rotations; and fertilizer, herbicide, and pesticide amendments. In most of these systems, there was a correlation between aggregate stability and glomalin concentration, with no-till or minimum-till systems having the highest values (Figure 6.5; Rillig et al., 2002; Wright and Anderson, 2000; Wright et al., 1999; Nichols and Wright, unpublished data). At a site in Maryland, different management systems [(1) no-till, synthetic amendments (NT); (2) conventional tillage, synthetic amendments (CT); and (3) minimum tillage, organic amendments (MT)] in the same field were examined to determine how management affects glomalin-C and POM-C. The MT treatment had the highest content of glomalin-C, POM-C, and aggregate stability, whereas the NT and CT systems did not differ (Figure 6.6; Nichols, Wright, and Cavigelli, unpublished data).

## SINGLE-SPECIES POT CULTURE EXPERIMENTS

Glomalin has been extracted from AM hyphae grown in single-species pot cultures with sterile sand media and crushed coal where roots are contained in a 38-μm nylon mesh bag that only hyphae can penetrate (Figure 6.2). Plants were watered with low-P nutrient solution (Millner and Kitt, 1992) and grown under metal halide and sodium vapor lights in a growth chamber or sodium vapor lights in the greenhouse. Several different pot culture experiments have been conducted to measure glomalin concentrations. In total, nine different species from four of five genera of AM fungi have been grown with up to five different host plants (i.e., corn, clover, sudangrass, sorghum, fescue). All these AM species produced glomalin in amounts that vary with culture conditions (i.e., light intensity, media, etc.) and plant host. Glomalin concentrations (TG) in the outer hyphae chamber typically range from 2 to 13 mg/pot, with values of 2 to 20 mg/g hyphae (Nichols and Wright, unpublished data) and 5 to 40 μg glomalin/cm$^2$ on horticultural mesh strips inserted into sand in the hyphae chamber (Wright and Upadhyaya, 1999).

In one experiment, TG was extracted from five AM species *(Glomus etunicatum, G. viscosum, G. caledonium, Gigaspora rosea,* and *Gi. gigantea)* produced on corn *(Zea mays)* and crimson clover *(Trifolium incarnatum* L.). For three of the fungi *(G. etunicatum, G. viscosum,* and *Gi. gigantea)*, two INVAM isolates that had been collected from the same state or country but from a

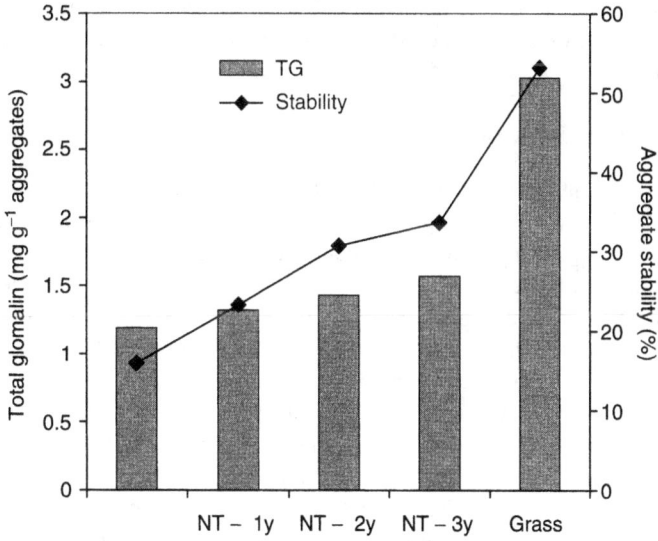

**FIGURE 6.5** Total glomalin concentration (mg/g aggregates) and aggregate stability (%) in 0- to 5-cm soil samples of plots in transition from plow tillage (PT) to no-till (NT) in 1-, 2-, and 3-year increments and a continuous grass (Grass) for ca. 15 years. (Adapted from Wright, S.F. et al. 1999. *Soil Sci. Soc. Am. J.* 63: 1825–1829. With permission.)

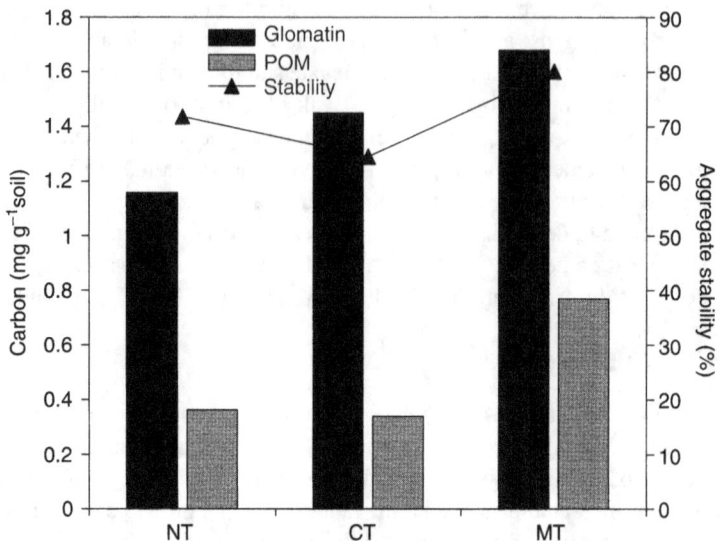

**FIGURE 6.6** Carbon in glomalin (glomalin-C, mg C glomalin/g soil) or particulate organic matter (POM) (POM-C, mg C POM/g soil) extracted from soil (1 g) with 50 m*M* citrate or by using density separation, respectively. Soil samples were collected from a Maryland field site under three different management treatments: (1) no-till, synthetic amendments (NT); (2) conventional tillage, synthetic amendments (CT); and (3) minimum tillage, organic amendments (MT).

different site were used. Glomalin concentrations (TG) in the hyphal compartment ranged from 0.5 to 2.7 mg/pot and in the root compartment from 1.2 to 13.8 mg/g hyphae (Table 6.1). *Glomus viscosum* had the lowest total protein values in both the hyphae and root chambers, but the % IR values in the root chamber were among the highest for both isolates and both hosts. One isolate of *G. etunicatum* (BR220) had total protein values almost twice that of the other isolates, but this was not true of the other *G. etunicatum* isolate (BR211). Although these plants were grown under the same conditions, overall variations existed among isolates, species, and hosts that do not follow any common trend.

## DEPTH AND DEPOSITION EXPERIMENT

A pot culture experiment was performed to determine the amount of glomalin produced by *Gigaspora rosea* (FL 224) colonizing single crimson clover (*Trifolium incarnatum* L.) plants. Figure 6.7 shows the experimental setup. Roots were confined within a nylon mesh bag, and hyphae grew into a root-free zone. Plants were supplied with a low-P nutrient solution (Millner and Kitt, 1992) and grown under sodium vapor lights. The following measurements were made in 8-cm-deep increments of the hyphae chamber after 3 months of plant growth: (1) glomalin on hyphae, (2) glomalin deposited on the horticultural mesh, (3) unattached glomalin (i.e., scum), and (4) percent colonization of roots. Hyphae and scum were obtained by immersing the sand from each depth increment in water, shaking the sand vigorously, and decanting the water into stacked sieves (150, 150, and 53 μm top to bottom). This was repeated four times. Hyphae and unattached glomalin on the 150 and 53 μm sieves were washed into petri dishes. Scum floats on water and was separated from hyphae by pipetting. The horticultural mesh was cut into small pieces for processing. Hyphae, scum, and horticultural mesh were extracted in 20 m*M* citrate, pH 7.0 at 121°C for 1 h. Glomalin was quantified by the Bradford assay (Wright and Upadhyaya, 1996). Percent colonization was determined by the method of Giovannetti and Mosse (1980).

**TABLE 6.1**
**Glomalin Extracted from Single-Species Pot Cultures of One or More Isolates of Five Arbuscular Mycorrhizal Fungal Species on Two Plant Hosts and Measured as Total Protein (TP) and Percentage Immunoreactive Protein (% IR)**

| Species[a] | Isolate | Host | Hyphae Chamber TP (mg/pot) | Hyphae Chamber % IR[b] | Root Chamber TP (mg/pot) | Root Chamber % IR[b] |
|---|---|---|---|---|---|---|
| G. etunicatum | BR211 | Clover | 1.90 | 52 | 6.57 | 44 |
| G. etunicatum | BR220 | Clover | 1.11 | 58 | 11.82 | 12 |
| Gi. rosea | FL224 | Clover | 2.05 | 52 | 5.25 | 12 |
| Gi. gigantea | MA401C | Clover | 1.63 | 48 | 5.45 | 46 |
| Gi. gigantea | MA453A | Clover | 1.28 | 50 | 4.08 | 47 |
| G. viscosum | MD215 | Clover | 0.99 | 31 | 1.98 | 156 |
| G. viscosum | MD216 | Clover | 1.17 | 34 | 1.55 | 116 |
| G. caledonium | UK301 | Clover | 1.44 | 35 | 4.53 | 28 |
| G. etunicatum | BR211 | Corn | 1.43 | 69 | 2.82 | 94 |
| G. etunicatum | BR220 | Corn | 1.70 | 57 | 13.80 | 22 |
| Gi. rosea | FL224 | Corn | 2.46 | 40 | 2.13 | 140 |
| Gi. gigantea | MA401C | Corn | 1.47 | 39 | 7.58 | 105 |
| Gi. gigantea | MA453A | Corn | 2.74 | 40 | 7.56 | 153 |
| G. viscosum | MD215 | Corn | 0.47 | 92 | 1.22 | 94 |
| G. viscosum | MD216 | Corn | 0.83 | 45 | 1.44 | 191 |
| G. caledonium | UK301 | Corn | 1.13 | 62 | 7.16 | 185 |

[a] Species from two genera *Glomus (G.)* and *Gigaspora (Gi.)*.

[b] % IR = (immunoreactive protein/total protein) × 100. Immunoreactive protein values were determined by comparison to 100% immunoreactive glomalin extracted from an undisturbed prairie soil. The scale for the standard curve used to measure immunoreactive protein concentrations was ca. 100 times less than the scale for the total protein standard curve. This gives very precise values for immunoreactive protein concentrations and might result in % IR values higher than 100, especially when measuring microgram quantities.

Note: n = 3 pots per isolate. In these pot cultures, the hyphae chamber was separated from the root chamber by a 38-μm nylon mesh bag that roots cannot penetrate.

Glomalin production varied greatly for three plants (Figure 6.8). This might have been due to factors controlled by individual plants, differences in light intensity, or a combination of factors, which will require further investigation.

Figure 6.9 examines more closely the distribution of glomalin produced by Plant 2. Hyphae grew into the root-free zone in the top half of the pot (Figure 6.9) and apparently released glomalin that was unattached to hyphae or adhered to the horticultural mesh. Movement of glomalin unattached to hyphae through coarse sand is suggested, because unattached and mesh-trapped glomalin were measured in the absence of detectable amounts of hyphae in the lower half of the pot. Sloughing of glomalin from hyphae and attachment to soil particles is also suggested by the results of this experiment.

## GLOMALIN AND AGGREGATE STABILITY

Loss of topsoil due to erosion is a serious consideration in agroecosystems. Pimentel et al. (1995) estimated that during the past 40 years nearly one third of the world's arable land was lost to erosion, with a current rate of 10 million ha/year. Soil aggregates are important to (1) maintain soil

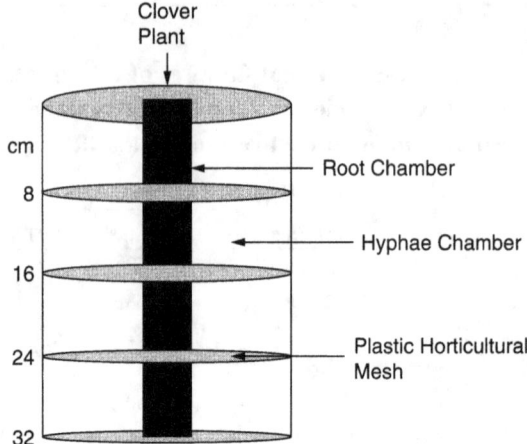

**FIGURE 6.7** Configuration of pots used to determine glomalin production and deposition. A single red clover plant was grown inside an 8-cm-diameter 38-µm nylon bag filled with coarse sand in a 25-cm-wide by 40-cm-deep pot. Outside the nylon bag, the pot was filled with coarse sand. Horticultural mesh disks were placed at 8-cm-deep intervals within the root-free hyphae chamber.

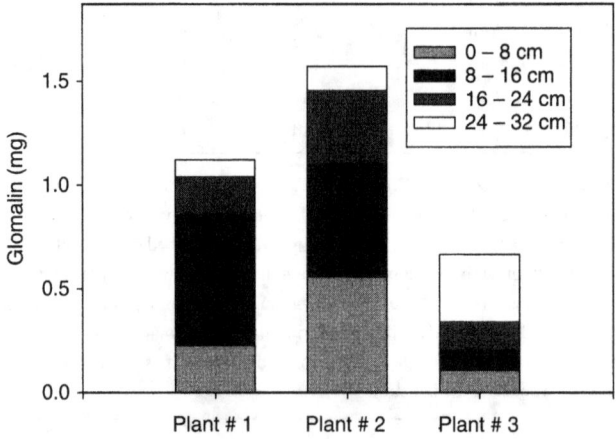

**FIGURE 6.8** Glomalin production by three individual clover plants. Production at each depth is the sum of glomalin on hyphae, attached to the horticultural mesh, and unattached scum.

porosity, which provides aeration and water infiltration rates favorable for plant and microbial growth, (2) increase stability against wind and water erosion, and (3) store carbon by protecting organic matter from microbial decomposition (Bird et al., 2002; Hassink and Whitmore, 1997). Because both aggregate stability and SOM decline on cultivation, it is possible that SOM (i.e., POM, humic substances, microbially produced molecules, and fungal hyphae) plays a role in aggregate formation, but the exact mechanism is not understood. Aggregate formation is a complex process of physical and chemical interactions.

Electron microscopy shows that aggregates are a conglomeration of soil minerals (clay particles, fine sand, and silt), small plant or microbial debris, bacteria, free amorphous organic matter, and organic matter strongly associated with clay coatings (Chenu et al., 2000). Fungal hyphae can initiate aggregate formation by providing the framework on which organic mater collects (Miller and Jastrow, 1990; Tisdall et al., 1997). Chemical processes then contribute to aggregate formation and stability by gluing with polysaccharides, coating with hydrophobic polymers, binding mineral

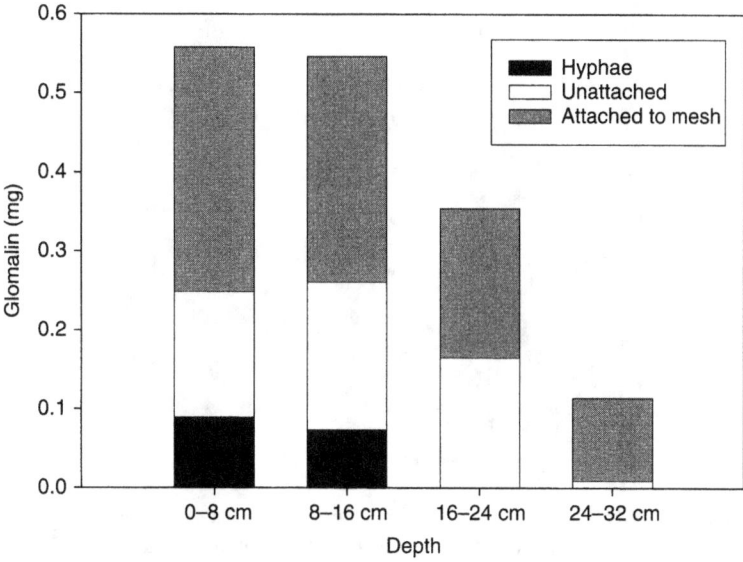

**FIGURE 6.9** Glomalin production on a clover plant by depth. Percent colonization of roots at each depth increment is shown above the bar.

particles with organic polymers, and bridging organic matter and clay particles by polyvalent cations (Degens, 1997; Piccolo and Mbagwu, 1999; Chenu et al., 2000). Drying and wetting actions, shrinking and swelling of clays, freeze–thaw cycles, compaction, and enmeshing by fungal hyphae and fine roots physically stabilize aggregates (Chaney and Swift, 1986a, 1986b; Degens, 1997).

Soil aggregates can be disrupted by rainfall because of slaking, differential swelling of clays, mechanical dispersion by the kinetic energy of raindrops, and physiochemical dispersion without the protection of hydrophobic coatings. The molecules involved in aggregate formation increase water stability and long-term survival of aggregates, because attractive forces between these molecules are much stronger internally than externally (Degens, 1997; Piccolo and Mbagwu, 1999; Chenu et al., 2000). In reduced or no-till systems, Chaney and Swift (1986a) found that the stubble and mulch litter promote aggregate formation, because fungal decomposition of organic matter produces gluing agents such as polysaccharides and mucigels. Caesar-TonThat and Cochran (2002) found that ligninolytic basidomycetes produce large quantities of polysaccharides, glycolipids, or glycoproteins that bind to and stabilize soil particles in water-stable aggregates. However, many of the polysaccharides produced by microbial degradation glue aggregates together quickly but are water-soluble and ephemeral and do not to contribute to the long-term stability of aggregates (Chaney and Swift, 1986a; Six et al., 2001). Soil organic matter containing high concentrations of aliphatic groups, such as HA, can increase aggregate stability and the long-term stabilization of organic materials (Piccolo and Mbagwu, 1999). These aliphatic hydrophobic groups and polymers are the major contributors to the water stability of aggregates. They increase the contact angle for water penetration, which restricts infiltration and slaking, lowers wettability, and increases the internal cohesion of aggregates (Chenu et al., 2000). Complexes between organic matter and amorphous Fe and Al compounds also decrease the wettability of aggregates (Chenu et al., 2000).

Glomalin contributes to the stabilization of aggregates because it sloughs off hyphae onto surrounding organic matter (Figure 6.10); binds to clays, probably via cation bridging by iron; and has hydrophobic character owing to a number of aliphatic groups (Nichols, 2003; Wright and Upadhyaya, 1999). This is demonstrated in a number of experiments in which total, and, especially, immunoreactive concentration, of glomalin are positively correlated with percent water-stable soil aggregates in both agricultural and native soils (Figure 6.5 and Figure 6.6; Bird et al., 2002; Rillig et al., in press; Wright and Anderson, 2000; Wright and Upadhyaya, 1998; Wright et al., 1999).

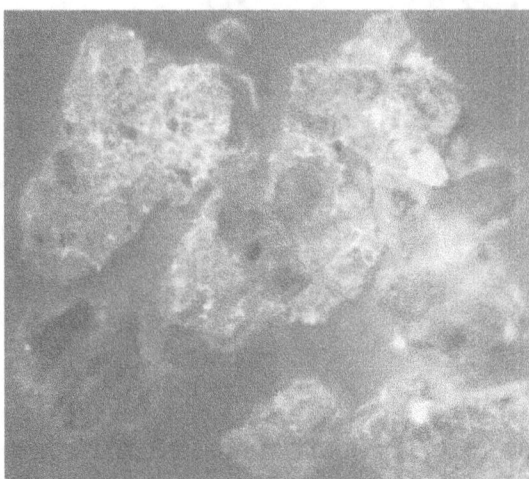

**FIGURE 6.10** Glomalin and arbuscular mycorrhizal hyphae on the surface of a 1- to 2-mm aggregate separated from a Philippine soil provided by Dr. Angela Almendras, Department of Agronomy and Soil Science, ViSCA. Glomalin is indicated by the bright spots, which are illuminated by an immunofluorescence assay using a monoclonal antibody against glomalin.

Figure 6.5 shows that both glomalin and aggregate stability can be used to quantify changes that occur in the soil, with a transition from continuous plow tillage to no till over a short period of time (1 to 3 years). Even though glomalin and aggregate stability increased significantly after only 2 years of no-till management, the 15 years of undisturbed grass site had much higher glomalin concentrations (TG) and aggregate stability than any other treatment. This study indicates that glomalin, POM, and aggregate stability all continue to increase with time within reduced tillage systems.

## GLOMALIN UNDER ELEVATED $CO_2$

Several studies were conducted to compare glomalin concentrations to aggregate stability under elevated $CO_2$ conditions. In a native grassland ecosystem in northern California, TG and IRTG concentrations increased with higher $CO_2$ concentrations, along with hyphal length at one site, and aggregate stability in 1–2 mm and 0.25–1 mm aggregate size fractions (Rillig et al., 1999). Long-term exposure to elevated atmospheric $CO_2$ conditions from a natural $CO_2$ spring in New Zealand resulted in a linear increase in percent root colonization by AM fungi, soil hyphal length, TG, and EEG along a $CO_2$ gradient (Rillig et al., 2000). In a sorghum field, aggregate stability, hyphal length, and EEG increased with elevated $CO_2$ (Rillig et al., 2001a). In both the grasslands (Rillig et al., 1999) and sorghum field (Rillig et al., 2001a), aggregate stability was correlated with glomalin concentrations. These studies show that under elevated $CO_2$ conditions, photosynthetic carbon is allocated belowground and glomalin might provide a significant sink to trap carbon in the soil.

## CONTRIBUTION OF SOIL FUNGI TO ORGANIC MATTER

Although hundreds of meters of hyphae can be found, fungal biomass is typically underestimated and the contribution of fungi to SOM on a mass basis is not quantified at present. Stevenson (1994) estimates that fungal numbers are 10 to 20 million/g of soil, whereas bacteria numbers are >1 billion/g of soil. Olsson et al. (1999) estimated the hyphal dry weight of AM fungi to be 0.03 to

0.35 mg/g soil according to phospholipid fatty acid analysis. Some of the largest organisms in the world are slow-growing soil fungi; for example, the basidiomycete *Armillaria bulbosa* has been found in the soil that covers 15 ha, weighs >10,000 kg, and is over 1500 years old (Paul and Clark, 1996).

Soil fungi contribute to the formation and function of SOM. Upon decomposition by saprophytic fungi, elements in POM, such as nitrogen and phosphorus, are transformed from unavailable organic compounds to available inorganic nutrient sources. The degraded plant material then becomes part of the humic fraction of soil. In addition, AM fungi and glomalin help stabilize SOM by contributing to aggregate formation.

Overall, the diversity of soil organisms is reduced by agricultural practices and mineralization rates increased, making C the limiting nutrient for soil fungi. Fungal biomass typically responds positively to no-till management; for example, Frey et al. (1999) found that fungal hyphae length was 2 to 2.5 times higher in no-till than conventionally tilled systems. Fungi are favored in no-till systems because (1) hyphal networks can be maintained, (2) fungi can bridge the soil–residue interface and utilize spatially separated nutrients, especially C and N; and (3) fungi can maintain activity, even in dry locations or across air-filled pores. With the identification of glomalin and the correlations between the immunoreactive fraction of glomalin and aggregate stability, this glycoprotein is proving useful as an indicator of soil quality and mycorrhizal input. As more information comes to light about the structure of this molecule and its different pools, the role of glomalin in agroecosystems will become more defined. However, from what is already known, glomalin is a major component of the SOM and important to sustainable agroecosystem functioning.

## MANAGING SOIL FUNGI TO INCREASE SOIL ORGANIC MATTER

A sustainable agroecosystem is one in which the system's internal mechanisms and resources can maintain productivity, recover quickly from disturbances (such as tillage), and keep pests and disease at tolerable levels with only minimal external inputs. Agricultural soils contain unnaturally high amounts of P, N, and K from fertilizers, are physically disrupted by tillage, and often are vegetated by only one or two plant species (Gliessman, 2000; Muramoto et al., 2000). To decrease pathogenic fungi and enhance the biomass, diversity, and function of mutualistic fungi, one or more of the following management options can be implemented (Carlile and Watkinson, 1996a; Horwath and Paul, 1994; Stenberg, 1999; Wright and Anderson, 2000):

1. Reduced fertilizer inputs (especially high-P fertilizers) will increase the need for mutualistic fungi to scavenge immobile nutrients.
2. Conservation or no-tillage systems will prevent disruption of hyphal networks.
3. Increasing the number of crops in the rotation, planting interrow crops, or using buffer strips or shelterbelts will increase aboveground diversity and decrease hosts for pathogenic fungi.
4. Planting cover crops instead of having fallow periods will maintain the presence of living roots as hosts for mutualistic fungi.
5. Use of biocontrol measures for weeds and pests will reduce the loss of beneficial fungi by fungicides and other pesticides.

These strategies can be used in agroecosystems to manage saprophytic, pathogenic, and mutualistic soil fungi in order to allow greater crop production at lower costs.

## REFERENCES

Abbott, L.K., and A.D. Robson. 1994. The impact of agricultural practices on mycorrhizal fungi. In C.E. Pankhurst et al. (Eds.), *Soil Biota: Management in Sustainable Farming*. CSIRO Melbourne, Australia.

Bird, S.B., J.E. Herrick, M.M. Wander, and S.F. Wright. 2002. Spatial heterogeneity of aggregate stability and soil carbon in semi-arid rangeland. *Environ. Pollut.* 116: 445–455.

Bolan, N.S. 1991. A critical review on the role of mycorrhizal fungi in the uptake of phosphorus by plants. *Plant Soil* 134: 189–207.

Borowicz, V.A. 2001. Do arbuscular mycorrhizal fungi alter plant-pathogen relations? *Ecology* 82(11): 3057–3068.

Bottomley, P.J. 1994. Light microscopic methods for studying soil microorganisms. In R.W. Weaver et al. (Eds.), *Methods of Soil Analysis. Part 2. Microbiological and Biochemical Properties*. SSSA Book Series, No. 5, Soil Science Society of America, Madison, WI, chap. 6.

Caesar-TonThat, T.-C., and V.L. Cochran. 2000. Soil aggregate stabilization by a saprophytic lignin-decomposing basidiomycete fungus. I. Microbiological aspects. *Biol. Fertil. Soils* 32: 374–380.

Carlile, M.J., and S.C. Watkinson. 1996a. Saprotrophs and ecosystems. In M.J. Carlile and S.C. Watkinson (Eds.), *The Fungi*. Academic Press, New York, chap. 6.

Carlile, M.J., and S.C. Watkinson. 1996b. Parasites and mutualistic symbionts. In M.J. Carlile and S.C. Watkinson (Eds.), *The Fungi*. Academic Press, New York, chap. 7.

Chaney, K., and R.S. Swift. 1986a. Studies on aggregate stability. I. Reformation of soil aggregates. *J. Soil Sci.* 37: 329–335.

Chaney, K., and R.S. Swift. 1986b. Studies on aggregate stability. II. The effect of humic substances on the stability of re-formed soil aggregates. *J. Soil Sci.* 37: 337–343.

Chenu, C., Y. Le Bissonnais, and D. Arrouays. 2000. Organic matter influence on clay wettability and soil aggregate stability. *Soil Sci. Soc. Am. J.* 64: 1479–1486.

Clark, R.B., and S.K. Zeto. 1996. Iron acquisition by mycorrhizal maize grown on alkaline soil. *J. Plant Nutr.* 19(2): 247–264.

Degens, B.P. 1997. Macro-aggregation of soils by biological bonding and binding mechanisms and the factors affecting these: A review. *Aust. J. Soil Res.* 35: 431–459.

Diaz, G., C. Azcon-Aguilar, and M. Honrubia. 1996. Influence of arbuscular mycorrhizae on heavy metal (Zn and Pb) uptake and growth of *Lygeum spartum* and *Anthyllis cytisoides*. *Plant Soil* 180: 241–249.

Douds, D.D., and P.D. Millner. 1999. Biodiversity of arbuscular mycorrhizal fungi in agroecosystems. *Agric. Ecosyst. Environ.* 74: 77–93.

Frey, S.D., E.T. Elliott, and K. Paustian. 1999. Bacterial and fungal abundance and biomass in conventional and no-tillage agroecosystems along two climatic gradients. *Soil Biol. Biochem.* 31: 573–585.

George, E., K. Haussler, S.K. Kothari, X.-L. Ki, and H. Marschner. 1992. Contribution of mycorrhizal hyphae to nutrient and water uptake by plants. In D.J. Read et al. (Eds.), *Mycorrhizas in Ecosystems*. CAB International, Cambridge, U.K.

Giovannetti, M, and B. Mosse. 1980. An evaluation of techniques for measuring vesicular arbuscular mycorrhizal infection in roots. *New Phytol.* 84: 489–500.

Gliessman, S.R. 2000. The ecological foundations of agroecosystem sustainability. In S.R. Gliessman (Ed.), *Agroecosystem Sustainability: Developing Practical Strategies*. CRC Press, Boca Raton, FL.

Goding, J.W. 1986. *Monoclonal Antibodies: Principles and Practice*. Academic Press, San Diego, 315 pp.

Gonzales-Chavez, C., J. D'Haen, J. Vangronsveld, and J.C. Dodd. 2002. Copper sorption and accumulation by the extraradical mycelium of different *Glomus* spp. (arbuscular mycorrhizal fungi) isolated from the same polluted soil. *Plant Soil* 240: 287–297.

Harley, J.L., and E.L. Harley. 1987. A check-list of mycorrhiza in the British flora: Addenda, errata, and index. *New Phytol.* 107: 741–749, 13–89.

Hassink, J., and A.P. Whitmore. 1997. A model of the physical protection of organic matter in soils. *Soil Sci. Soc. Am. J.* 61: 131–139.

Hooker, J.E., and K.E. Black. 1995. Arbuscular mycorrhizal fungi as components of sustainable soil-plant systems. *Crit. Rev. Biotechnol.* 15 (3/4): 201–212.

Horwath, W.R., and E.A. Paul. 1994. Microbial biomass. In R.W. Weaver et al. (Eds.), *Methods of Soil Analysis. Part 2. Microbiological and Biochemical Properties*. SSSA Book Series No. 5, Soil Science Society of America, Madison, WI, chap. 36.

Klein, D.A., T. McLendon, M.W. Paschke, and E.F. Redente. 1995. Saprophytic fungal-bacterial biomass variations in successional communities of a semi-arid steppe ecosystem. *Biol. Fertil. Soils* 19: 253–256.

Miller, R.M., and J.D. Jastrow. 1990. Hierarchy of root and mycorrhizal fungal interactions with soil aggregation. *Soil Biol. Biochem.* 22(5): 579–584.

Miller, R.M., D.R. Reinhardt, and J.D. Jastrow. 1995. External hyphal production of vesicular-arbuscular mycorrhizal fungi in pasture and tallgrass prairie communities. *Oecologia* 103: 17–23.

Millner, P.D., and D.G. Kitt. 1992. The Beltsville method for soilless production of vesicular-arbuscular mycorrhizal fungi. *Mycorrhiza* 2:9–15.

Millner, P.D., and S.F. Wright. 2002. Tools for support of ecological research on arbuscular mycorrhizal fungi. *Symbiosis* 33: 101–123.

Muramoto, J., E.C. Ellis, Z. Li, R.M. Machado, and S.R. Gliessman. 2000. Field-scale nutrient cycling and sustainability: Comparing natural and agricultural ecosystems. In S.R. Gliessman (Ed.), *Agroecosystem Sustainability: Developing Practical Strategies*. CRC Press, Boca Raton, FL.

Nichols, K.A. 2003. Characterization of glomalin: A glycoprotein produced by arbuscular mycorrhizal fungi. Ph.D. thesis, University of Maryland, College Park.

Olsson, P.A., I. Thingstrup, I. Jakobsen, and E. Baath. 1999. Estimation of the biomass of arbuscular mycorrhizal fungi in linseed field. *Soil Biol. Biochem.* 31(13): 1879–1887.

Paul, E.A., and F.E. Clark. 1996. *Soil Microbiology and Biochemistry*, 2nd ed. Academic Press, New York.

Pawlowska, T.E., R.L. Chaney, M. Chin, and I. Charvat. 2000. Effects of metal phytoextraction practices on the indigenous community of arbuscular mycorrhizal fungi at a metal-contaminated landfill. *Appl. Environ. Microbiol.* 66(6): 2526–2530.

Piccolo, A., and J.S.C. Mbagwu. 1999. Role of hydrophobic components of soil organic matter in soil aggregate stability. *Soil Sci. Soc. Am. J.* 63: 1801–1810.

Pimentel, D., C. Harvey, P. Resosudarmo, K. Sinclair, D. Kurz, M. McNair, S. Crist, L. Shpritz, L. Fitton, R. Saffouri, and R. Blair. 1995. Environmental and economic costs of soil erosion and conservation benefits. *Science* 267: 1117–1123.

Rillig, M.C., G.Y. Hernandez, and P.C.D. Newton. 2000. Arbuscular mycorrhizae respond to elevated atmospheric $CO_2$ after long-term exposure: Evidence from a $CO_2$ spring in New Zealand supports the resource balance model. *Ecol. Lett.* 3: 475–478.

Rillig, M.C., P.W. Ramsey, S. Morris, and E.A. Paul. 2003. Glomalin, an arbuscular-mycorrhizal fungal soil protein, responds to land-use change. *Plant Soil* 253: 293–299.

Rillig, M.C., and P.D. Steinberg. 2002. Glomalin production by an arbuscular mycorrhizal fungus: A mechanism of habitat modification? *Soil Biol. Biochem.* 34: 1371–1374.

Rillig, M.C., S.F. Wright, M.F. Allen, and C.B. Field. 1999. Rise in carbon dioxide changes soil structure. *Nature* 400: 628.

Rillig, M.C., S.F. Wright, and V.T. Eviner. 2002. The role of arbuscular mycorrhizal fungi and glomalin in soil aggregation: Comparing effects of five plant species. *Plant Soil* 238: 325–333.

Rillig, M.C., S.F. Wright, B.A. Kimball, P.J. Pinter, G.W. Wall, M.J. Ottman, and S.W. Leavitt. 2001a. Elevated carbon dioxide and irrigation effects on water stable aggregates in a *Sorghum* field: A possible role for arbuscular mycorrhizal fungi. *Glob. Change Biol.* 7: 333–337.

Rillig, M.C., S.F. Wright, K.A. Nichols, W.F Schmidt, and M.S. Torn. 2001b. Large contribution of arbuscular mycorrhizal fungi to soil carbon pools in tropical forest soils. *Plant Soil* 233: 167–177.

Six, J., A. Carpenter, C. van Kessel, R. Merck, D. Harris, W.R. Horwath, and A. Lüscher. 2001. Impact of elevated $CO_2$ on soil organic matter dynamics as related to changes in aggregate turnover and residue quality. *Plant Soil* 234: 27–36.

Stahl, P.D, T.B. Parkin, and N.S. Eash. 1995. Sources of error in direct microscopic methods for estimation of fungal biomass in soil. *Soil Biol. Biochem.* 27(8): 1091–1097.

Steinberg, P.D., and M.C. Rillig. 2003. Differential decomposition of arbuscular mycorrhizal fungal hyphae and glomalin. *Soil Biol. Biochem.* 35: 191–194.

Stenberg, B. 1999. Monitoring soil quality of arable land: microbiological indicators. *Acta Agric. Scand., B: Soil and Plant Sci.* 49: 1–24.

Stevenson, F.J. 1994. *Humus Chemistry: Genesis, Composition, Reactions*, 2nd ed. John Wiley & Sons, New York.

Tinker, P.B., D.M. Durall, and M.D. Jones. 1994. Carbon use efficiency in mycorrhizas: Theory and sample calculations. *New Phytol.* 128: 115–122.

Trappe, J.M. 1987. Phylogenetic and ecological aspects of mycotrophy in the angiosperms from an evolutionary standpoint. In G.R. Safir (Ed.), *Ecophysiology of VA Mycorrhizal Plants*. CRC Press, Boca Raton, FL, pp. 5–25.

Tisdall, J.M., S.E. Smith, and P. Rengasamy. 1997. Aggregation of soil by fungal hyphae. *Aust. J. Soil Res.* 35: 55–60.

Vieira, R.F., C.M.M.S. Silva, A.H.N. Maia, E.F. Fay, and K.C. Coelho. 2000. An appraisal of five methods for the measurement of the fungal population in soil treated with chlorothalonil. *Pest Manage. Sci.* 56: 431–440.

Wessels, J.G.H. 1997. Hydrophobins: Proteins that change the nature of the fungal surface. *Adv. Microb. Physiol.* 38: 1–45.

Wright, S.F. 2000. A fluorescent antibody assay for hyphae and glomalin from arbuscular mycorrhizal fungi. *Plant Soil* 226: 171–177.

Wright, S.F., and R.L. Anderson. 2000. Aggregate stability and glomalin in alternative crop rotations for the central Great Plains. *Biol. Fertil. Soils* 31: 249–253.

Wright, S.F., M. Franke-Snyder, J.B. Morton, and A. Upadhyaya. 1996. Time-course study and partial characterization of a protein on arbuscular mycorrhizal hyphae during active colonization of roots. *Plant Soil* 181: 193–203.

Wright, S.F., J.L. Starr, and I.C. Paltineanu. 1999. Changes in aggregate stability and concentration of glomalin during tillage management transition. *Soil Sci. Soc. Am. J.* 63: 1825–1829.

Wright, S.F., and A. Upadhyaya. 1996. Extraction of an abundant and unusual protein from soil and comparison with hyphal protein of arbuscular mycorrhizal fungi. *Soil Sci.* 161: 575–585.

Wright, S.F., and A. Upadhyaya. 1998. A survey of soils for aggregate stability and glomalin, a glycoprotein produced by hyphae of arbuscular mycorrhizal fungi. *Plant Soil* 198: 97–107.

Wright, S.F., and A. Upadhyaya. 1999. Quantification of arbuscular mycorrhizal fungi activity by the glomalin concentration on hyphal traps. *Mycorrhiza* 8: 283–285.

Wright, S.F., A. Upadhyaya, and J.S. Buyer. 1998. Comparison of N-linked oligosaccharides of glomalin from arbuscular mycorrhizal fungi and soils by capillary electrophoresis. *Soil Biol. Biochem.* 30(13): 1853–1857.

Zak, J.C., and B. McMichael. 2001. Agroecology of arbuscular mycorrhizal activity. In M. Shiyomi and H. Koizumi (Eds.), *Structure and Function in Agroecosystem Design and Management*. CRC Press, Boca Raton, FL.

# 7 Connecting Belowground and Aboveground Food Webs: The Role of Organic Matter in Biological Buffering

*P. Larry Phelan*

## CONTENTS

## INTRODUCTION

In unmanaged temperate ecosystems, cycling of mineral nutrients and carbon, development of soil structure, and establishment of soil moisture levels are as much biological as chemical processes. These processes are regulated by the interactions of a highly diverse and complex web of soil flora and fauna that is sustained by the influx of organic matter into the soil. As generally practiced at present, agriculture simplifies the natural system by not only reducing plant and animal diversity but also taking control of many biological functions normally modulated by the soil community. Following the industrial model, modern agriculture attempts to maximize yield by identifying and removing key restraints to productivity through inputs of chemicals and fossil fuel energy. Thus, driven by industrial precepts, agricultural management

0-8493-1294-9/04/$0.00+$1.50
© 2004 by CRC Press LLC

employs a series of discrete problem-solving techniques: loosening the soil matrix for more efficient growth; adding raw materials (water and minerals) as required by the crop; making the plant machine more productive through breeding programs that select for higher yield potential; and using pesticides to eliminate insects, pathogens, and weeds that potentially compete with the crop. Initially these practices increased production levels, but over time they have also caused changes in the functioning of the agroecosystem with impacts that only gradually have become manifest. Increased soil compaction and erosion, reduced levels or shifts in the species composition of soil flora and fauna, and increased crop susceptibility to biotic and abiotic stress can be traced to industrialization of agriculture.

The harmful impacts and their root causes were not initially apparent while the industrial model and its reductionist underpinnings first led to improvements in agricultural productivity. However, we are now beginning to understand the limitations of the industrial model when applied to the natural world. As we take a more inclusive ecological approach and compare natural vs. managed soil ecosystems and chemically intensive vs. organically farmed soils, our knowledge of the connections between soil organic matter (SOM) and biological function in agroecosystems is starting to take shape. This chapter highlights the connection between the belowground and above-ground domains. In so doing, it illustrates the wonderful complexity of the soil ecosystem and emphasizes the interactions of SOM with both soil-dwelling and herbivorous arthropods.

# CONNECTING ABOVEGROUND AND BELOWGROUND ECOSYSTEMS

One of the central tenets of the ecosystem theory is that changes in one component of the system can produce unanticipated effects in other compartments. However, ecological research has gener-ally dichotomized the aboveground and belowground portions of the terrestrial ecosystem, with little consideration of their interdependence (Wardle, 1999, 2002). As discussed later, assertions have been made that some fundamental ecological principles developed from aboveground studies do not apply to interactions in the soil. Although this question remains to be resolved, recent studies have demonstrated important connections between the two worlds.

## UPWARD REGULATION: SOM AND SOIL COMMUNITY MEDIATION OF PLANT HEALTH

The plant is the primary agent connecting the belowground and aboveground universes, acting as the main conduit of nutrient, energy, and information flows (Figure 7.1). In broad terms, mineral nutrients flow up through the plant to the aboveground food web, which consists of plant shoots, invertebrate and vertebrate herbivores, and their predators. Conversely, carbon from the atmosphere is fixed by the plant and flows, via plant litter, root herbivory, and root exudates, into the soil food web, which consists of plant roots, microflora, microfauna, and soil invertebrates. Although the C and N cycles include many organisms, in terms of relative biomass, plants play the central role in connecting above- and belowground ecosystems.

One linkage garnering recent attention, particularly in agroecosystems, is that of SOM and aboveground plant susceptibility to insects and disease. Given that the soil decomposer food web is the primary source for plant N in natural systems, it is logical to expect belowground biological interactions to profoundly impact aboveground ecology. However, beyond the chemical and physical benefits of SOM, such as better moisture retention, improved soil structure, and decreased erosion (Chapter 1), only a limited set of empirical studies has attempted to define the relationship between SOM and the ability of the crop to resist losses to insects and disease.

In contrast, a long-held maxim of organic farming is that soil organic matter management can hold diseases and foliage-feeding insects in check without the intervention of pesticides. Howard (1943) states:

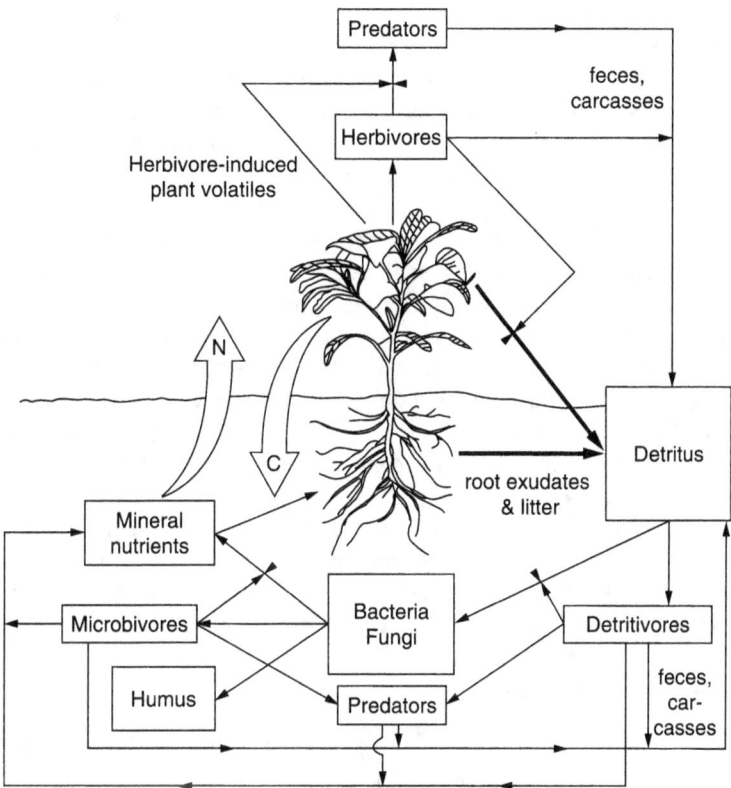

**FIGURE 7.1** With its roots belowground and its shoots aboveground, plants provide the primary conduit connecting these two ecosystems. Having evolved in an environment dominated by soil microbes and acting as the basis of the aboveground food web, plants mediate interactions between the two ecosystems. Arrows represent flows of energy or mineral nutrients and the four double arrowheads straddling a line indicate points of regulation and indirect effects.

*Insects and fungi are not the real cause of plant diseases, and only attack unsuitable varieties or crops improperly grown. Their true role in agriculture is that of censors for pointing out the crops which are imperfectly nourished. Disease resistance seems to be the natural reward of healthy and well-nourished protoplasm. The first step is to make the soil live by seeing that the supply of humus is maintained.*

Mainstream agricultural research took a different course over the past half century, and such thinking was relegated to the backwaters. However, recently a number of studies have provided evidence for lower pest pressure in agricultural systems that maintain high levels of SOM. Kowalski and Visser (1979) found significantly lower aphid populations in organically grown winter wheat compared with that on a neighboring conventional farm. Andow and Hidaka (1989) compared the effect of production-management systems on pest outbreaks for adjacent conventional and shizen (natural) rice farms in Japan. Populations of a chrysomelid beetle as well as delphacid pests were higher on the conventional farm and required intervention with pesticides, whereas intervention was not needed on the shizen farm. In addition, shizen plants suffered lower losses by pathogens despite the application of fungicides to conventional fields. Subsequent laboratory choice tests demonstrated that two delphacid pests, *Nilaparvata lugens* and *Nephotettix cincticeps*, strongly preferred conventionally grown rice plants to shizen-grown rice plants.

As part of one of the most extensive multidisciplinary comparisons of organic and conventional management systems, Drinkwater et al. (1995) sampled insect pests in commercial California tomato fields. They found that organic and conventional production systems could not be distinguished based on either crop yields or losses by foliage- or fruit-feeding insects. However, the conventional fields required a greater reliance on insecticides to suppress pest damage levels. They concluded that in organic fields, "biological processes compensated for reductions in synthetic fertilizers and pesticides." In matched corn fields at three locations in Guatemala, Morales et al. (2001) recorded lower aphid populations in fields receiving organic fertilizers for at least 2 years compared with those treated with synthetic NPK without significant differences in corn yields. Aphids were not significantly reduced in fields receiving organic fertilizer for only 1 year, suggesting that changes were not immediately associated with the use of organic-based fertilizers. Bedet (2000) compared corn pests for paired organic and conventional farms in three Ohio counties across two growing seasons. Substantial differences were recorded in plants damaged by the European corn borer (*Ostrinia nubilalis*), with conventional farms averaging 63% damage in first-year corn across the two years compared with organic farms, which suffered only 21% damage. Conventional farms exceeded the economic threshold for this pest four of six times, whereas organic farms never reached the threshold. Although billbugs were statistically more abundant on organic farms than on conventional farms — four of six times — and armyworms were higher on organic farms — two of six times — these levels never approached economic thresholds. It was concluded that *O. nubilalis* damage corresponded to the higher spring plant N levels of conventional farms, whereas billbugs and armyworms were associated with grassy weeds, which were in far greater abundance on the organic farms. Once again, no consistent differences in crop yield between farming systems were measured.

## POSSIBLE MECHANISMS OF MEDIATION

Although there are numerous distinctions between organic and conventional farms that might contribute to the observed differences in pest populations, there is good reason to deduce that belowground processes play a role, and that the plants themselves differ biochemically. First, in most of the field studies cited previously, pest differences appeared not to be due to differential predator densities, because they were either similar on the two farms or higher in conventional fields. Second, most studies either directly measured biochemical differences (e.g., amino acid levels) in the plants or demonstrated herbivore behavioral preferences that indirectly indicated such differences. Unfortunately, beyond these few examples, there has been little effort to elucidate plant mechanisms underlying such field observations. To test whether the lower *O. nubilalis* damage reported for organic corn was due to a behavioral preference, Phelan et al. (1995, 1996) conducted greenhouse studies in which the short-term effect of the form of fertilizer was statistically separated from the long-term effect of soil management history by using a factorial design. Soils collected from a series of neighboring organic and conventional farms were placed in pots, each of which was amended with mineral fertilizers, animal manures, or left unamended and planted to corn. In each of four experiments, females of *O. nubilalis* released into the greenhouse showed egg-laying preferences among the plants, such that relatively few eggs were laid on corn plants in soil from organic farms, irrespective of the fertilizer added. In contrast, plants in conventional soil varied significantly among fertilizer treatments in the number of eggs received, the variance in egg laying across all experiments being ca. 18 times higher than that among plants in organically managed soil. In addition, although the fertilizer treatment affected plant growth, no growth differences were attributable to soil management history. These results led to the concept of biological buffering (Phelan et al., 1995) for organically managed soil, which asserts that the sustained influx of SOM provides the resource base for the soil community, whose interactions act to lessen changes in the soil environment. As such, the concept extends the well-established potential for organic matter to buffer physicochemical soil parameters (Chapter 1) to higher-order buffering, i.e., belowground and aboveground biological interactions. Biological buffering then predicts that SOM can impart greater stability in population levels and resistance to disturbances by dampening fluctuations in the flow of nutrients, moisture, and energy.

A subsequent test of the biological buffering concept (Phelan et al., 1996), repeating the experimental design given previously, determined that corn grown in organic soil showed significantly less variation among fertility treatments in *O. nubilalis* egg laying, rates of photosynthesis, leaf-mineral profiles, protein content, and final shoot weight compared with plants grown in soil from a conventional farm. Compared with plants in organic soil, plants grown in conventional soil that either was unfertilized or amended with composted cow manure elicited greater egg laying, showed lower rates of photosynthesis and growth, and possessed lower levels of true leaf protein.

## MINERAL BALANCE HYPOTHESIS

I have previously asserted that the primary mechanism for biological buffering of plant health is the modulation of mineral availability in the soil solution by the SOM-decomposer food web (Phelan, 1997). An enormous amount of research has been conducted on the effects of plant mineral nutrition on insect herbivory, focusing almost exclusively on the role of N (Mattson, 1980; Scriber 1984a, 1984b; Waring and Cobb, 1992). A generally positive correlation between plant N and insect performance has been attributed to the disparity between the levels of N in plants (2–4%) and that of insect tissues (7–14%; Scriber, 1984a; White, 1984). Because high N fertility is considered necessary to maximize agricultural yields, increased crop susceptibility to insects is often viewed as a necessary trade-off. The current ecological theory of plant defense also seems to reinforce this idea. Plants have evolved a diverse collection of compounds that inhibit feeding, larval development, and survival of herbivores. However, because production of these compounds comes at the expense of growth, their concentrations represent a resource allocation problem for the plant. A number of conceptual models have been proposed to explain the genetic (population-level) and phenotypic variability in plant allocations to primary chemistry (growth) and secondary chemistry (defense; reviewed by Herms and Mattson, 1992). The common element in these models is the prediction that high resource availability (minerals and water) will result in preferential allocation priority to growth over defense. Therefore, with the conditions found in the agricultural ecosystem, plants generally would be expected to exhibit high growth rates but low resistance to insect attack.

Although more research is needed to determine the generality of the results, the field and controlled studies cited previously suggest that plants grown in soils with improved organic matter management do not fit the current ecological paradigm for growth and defense against insects. Therefore, other factors might be operating that are not being considered. Although optimal defense theories have dominated conceptual development in the area of insect–plant interactions and particularly herbivore resistance, they are better thought of as models of plant-resource allocation, not models of herbivore performance, as they address only one aspect of host quality. Host-plant quality includes the relative levels of behavioral stimulants and repellents, essential nutrients, and nonnutritional or antinutritional (defensive) chemical factors, as well as plant structural traits. By focusing on only one component, the models fail to address how phenotypic variation in primary chemistry impacts herbivory.

The mineral balance hypothesis was proposed to explain the lower pest pressures observed in organic systems (Phelan et al., 1996; Phelan, 1997) and shares key components with the plant stress hypothesis (White, 1984). The hypothesis asserts that plants require an optimal balance of mineral nutrients, which is defined not only by absolute levels of each essential mineral but also by proportions among the minerals. This optimal balance is genetically determined, but differs among species and varies because of abiotic factors, e.g., moisture availability, temperature, and light. When provided its optimal mineral balance in the soil solution, the plant achieves maximum health, characterized by both strong growth and resistance to insects and disease. Graphically, the optimal balance for a plant occupies a multidimensional space, with each axis symbolizing levels of a nutrient (Figure 7.2). If we could envision this nutrient hyperspace, we would find it elongated along certain axes, representing those nutrients for which the plant can tolerate a wide range of concentrations, and narrow along other dimensions, indicating those nutrients to which the plant

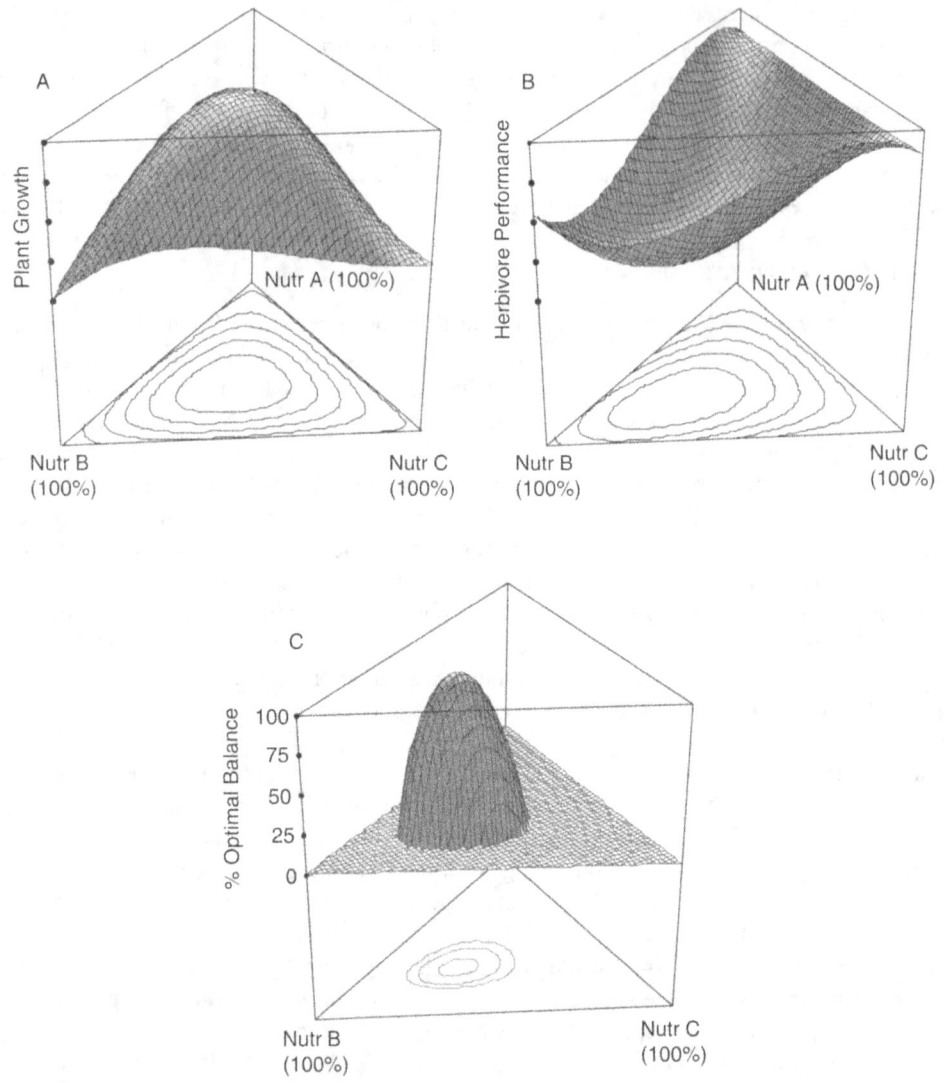

**FIGURE 7.2** Graphical representation of the mineral balance hypothesis in response to mixtures of three plant minerals. Optimal mineral balance (c) is the range of mineral proportions characterized by the overlap between strong plant growth (a) and relatively poor performance by herbivorous insects (b). The proportion of each mineral in the mixture ranges from 0% along the side of the triangle to 100% at the opposing vertex. Note that certain minerals are expected to have greater impact than others, and the effect of changing levels of one nutrient depends on the levels of others. As illustrated in (b), replacing Nutrient B primarily with Nutrient A in the mixture results in a substantial drop in herbivore performance, whereas its replacement primarily with Nutrient C causes relatively little change.

is sensitive. Occupying a habitat that provides this optimal blend of nutrients is probably the exception for plants, and therefore they have evolved a number of physiological mechanisms or biological associations to better meet their nutrient requirements. Examples of such mechanisms include active membrane transport for a number of minerals, release of phytosiderophores to increase availability of Fe, symbiotic associations with mycorrhizal fungi to enhance uptake of P and other nutrients, and active exclusion by roots or vacuolar sequestration to reduce plant concentration of some minerals to below harmful levels (reviewed by Marschner, 1995). Although these mechanisms help the plant ameliorate nonoptimal nutrient conditions, costs are generally associated with these mechanisms, investments that might require trade-offs with other aspects of plant growth or differentiation.

Given the manifold roles minerals play in every metabolic pathway of plants, all aspects of host-plant quality are impacted to varying degrees. Because the mineral balance hypothesis considers effects on both primary and secondary chemistry, it is more inclusive than current resource allocation models and can be differentiated experimentally by its predictions. Although plant investment in defensive chemistry is constrained by a physiological trade-off with plant growth, under favorable mineral conditions, growth and resistance can actually be positively correlated. When mineral proportions and levels are optimized, both primary and secondary metabolic pathways are able to operate most efficiently. In contrast, when imbalances exist, blockages occur in metabolic paths, potentially resulting in slower growth, levels of secondary metabolites below the plant's genetic optimal, and elevated levels of soluble precursors such as sugars and amino acids. Figure 7.3 shows the relationship among plant resistance, secondary chemistry, and mineral balance. Diverting photosynthate into secondary metabolism might increase resistance to herbivory, but only at the cost of plant growth or reproduction. When all plant resource requirements are met, metabolic processes are optimized, allowing the plant to realize growth and reproductive potentials. However, a blockage in metabolic pathways because of mineral deficiencies or nonoptimal mineral ratios can both reduce growth and increase susceptibility to herbivores. Figure 7.3 specifically illustrates what might occur if the efficient functioning of Pathway C requires higher levels of a Mineral Y than does Pathway A, which is more dependent on Mineral X. If the ratio of Mineral X to Mineral Y is relatively high, Pathway A produces a large amount of intermediate metabolites, but Pathway C is not able to keep up and assimilate all the intermediates into larger and more complex compounds.

Therefore, growth slows and host quality increases for insect herbivores. For example, N is utilized in early pathways, primarily in the synthesis of amino acids and other low-molecular-weight N-containing compounds, whereas K is required for every major step of protein synthesis. Plants with a higher than optimal N:K ratio are impaired in their ability to produce protein, resulting in the accumulation of amino acids, amides, and nitrates (Mengel and Kirkby, 1987). Similarly, K also activates the enzyme responsible for synthesis of starch (starch synthetase). Thus, with inad-

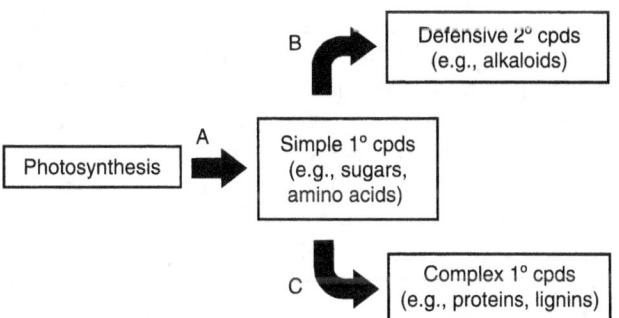

**FIGURE 7.3** Simplified plant metabolic paths showing the trade-offs between primary (structural and regulatory) and secondary (defensive) chemistry and the role of mineral balance in resistance to insect herbivory. Large pools of available simple primary compounds enhance feeding and development by many herbivores. Herbivory can be reduced by investment in secondary chemistry, however, generally at the expense of primary chemistry and therefore growth and reproduction. Partitioning of resources between these two main paths is determined by genetic and environmental factors. In contrast, an optimal balance of nutrients essential for Pathway C relative to those for A facilitates plant growth, but also reduces herbivory by reducing the pool of low-molecular-weight compounds. Additionally, sufficient levels of nutrients required for B are needed for the plant to realize its genetically optimal level of chemical defense.

equate K, the rates of starch and structural carbohydrate synthesis decline, whereas soluble carbohydrates accumulate. Free amino acid levels also increase in response to .other mineral nutrient imbalances (Court et al., 1972), including sulfur deficiency (Amancio et al., 1997), high N:P ratios (McLaughlin et al., 1994), and Fe deficiency (Alam et al., 2001)

In addition to reducing plant growth, the accumulation of simple biochemical components can promote insect herbivory in a number of ways (Cockfield, 1988): (1) amino acids and sugars are metabolically more accessible to insect herbivores than are proteins, nucleic acids, and structural carbohydrates; (2) these compounds might be quantitatively more available because of their greater solubility and mobility in plant tissues than those of larger molecules; (3) these compounds reduce the effectiveness of some plant defensive compounds, such as proteinase inhibitors; and (4) many amino acids and sugars act as feeding and oviposition stimulants for insects. Also, insects and mites are known to contain unusually high levels of free amino acids, more than 30 times higher than in other animal groups (Chen, 1966). For these reasons, developmental performance of most arthropods is enhanced by conditions that increase soluble N in plants. Thus, population growth rates for the twospotted spider mite (*Tetranychus urticae*) correlate better with the N:K ratio of plant tissue than with absolute plant levels of either nutrient (Suski and Badowski, 1975). Herbivorous insects commonly show a greater preference for and enhanced developmental performance on young plant parts, an association probably related to the higher pools of free amino acids concentrated in the growing tissue (White, 1984).

Similarly, the energy requirements of herbivorous insects are met predominantly through consumption of simple plant sugars (Waldbauer, 1968), whereas insect utilization of starch is less widespread and digestion of complex structural chemicals such as cellulose, hemicellulose, and lignin is rare (Herms and Mattson, 1992). Feeding preferences and development performance of gypsy moth (*Lymantria dispar*) on American basswood is positively correlated with levels of soluble carbohydrates and is not related to secondary metabolites (Rieske and Raffa, 1998). Oviposition by *O. nubilalis* is highly correlated with leaf fructose, and to a lesser extent with proline, glucose, and sucrose (Derridj et al., 1989), and feeding by the Mexican bean beetle (*Epilachna varivestis*) also is stimulated by simple sugars (Augustine et al., 1964). Thus, any plant mineral imbalances that increase levels of these simple components should be expected to increase oviposition, larval feeding, and developmental performance.

Minerals also play important roles in secondary metabolic pathways, providing another way in which nutrient deficiencies or imbalances (e.g., suboptimal ratio of Mineral B to Mineral A in Figure 7.3) could increase plant susceptibility to herbivory or disease. Boron deficiency inhibits lignin synthesis and increases levels of free phenolics (Marschner, 1995). As a direct test of the mineral balance hypothesis, Beanland et al. (2003) used a mixture-model experimental design to compare soybean growth and insect herbivore response to different proportions of B, Fe, and Zn in hydroponic solution. Plants grew ca. four times larger at intermediate proportions (B:Fe:Zn 2:2:1) than plants receiving no B (B:Fe:Zn 0:6:1), whereas Mexican bean beetles (*E. varivestis*) developing on plants with balanced nutrition were 45% smaller than those fed B-deficient plants, and soybean loopers (*Pseudoplusia includens*) were 20% smaller (Figure 7.4). Although lignin and phenol concentrations were not measured, levels of amino acids were five to ten times higher in the unbalanced plants (Phelan et al., unpublished data), suggesting one mechanism underlying greater insect performance.

I suggest that organically managed soils are better able to approximate the optimal plant nutrient balance because they are closer to the soil environment in which plants evolved and because of their capacity to buffer mineral availability. The most striking difference in nutrient dynamics between organic and conventional soils relates to N availability. Given that microbes generally outcompete plants for nutrients in the presence of sufficient carbon sources (Jackson et al., 1989), N becomes limiting as it is immobilized by the microbial community. Some carbon is lost through respiration as organic matter is decomposed, and N is concentrated in the microbial soil fraction. However, a portion of this N is continuously mineralized by death of the microbes, grazing by soil

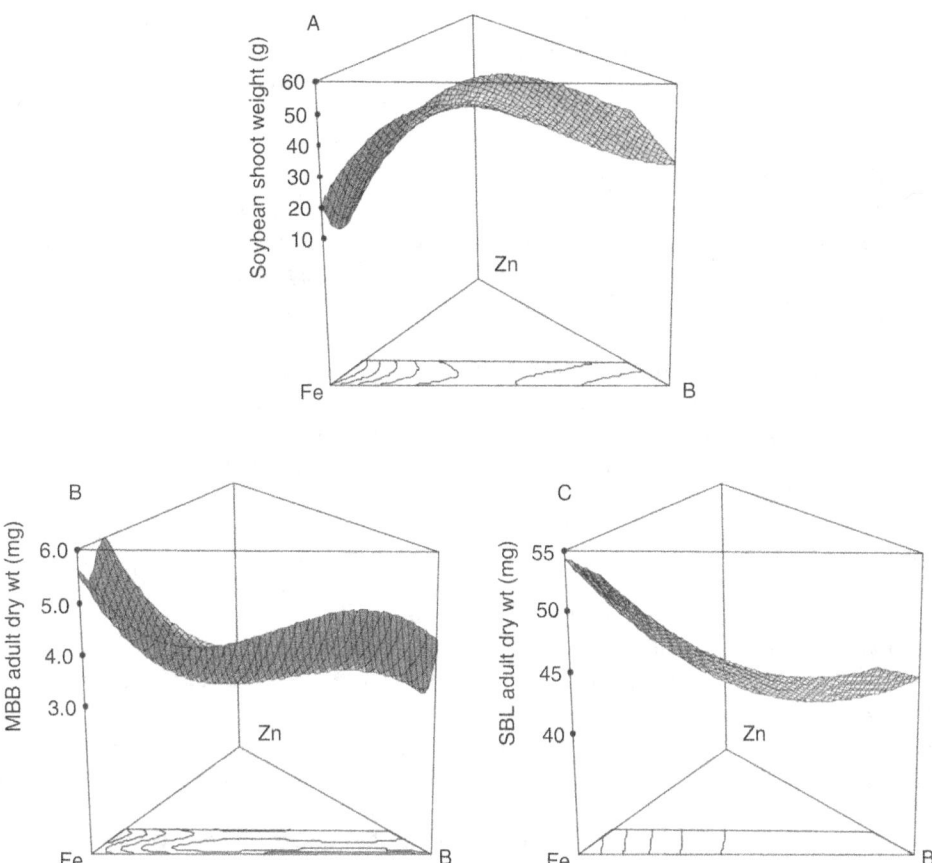

**FIGURE 7.4** Results of test of mineral balance hypothesis for the plant micronutrients Fe, Zn, and B. By using a mixture model experimental design, soybean plants were grown in hydroponic nutrient solutions in which the concentrations of Fe, Zn, and B were varied 0–0.05 mM, 0–0.05 mM, and 0–0.01 mM, respectively, and fed to two soybean herbivores. Graphic representation of resulting Scheffé polynomial models for (a) soybean shoot growth, (b) Mexican bean beetle (*Epilachna varivestis*) adult weight, and (c) soybean looper (*Pseudoplusia includens*) adult weight. (From Beanland, L. et al. 2003. *Environ. Entomol.* 32: 641–651. With permission.)

fauna on microbial communities, and predation on micro- and mesofauna. This cyclic flux between immobile and mineralized forms provides a continuous low-level supply of N to plants with minimal losses to leaching. Thus, the complex interactions of the soil food web buffer extremes in mineralized N levels and probably other nutrients.

By contrast, in agricultural soils where N inputs are primarily applied as large pulses of inorganic compounds and organic matter inputs are low, carbon is limiting for microbes and N remains mobile in the soil. This high availability stimulates plant growth along with rapid uptake in excess of metabolic needs, likely putting the plant in a state of nutrient imbalance. Studies of nutrition have revealed variation in plants' capacity for regulating tissue-mineral levels (Marschner, 1995). Some plant-essential nutrients are tightly regulated whereas others are taken up via passive transfer, so that tissue levels generally reflect those available in the soil. This differential response to minerals reflects the plant's evolutionary history and the range of nutrient environments relative to its requirements that it encountered during this time. Because N is generally considered the most limiting mineral for many organisms, most plants probably have not encountered N-excessive environments in their evolutionary history. It is possible that only with the advent of mineral fertilizers in the agricultural ecosystem have plants encountered chronically or spatially pervasive N excess, and have therefore not evolved mechanisms for moderating its uptake. Uptake of N beyond plant metabolic requirements has come to be termed *luxury consumption*. The positive connotation of this term obscures possible costs of the resulting nutrient imbalance in terms of susceptibility to insects and pathogens. Also, higher levels of N can elevate or inhibit uptake of other nutrients, potentially leading to other imbalances (Marschner, 1995).

That plants grown with inorganic fertilizers show more seasonal variation in N content than those getting their nutrients from organic sources was affirmed by Bedet (2000), who measured a four- to sixfold greater variation in corn-leaf amino-N levels across the growing season in conventional fields than in organic fields. The ability of organically managed soils to buffer nutrient uptake was also demonstrated by a principal-component mapping of eight mineral profiles of corn plants grown in soil from organic vs. conventional farms (Figure 7.5; Phelan, 1997). When either compost or $NH_4NO_3$ was added to the organic soil, the plant mineral profile showed little change compared to unfertilized plants, but plants in conventional soil showed dramatic shifts, particularly when amended with $NH_4NO_3$.

## PLANT SIGNALING AND INDUCED RESISTANCE

In addition to constitutive chemical defense, plants can respond to attack by herbivores and pathogens with a rapid accumulation of secondary metabolites. Inducible plant resistance represents a just-in-time strategy that might provide an effective defense at a lower cost than continuous maintenance of defensive chemical levels. Recent research in plant resistance has focused on the two biochemical pathways that signal attack to other parts of the plant: jasmonic acid and salicylic acid. In general terms, jasmonic acid induces defense in response to insect feeding, elevating levels of proteinase inhibitors and various oxidative enzymes (Thaler et al., 2002), whereas salicylate promotes protection against plant pathogens (Chapter 5). Numerous studies are unveiling great complexity in these pathways, with evidence that each triggers an array of biochemical responses, including downreg-

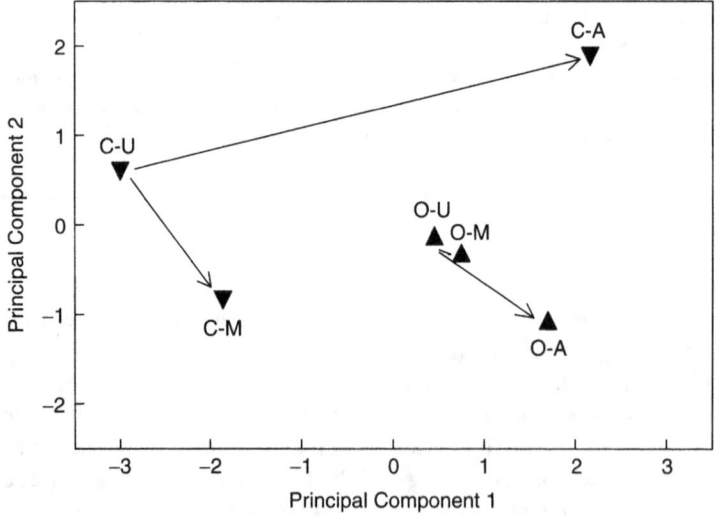

**FIGURE 7.5** Principal components analysis of corn grown in soil collected either from organically (O) or conventionally (C) managed farms and receiving $NH_4NO_3$ (A), composted manure (M), or no fertilizer (U), based on eight leaf nutrients (N, Mg, Ca, Fe, Al, Mn, Cu, and Zn). For clarity, only the centroid values for each treatment are plotted with arrows indicating the degree of change in leaf-mineral profiles resulting from each fertilizer relative to the unfertilized plants in each soil (*n* = 5). (From Phelan, P.L. 1997. *Biol. Agric. Hortic.* 15: 25–34. With permission.)

ulation of basic processes such as photosynthesis and indications of extensive cross-talk between the signaling pathways (Thaler et al., 2002). This cross-talk is usually negative; for example, pathogen induction of the salicylate pathway can result in a plant with greater resistance to other pathogens but also increased susceptibility to insect attack. In other cases, there can be cross-potentiation between the pathways. The complexity of these interactions has led to the realization that rather than viewing these two systems as dichotomous channels, they should be considered components of a defense-signal network (Felton and Korth, 2000). Although most of the research in plant signaling has been conducted at the molecular level, it is not hard to imagine the potential significance of this network for mediating belowground–aboveground interactions. For example, if a root pathogen triggers the salicylate pathway, which induces biochemical changes aboveground, how does this change the plant shoot susceptibility to insect herbivores? The same question can be asked for plant-beneficial soil microflora, such as mycorrhizae or rhizosphere microbes. Shaul et al. (1999) found that leaves of mycorrhizal-colonized tobacco plants showed a higher incidence and severity of necrotic lesions caused by *Botrytis cinerea* or tobacco mosaic virus than those of nonmycorrhizal plants. Both negative and positive effects of vesicular-arbuscular mycorrhizae (VAM) colonization on herbivory have been demonstrated, including reduced performance of leaf-chewing caterpillars (Rabin and Pacovsky, 1985; Gange and West, 1994) and root-feeding weevil larvae (Gange et al., 1994), no effect on aphids (Pacovsky et al., 1985), and enhanced development of Mexican bean beetles (Borowicz, 1997). To the extent that activity of VAM, plant-beneficial, or plant-pathogenic fungi and bacteria are impacted by SOM, cross-talk in plant-signaling systems represents another mechanism by which SOM might modulate aboveground plant–insect interactions.

## IMPACTS ON THE THIRD TROPHIC LEVEL

The effects of SOM on plant physiology persist up the trophic chain to the predators and parasites of herbivorous insects through two possible mechanisms: host-plant quality for herbivores and plant volatiles attractive to predators and parasites. Low host quality through biological buffering can slow the developmental rate of herbivorous insects, extending their period of susceptibility to predation (Benrey and Denno, 1997). Low host quality can also increase the amount of feeding required by herbivorous insects to complete development and thus increase their detection by natural enemies (Stiling et al., 1999). In this scenario, lower host-plant quality has multiple negatives for herbivore fitness: slower growth, lower resulting adult weight, reduced survival due to suboptimal nutrition or higher plant defenses, and higher mortality through increased predation. On the other hand, higher host-plant quality for herbivores can produce prey with greater nutritional quality for predators. For example, the parasitoid wasp, *Diadegma insulare*, showed higher rates of parasitism on diamondback caterpillars, *Plutella xylostella*, reared on N-fertilized host plants than on those feeding on unfertilized plants (Fox et al., 1990). Similarly, parasitism of the silverleaf whitefly, *Bemesia argentifolii*, by *Encarsia formosa* increases significantly with fertilization of their poinsettia host (Bentz et al., 1996).

Because the cues produced by herbivorous insects are difficult for predators to detect from long range, they have come to depend more on cues from the prey's habitat. It has recently been demonstrated that many plant species respond to insect attack with an increased production of specific volatiles (reviewed by Cortesero et al., 2000). This herbivore-induced synamone benefits predators and parasitoids by increasing their finding rate of prey items, and plants obviously benefit from the reduction in herbivore populations attacking them. Because mineral nutrition impacts so many aspects of plant chemistry, including plant defenses, it is a reasonable expectation that SOM-mediated changes in plant nutrient availability will impact this external communication system as well. Although studies of SOM and herbivore-specific plant volatiles have yet to be conducted, Gouinguene and Turlings (2002) demonstrated effects of a number of abiotic factors on the levels of these volatile emissions, including increased emissions with N fertility and reduced emissions with moisture level.

## Downward Regulation: Aboveground Herbivory Modulates Soil Communities

Although the aboveground food web has its roots, both literally and figuratively, in the belowground food web, regulatory connections between the two flow in both directions. Bardgett et al. (1998) identified two conduits for the effects that aboveground herbivory can have on soil communities. Immediate effects can be realized by changes in the amount and type of root exudates and shifts in the allocation of carbon and nutrients to roots vs. shoots due to the removal of foliage, whereas more long-term changes are caused by qualitative changes in the resulting leaf litter. It is not widely recognized that root exudations and tissue slough can be significant sources of labile soil carbon for microbial communities, with some plant species releasing up to 20% of photosynthate in this manner (Huetsch et al., 2002). Bardgett et al. (1996) demonstrated that changes in root exudations caused by vertebrate grazing on shoots can translate to alterations of the soil microbial community in grasslands, with heavy grazing (defoliation) resulting in communities dominated by bacteria compared to lightly grazed pastures, where decomposition processes are dominated by fungi. In this same system, Bardgett et al. (1993) also demonstrated that herbivore effects continue at the next soil trophic level, with populations of microbivorous Collembola enhanced in moderately grazed grasslands.

Long-term changes in leaf litter can be caused by selective herbivory among plant species, altering the make-up of leaf litter, which can lead to shifts in the soil community composition. Additionally, litter quality changes can result from plant response to herbivory, which can increase allocations to secondary (defensive) compounds or change levels or availability of leaf N. Plants commonly respond to attack by arthropods and pathogens by increasing production of defensive compounds. These chemicals can have toxic/antibiotic effects or they might act by increasing leaf toughness and inhibition of digestive enzymes. The increased tannin production of poplar seedlings responding to spider mite feeding led to a 50% reduction in the rate of decomposition in the resulting leaf litter (Findlay et al., 1996).

Episodic outbreaks of insect herbivores, such as gypsy moth, can lead to severe or complete defoliation of forests. This results in a dramatic change in nutrient flow, with a large portion of N diverted from the internal N cycle of the tree to the soil in the form of insect frass and green leaf fragments with a lower C:N ratio than that of normal leaf fall. This influx results in an increase in both soil microbial and microarthropod activity (Hunter, 2001). In stable isotope studies of leaf litter and insect frass, Lovett et al. (2002) reported <0.01% loss of inorganic N by leaching, although other field studies have shown significant increases in leached N after defoliation (reviewed by Hunter, 2001). Nonetheless, if sufficient labile soil carbon is available, the increased activity of the decomposer food web provides a feedback loop that retains a large portion of the N influx for subsequent plant uptake.

On the other hand, severe aboveground herbivory can decrease the allocation of carbon to the roots, causing the soil microbial community to be carbon limited, which can reduce immobilization of N and increase N availability to plants (Holland and Detling, 1990). As Holland and Detling (1990) assert:

> *Although herbivores represent a small fraction of total biomass in grasslands, they have a disproportionate influence on regulation of nutrient and energy flow. Indirect effects of herbivory on nutrient flow in grasslands include nutrient redistribution within landscapes, changes in nutrient recycling rates, increased nutrient uptake by grazed plants, and increased loss of nitrogen via ammonia volatilization.*

Although they were speaking of vertebrates, the same holds for invertebrate herbivores.

# BELOWGROUND INTERACTIONS

At the core of the biological buffering concept is the assertion that SOM generates a higher functioning soil community, characterized by more interactions and feedbacks. Other chapters in the book address the interactions between organic matter and soil microbes and earthworms; here I focus on the roles of soil microarthropods. Many excellent reviews are available on soil arthropods and their ecological roles (Seastedt, 1984; Moore et al., 1988; Coleman and Crossley, 1996). Rather than provide an exhaustive review of this literature, the objective here is to provide examples to illustrate how microarthropods can act as intermediaries in the linkage between organic matter and plant health. Arthropods comprise only a small fraction of the total soil biomass; however, they are unparalleled by any other soil fauna in their diversity of size, morphology, and ecological function. Arthropods associated with the soil habitat occupy the categories of macrofauna and mesofauna. The former group (animals >2 mm) includes macroarthropods such as adult and larval beetles, ants, fly larvae, termites, spiders, millipedes, centipedes, and isopods (Brady and Weil, 1999). Mesofauna (0.2 to 2 mm) include the microarthropods, primarily Collembola (springtails) and mites of various orders, and, to a lesser extent, Protura and Pauropoda. Functionally, soil macroarthropods occupy most feeding guilds: predators, phytophages, and saprophages. Among the microarthropods, Collembola and cryptostigmatid and astigmatid mites generally feed on detritus and microflora, whereas mesostigmatid and prostigmatid mites are primarily predatory, feeding on nematodes as well as other microarthropods (Ingham et al., 1986; Moore et al., 1988).

## SOIL QUALITY AND ARTHROPODS

As natural grasslands and forestlands are converted to agriculture, the roles of nutrient and energy cycling played by the complex web of interactions between soil flora and fauna are largely replaced by the use of highly mobile inorganic fertilizers, accompanied by a reduction in OM influx. This change in the soil environment generally leads to a reduction in the activity level or a change in the species composition of the soil food webs, or both. However, the loss of soil fauna can be ameliorated by reintroducing organic matter inputs. Scholte and Lootsma (1998) measured significantly larger spring populations of mycophagous Collembola in fields amended with farmyard manure the previous autumn. A positive relationship between organic matter and the distribution of soil microarthropods has also been registered in woodland habitats (Poole, 1964; Mitchell, 1978).

One of the major precepts of organic farming is that this system promotes biological activity and soil health by maintaining high levels of fresh organic matter. Many comparative studies of soil biota have supported this claim, finding a greater number or species diversity of soil arthropods in organic than in conventional farms. El Titi and Ipach (1989) compared replicated plots under different long-term crop management and found soil fauna to be higher overall in integrated plots, whose fertility input was almost solely manure, compared with conventional plots, which received only mineral fertilizers. Of soil invertebrates in the integrated plots, Collembola levels were higher, and predatory mite populations were both greater and more species diverse. Saprophytic and predatory nematodes were more numerous, whereas plant-parasitic nematodes were less numerous. In Denmark, Krogh (1994) determined levels of soil microarthropods to be twice as high in organically managed fields as in conventionally managed ones, and these levels were also higher than those under integrated production. Paoletti et al. (1992) and Moreby et al. (1994) found consistently higher levels of Collembola in organic farms. Similarly, Yeates et al. (1997) found significantly higher populations of mites in grasslands under organic management. Doles et al. (2001) found that organic management in apple orchards increased soil microarthropod densities and diversities, although only on certain sampling dates, possibly reflecting the dynamics of organic matter accumulation and loss. Surface-dwelling predators, such as carabid beetles, might also be more in organic systems (Kromp, 1989).

SOM impacts soil arthropods in two ways: through its effect on soil structure and the physical environment and as the resource base that drives the soil food web. By increasing moisture retention, organic matter in or on the soil surface improves the soil habitat for many microarthropods, such as springtails (Steinberger et al., 1984). Even so, the response of mites to soil moisture is group dependent, as cryptostigmatid and mesostigmatid mites have high moisture requirements (Cepeda and Whitford, 1989), whereas members of the Prostigmata can tolerate lower moisture levels (Cepeda-Pizzaro and Whitford, 1989). In addition to improving the soil environment for arthropods, organic matter provides a direct food source for detritivorous microarthropods. It also has indirect effects on microbivorous and predatory species by increasing fungal activity and populations of other microarthropods and nematodes.

Decomposer arthropod population dynamics are impacted by not only the quantity but also the quality of organic matter. Studying the effects of cover crops on soil arthropods, House and Del Rosario Alzugaray (1989) determined that although decomposers increased initially with destruction of all cover crops, later in the season population levels remained similar or declined with legume covers (low C:N ratio), but increased further with wheat (higher C:N ratio). In addition, there is evidence of niche separation among some microarthropods. Scholte and Lootsma (1998) measured increases in saprophagous Collembola populations, along with significantly higher populations of saprophagous nematodes, in fields amended with a mixture of manure and mustard compared with those receiving either amendment alone. Other studies have found similar stimulatory effects of green or animal manures on Collembola and non-plant-parasitic nematodes (Van de Bund, 1970; Curry and Purvis, 1982). Hansen and Coleman (1998) found different species composition of oribatids in microplots with pure birch, maple, or oak litters, whereas mixed-species leaf litters sustained a more diverse oribatid community and higher decomposition rates. Moreover, the community structure in mixed litters paralleled the species found in single-litter leaf species making up the mixture.

Comparative farming studies have highlighted a number of edaphic factors in addition to organic matter to be important to invertebrate communities. For example, the higher levels of weeds often associated with organic fields provide a microclimate beneficial to both euedaphic (soil-dwelling) and epigeic (surface-litter-dwelling) arthropods (Pfiffner and Niggli, 1996). On the other hand, tillage generally has a disruptive effect (Wardle, 1995), reducing both numbers and genetic diversity of microarthropods (Rockett, 1986; House and Del Rosario Alzugaray, 1989). The negative impact of tillage on arthropod populations might be due to inversion of the soil layers and physical injury during tillage, the disruption of soil structure and thus habitat for soil fauna, and changes in soil temperature and moisture that result. Tillage generally depresses microarthropods to a greater degree than lower trophic groups, indicating either a greater susceptibility of this group or an amplification of negative effects as one moves up the food chain (Wardle, 1995). Although the general effect of tillage is negative on microarthropods, different taxa vary in their sensitivity. Hülsmann and Wolters (1998) measured a 50% reduction in total soil mites because of spring cultivation, but the intensity of effects differed not only in the type of tillage employed but also among mite species. Most sensitive were the fungivorous Oribatida, which probably reflects the negative impact of tillage on soil fungi. Soil compaction has a similar long-term detrimental effect on microarthropod populations because of a reduction in soil porosity, whether caused by farm machinery (Heisler 1994, 1995a, 1995b) or high density of livestock (King et al., 1976). This suppression is less pronounced in shallow-tilled than deep-tilled soils (Schrader and Lingnau, 1997), and the effect is reduced in agricultural soils under conservation tillage (Kracht and Schrader, 1997), which conserves organic matter and generally reduces the bulk density of soil. However, Schjønning et al. (2002) demonstrated that even in fields under organic management, heavy machinery traffic could cause compaction. Thus, although organic matter is the foundation of the soil food web, it alone does not guarantee a fully functioning system. The requirement of good aeration for a biologically active soil has been recognized by organic farming and underlies such practices as shallow tillage. As one farmer expressed it, "you can go without food for a couple of weeks, but without air, you will only last a few minutes."

Not all studies indicate a clear-cut relationship between farm management system and soil arthropod biodiversity. Hadjicharalampous et al. (2002) found no management differences in soil arthropods, whereas others have shown significant differences that are site-specific or inconsistent in direction (Alvarez et al., 2001). Given that farming systems do not represent monolithic management practices, such discrepancies undoubtedly reflect the collective outcome of the different parameters with positive and negative effects on soil microarthropods discussed previously. Although it appears that increasing organic matter influx generally has a positive effect on microarthropods, the effect can be diminished by other soil environmental factors. These findings also point to the limitations of organic vs. conventional comparisons. A more informative experimental design includes a range of farm-management practices, employing multivariate statistics to determine the relative importance of individual parameters to soil health. Interpretation of such comparative studies must also consider issues of space and time: the scale of experimental plots relative to the immigration potential of arthropods from nearby unmanaged areas, the duration of time over which different management practices have been maintained, and the seasonal dynamics of soil communities.

## CAN SOIL INVERTEBRATES ACT AS BIOINDICATORS OF BIOLOGICAL BUFFERING?

With the mounting interest in measuring soil quality, many studies have been designed to identify bioindicators of soil status. Given the effects of soil parameters on arthropods outlined previously, analysis of this community would seem to tell us something about the condition of the soil. van Straalen (1997) contends that microarthropod community structure might not only indicate soil quality but also ecosystem health. Three groups of mesofauna have been proposed as candidate bioindicators of soil conditions: nematodes (Bongers, 1990; Neher, 2001), mites (Kay et al., 1999), and Collembola (Frampton, 1997). Attributes that make these groups candidates for bioindication are that (1) they are among the most numerically dominant of soil fauna, (2) they include species that occupy various levels of the soil food chain and perform a range of ecological functions, and (3) within functional groups, they might contain multiple species with different niche requirements. The last aspect means that absolute numbers might not be as informative to soil condition as species composition. Other researchers have found invertebrates to be less useful tools for ecological analysis. Doube and Schmidt (1997) concluded that soil macrofauna are relatively poor indicators of functional aspects of soil quality, although they might act as bioaccumulators of xenobiotics because they occupy the top of the soil food chain. This fact would increase the sensitivity of analysis for contaminants.

Such disparate findings point out a number of issues: (1) the usefulness of fauna for bioindication will vary with the species or groups considered, (2) groups must be analyzed from a wide range of habitats or soil conditions to be considered robust bioindicators, and (3) methodological differences can lead to conflicting conclusions about the same group. There appear to be no examples in the literature of single species acting as effective soil indicators, and total densities or even diversity indices such as Shannon–Weiner have fallen short of the goal. van Straalen (1997) evaluated a variety of approaches for characterizing soil microarthropod community structure with regard to specificity of response to soil factors and resolution of response, i.e., sensitivity to changes in level of the factor (Table 7.1). Methodologies with high specificity respond to one factor more strongly than to others, making them a good signal for changes in specific soil parameters, but probably not very useful for assessing more general characteristics such as soil health. Measurements of soil communities with high resolution (such as multivariate statistics) are sensitive to soil changes, but might not explain which soil factors affect the change in community structure. Thus, it was concluded that the most promising approach for characterizing soil conditions is one that combines multivariate analysis of community composition with a functional classification of species.

**TABLE 7.1**
**Evaluation of Soil Microarthropod Community Bioindicators with Regard to the Extent to Which They Respond to a Single Soil Factor (Specificity +) or to Many Different Factors (Specificity –), and with Regard to Their Ability to Detect Small Changes (Resolution +) or Only Large Changes (Resolution –)**

| Microarthropod Community | Specificity | Resolution |
|---|:---:|:---:|
| Species diversity indices | – – | – |
| Single indicator species | + | – – |
| Ratios between species | + | – |
| Dominance structure | – | + |
| Multivariate statistics | – | ++ |
| Life-history patterns | – | + |
| Feeding types | ++ | + |
| Functional groupings | ++ | + |
| Food-web parameters | ? | ? |
| Ecophysiological classifications | ++ | + |

*Source:* From van Straalen, N.M. 1997. In Pankhurst, C., Doube, B.M., and Gupta, V.V.S.R. (Eds.), *Biological Indicators of Soil Health*. CAB International, New York, pp. 235–264. With permission.

Given the complexity of microarthropod response to soil parameters, their use as a simple tool to quantitatively compare soil condition might not occur on a widespread basis. However, in a qualitative manner, it is clear that a healthy soil that is functioning well will include a large and diverse microarthropod community.

## Effects of Arthropods on SOM

Most general discussions of decomposition of organic matter in soil focus on the roles played by microbes and earthworms. The significance of microarthropods to soil ecology has been undervalued to a large degree (Crossley et al., 1992). A number of factors could have contributed to this oversight. First, members of this group are minute organisms living in a cryptic environment, which makes study of the behavior and bionomics of soil microarthropods difficult. Thus, the soil system is often treated as a black box, and conclusions about the roles of organisms must be indirectly derived. Second, the role of soil microarthropods in organic matter decomposition has been obscured by measurements of respiration and relative biomass, which are both very low for soil microarthropods relative to microbes or even earthworms, causing many soil ecologists to overlook their functional importance. For example, Reichle (1977) determined that microflora were responsible for 90% of the respiration from decomposition in a southeastern U.S. forest, with the remaining 10% divided among all soil invertebrates. However, assigning ecological significance by comparing respiration masks the true role of soil fauna, as it has become increasingly clear that arthropods along with other soil fauna have their greatest impact as regulatory agents in soil decomposition processes: directly through feeding and comminution of organic matter and indirectly through microbial grazing.

## A Diversity of Ecological Functions

Also obscuring the total contribution of arthropods in the soil is their functional diversity. Whereas discrete functions can be assigned to most other soil taxa, allowing them to be assigned to separate boxes in a soil food web, microarthropods play a diverse array of ecological functions occurring at all trophic levels (Figure 7.6). Moore et al. (1988) categorize soil arthropods according to six

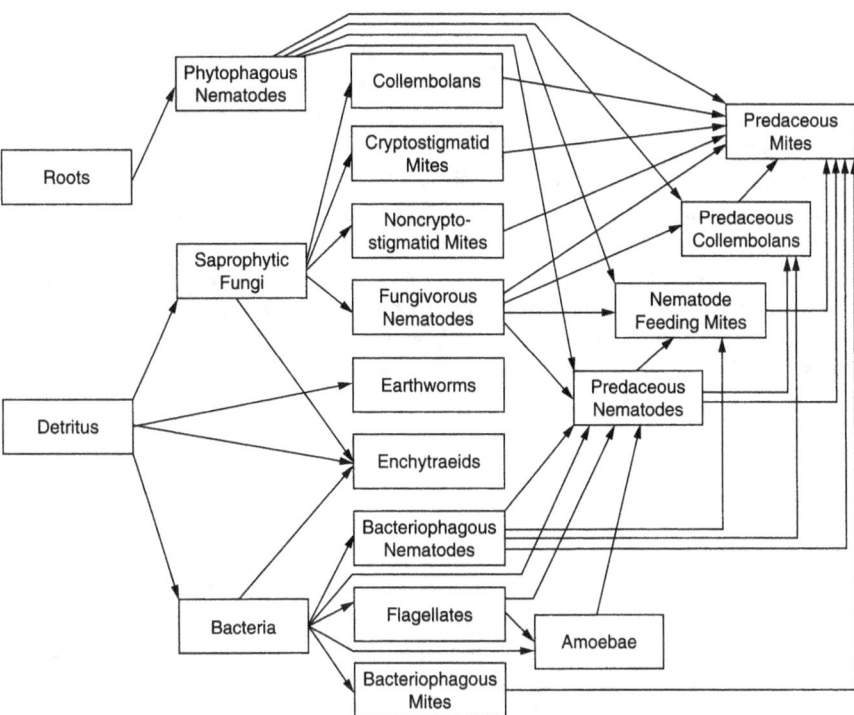

**FIGURE 7.6** Soil food web from a farm in which >50% of nitrogen was provided by manure application. Although soil invertebrates contribute relatively little to soil biomass and respiration, this diagram illustrates the variety of ecological boxes that they fill. (From De Ruiter, P.C. et al. 1993. *J. Appl. Ecol.* 30: 95–106. With permission.)

trophic functions: detritivores, fungivores, bacteriovores, herbivores, omnivores, and predators. As such, soil arthropods in total directly and indirectly impact SOM dynamics in four main ways: direct roles include litter reduction through feeding and transformation through production of fecal matter and ultimately their own death, and indirect effects include facilitation of microbial action and feeding on detritivorous microbes. Thus, the importance of microarthropods on belowground food webs comes not through mass but through their interactions at a number of key junctures in these webs. Quantifying the full impact of arthropods on the soil ecosystem is further complicated by the unwillingness of a number of taxa to stay in their trophic boxes, with some species showing omnivorous feeding and other species switching opportunistically among feeding habits as prevailing resources dictate (Wardle, 1995). Although not as morphologically diverse as soil arthropods, this functional diversity is also characteristic of the nematodes, which as a group impact the soil ecosystem at a number of trophic levels (Yeates et al., 1993).

## COMMINUTION AND DIGESTION OF ORGANIC LITTER

Detritivorous arthropods feed on organic matter and litter. Respiration measurements do not give a true picture of the arthropod contribution to decay. A more accurate assessment of the impact of detritivorous arthropods is provided by exclusion studies, which compare decay rates in the presence and absence of these organisms. A review of such studies (Seastedt, 1984) indicates a wide range of impact, with an enhanced decomposition rate of 0 to 70%, and averaging 23%. This variability reflects interactions with a number of other factors, including climate, faunal species present, and nature of organic matter substrate, with microarthropods possibly having a greater impact on the decomposition of more recalcitrant substrates (Seastedt, 1984). It is also suggested that conclusions can vary with the experimental methods employed, with shorter-term studies tending to underestimate the impact of microarthropods on decomposition rate. Probably more important than direct contributions to organic matter decomposition are the indirect effects these detritivores have through increasing the surface area of detritus, which enhances the rate of microbial digestion, and in some cases by redistribution of both the organic matter and microbes and protozoa. Incidental or adaptive ingestion of microbes while feeding on detritus also accelerates the process. The processes of microbial transport and ingestion are largely passive, but mutualisms are also evident, as in the heterostigmatid mite, *Trochometridium*, which possesses sporothecae, structures specialized for transport of fungal spores (Lindquist, 1985). Also like earthworms, detritivorous microarthropods play a significant role in establishing a plant-beneficial soil structure. Feeding on surface litter, these microarthropods produce particulate organic matter and fecal aggregates that improve the structure of soils, increasing porosity, as well as creating hotspots of microbial activity (Rusek, 1985).

## REGULATION OF MICROBIAL POPULATIONS

Even though microarthropods encompass a range of roles, it is generally agreed that as a group they have their greatest impact on regulating soil nutrient cycling via microbial grazing. Microarthropods reverse the process of nutrient immobilization from microbial decomposition of organic matter. Mineralization of N by microbial grazing is derived from the fact that the C:N ratio of microarthropods is typically higher than that of the microbes on which they feed.

Often thought to be general in their feeding, two main groups of microbivorous invertebrates are evident, preferring either fungi or bacteria. Beyond that, fungivores show feeding preferences among fungal species and typically preferences for different fungal structures (Griffiths and Bardgett, 1986). These feeding preferences can significantly alter microbial activity levels and community structure. Moderate levels of grazing on senescent hyphae actually stimulate fungal activity through compensatory growth of metabolically active hyphae (Hanlon, 1981; Bardgett et al., 1993). Similarly, bacteria-feeding invertebrates can increase bacterial respiration by keeping microbial populations in the log phase of growth (Bloem et al., 1986). On the other hand, high levels of selective fungal grazing can reduce biomass and respiration of fungal species (Bardgett et al., 1993). In a natural system, selective feeding can lead to either changes in the fungal species composition or shifts in the fungal–bacterial balance (Hanlon and Anderson, 1979; Parker et al., 1984), with implications for decomposition and nutrient cycling. Heavy preferential feeding by the springtail *Onychiurus armatus* on the fungus *Marasmius androsaceus* in conifer litter shifted population dominance to the competing and nonpalatable fungus *Mycena galopus*, resulting in reduced rates of litter decomposition due to the slower growth of *M. galopus* (Newell, 1984a, 1984b). Plants in soil microcosms inoculated with bacteria and a bacterial-feeding nematode grew larger than those in soil with bacteria alone because of increases in $NH_4^+$ production (Ingham et al., 1985). On the other hand, the addition of fungi and a fungivorous nematode to microcosms did not enhance plant growth relative to fungi alone, because these nematodes excreted 2.5 times lesser $NH_4^+$ and because the N mineralized by the fungi alone was enough for plant growth.

The suppression of plant pathogens by organic matter has been largely attributed to an interaction among microbes (Chapter 5); however, soil fauna also contribute significantly to this beneficial effect of organic matter. The Collembola *Onychiurus fimatus* and *Folsomia fimetaria* reduced the infection potential of *Pythium* in sugar beets (Ulber, 1983), and Curl et al. (1985) demonstrated that a mixture of *Proisotoma* and *Onychiurus* springtails rapidly consumed colonies of *Fusarium* and *Rhizoctonia* and reduced spores of *Fusarium* and *Trichoderma* in soil by 92% and 75%, respectively. Because Collembola tend to be concentrated in the rhizosphere, they have the potential to reduce plant-disease severity, particularly in the competitive milieu of other soil fungi. Also, the

Collembola in this study showed preferences for certain fungi and were repelled by others; for some fungal species, they would feed on certain structures only. Subsequently, Curl et al. (1988) found that this feeding preference for *Rhizoctonia solani* could result in a reduction in infection of cotton seedlings, and Scholte and Lootsma (1998) recorded a reduction in *Rhizoctonia* stem canker of potato where populations of Collembola and mycophagous nematodes were enhanced by the additions of farmyard and green manures. When insect and nematode populations were suppressed with insecticide, disease severity increased. In a similar manner, plant interactions with mutualistic microbes can also be impacted; the selective grazing habit of some collembolans and mites on VAM fungi generally results in reduced nutrient flow to plants (Finlay, 1985; Lussenhop, 1992).

Finally, other microarthropod species can impact microbial communities by predation on microbivorous arthropods and nematodes. Microarthropod predation on nematodes can depress populations of the latter groups (Santos and Whitford, 1981), resulting in increases in bacterial populations (Santos et al., 1981; Seastedt, 1984; Moore et al., 1988). If soil bacteria are primarily predator-limited (Wardle and Yeates, 1993), the cascading effects of predatory microarthropods on lower trophic levels can contribute to the shift in balance of bacteria and fungi populations in conventional vs. reduced-tillage systems.

## IMPACT ON NUTRIENT CYCLING AND PLANT PRODUCTIVITY

N is generally one of the most limiting elements for most organisms. N from organic sources must pass through the detrital food web before it becomes available to plants. As microbes feed on organic matter, they respire carbon while immobilizing N. Microbivorous microarthropods reverse this process, mineralizing N for plant availability. Even for SOM of relatively high C:N ratio, N mineralization can result from microbial grazing by microarthropods (Groffman, 1999). In an analysis of belowground food webs, Hunt et al. (1987) estimated that although the total biomass of soil fauna was less than 3% of that of soil microbes, they could be responsible for almost 40% of the N mineralized in a temperate grassland system. As stated by Luxton (1982):

> The relationship between the fauna and microflora is close, complex, and imperfectly understood, but a full comprehension of decomposition processes requires a study of this biological concert rather than a consideration of the activities of fauna or of microflora separately.

Just as plant productivity and species composition can impact the make-up of the belowground food web, the reverse also holds true. Setälä and Huhta (1991) demonstrated that microcosms that included nematodes, enchytraeids, Collembola, and mites had higher mineralization rates and plant uptake for N and P, resulting in faster growth of birch seedlings than microcosms without soil fauna. Although mineralization was increased, leaching, particularly of N, was reduced with soil fauna. Subsequently, Laakso and Setälä (1999) demonstrated that this beneficial effect of soil fauna on plant productivity was dependent not only on abundance but also on position of the species in the food web. They determined that impact on mineralization and plant growth inversely related to trophic level; although higher-level predators could reduce prey biomass, their impact on detrital food-web functioning was small.

Belowground herbivory represents another avenue for arthropod effects on soil nutrient flux. Legume roots, and particularly nodules with their N-fixing symbionts, are concentrated sources of N. *Sitona* weevils are significant pests of clover, with adults feeding on foliage and larvae feeding belowground on the roots, which can result in the release of N. Murray and Hatch (1994) found that ryegrass interplanted with white clover had a 75% increase in biomass and 74% increase in N content when the clover was infested with *Sitona* larvae, compared with uninfested plants.

Microarthropods also impact the cycling of other soil nutrients, although studies have been limited, and where measured, the effects were variable (Seastedt, 1984). For example, microarthropods accelerated the loss of phosphorus from leaf litter on forest soil, while magnesium and calcium

concentrations declined initially but then increased, and potassium was relatively unaffected (Seastedt and Crossley, 1980). These variable effects on nutrient flux were attributed to the combined
result of comminution of litter (leading to nutrient loss) and stimulation of microbial respiration
(tending to concentrate nutrients).

In summary, soil microarthropods through their functional diversity can impact plant development, productivity, health, and aboveground interactions by a multitude of mechanisms (Scheu et
al., 1999):

1. Nutrient mineralization by detritivory and microbial grazing
2. Improvements in soil structure
3. Grazing on mycorrhizal fungi
4. Grazing on plant pathogens
5. Dispersal of plant growth-stimulating microorganisms (Stephens et al., 1995; Lussenhop,
   1996)
6. Dispersal of microorganisms antagonistic to root pathogens (Stephens and Davoren 1997)
7. Enhancement of the phytohormonal activity of microbial decay products via their roles
   in organic matter decomposition (Jentschke et al., 1995; Muscolo et al., 1996)

## FUNCTIONAL REDUNDANCY AND THE NICHE CONCEPT

The coexistence of hundreds of species of soil microarthropods (Giller, 1996) sustained by the
single resource of organic matter has caused some difficulty for the classic equilibrium theory,
which predicts that interspecific competition will eliminate niche sharing (Anderson 1975a, 1975b).
The appearance of species niche overlap or functional redundancy has led some to conclude that
soil ecosystems function differently than do aboveground systems, on which most ecological
theories are based. In discussing the species diversity of Collembola in tropical mountain forest
soils, Deharveng and Bedos (1993) argue that besides the broad trophic categories of sucking,
bacterial feeding, and fungal feeding, there is little evidence for trophic specialization. Within each
of these categories, analysis of guts showed a wide variety of contents for individuals, with
considerable overlap in contents among Collembola species, leading them to conclude that resource
heterogeneity plays only a minor role in species diversity. However, other researchers have found
considerable evidence for fine-scale feeding specialization. Abundance and species richness of
oribatid mites increase with level of organic matter decomposition (Anderson, 1975b). There is
even evidence of niche separation within species, with juveniles of the Collembola *Folsomia candida*
feeding primarily on bacteria and adults primarily utilizing fungi (Hanlon and Anderson, 1979;
Bakonyi, 1989).

Beare et al. (1995) challenged the notion that soil biota are highly functionally redundant.
Instead, they argue for a hierarchical view of soil biodiversity that explains species richness as a
function of the scale of resolution used. This view emphasizes the temporal and spatial diversity
of resources that forms different spheres of influence on the soil ecosystem. On the other hand,
Laakso and Setälä (1999) provided evidence for functional redundancy in soil animals by measuring
ecosystem productivity. They manipulated faunal composition in boreal forest miniecosystems to
distinguish the effects of functional and species diversity on soil nutrient cycling. They concluded
that the link between decomposition and plant production was driven primarily by soil invertebrate
effects on nutrient dynamics. When members of lower trophic levels (i.e., bacteriovores, fungivores,
detritivores) were introduced to these microcosms, nutrient mobilization increased. However, little
effect on plant productivity was noted when the number of species within each of these functional
groups was reduced. The authors interpreted the results as an indication of ecosystem robustness
in the face of lower species diversity, as long as broad functional roles were represented. These
results were consistent with a mechanism of density compensation suggested by others (Walker,
1992). In a recent review of decomposer food-web studies, Setälä (2002) further emphasized the

need to define diversity and concluded that diversity in terms of faunal species number was less important than species-specific properties of individual taxa to plant productivity. Thus, within a trophic level, species vary in their influence on soil processes based on such characteristics as efficiency of resource utilization and vulnerability to predation. Also, a greater heterogeneity among feeding guilds (e.g., bacterivores vs. fungivores) is more important than heterogeneity within a guild on the rate of decomposition and plant productivity.

## CONCLUSIONS AND FUTURE DIRECTIONS

Other chapters discuss the various benefits of organic matter to soil quality and soil ecosystem function. The primary focus of this chapter is on how these benefits cascade through various trophic levels, particularly aboveground components of the terrestrial ecosystem. Much of the ecological theory has been developed from the study of aboveground associations in isolation from what transpires belowground. However, there is a growing recognition of the importance of the interface between these two domains to understanding ecosystem function. Contrasting chemically intensive agricultural with undisturbed ecosystems or with organically managed farms provides an opportunity for understanding whole-system function and elucidates how distant components of this system are interdependent. As the complexity of an ecosystem increases, so does the importance of indirect effects on ecosystem regulation, "for there are so many alternative pathways along which interactions may develop" (Price, 1988).

Most pertinent to agricultural management systems, we must recognize that in bypassing the natural nutrient cycles of the soil food web, we are superseding a community of organisms with short generation times, whose origin predates plants by billions of years. This long evolutionary history and great evolutionary potential has resulted in an enormous capacity to answer ecological challenges. First primitive invertebrate taxa, and later higher plants and animals, evolved to exploit this microbial resource (Price, 1988). In this context, it is easy to see that terrestrial trophic interactions are built on the soil decomposer web. Is it not unrealistic to think that a soil management approach that ignores these associations would not have other consequences unforeseen by reductionistic analysis?

Having stated the importance of connecting above- and belowground components, there still remain a number of unresolved issues. Moving away from the industrial model of agriculture to an ecologically based one is a prerequisite to developing sustainable farm-management systems for the future. Central to the latter model is a greater reliance on biological functionality in agroecosystems to provide key ecological services and internal regulation, resulting in a system that requires fewer external inputs and possesses greater resistance to biotic and abiotic disturbances. However, for such adoption to be widespread, we need to understand how to optimize our management of individual cropping systems so as to achieve parity in productivity and economics with chemical-intensive management. Specifically, more studies are needed to detail the mechanisms mediating the impacts soil processes have on plant growth and aboveground interactions.

A number of fundamental ecological questions also remain. Resolving the large number of soil microarthropods of seemingly similar resource and habitat requirements with the fundamental concept of the niche by Hutchinson (1957) and understanding how this great diversity relates to soil ecosystem function are interrelated and represent some of the most hotly debated issues in soil biology. There has been an explosion of studies related to this debate in the past few years, driven both by the broader interest in biodiversity and environmental concerns and by the need to understand how anthropogenic activities affect soil biodiversity and soil health. Some of the disparate conclusions of these studies can be traced to methodological differences, differences in the taxa studied and conditions used, and particularly issues of spatial and temporal scale. These studies are made difficult by the cryptic soil environment and the complexity of biotic interactions.

One concern in this debate is that stating that there exists a high level of species redundancy in the belowground food web and that species diversity is unimportant to ecosystem productivity can be misconstrued by some as meaning that we do not need to preserve the biological richness of the soil. Clearly, there is strong evidence of a number of connections within the soil community that impact carbon and nutrient cycling and that there is a need for functional diversity in the soil. Species diversity simply means that not all species found in the soil impact ecosystem function equally and that there does not exist a one-for-one connection between species number and either productivity or stability.

Another concern is that conclusions from most microcosm studies are drawn from the short-term changes in plant productivity under controlled conditions. This approach might cause us to overlook certain ecological benefits. Most notable is the paucity of studies that attempt to relate soil faunal diversity to the resistance and resilience of the soil to perturbations (i.e., biological buffering capacity). More studies are needed that address these issues and that allow testing under a broader range and greater fluctuation of abiotic conditions over longer periods. Measuring responses under conditions that vary widely with natural perturbations will provide a more robust picture of what defines a truly healthy soil ecosystem.

## REFERENCES

Alam, S., Kamei, S., and Kawai, S. 2001. Effect of iron deficiency on the chemical composition of the xylem sap of barley. *Soil Sci. Plant Nutr.* 47: 643–649.

Amancio, S., Clarkson, D.T., Diogo, E., Lewis, M., and Santo, H. 1997. Assimilation of nitrate and ammonium by sulfur deficient *Zea mays* cells. *Plant Physiol. Biochem.* 35: 41–48.

Anderson, J.M. 1975a. The enigma of soil animal species diversity. In Vanek, J. (Ed.), *Progress in Soil Zoology.* Dr. W. Junk, The Hague, the Netherlands, pp. 51–58.

Anderson, J.M. 1975b. Succession, diversity and trophic relationships of some soil animals in decomposing leaf litter. *J. Anim. Ecol.* 44: 475–495.

Andow, D.A., and Hidaka, K. 1989. Experimental natural history of sustainable agriculture: Syndromes of production. *Agric. Ecosyst. Environ.* 27: 447–462.

Alvarez, T., Frampton, G. K., and Goulson, D. 2001. Epigeic Collembola in winter wheat under organic, integrated, and conventional farm management regimes. *Agric. Ecosyst. Environ.* 83: 95–110.

Augustine, M.G., Fisk, F.W., Davidson, R.H., Lapidus, J.B., and Cleary, R.W. 1964. Host-plant selection by the Mexican bean beetle, *Epilachna varivestis. Ann. Entomol. Soc. Am.* 57: 127–134.

Bakonyi, G. 1989. Effects of *Folsomia candida* (Collembola) on the microbial biomass in a grassland soil. *Biol. Fertil. Soils* 7: 138–141.

Bardgett, R.D., Frankland, J.C., and Whittaker, J.B. 1993. The effects of agricultural practices on the soil biota of some upland grasslands. *Agric. Ecosyst. Environ.* 45: 25–45.

Bardgett, R.D., Hobbs, P.J., and Frostegård, Å. 1996. Changes in fungal:bacterial biomass ratios following reductions in the intensity of management on an upland grassland. *Biol. Fertil. Soils* 22: 261–264.

Bardgett, R.D., Wardle, D.A., and Yeates, G.W. 1998. Linking aboveground and belowground interactions: How plant responses to foliar herbivory influence soil organisms. *Soil Biol. Biochem.* 30: 1867–1878.

Bardgett, R.D., Whittaker, J.B., and Frankland, J.C. 1993. The effect of Collembolan grazing on fungal activity in differently managed upland pastures: A microcosm study. *Biol. Fertil. Soils* 16: 255–262.

Beanland, L., Phelan, P.L., and Salminen, S. 2003. Micronutrient interactions on soybean growth and the developmental performance of three insect herbivores. *Environ. Entomol.* 32: 641–651.

Beare, M.H., Coleman, D.C., Crossley, Jr., D.A., Hendrix, P.F., and Odum, E.P. 1995. A hierarchical approach to evaluating the significance of soil biodiversity to biogeochemical cycling. *Plant Soil* 170: 5–22.

Bedet, C. 2000. Soil fertility, crop nutrients, weed biomass and insect populations in organic and conventional field corn (*Zea mays* L.) agroecosystems. Ph.D. Dissertation, The Ohio State University, Columbus, OH.

Benrey, B., and Denno, R.F. 1997. The slow-growth-high-mortality hypothesis: A test using the cabbage butterfly. *Ecology* 78: 987–1119.

Bentz, J., Reeves, J., Barbosa, P., and Francis, B. 1996. The effect of nitrogen fertilizer applied to *Euphoria pulcherrima* on the parasitization of *Bemisia argentifolii* by the parasitoid *Encarsia formosa. Entomol. Exp. Appl.* 78: 105–110.

Bloem, J., de Ruiter, P., and Bouwman, L. 1986. Soil food webs and nutrient cycling in agroecosystems. In, van Elsas, J.D., Trevors, J.T., and Wellington, E.M.H. (Eds.), *Modern Soil Microbiology*. Marcel Dekker, New York.

Bongers, T. 1990. The maturity index: An ecological measure of environmental disturbance based on nematode species composition. *Oecologia* 83: 14–19.

Borowicz, V.A. 1997. A fungal root symbiont modifies plant resistance to an insect herbivore. *Oecologia* 112: 534–542.

Brady, N.C., and Weil, R.R. 1999. *The Nature and Properties of Soils*, 12th ed. Prentice-Hall, Upper Saddle River, NJ.

Cepeda, J.G., and Whitford, W.G. 1989. The relationships between abiotic factors and the abundance patterns of soil microarthropods on a desert watershed. *Pedobiology* 33: 79–86.

Cepeda-Pizzaro, J.G., and Whitford, W.G. 1989. Spatial and temporal variability of higher soil microarthropod taxa along a transect in a northern Chihuahuan Desert watershed. *Pedobiology* 33: 101–111.

Cockfield, S.D. 1988. Relative availability of nitrogen in host plants of invertebrate herbivores: Three possible nutritional and physiological definitions. *Oecologia* 77: 91–94.

Chen, P.S. 1966. Amino acid and protein metabolism in insect development. *Adv. Insect Physiol.* 3: 53–132.

Coleman, D.C., and Crossley, D.A. 1996. *Fundamentals of Soil Ecology*. Academic Press, San Diego, 205 pp.

Cortesero, A.M., Stapel, J.O., and Lewis, W.J. 2000. Understanding and manipulating plant attributes to enhance biological control. *Biol. Contr.* 17: 35–49.

Court, R.D., Williams, W.T., and Megarty, M.P. 1972. The effect of mineral nutrient deficiency on the content of free amino acids in *Setaria sphacelata*. *Aust. J. Biol. Sci.* 25: 77–87.

Crossley, D.A., Jr., Mueller, B.R., and Perdue, J.C. 1992. Biodiversity of microarthropods in agricultural soils: Relations to functions. *Agric. Ecosyst. Environ.* 40: 37–46.

Curl, E.A., Gudauskas, R.T., Harper, J.D., and Peterson, C.M. 1985. Effects of soil insects on populations and germination of fungal propagules. In Parker, C.A., Rovira, A.D., Moore, K.J., Wong, P.T.W., and Kollmorgan, J.F. (Eds.), *Ecology and Management of Soilborne Plant Pathogens*. American Phyto-pathological Society, St. Paul, MN, pp. 20–23.

Curl, E.A., Lartey, R., and Peterson, C.M. 1988. Interactions between root pathogens and soil microarthropods. *Agric. Ecosyst. Environ.* 24: 249–261.

Curry, J.P., and Purvis, G. 1982. Studies on the influence of weeds and farmyard manure on the arthropod fauna of sugar beet. *J. Life Sci.* 3: 397–408.

Deharveng, L., and Bedos, A. 1993. Factors influencing diversity of soil Collembola in a tropical mountain forest (Dol Inthanon, Northern Thailand). In Paoletti, M.G., Foissner, W., and Coleman, D. (Eds.), *Soil Biota, Nutrient Cycling, and Farming Systems*. Lewis Publishers, Boca Raton, FL, pp. 91–111.

Derridj, S., V. Gregoire, J.P. Boutin, and V. Fiala. 1989. Plant growth stages in the interspecific oviposition preference of the European corn borer and relations with chemicals present on the leaf surfaces. *Entomol. Exp. Appl.* 53: 267–276.

De Ruiter, P.C., Moore, J.C., Zwart, K.B., Bouwman, L.A., Hasink, J., Bloem, J., De Vos, J.A., Marinissen, J.C.Y., Didden, W.A.M., Lebbink, G., and Brussaard, L. 1993. Simulation of nitrogen mineralization in the belowground food web of two winter wheat fields. *J. Appl. Ecol.* 30: 95–106.

Doles, J.L., Zimmerman, R.J., and Moore, J.C. 2001. Soil microarthropod community structure and dynamics in organic and conventionally managed apple orchards in western Colorado, USA. *Appl. Soil Ecol.* 18: 83–96.

Doube, B.M., and Schmidt, O. 1997. Can the abundance or activity of soil macrofauna be used to indicated the biological health of soils? In Pankhurst, C.E., et al. (Eds.), *Biological Indicators of Soil Health*. CAB International, London, pp. 265–295.

Drinkwater, L.E., Letourneau, D.K., Workneh, F., van Bruggen, H.C., and Shennan, C. 1995. Fundamental differences between conventional and organic tomato agroecosystems in California. *Ecol. Appl.* 5: 1098–1112.

El Titi, A., and Ipach, U. 1989. Soil fauna in sustainable agriculture: Results of an integrated farming system at Lautenbach, F.R.G. *Agric. Ecosyst. Environ.* 27: 561–572.

Felton, G.W., and Korth, K.L. 2000.Trade-offs between pathogen and herbivore resistance. *Curr. Opin. Plant Biol.* 3: 309–314.

Findlay, S., Carreiro, M., Krishik, V., and Jones, C.G. 1996. Effects of damage to living plants on leaf litter quality. *Ecol. Appl.* 6: 269–275.

Finlay, R.D. 1985. Interactions between soil micro-arthropods and endomycorrhizal associations of higher plants. *Spec. Publ. Brit. Ecol. Soc. 1985:* 319–332.

Fox, L.R., Letourneau, D.K., Eisenbach, J., and van Nouhuys, S. 1990. Parasitism rates and sex ratios of a parasitic wasp: Effects of herbivore and plant quality. *Oecologia* 83: 414–419.

Frampton, G.K. 1997. The potential of Collembola as indicators of pesticide usage: Evidence and methods from the UK arable ecosystem. *Pedobiology* 41: 179–184.

Gange, A.C., Brown, V.K., and Sinclair, G.S. 1994. Reduction of black vine weevil larval growth by vesicular-arbuscular mycorrhizal infection. *Entomol. Exp. Appl.* 70: 115–119.

Gange, A.C., and West, H.M. 1994. Interactions between arbuscular mycorrhizal fungi and foliar-feeding insects in *Plantago lanceolata* L. *New Phytol.* 128: 79–87.

Giller, P.S. 1996. The diversity of soil communities, the "poor man's tropical rainforest." *Biodivers. Conserv.* 5: 135–168.

Gouinguene, S.P., and Turlings, T.C.J. 2002. The effects of abiotic factors on induced volatile emissions in corn plants. *Plant Physiol.* 129: 1296–1307.

Griffiths, B.S., and Bardgett, R.D. 1986. Interactions between microbe-feeding invertebrates and soil micro-organisms. In van Elsas, J.D., Trevors, J.T., and Wellington, E.M.H. (Eds.), *Modern Soil Microbiology.* Marcel Dekker, New York.

Groffman, P.M. 1999. Carbon additions increase nitrogen availability in northern hardwood forest soils. *Biol. Fertil. Soils* 29: 430–433.

Hadjicharalampous, E., Kalburtji, K.L., and Mamolos, A.P. 2002. Soil arthropods (Coleoptera, Isopoda) in organic and conventional agroecosystems. *Environ. Manage.* 29: 683–690.

Hanlon, R.D. 1981. Influence of grazing by Collembola on the activity of senescent fungal colonies grown on media of different nutrient concentration. *Oikos* 36: 362–367.

Hanlon, R.D., and Anderson, J.M. 1979. The effect of Collembola grazing on microbial activity in decomposition of leaf litter. *Oecologia* 38: 93–99.

Hansen, R.A., and Coleman, D.C. 1998. Litter complexity and composition are determinants of the diversity and species composition of oribatid mites (Acari: Oribatida) in litterbags. *Appl. Soil Ecol.* 9: 17–23.

Heisler, C. 1994. Effects of soil compaction on the soil mesofauna: Collembola and Gamasina: A three-year field experiment. *Pedobiology* 38: 566–576.

Heisler, C. 1995a. Collembola and Gamasina: Bioindicators for soil compaction. *Acta Zool. Fenn.* 196: 2229–2231.

Heisler, C. 1995b. Influence of agricultural traffic and crop management on Collembola and microbial biomass in arable soil. *Biol. Fertil. Soils* 19: 159–165.

Herms, D.A., and Mattson, W.J. 1992. The dilemma of plants: To grow or defend. *Q. Rev. Biol.* 67: 283–335.

Holland, E.A., and Detling. J.K. 1990. Plant response to herbivory and belowground nitrogen cycling. *Ecology* 71: 1040–1049.

House, G.J., and Del Rosario Alzugaray, M. 1989. Influence of cover cropping and no-tillage practices on community composition of soil arthropods in a North Carolina agroecosystems. *Environ. Entomol.* 18: 302–307.

Howard, A. 1943. *An Agricultural Testament.* Oxford University Press, Oxford, 253 pp.

Huetsch, B.W., Augustin, J., Merbach, W. 2002. Plant rhizodeposition: An important source for carbon turnover in soils. *J. Plant Nutr. Soil Sci.* 165: 397–407.

Hülsmann, A., and Wolters, V. 1998. The effects of different tillage practices on soil mites, with particular reference to Oribatida. *Appl. Soil Ecol.* 9: 327–332.

Hunt, H.W., Coleman, D.C., Ingham, E.R., Elliott, E.T., Moore, J.C., Rose, S.L., Reid, C.P.P., and Morley, C.R. 1987. The detrital food web in a short-grass prairie. *Biol. Fertil. Soils* 3: 57–68.

Hunter, M.D. 2001. Insect population dynamics meets ecosystem ecology: Effects of herbivory on soil nutrient dynamics. *Agric. Forest Entomol.* 3: 77–84.

Hutchinson, G.E. 1957. Concluding remarks. *Cold Spring Harb. Symp. Quant. Biol.* 22: 415–427.

Ingham, E.R., Trofymow, J.A., Ames, R.N., Hunt, H.W., Morley, C.R., Moore, J.C., and Coleman, D.C. 1986. Trophic interactions and nitrogen cycling in a semi-arid grassland soil. II. System responses to removal of different groups of soil microbes or fauna. *J. Appl. Ecol.* 23: 615–630.

Ingham, R.E., Trofymow, J.A., Ingham, E.R., and Coleman, D.C. 1985. Interactions of bacteria, fungi, and their nematode grazers: Effects on nutrient cycling and plant growth. *Ecol. Monogr.* 55: 119–140.

Jackson, L.E., Schimel, J.P., and Firestone, M.K. 1989. Short-term partitioning of ammonium and nitrate between plants and microbes in an annual grassland. *Soil Biol. Biochem.* 21: 409–415.

Jentschke, G., Bonkowski, M., Godbold, D.L., and Scheu, S. 1995. Soil protozoa and forest tree growth: Non-nutritional effects and interaction with mycorrhizae. *Biol. Fertil. Soils.* 20: 263–269.

Kay, F.R., Sobhy, H.M., Whitford, W.G. 1999. Soil microarthropods as indicators of exposure to environmental stress in Chihuahuan Desert rangelands. *Biol. Fertil. Soils* 28: 121–128.

King, K.L., Hutchinson, K.J., and Greenslade, P. 1976. The effects of sheep numbers on associations of Collembola in sown pastures. *J. Appl. Ecol.* 13: 731–739.

Kowalski, R., and Visser, P.E. 1979. Nitrogen in a crop-pest interaction: Cereal aphids 1979, In Lee, J.A., McNeill, S., and Rorison, L.H. (Eds.), *Nitrogen as an Ecological Parameter.* Blackwell Scientific, Oxford, pp. 283–300.

Kracht, M.S., and Schrader, S. 1997. Collembola and Acari in compacted soil of agricultural land under different soil tillage systems. *Braun. Natur. Schrift.* 5: 425–440.

Krogh, P.H. 1994. Microarthropods as bioindicators. A study of disturbed populations. In *Terrestrial Ecology.* Natural Environmental Research Institute, Silkeborg.

Kromp, B. 1989. Carabid beetles communities (Carabidae, Coleoptera) in biologically and conventionally farmed agroecosystems. *Agric. Ecosyst. Environ.* 27: 241–251.

Laakso, J., and Setälä, H. 1999. Sensitivity of primary production to changes in the architecture of belowground food webs. *Oikos* 87: 57–64.

Lindquist, E.E. 1985. Discovery of sporothecae in adult female *Trochometridium* Cross, with notes on analogous structures in *Siteroptes amerling* (Acari: Heterostigmata). *Exp. Appl. Acarol.* 1: 73–85.

Lovett, G.M., Christenson, L.M., Groffman, P.M., Jones, C.G., Hart, J.E., Mitchell, M.J. 2002. Insect defoliation and nitrogen cycling in forests. *Bioscience* 52: 335–341.

Lussenhop, J. 1992. Mechanisms of microarthropod-microbial interactions in soil. *Adv. Ecol. Res.* 23: 1–33.

Lussenhop, J. 1996. Collembola as mediators of microbial symbiont effects upon soybean. *Soil Biol. Biochem.* 28: 363–369.

Luxton, M. 1982. General ecological influence of the soil fauna on decomposition and nutrient circulation. *Oikos* 39: 355–357.

Marschner, H. 1995. *Mineral Nutrition of Higher Plants*, 2nd ed., Academic Press, San Diego, 889 pp.

Mattson, W.J. 1980. Herbivory in relation to plant nitrogen content. *Annu. Rev. Ecol. Syst. 11:* 119–162.

McLaughlin, J.W., Reed, D.D., Bagley, S.T., Jurgensen, M.F., and Mroz, G.D. 1994. Foliar amino acid accumulation as an indicator of ecosystem stress for first-year sugar maple seedlings. *J. Environ. Qual.* 23:154–161.

Mengel, K., and Kirkby, E.A. 1987. *Principles of Plant Nutrition*, 4th ed. International Potash Institute, Bern, Switzerland, 685 pp.

Mitchell, M. 1978. Vertical and horizontal distributions of Oribatid mites (Acari: Cryptostigmata) in an Aspen woodland soil. *Ecology* 59: 516–525.

Moore, J.C., Walter, D.E., and Hunt, H.W. 1988. Arthropod regulation of micro- and mesobiota in belowground detrital food webs. *Annu. Rev. Entomol.* 33: 419–439.

Morales, H., Perfecto, I., and Ferguson, B. 2001. Traditional fertilization and its effect on corn insect populations in the Guatemalan highlands. *Agric. Ecosyst. Environ.* 84: 145–155.

Moreby, S.J., Aebischer N.J., Southway, S.E., and Sotherton, N.W. 1994. A comparison of the flora and arthropod fauna of organic and conventionally grown winter wheat in southern England. *Ann. Appl. Biol.* 125: 13–27.

Murray, P. J., and Hatch, D. J. 1994. *Sitona* weevils (Coleoptera: Curculionidae) as agents for rapid transfer of nitrogen from white clover (*Trifolium repens* L.) to perennial ryegrass (*Lolium perenne* L.). *Ann. Appl. Biol.* 125: 29–33.

Muscolo, A., Panuccio, M.R., Abenavoli, M.R., Concheri, G., and Nardi, S. 1996. Effect of molecular complexity and acidity of earthworm faeces humic fractions on glutamate dehydrogenase, glutamine synthetase, and phosphoenolpyruvate carboxylase in *Daucus carota* alpha II cells. *Biol. Fertil. Soils* 22: 83–88.

Neher, D.A. 2001. Role of nematodes in soil health and their use as indicators. *J. Nematol.* 33: 161–168.

Newell, K. 1984a. Interaction between two decomposer Basidiomycetes and a collembolan under Sitka spruce: Distribution, abundance, and selective grazing. *Soil Biol. Biochem.* 16: 227–233.

Newell, K. 1984b. Interaction between two decomposer Basidiomycetes and a collembolan under Sitka spruce: Grazing and its potential effects on fungal distribution and litter decomposition. *Soil Biol. Biochem.* 16: 235–239.

Pacovsky, R.S., Rabin, L.B., Montllor, C.B., and Waiss, A.C., Jr. 1985. Host-plant resistance to insect pests altered by *Glomus fasciculatum* colonization. In Molina, R. (Ed.), *Proceedings of the 6th North American Conference on Mycorrhizae*, Oregon State University, Corvallis, OR, p. 288.

Paoletti, M.G., Favretto, M.R., Bressan, M., Marchiorato, A., and Babetto, M. 1992. Biodiversita in pescheti forlivesi. In Paoletti, M.G., Favretto, M.R., Nasolini, T., Scaravelli, D., and Zecchi, G. (Eds.), *Biodiversita Negli Agroecosystemi*. Wafra Litografica, Cesena, pp. 30–80.

Parker, L.W., Santos, P.F., Phillips, J., and Whitford, W.G. 1984. Carbon and nitrogen dynamics during decomposition of litter and roots of a Chihuahuan desert annual, *Lepidium lasioccurpum. Ecol. Monogr.* 54: 339–360.

Pfiffner, L., and Niggli, U. 1996. Effects of biodynamic, organic, and conventional farming on ground beetles (Col. Carabidae) and other epigaeic arthropods in winter wheat. *Biol. Agric. Hortic.* 12: 353–364.

Phelan, P.L. 1997. Soil-management history and the role of plant mineral balance as a determinant of maize susceptibility to the European corn borer. *Biol. Agric. Hortic.* 15: 25–34.

Phelan, P.L., Mason, J.R., and Stinner, B.R. 1995. Soil-fertility management and host preference by European corn borer, *Ostrinia nubilalis* (Hübner), on *Zea mays* L.: A comparison of organic and conventional chemical farming. *Agric. Ecosyst. Environ.* 56: 1–8.

Phelan, P.L., Norris, K., and Mason, J.R. 1996. Soil-management history and host preference by *Ostrinia nubilalis* (Hübner): Evidence for plant mineral balance as a mechanism mediating insect/plant interactions. *Environ. Entomol.* 25: 1329–1336.

Poole, T.B. 1964. A study of the distribution of soil Collembola in three small areas in a coniferous woodland. *Pedobiology* 4: 35–42.

Price, W.P. 1988. An overview of organismal interactions in ecosystems in evolutionary and ecological time. *Agric. Ecosyst. Environ.* 24: 369–377.

Rabin, L.B., and Pacovsky, R.S. 1985. Reduced larval growth of two Lepidoptera (Noctuidae) on excised leaves of soybean infected with a mycorrhizal fungus. *J. Econ. Entomol.* 78: 1358–1363.

Reichle, D.E. 1977. The role of soil invertebrates in nutrient cycling. *Ecol. Bull. Stockholm* 25: 145–156.

Rieske, L.K., and Raffa, K.F. 1998. Interactions among insect herbivore guilds: Influence of thrips bud injury on foliar chemistry and suitability to gypsy moths. *J. Chem. Ecol.* 24: 501–523.

Rusek, J. 1985. Soil microstructures: Contributions of specific organisms. *Quaest. Entomol.* 21: 497–514.

Santos, P.F., Phillips, J., and Whitford, W.G. 1981. The role of mites and nematodes in early stages of buried litter decomposition in a desert. *Ecology* 62: 664–669.

Santos, P.F., and Whitford, W.G. 1981. The effects of microarthropods in litter decomposition in a Chihuahuan desert ecosystem. *Ecology* 62: 654–663.

Scheu, S., Theenhaus, A., and Jones, T.H. 1999. Links between the detritivore and the herbivore system: Effects of earthworms and Collembola on plant growth and aphid development. *Oecologia* 119: 541–551.

Schjonning, P., Elmholt, S., Munkholm, L.J., and Debosz, K. 2002. Soil quality aspects of humid sandy loams as influenced by organic and conventional long-term management. *Agric. Ecosyst. Environ.* 88: 195–214

Scholte, K., and Lootsma, M. 1998. Effect of farmyard manure and green manure crops on populations of mycophagous soil fauna and *Rhizoctonia* stem canker of potato. *Pedobiology* 42: 223–231.

Schrader, S., and Lingnau, M. 1997. Influence of soil tillage and soil compaction on microarthropods in agricultural land. *Pedobiology* 41: 202–209.

Scriber, J.M. 1984a. Nitrogen nutrition in plants and insect invasion. In *Nitrogen in Crop Production*, American Society of Agronomy, Madison, WI, pp. 441–460.

Scriber, J.M. 1984b. Host-plant suitability. In Bell, W.J., and Cardé, R.T. (Eds.), *Chemical Ecology of Insects*. Chapman & Hall, New York, pp. 159–202.

Seastedt, T.R. 1984. The role of microarthropods in decomposition and mineralization processes. *Annu. Rev. Entomol.* 29: 25–46.

Seastedt, T.R., and Crossley, D.A., Jr. 1980. Effects of microarthropods on the seasonal dynamics of nutrients in forest litter. *Soil Biol. Biochem.* 12: 337–342.

Setälä, H. 2002. Sensitivity of ecosystem functioning to changes in trophic structure, functional group composition and species diversity in belowground food webs. *Ecol. Res.* 17: 207–215.

Setälä, H., and Huhta, V. 1991. Soil fauna increase *Betula pendula* growth: Laboratory experiments with coniferous forest floor. *Ecology* 72: 665–671.

Shaul, O., Galili, S., Volpin, H., Ginzberg, I., Elad, Y., Chet, I., and Kapulnik, Y. 1999. Mycorrhiza-induced changes in disease severity and PR protein expression in tobacco leaves. *Mol. Plant-Microbe Interact.* 12: 1000–1007.

Steinberger, Y., Freckman, D.W., Parker, L.W., and Whitford, W.G. 1984. Effects of simulated rainfall and litter quantities on desert soil biota: Nematodes and microarthropods. *Pedobiology* 26: 267–274.

Stephens, P.M., and Davoren, C.W. 1997. Influence of earthworms *Aporrectodea trapezoides* and *A. rosea* on the disease severity of *Rhizoctonia solani* on subterranean clover and ryegrass. *Soil Biol. Biochem.* 29: 511–516.

Stephens, P.M., Davoren, C.W., and Hawke, B.G. 1995. Influence of barley straw and the lumbricid earthworm *Aporrectodea trapezoids* on *Rhizobium meliloti* L5-30R, *Pseudomonas corrugata* 2140R, microbial biomass and microbial activity in a red-brown earth soil. *Soil Biol. Biochem.* 27: 1489–1497.

Stiling, P., Rossi, A., Hungate, B., Dijkstra, P., Hinkle, C.R., Knott, W.M., III, and Drake, B. 1999. Decreased leaf-miner abundance in elevated $CO_2$: Reduced leaf quality and increased parasitoid attack. *Ecol. Appl.* 9: 240–244.

Suski, A.W., and Badowski, T. 1975. Effect of the host plant nutrition on the population of the twospotted spider mite, *Tetranychus urticae* Koch (Acarina, Tetranychidae). *Ekol. Pol.* 23: 185–209.

Thaler, J.S.. Karban, R., Ullman, D.E., Boege, K., Bostock, R.M. 2002. Cross-talk between jasmonate and salicylate plant defense pathways: Effects on several plant parasites *Oecologia* 131: 227–235.

Ulber, B. 1983. Einfluss von *Onychiurus fimatus* Gisin (Collembola: Onychiuridae) und *Folsomia fimetaria* (L.) (Collembola: Isotomidae) auf *Pythium ultimum* Trow., einem Erreger des Wurzelbrandes der Zuckerrübe. In Lebrun, P., Andre, H.M., deMedts, A., Grigoire-Wibo, C., and Wauthy, G. (Eds.), *New Trends in Soil Biology.* Dieu-Brichart, Ottignies, Louvain-la-Neuve, Belgium, pp. 262–268.

Van de Bund, C.F. 1970. Influence of crop and tillage on mites and springtails in arable soil. *Netherl. J. Agric. Sci.* 18: 308–314.

van Straalen, N.M. 1997. Community structure of soil arthropods as a bioindicator of soil health. In Pankhurst, C., Doube, B.M., and Gupta, V.V.S.R. (Eds.), *Biological Indicators of Soil Health.* CAB International, New York, pp. 235–264.

Waldbauer, G.P. 1968. The consumption and utilization of food by insects. *Adv. Insect Physiol.* 5: 229–288.

Walker, B.H. 1992. Biodiversity and functional redundancy. *Conserv. Biol.* 6:18–23.

Wardle, D.A. 1995. Impacts of disturbance on detritus food webs in agro-ecosystems of contrasting tillage and weed management practices. *Adv. Ecol. Res.* 26: 105–185.

Wardle, D.A. 1999. How soil food webs make plants grow. *TREE* 14: 418–420.

Wardle, D.A. 2002. *Communities and Ecosystems: Linking the Aboveground and Belowground Components.* Princeton University Press, Princeton, NJ.

Wardle, D.A., and Yeates, G.W. 1993. The dual importance of competition and predation as regulatory forces in terrestrial ecosystems: Evidence from decomposer food-webs. *Oecologia* 93: 303–306.

Waring, G.L., and Cobb, N.S. 1992. The impact of plant stress on herbivore population dynamics. In Bernays, E. (Ed.), *Insect–Plant Interactions,* Vol. IV. CRC Press, Boca Raton, FL, pp. 167–226.

White, T.C.R. 1984. The abundance of invertebrate herbivores in relation to the availability of nitrogen in stressed food plants. *Oecologia* 63: 90–105.

Yeates, G.W., Bardgett, R.D., Cook, R., Hobbs, P.J., Bowling, P.J., and Potter, J.F. 1997. Faunal and microbial diversity in three Welsh grassland soils under conventional and organic management regimes. *J. Appl. Ecol.* 34: 453–470.

Yeates, G.W., Bongers, T., de Goede R.G.M., Freckman, D.W., and Georgieva S.S. 1993. Feeding habits in soil nematode families and genera: An outline for soil ecologists. *J. Nematol.* 25: 315–331.

# 8 Tillage and Residue Management Effects on Soil Organic Matter

*Alan J. Franzluebbers*

## CONTENTS

## TYPES OF TILLAGE

Soil tillage is an ancient practice that was originally used to eradicate weeds and loosen the soil for planting seeds (Lal, 2001). In modern agriculture, tillage is still performed for controlling weeds, insects, and diseases; improving the soil's physical condition by loosening compacted layers and enhancing soil warming in spring; incorporating fertilizer, herbicide, and plant residues; conserving soil and water; and preparing a quality seedbed (Jones et al., 1990). The type of tillage employed should be designed to achieve a specific set of goals. During the past several decades, conservation tillage, and, particularly, no tillage have been increasingly utilized, as the need for inversion tillage has been reevaluated. The susceptibility of inverted soil to wind and water erosion has highlighted the environmental and production threats to sustainability (Figure 8.1). The term *conservation tillage* includes a variety of systems, all designed to minimize residue incorporation with the intent of abating soil erosion. According to the definition of the term by the United States Department of Agriculture (USDA), >30% residue cover must be on the soil surface immediately after planting (Figure 8.2). A major part of this chapter compares the influences of conservation and inversion tillage on soil organic matter.

Tillage practices range from the very simple to the very complex. Buckingham (1976) and Swinford (1994) give excellent descriptions of the types of tillage operations and their intended use. This chapter focuses on four groups of tillage practices affecting soil organic matter dynamics:

**FIGURE 8.1** Wind and water erosion are serious threats to the sustainability of agriculture. Both these erosive forces preferentially displace the lighter organic matter fraction from the soil surface, resulting in a decline of long-term productivity. Photos depict water erosion in the Georgia Piedmont and wind erosion in the loess hills of Nebraska.

**FIGURE 8.2** Alfalfa is an excellent sod component of long-term rotations that can help abate erosion. Traditionally, sod was broken by plowing and smoothing before planting maize, leaving the soil surface exposed to erosive forces (left). With no tillage, sod is killed with herbicides and maize can be grown without soil disturbance (right). Photos from eastern Nebraska.

moldboard plow, shallow, ridge, and no tillage. The moldboard plow was perhaps the most widely used primary tillage implement during the early part of the 20th century (Allmaras et al., 2000). The moldboard plow inverts soil to a depth of usually 15 to 30 cm, resulting in complete burial of aboveground crop residues. Secondary tillage operations of disking or harrowing, or both, are often needed to prepare a suitable seedbed following plowing.

Shallow tillage is accomplished by using a wide diversity of implements to scarify the soil surface. One primary tillage tool that has replaced the moldboard plow in some regions is the chisel plow. Although the working depth of the chisel plow might be similar to that of the moldboard plow, the degree of soil inversion with the chisel plow is much less. In semiarid regions with small grains as the main crop, primary tillage operations can be accomplished with an offset disk or field cultivator. Working depth of these implements is often less than that with plow tools, e.g., 10 to 15 cm depth. The extent of residue incorporation depends on the number of passes performed.

A conservation-tillage method with greater opportunities for controlling traffic is ridge tillage. The extent of soil disturbance varies greatly with the type of equipment and number of cultivations with this system. Ridges are typically formed, the tops scraped off to create a clean seedbed, and ridges reshaped during summer cultivation. The negative effects of machinery traffic can be limited to the same rows year after year so that the majority of the field is not compacted.

No tillage relies completely on herbicides and management to control weeds. Planting operations are typically the only disturbance to the soil surface.

## TYPES OF RESIDUE MANAGEMENT

If residues of various crops are considered a by-product without much value and a hindrance to future production, they can be removed from the field by burning. Residues can also be removed from the field as valuable fodder for animals or as materials for construction. Removal of residues either by burning or by harvest has important implications for soil organic matter dynamics. Crop residues are rich in organic C and N, and therefore their removal is a loss of input to the soil, resulting almost always in a decline in soil organic matter compared with retention of residues (Saffigna et al., 1989; Dalal et al., 1991; Kapkiyai et al., 1999).

Residues left in the field ultimately undergo decomposition with a majority of the C respired back to the atmosphere as $CO_2$ and a smaller fraction retained as soil organic matter. The rate and extent of residue transformation into soil organic matter depends on the type, quantity, and quality of residues produced and how and when residues are manipulated. The quantity of residues depends on climatic, soil, and fertility variables. The quality of residues depends on the plant species (e.g., small grain straw low in N vs. legume cover crop forage rich in N) and developmental stage when killed. Residues of primary crops can be cut, shredded, or left standing in the field. Cover crops can be allowed to mature, mowed, rolled, or terminated with herbicides. No-tillage management with a dense mat of previous crop residues can be effective at controlling erosion and weeds and moderate temperature and moisture fluctuations (Figure 8.3).

**FIGURE 8.3** Cotton planted with no tillage following harvest of barley in the Georgia Piedmont.

## EFFECT OF TILLAGE ON PLANT GROWTH

Agronomic production of food and fiber is a vocation that brings both joys and challenges to those called to be stewards of the land (Figure 8.4). For those who farm the land, nature can be both friend and foe. With care and management, the fruits of the earth can be harvested in bounty. However, the desire to obtain more from the land is often limited by the harsh realities of weather and pestilence. Those who believe that they know their system are often taught new lessons by nature, neighbors, accountants, or the government. Modern agricultural production is a complicated system involving natural resources, technology, finance, ingenuity, labor, and social fabric. There will always be different systems of agricultural production requiring different solutions to problems.

Soil erosion is a natural disaster that damages resources in a slow but continuous, and, occasionally, dramatic manner. Exposure of the fragile surface soil to the erosive forces of wind and water without protective cover has led to long-term soil, water, and air degradation (Trimble, 1974). Conservation tillage systems attempt to mimic nature by allowing residues that fall to the surface to remain there without mechanical incorporation. Seeds can then be planted directly through this mulch layer with minimal disturbance to the protective surface cover. This approach was partly made possible with the development of herbicides, which reduced one of the greatest needs for tillage, i.e., weed control.

Changes in microclimate under conservation compared with inversion tillage systems result in more water available for crop uptake by (1) getting more precipitation to infiltrate soil rather than run off of the land and (2) reducing evaporation of water from the soil surface during intervals between precipitation events (Lascano et al., 1994). Lack of tillage, however, could result in excessive compaction of soil, especially in systems with heavy equipment and random traffic patterns. In many studies, soil immediately below the surface becomes compacted during early adoption of no tillage, a process that could limit root growth and development. In the long term, however, freezing–thawing and bioperturbation loosen soil under no tillage compared with plow tillage (Voorhees and Lindstrom, 1984). It is also possible that old root channels and worm holes that remain intact without soil disturbance enhance water infiltration and root growth without a major change in bulk density.

**FIGURE 8.4** Statue of St. Isidore, the patron saint of farmers, in Bow Valley, Nebraska.

**FIGURE 8.5** Side-by-side long-term experiment near College Station, TX, comparing conventional disk-and-bed tillage of sorghum on left with no tillage of sorghum on right.

Many short-term studies and a few long-term studies have evaluated the effect of tillage system on plant productivity (Figure 8.5). In 33 comparisons with small grains, yield under no-tillage systems was equivalent to that under shallow-tillage systems (Table 8.1). However, yield under plow tillage was, on average, lower than under shallow or no tillage. At many of the semiarid locations, water conservation with shallow- or no-tillage management probably contributed to improved yield. From a compilation of studies with various crops other than maize or small grains, similar effects of tillage systems on yield occurred (Table 8.2). However, from a compilation of studies with maize, tillage system had no overall effect on yield (Table 8.3). Individual experiments might have shown significant reductions or increases in yield with adoption of conservation tillage, but on average there was no negative or positive effect of conservation tillage on maize yield. The lack of tillage system effect on yield might be important in promoting conservation tillage to control soil erosion and improve water quality in a particular watershed or region. No yield reduction can make conservation tillage attractive because, other than the initial investment in modifying or purchasing a conservation-tillage planter, operating costs are often lower with conservation-tillage systems than with conventional-tillage systems (Jones et al., 1990).

In the long term, accumulation of soil organic matter under conservation-tillage systems should lead to an increase in the storage and potential availability of nutrients. On a Fluventic Ustochrept in Texas, the N fertilizer required to achieve 95% of maximum sorghum grain yield was 40 to 60% higher during the first year of no-tillage management compared with conventional tillage (Figure 8.6). With time, however, the N fertilizer required became similar between tillage systems. It could be expected that during the second decade of no-tillage management, N fertilizer requirement would be lower than under conventional tillage. Although higher initial fertilizer expenditures might be needed to achieve optimum yield with no-tillage management because of sequestration of nutrients into organic matter, the long-term benefits of sustained nutrient storage, enhanced water infiltration and retention, improved soil biological activity, and more stable production can more than offset the initial costs. Cropping systems that include legumes with substantial biological N-fixation could help offset any additional requirement for N fertilizer inputs in conservation-tillage systems. In a long-term tillage study on a Typic Fragiudalf in Ohio, maize and soybean yields tended to increase with time (18 years) under no tillage compared with conventional tillage (Dick et al., 1991). On a very poorly drained Mollic Ochraqualf, yields were lower under no tillage than under conventional tillage during early years, but became similar between tillage systems with time. Similar positive changes in yield under no tillage compared with conventional tillage occurred with time in long-term studies in Maryland and Kentucky (Bandel and Meisinger, 1993; Ismail et al., 1994). Other studies that indicate negative yield effects of conservation tillage compared with conventional tillage have often been limited by weed control (Brandt, 1992), diseases due to crop sequencing (Dick et al., 1991), or poor seedling establishment due to straw management (Cannell and Hawes, 1994).

## TABLE 8.1
### Comparison of Small Grain Yield (g m$^{-2}$) among Tillage Systems

| Crop/Component | Location | Observations | Conditions | Soil | Plow | Shallow | Ridge | No. | Reference |
|---|---|---|---|---|---|---|---|---|---|
| Spr wheat – grain | Montana | 10 | Continuous spring wheat | Argiboroll | — | 155 | — | 176 | Aase et al. (1995) |
| Spr wheat – grain | North Dakota | 48 | Spring wheat–fallow | Argiboroll | 119 | 118 | — | 116 | Black and Tanaka (1997) |
| Spr wheat – grain | North Dakota | 48 | Wheat–wheat–sunflower | Argiboroll | 131 | 145 | — | 145 | Black and Tanaka (1997) |
| Spr wheat – grain | SK, Canada | 2 | Continuous wheat | Haploboroll | — | 256 | — | 245 | Curtin et al. (2000) |
| Spr wheat – grain | SK, Canada | 2 | Wheat–fallow | Haploboroll | — | 280 | — | 256 | Curtin et al. (2000) |
| Win wheat – grain | North Dakota | 48 | Wheat–wheat–sunflower | Argiboroll | 169 | 188 | — | 199 | Black and Tanaka (1997) |
| Win wheat – grain | New Mexico | 1 | 2-year sorghum–fallow–wheat | Paleustoll | — | 202 | — | 271 | Christensen et al. (1994) |
| Win wheat – grain | Texas | 12 | Continuous wheat | Ustochrept | — | 277 | — | 247 | Franzluebbers et al. (1995a) |
| Win wheat – grain | Texas | 12 | Wheat/soybean | Ustochrept | — | 371 | — | 361 | Franzluebbers et al. (1995a) |
| Win wheat – grain | Texas | 12 | Sorghum–wheat/soybean | Ustochrept | — | 385 | — | 407 | Franzluebbers et al. (1995a) |
| Win wheat – grain | Colorado | 12 | Wheat–fallow | Paleustoll | — | 289 | — | 279 | Halvorson et al. (1997) |
| Win wheat – grain | Austria | 5 | 9-year rotation | Chernozem | 498 | 493 | — | 114 | Kandeler et al. (1999) |
| Win wheat – grain | Texas | 3 | 0 g N m$^{-2}$ | Haplustoll | — | 218 | — | 134 | Knowles et al. (1993) |
| Win wheat – grain | Texas | 3 | 5 g N m$^{-2}$ | Haplustoll | — | 277 | — | 221 | Knowles et al. (1993) |
| Win wheat – grain | Texas | 3 | 9 g N m$^{-2}$ | Haplustoll | — | 328 | — | 272 | Knowles et al. (1993) |
| Win wheat – grain | Texas | 3 | 14 g N m$^{-2}$ | Haplustoll | — | 349 | — | 331 | Knowles et al. (1993) |
| Win wheat – grain | Texas | 10 | Continuous wheat | Paleustoll | — | 95 | — | 114 | Schomberg and Jones (1999) |
| Barley – grain | BC, Canada | 10 | Continuous barley | Cryoboralf | — | 300 | — | 309 | Arshad et al. (1999a) |
| Barley – grain | AB, Canada | 3 | 2-year barley with canola/pea | Cryoboralf | — | 342 | — | 370 | Arshad et al. (1999b) |
| Barley – grain | U.K. | 3 | Continuous barley | Cambisol | 624 | 640 | — | 668 | Ball et al. (1989) |
| Barley – grain | U.K. | 3 | Continuous barley | Gleysol | 611 | 625 | — | 680 | Ball et al. (1989) |
| Spr wheat – straw | North Dakota | 48 | Spring wheat–fallow | Argiboroll | 177 | 176 | — | 176 | Black and Tanaka (1997) |
| Spr wheat – straw | North Dakota | 48 | Wheat–wheat–sunflower | Argiboroll | 224 | 248 | — | 266 | Black and Tanaka (1997) |
| Spr wheat – straw | SK, Canada | 12 | Continuous wheat | Haploboroll | — | 191 | — | 191 | Campbell et al. (1999) |
| Spr wheat – straw | SK, Canada | 12 | Wheat–fallow | Haploboroll | — | 248 | — | 235 | Campbell et al. (1999) |
| Spr wheat – straw | SK, Canada | 12 | Continuous wheat | Haploboroll | — | 288 | — | 287 | Campbell et al. (1995) |
| Spr wheat – straw | SK, Canada | 6 | Wheat–fallow | Haploboroll | — | 364 | — | 348 | Campbell et al. (1995) |
| Spr wheat – straw | SK, Canada | 11 | Continuous wheat | Haploboroll | — | 158 | — | 163 | Campbell et al. (1996) |

| Crop/Component | Location | Observations | Conditions | Soil | Plow | Shallow | Ridge | No. | Reference |
|---|---|---|---|---|---|---|---|---|---|
| Spr wheat – straw | SK, Canada | 6 | Wheat-fallow | Haploboroll | — | 300 | — | 314 | Campbell et al. (1996) |
| Spr wheat – straw | SK, Canada | 2 | Continuous wheat | Haploboroll | — | 464 | — | 431 | Curtin et al. (2000) |
| Spr wheat – straw | SK, Canada | 2 | Wheat-fallow | Haploboroll | — | 514 | — | 440 | Curtin et al. (2000) |
| Win wheat – straw | North Dakota | 8 | Wheat-wheat-sunflower | Argiboroll | 287 | 324 | — | 346 | Black and Tanaka (1997) |
| Win wheat – straw | Texas | 10 | Continuous wheat | Paleustoll | — | 244 | — | 310 | Schomberg and Jones (1999) |
| Barley – straw | AB, Canada | 3 | 2-year barley with canola/pea | Cryoboralf | — | 551 | — | 551 | Arshad et al. (1999b) |
| Plow vs. shallow tillage (Pr > F = 0.02) | | | | | 316 | 329 | — | — | $n = 9$ |
| Plow vs. no tillage (Pr > F = 0.01) | | | | | 293 | — | — | 325 | $n = 8$ |
| Shallow vs. no tillage (Pr > F = 0.80) | | | | | — | 300 | — | 299 | $n = 33$ |

*Note:* Spr wheat, spring wheat; Win wheat, winter wheat.

**TABLE 8.2**
**Comparison of Various Other Crop Yields (g m⁻²) among Tillage Systems**

| Crop/Component | Location | Observations | Conditions | Soil | Plow | Shallow | Ridge | No. | Reference |
|---|---|---|---|---|---|---|---|---|---|
| Sorghum – grain | New Mexico | 3 | 2-year sorghum–fallow–wheat | Paleustoll | — | 297 | — | 314 | Christensen et al. (1994) |
| Sorghum – grain | Texas | 12 | Continuous sorghum | Ustochrept | — | 503 | — | 468 | Franzluebbers et al. (1995a) |
| Sorghum – grain | Texas | 12 | Continuous sorghum | Ustochrept | — | 519 | — | 453 | Franzluebbers et al. (1995a) |
| Sorghum – grain | Georgia | 2 | Sorghum–soybean | Rhodudult | 318 | — | — | 379 | Groffman et al. (1987) |
| Sorghum – grain | Austria | 2 | 9-year rotation | Chernozem | 600 | 586 | — | — | Kandeler et al. (1999) |
| Sorghum – grain | Texas | 10 | Continuous sorghum | Paleustoll | — | 293 | — | 293 | Schomberg and Jones (1999) |
| Sorghum – grain | Texas | 3 | Wheat–sorghum–sunflower | Paleustoll | 256 | 244 | — | 334 | Unger (1984) |
| Sorghum – straw | Georgia | 2 | Sorghum–soybean | Rhodudult | 783 | — | — | 762 | Groffman et al. (1987) |
| Sorghum – straw | Texas | 10 | Continuous sorghum | Paleustoll | — | 413 | — | 431 | Schomberg and Jones (1999) |
| Sorghum – straw | Texas | 3 | Wheat–sorghum–sunflower | Paleustoll | 394 | 501 | — | 469 | Unger (1984) |
| Soybean – grain | Ohio | 20 | Maize–soybean | Fragiudalf | 163 | — | — | 193 | Dick et al. (1991) |
| Soybean – grain | Ohio | 20 | Maize–soybean | Ochraqualf | 230 | — | — | 199 | Dick et al. (1991) |
| Soybean – grain | Alabama | 4 | Continuous soybean | Hapludult | 164 | 203 | — | 239 | Edwards et al. (1988) |
| Soybean – grain | Alabama | 4 | Maize–soybean | Hapludult | 222 | 263 | — | 266 | Edwards et al. (1988) |
| Soybean – grain | Alabama | 4 | Maize–wheat/soybean | Hapludult | 241 | 245 | — | 216 | Edwards et al. (1988) |
| Soybean – grain | Texas | 12 | Continuous soybean | Ustochrept | — | 176 | — | 145 | Franzluebbers et al. (1995a) |
| Soybean – grain | Iowa | 1 | Maize–soybean: 10 years | Haplaquoll | 324 | 283 | 299 | — | Singh et al. (1992) |
| Soybean – grain | Minnesota | 1 | Maize–soybean: 10 years | Hapludoll | 239 | 273 | — | 218 | Singh et al. (1992) |
| Soybean – grain | North Carolina | 5 | Maize–soybean | Kanhapludult | — | 235 | — | 254 | Wagger and Denton (1992) |
| Soybean – grain | North Carolina | 5 | Maize–soybean | Hapludult | — | 245 | — | 245 | Wagger and Denton (1992) |
| Sunflower – seed | North Dakota | 8 | Wheat–wheat–sunflower | Argiboroll | 131 | 140 | — | 138 | Black and Tanaka (1997) |
| Sunflower – seed | Texas | 3 | Wheat–sorghum–sunflower | Paleustoll | 155 | 163 | — | 154 | Unger (1984) |
| Sunflower – straw | North Dakota | 8 | Wheat–wheat–sunflower | Argiboroll | 296 | 312 | — | 319 | Black and Tanaka (1997) |
| Pea – grain | Austria | 1 | 9-year rotation | Chernozem | 508 | 453 | — | — | Kandeler et al. (1999) |
| Sugar beet | Austria | 1 | 9-year rotation | Chernozem | 5170 | 5375 | — | — | Kandeler et al. (1999) |
| | | | | | | | | | |
| Plow vs. shallow tillage (Pr > F = 0.19) | | | | | 669 | 695 | | | $n = 13$ |
| Plow vs. no tillage (Pr > F = 0.07) | | | | | 276 | 299 | | | $n = 13$ |
| Shallow vs. no tillage (Pr > F = 0.66) | | | | | — | 296 | | 292 | $n = 17$ |

**TABLE 8.3**
**Comparison of Maize Yield (g m$^{-2}$) among Tillage Systems**

| Crop/Component | Location | Observations | Conditions | Soil | Plow | Shallow | Ridge | No. | Reference |
|---|---|---|---|---|---|---|---|---|---|
| Maize – grain | Ohio | 20 | Continuous maize | Fragiudalf | 545 | — | — | 672 | Dick et al. (1991) |
| Maize – grain | Ohio | 20 | Maize–soybean | Fragiudalf | 656 | — | — | 778 | Dick et al. (1991) |
| Maize– grain | Ohio | 20 | Continuous maize | Ochraqualf | 702 | — | — | 686 | Dick et al. (1991) |
| Maize – grain | Ohio | 20 | Maize–soybean | Ochraqualf | 603 | — | — | 515 | Dick et al. (1991) |
| Maize – grain | Alabama | 4 | Continuous maize | Hapludult | 850 | 803 | — | 810 | Edwards et al. (1988) |
| Maize – grain | Alabama | 4 | Maize–soybean | Hapludult | 819 | 897 | — | 857 | Edwards et al. (1988) |
| Maize – grain | Alabama | 4 | Maize–wheat/soybean | Hapludult | 819 | 899 | — | 872 | Edwards et al. (1988) |
| Maize – grain | Indiana | 7 | Continuous maize | Haplaquoll | 1092 | 1047 | 1047 | 947 | Griffith et al. (1988) |
| Maize – grain | Indiana | 7 | Maize–soybean | Haplaquoll | 1182 | 1163 | 1191 | 1136 | Griffith et al. (1988) |
| Maize – grain | Indiana | 7 | Continuous maize | Ochraqualf | 825 | 846 | 824 | 882 | Griffith et al. (1988) |
| Maize – grain | Indiana | 7 | Maize–soybean | Ochraqualf | 821 | 837 | 876 | 933 | Griffith et al. (1988) |
| Maize – grain | Kentucky | 20 | Continuous: 0 g N m$^{-2}$ | Paleudalf | 477 | — | — | 429 | Ismail et al. (1994) |
| Maize – grain | Kentucky | 20 | Continuous: 8 g N m$^{-2}$ | Paleudalf | 682 | — | — | 677 | Ismail et al. (1994) |
| Maize – grain | Kentucky | 20 | Continuous: 17 g N m$^{-2}$ | Paleudalf | 711 | — | — | 750 | Ismail et al. (1994) |
| Maize – grain | Kentucky | 20 | Continuous: 34 g N m$^{-2}$ | Paleudalf | 732 | — | — | 757 | Ismail et al. (1994) |
| Maize – grain | Iowa | 12 | Continuous maize | Hapludoll | 805 | 782 | 752 | 741 | Karlen et al. (1991) |
| Maize – grain | Iowa | 12 | Maize–soybean | Hapludoll | 876 | 881 | 856 | 863 | Karlen et al. (1991) |
| Maize – grain | Pennsylvania | 3 | Alfalfa–maize: 0 g N m$^{-2}$ | Hapludult | 674 | — | — | 707 | Levin et al. (1987) |
| Maize– grain | Pennsylvania | 3 | Alfalfa–maize: 5 g N m$^{-2}$ | Hapludult | 701 | — | — | 771 | Levin et al. (1987) |
| Maize – grain | Pennsylvania | 3 | Alfalfa–maize: 9 g N m$^{-2}$ | Hapludult | 727 | — | — | 798 | Levin et al. (1987) |
| Maize – grain | Pennsylvania | 3 | Alfalfa–maize: 14 g N m$^{-2}$ | Hapludult | 756 | — | — | 802 | Levin et al. (1987) |
| Maize – grain | Pennsylvania | 3 | Alfalfa–maize: 18 g N m$^{-2}$ | Hapludult | 732 | — | — | 817 | Levin et al. (1987) |
| Maize – grain | Texas | 1 | Continuous maize | Ustochrept | — | 978 | 1021 | — | McFarland et al. (1991) |
| Maize – grain | New York | 2 | Various cover and nitrogen | Hapludalf | 542 | — | — | 485 | Sarrantonio and Scott (1988) |
| Maize – grain | Iowa | 1 | Continuous maize: 10 years | Haplaquoll | 941 | 799 | 700 | — | Singh et al. (1992) |
| Maize – grain | Minnesota | 1 | Maize–soybean: 10 years | Haplaquoll | 824 | 876 | — | 795 | Singh et al. (1992) |
| Maize – grain | North Carolina | 5 | Continuous maize | Kanhapludult | — | 403 | — | 593 | Wagger and Denton (1992) |

*continued*

**TABLE 8.3 (continued)**
**Comparison of Maize Yield (g m⁻²) among Tillage Systems**

| Crop/Component | Location | Observations | Conditions | Soil | Plow | Shallow | Ridge | No. | Reference |
|---|---|---|---|---|---|---|---|---|---|
| Maize – grain | North Carolina | 4 | Maize–soybean | Kanhapludult | — | 317 | — | 474 | Wagger and Denton (1992) |
| Maize – grain | North Carolina | 5 | Continuous maize | Hapludult | — | 797 | — | 845 | Wagger and Denton (1992) |
| Maize – grain | North Carolina | 4 | Maize–soybean | Hapludult | — | 628 | — | 690 | Wagger and Denton (1992) |
| Maize – straw | New York | 2 | Various cover and nitrogen | Hapludalf | 529 | — | — | 491 | Sarrantonio and Scott (1988) |
| Maize – silage | QC, Canada | 11 | Continuous maize silage | Haplaquept | 981 | 975 | 1029 | — | Angers et al. (1995) |
| Maize – silage | New York | 2 | Continuous maize | Haplaquept | 1473 | — | 1538 | — | Mataruka et al. (1993) |
| Maize – silage | Michigan | 2 | Alfalfa–maize: 0 g N m⁻² | Hapludalf | 518 | — | — | 510 | Rasse and Smucker (1999) |
| Maize – silage | Michigan | 2 | Alfalfa–maize: 12 g N m⁻² | Hapludalf | 876 | — | — | 704 | Rasse and Smucker (1999) |
| Plow vs. shallow tillage (Pr > F = 0.89) | | | | | 903 | 900 | — | — | n = 12 |
| Plow vs. ridge tillage (Pr > F = 0.53) | | | | | 1000 | — | 979 | — | n = 9 |
| Plow vs. no tillage (Pr > F = 0.78) | | | | | 744 | — | — | 748 | n = 27 |
| Ridge vs. no tillage (Pr > F = 0.79) | | | | | — | — | 924 | 917 | n = 6 |
| Shallow vs. no tillage (Pr > F = 0.43) | | | | | — | 798 | — | 817 | n = 14 |

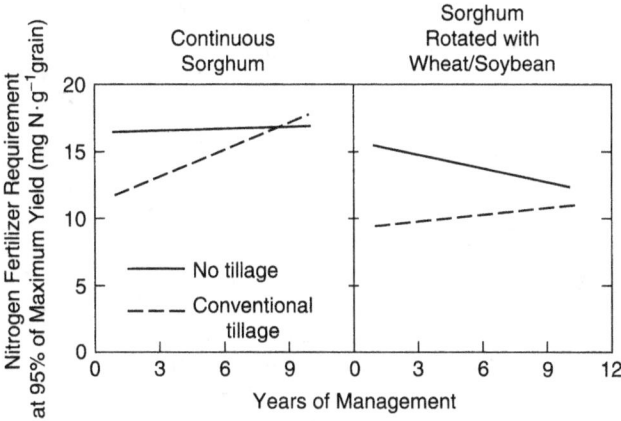

**FIGURE 8.6** Calculated N fertilizer requirement to achieve 95% of maximum sorghum grain yield each year during the first 10 years of a long-term tillage study in southcentral Texas. Yield response was derived from N fertilizer application rates of 0, 4.5, 9.0, and 13.5 g m$^{-2}$ year$^{-1}$. (Data from Franzluebbers, A.J. et al. 1995a. *Plant Soil* 173:55–65.)

The implication from this compilation of studies is that higher or equal crop yield under conservation tillage compared with inversion tillage will lead to, on average, higher or equal C inputs into the soil system.

## EFFECTS OF DISTURBANCE/TILLAGE ON SOIL ORGANIC MATTER

### DEPTH DISTRIBUTION OF ORGANIC MATTER

Inversion tillage mixes organic residues with soil at deeper depths. The type of tillage tool greatly influences the eventual location of aboveground residues within the soil profile. Allmaras et al. (1996) showed that moldboard plowing to a depth of 25 cm buried 70% of the aboveground oat residue at a depth of 12–24 cm, whereas chisel plowing to a depth of 15 cm left nearly 60% of the residue at a depth of 0–6 cm (Figure 8.7). Obviously, no tillage would leave nearly all of the residue at or above the soil surface. Because plant residues contribute greatly to subsequent soil organic matter formation, the placement of plant residues with different tillage practices is of utmost importance for understanding the depth distribution of soil organic matter.

With repeated inversion tillage, soil organic matter becomes uniformly distributed within the plowed layer (Figure 8.8). The fate of organic matter mixed into soil vs. that left on the soil surface depends on the prevailing climatic conditions. In general, however, the environment within soil is more buffered against extremes in moisture and temperature than that at the soil surface. Higher moisture content in soil than on the soil surface is probably the biggest factor that leads to greater decomposition of organic matter in tilled soil (Franzluebbers et al., 1996a).

Surface-placed crop residues under conservation tillage systems experience frequent drying and rewetting, depending on precipitation events. Although decomposition of surface-placed residues is slower than that of buried residues (Brown and Dickey, 1970; Douglas et al., 1980; Wilson and Hargrove, 1986; Ghidey and Alberts, 1993), N concentration of remaining residues can increase with time relative to buried residues (Varco et al., 1993; Franzluebbers et al., 1994c). Typically, higher N concentration residues leads to faster decomposition of residues (Vigil and Kissel, 1991). This contradiction suggests that frequent drying and rewetting of surface-placed residues increases the resistance of certain N compounds to microbial decomposition (Franzluebbers et al., 1994c), which leads to higher total N accumulation in the surface of no-tillage soils. However, during

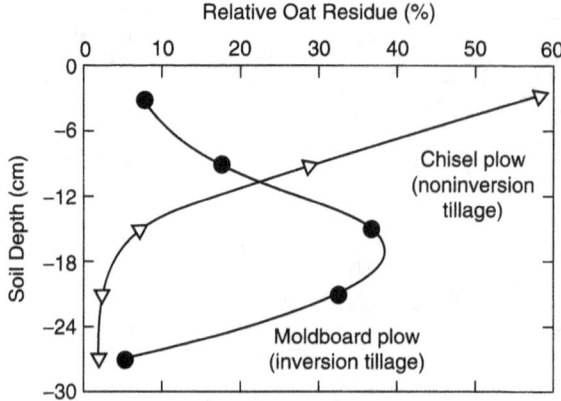

**FIGURE 8.7** Oat residue distribution in soil following moldboard plow and chisel plow in Minnesota. (Data from Allmaras, R.R. et al. 1996. *Soil Sci. Soc. Am. J.* 60:1209–1216.)

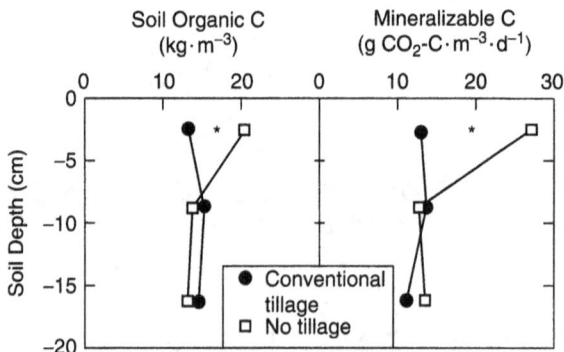

**FIGURE 8.8** Depth distribution of soil organic C and mineralizable C at the end of 9 years under conventional disk-and-bed tillage and no tillage in southcentral Texas. * indicates significance between tillage systems at $p < 0.1$. (Data from Franzluebbers, A.J. et al. 1994a. *Soil Sci. Soc. Am. J.* 58:1639–1645.)

decomposition of buried and surface-placed canola residues, the portion of total N remaining as lignin-bound N increased, but was not different between the two environments (Figure 8.9). More work is needed to understand the transformations that occur during decomposition of various crop residues under different micro- and macroclimatic conditions.

Organic matter has a direct impact on the density of soil and therefore on the content of organic matter within a given volume of soil. Because conservation tillage systems leave residues near the soil surface, most investigations report a substantial change in soil organic matter in surface soil as compared with inversion-tillage systems. However, calculation of net change in soil organic matter with a change in tillage management should be made to at least the depth of the tillage tool in both systems. At the end of 4 years of management in a Typic Kanhapludult in Georgia, soil organic C under no tillage was higher than under disk tillage (15-cm depth) at a depth of 0 to 2.5 cm, but not different at lower depths (Figure 8.10). C content was 81% higher (although C concentration was 95% higher) with no tillage than with disk tillage at a depth of 0 to 2.5 cm. Similarly, C content was only 2% lower with no tillage (C concentration was 14% lower) than with disk tillage at a depth of 2.5 to 7.5 cm. Summation of C content to a depth of 15 cm indicated no difference between tillage systems because of counteracting effects of residue placement at lower

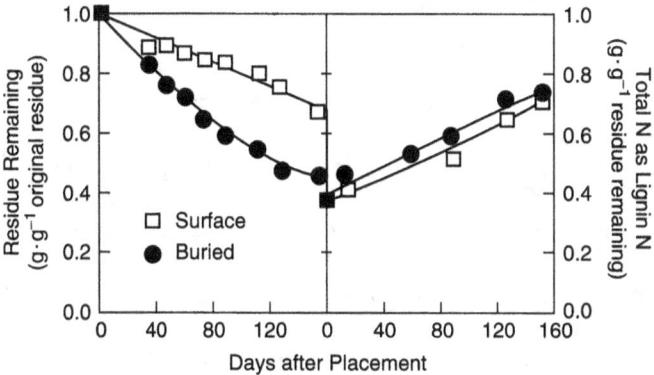

**FIGURE 8.9** Canola residue mass and the fraction of total N in remaining residue as lignin-bound N during field incubation in Alberta, Canada, when placed on the surface or buried at 10 cm. (Data from Franzluebbers, A.J., and Arshad, M.A. 1996a. *Soil Sci. Soc. Am. J.* 60:1422–1427.)

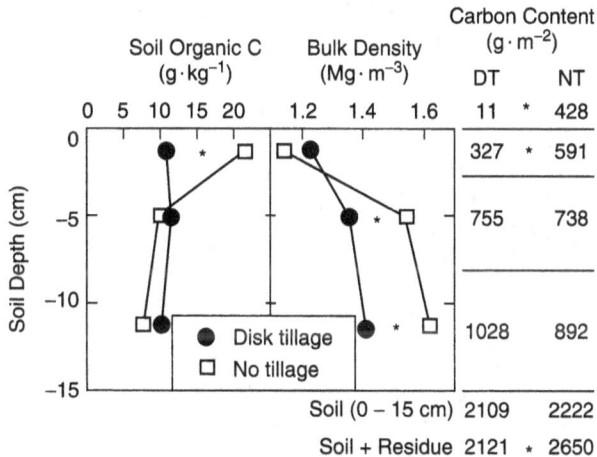

**FIGURE 8.10** Depth distribution of soil organic C concentration, soil bulk density, and C content at the end of 4 years under conventional disk tillage and no tillage in the Georgia Piedmont. * indicates significance between tillage systems at $p < 0.1$. (Data from Franzluebbers, A.J. et al. 1999. *Soil Sci. Soc. Am. J.* 63:349–355.)

depths with disk tillage. However, including surface residue C with soil organic C to a depth of 15 cm resulted in significantly greater storage of C under no tillage compared with disk tillage.

Stratification of soil organic matter pools with depth under conservation tillage systems has consequences on soil functions beyond that of potentially sequestering more C in soil. The soil surface is the vital interface that receives much of the fertilizers and pesticides applied to cropland, receives the intense impact of rainfall, and partitions the fluxes of gases and water into and out of soil. Surface organic matter is therefore essential to erosion control, water infiltration, and conservation of nutrients, all important soil functions. No-tillage management of a 2.7-ha cropped watershed for 24 years on a Typic Kanhapludult in Georgia reduced water runoff to 22 mm year⁻¹ compared with 180 mm year⁻¹ under previous management of the watershed under conventional inversion tillage (Endale et al., 2000). Soil loss was even more dramatically reduced with no-tillage management (3 vs. 129 kg ha⁻¹ mm⁻¹ runoff). A greenhouse study to separate the short- and long-term effects of disturbance on soil hydraulic properties of the same soil revealed that doubling soil organic C content in freshly tilled soil improved water infiltration by 27% (Franzluebbers, 2002b). However, water infiltration was 3.3 times higher in intact cores from long-term conservation tillage with a high degree of soil organic matter stratification compared with intact cores from a long-term conventionally tilled soil (but untilled during the previous 14 months) with a low degree of soil organic matter stratification.

Stratification of soil organic matter with conservation tillage depends on the inherent level of soil organic matter, intensity of disturbance, type of cropping system, and length of time. In an analysis of stratification ratios (soil organic C in the surface 5 cm divided by that at 12.5- to 20-cm depth) under no tillage in three different ecoregions, there were greater differences in the stratification of soil organic C between tillage systems in hot, wet, low soil organic matter environments than in cold, dry, high soil organic matter environments (Figure 8.11). Soils with low inherent levels of organic matter can be the most functionally improved with conservation tillage, despite modest or no change in total standing stock of soil organic C within the rooting zone. Alternatively, soils with inherently high soil organic matter even under conventional-tillage management would likely obtain relatively little additional soil functional benefit with adoption of conservation tillage, because inherent soil properties would be at a high level.

Stratification ratio of particulate organic C in a Typic Kanhapludult in Georgia decreased along a disturbance gradient created by tillage tools with different inversion characteristics (Figure 8.12). Less intensive mixing of soil preserves crop residues and soil organic matter near the soil surface, where it has the most beneficial impact. Stratification of mineralizable C in a Fluventic Ustochrept in Texas increased with increasing cropping intensity under conventional tillage, but was always higher under no tillage (Figure 8.13). More intensive cropping increases the quantity of residues produced, which can lead to higher soil organic matter. Stratification ratio of soil organic C (0- to 2.5-cm divided by 12.5- to 20-cm depth) in an Aquic Hapludult in Maryland was 1.0 under plow tillage and increased with time under no tillage to 1.1 at 1 year, 1.4 at 2 years, and 1.5 at 3 years (McCarty et al., 1998).

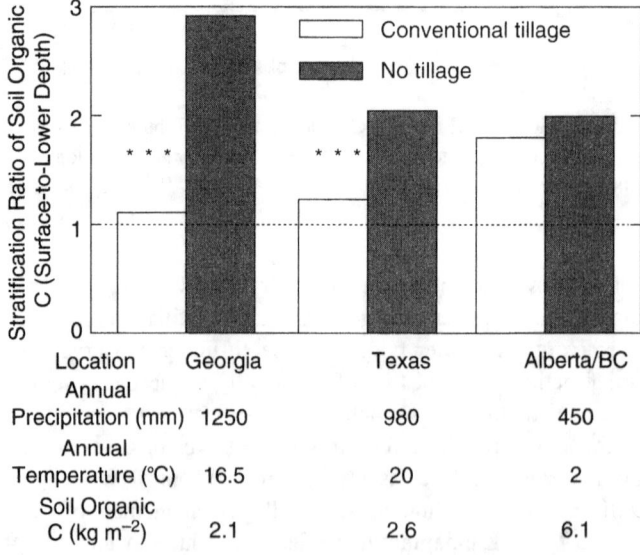

**FIGURE 8.11** Stratification ratio of soil organic C under conventional and no tillage at three locations differing in climatic characteristics and standing stock of soil organic C. *** indicates significance between tillage systems at $p < 0.001$. (Data from Franzluebbers, A.J. 2002a. *Soil Tillage Res.* 66:95–106.)

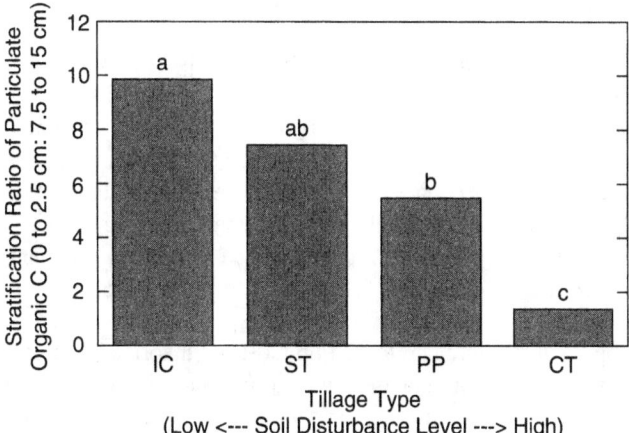

**FIGURE 8.12** Stratification ratio of particulate organic C at the end of 4 years under four tillage systems in the Georgia Piedmont. IC, in-row chisel at planting; ST, shallow tillage with sweeps during the growing season; PP, paraplow following harvest; and CT, conventional disk tillage. Bars labeled with different letters are significantly different at $p < 0.1$. (Data from Franzluebbers, A.J. 2002a. *Soil Tillage Res.* 66:95–106.)

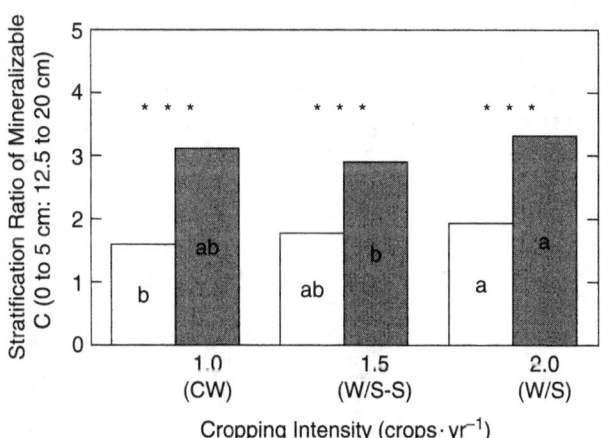

**FIGURE 8.13** Stratification ratio of mineralizable C at the end of 9 years under conventional (open bars) and no tillage (shaded bars) in three wheat rotation systems in southcentral Texas. CW, continuous wheat; W/S–S, wheat/soybean–sorghum; and W/S, continuous wheat/soybean double crop. *** indicates significance between tillage systems at $p < 0.001$. Within a tillage system, bars labeled with different letters are significantly different at $p < 0.1$. (Data from Franzluebbers, A.J. 2002a. *Soil Tillage Res.* 66:95–106.)

Stratification ratios of soil organic matter pools can be good indicators of soil quality, because surface soil properties are responsive to management, inherent levels of soil organic C are normalized in the calculation, and high stratification ratios are uncommon under degraded conditions (Franzluebbers, 2002a).

The effect of tillage/disturbance on soil organic matter is not equal among the components of organic matter. The following sections describe how tillage impacts total, particulate, and biologically active fractions of organic C and N.

## AGGREGATE-SIZE DISTRIBUTION OF ORGANIC MATTER

Organic matter is not only stratified with depth but can also be stratified three-dimensionally according to soil aggregation. Soil disturbance results in a more uniform distribution of organic substrates within soil (Figure 8.14). Lack of soil disturbance leads to concentration of organic matter within soil macroaggregates, which protect and isolate soil organic matter from consumption by soil fauna and microorganisms (Beare et al., 1994a, 1994b; Franzluebbers and Arshad, 1997b; Six et al., 2000c). Soil disturbance with tillage breaks apart macroaggregates and allows organic matter, once protected from decomposition, to be exposed to new environments and communities of organisms. Mineralization of organic C following disruption of soil macroaggregates is rapid, suggesting that this organic matter is highly labile on exposure (Figure 8.15).

A hierarchical approach to aggregate formation has been theorized, such that macroaggregates (>0.25 mm) form as a result of root entanglement and polysaccharides produced by heterotrophic microorganisms decomposing particulate organic matter glue together microaggregates (0.05 to 0.25 mm; Tisdall and Oades, 1982). A compilation of studies from the literature report that water-stable macroaggregation of surface soil is higher under no-tillage compared with inversion-tillage systems (Table 8.4). Available data suggests that macroaggregates under no tillage have a slower turnover time than under conventional tillage because of less physical perturbation, resulting in macroaggregates under no tillage that are enriched in fine particulate organic matter, which is more resistant to decomposition (Six et al., 2000b).

No tillage often leads to an improvement in soil structure because of reduced mechanical disturbance and greater reliance on soil organisms that deposit enriched organic debris along permanent soil pores. However, the depth to which changes in soil aggregation occurs might be limited, at least in the first decade. From a set of four soils in northern Alberta and British Columbia, the fraction of soil as water-stable macroaggregates (>0.25 mm) was higher under no tillage than under conventional tillage to a depth of 12.5 cm, but not below this depth (Figure 8.16). Enrichment

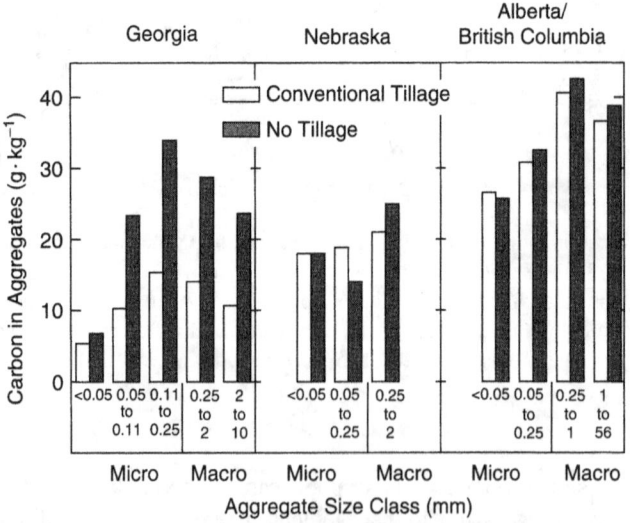

**FIGURE 8.14** Carbon concentration in water-stable aggregate fractions under conventional tillage and no tillage from Georgia, Nebraska, and Alberta/British Columbia, Canada. In general, soil organic C becomes enriched in macroaggregates (>0.25 mm) under no tillage. (Data for Georgia from Beare, M.H. et al. 1994b. *Soil Sci. Soc. Am. J.* 58:777–786; for Nebraska from Cambardella, C.A., and Elliott, E.T. 1993. *Soil Sci. Soc. Am. J.* 57:1071–1076; and for Alberta/British Columbia, Canada, from Franzluebbers, A.J., and Arshad, M.A., 1996c. *Can. J. Soil Sci.* 76:387–393.)

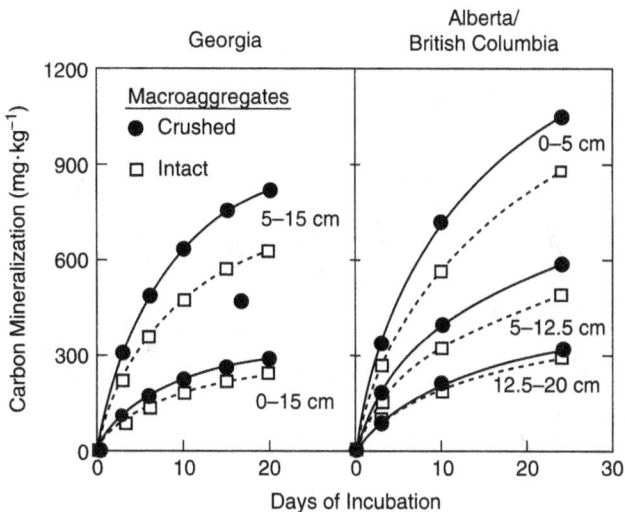

**FIGURE 8.15** Carbon mineralization during incubation of intact and crushed macroaggregates (>0.25 mm) from different soil depths in Georgia and in Alberta/British Columbia, Canada. Labile C protected within macroaggregates declines with soil depth. (Data for Georgia from Beare, M.H. et al. 1994b. *Soil Sci. Soc. Am. J.* 58:777–786; and for Alberta/British Columbia, Canada, from Franzluebbers, A.J., and Arshad, M.A. 1996c. *Can. J. Soil Sci.* 76:387–393.)

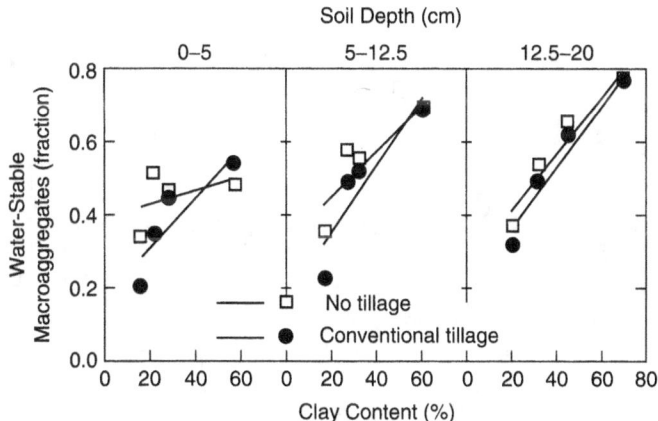

**FIGURE 8.16** Water-stable macroaggregates from three soil depths in four soils varying in soil texture under conventional and no tillage in Alberta/British Columbia, Canada. The positive effect of no tillage on macroaggregation was highest in coarse-textured soils and diminished with soil depth. (Data from Franzluebbers, A.J., and Arshad, M.A. 1996c. *Can. J. Soil Sci.* 76:387–393.)

of the soil surface with crop residues under no tillage led to significantly greater macroaggregation, especially in soils with coarse texture because their level of macroaggregation was lower than that in soils with fine texture. Fine-textured soils have a higher inherent level of macroaggregation even with soil disturbance because of the cohesive nature of highly reactive clays. This higher inherent level of aggregation can prevent further improvement with adoption of conservation tillage.

**TABLE 8.4**
**Comparison of Percent Water-Stable Macroaggregation among Tillage Systems**

| Soil Type | Texture | Location | Years | Depth (cm) | Plow | Shallow | Ridge | No. | Reference |
|---|---|---|---|---|---|---|---|---|---|
| Rhodic Kanhapludult | SCL | Georgia | 13 | 5 | — | 63 | — | 83 | Beare et al. (1994b) |
| Pachic Haplustoll | L | Nebraska | 20 | 20 | 23 | 29 | — | 39 | Cambardella and Elliott (1993) |
| Typic Cryoboralf | L | BC, Canada | 7 | 5 | — | 21 | — | 34 | Franzluebbers and Arshad (1997a) |
| Typic Cryoboralf | SiL | BC, Canada | 16 | 5 | — | 34 | — | 51 | Franzluebbers and Arshad (1997a) |
| Mollic Cryoboralf | CL | AB, Canada | 4 | 5 | — | 45 | — | 46 | Franzluebbers and Arshad (1997a) |
| Typic Natriboralf | C | AB, Canada | 6 | 5 | — | 54 | — | 48 | Franzluebbers and Arshad (1997a) |
| Typic Kanhapludult | SL | Georgia | 4 | 2.5 | — | 69 | — | 79 | Franzluebbers et al. (1999) |
| Pachic Haplustoll | L | Nebraska | 27 | 5 | 14 | — | — | 30 | Six et al. (2000a) |
| Typic Fragiudalf | SiL | Ohio | 34 | 5 | 15 | — | — | 56 | Six et al. (2000a) |
| Typic Hapludalf | SL | Michigan | 10 | 5 | 43 | — | — | 56 | Six et al. (2000a) |
| Typic Paleudalf | SiCL | Kentucky | 25 | 5 | 47 | — | — | 73 | Six et al. (2000a) |
| Vertisol | C | Qld, Australia | 8 | 10 | — | 25 | — | 32 | Hamblin (1980) |
| Alfisol | SL | NSW, Australia | 7 | 10 | — | 22 | — | 25 | Hamblin (1980) |
| Entisol | SL | WA, Australia | 5 | 10 | — | 9 | — | 17 | Hamblin (1980) |
| Spodosol | SL | Vic, Australia | 3 | 10 | — | 11 | — | 14 | Hamblin (1980) |
| Alfisol | SL | SA, Australia | 3 | 10 | — | 13 | — | 15 | Hamblin (1980) |
| Plow vs. no tillage (Pr > F = 0.01) | | | | | 28 | — | — | 51 | $n = 5$ |
| Shallow vs. no tillage (Pr > F = 0.01) | | | | | — | 33 | — | 40 | $n = 12$ |

## Total Organic C and N

Numerous reports are now available comparing the effect of conservation tillage with conventional inversion tillage on soil organic C and N. Although estimates of soil organic C and N were not always available at the initiation of long-term studies, relative changes in soil organic C and N between tillage systems can provide useful information on the fate of organic matter. Soil organic C in the Ap horizon (0- to 20-cm depth) of a Dark Brown Chernozemic clay loam in Alberta increased at 0.17 to 0.20 mg g$^{-1}$ soil year$^{-1}$ in two studies conducted for 9 and 19 years under no tillage compared with shallow disk tillage (Dormaar and Lindwall, 1989). In contrast, soil organic C at a depth of 0 to 7.5 cm during 4 years under no tillage compared with plowing increased at 0.69 mg g$^{-1}$ soil year$^{-1}$ on a Waukegon silt loam in Minnesota (Hansmeyer et al., 1997) and at ~1.15 mg g$^{-1}$ soil year$^{-1}$ on a Kamouraska clay in Quebec (Angers et al., 1993a). Incorporation of residues below 7.5 cm with plowing would likely reduce this effect when considering the entire plow depth. Soil organic C accumulation rates between these extremes have also been observed. At a depth of 0 to 5 cm, soil organic C increased at 0.42 mg g$^{-1}$ soil year$^{-1}$ during 14 years under no tillage compared with multiple-disk tillage on a Norfolk loamy sand in the South Carolina Coastal Plain (Hunt et al., 1996) and at 0.28 to 0.42 mg g$^{-1}$ soil year$^{-1}$ during more than 20 years under no tillage compared with plowing on a Bertie silt loam in the Maryland Coastal Plain (McCarty and Meisinger, 1997). On a Hoytville silty clay loam in Ohio, soil organic C of the 0- to 10-cm depth increased at 0.66 mg g$^{-1}$ soil year$^{-1}$ during 12 years under no tillage compared with plowing (Lal et al., 1990). The large range of changes in soil organic C with no tillage compared with inversion tillage among the aforementioned studies can be related to differences in cropping system, fertilization, depth of tillage tool, numerous soil characteristics, climatic conditions, and depth of sampling.

In general, compilation studies looking at the effect of conservation tillage on soil organic matter indicate that soil under long-term no tillage accumulates organic C to a greater extent than under inversion tillage (Kern and Johnson, 1991; Rasmussen and Collins, 1991; Reicosky et al., 1995; Paustian et al., 1997; Lal et al., 1998). The magnitude of difference between no tillage and conventional tillage can be as high as 2 kg m$^{-2}$ (Dick et al., 1998), but more typical differences center around 30 g m$^{-2}$ year$^{-1}$ (Figure 8.17). There are a number of cases where the total stock of soil organic C and N in the upper 20 to 30 cm does not change with adoption of conservation tillage compared with conventional tillage (Carter and Rennie, 1982; Franzluebbers and Arshad, 1996c; Angers et al., 1997; Wander et al., 1998). Although the C and N content in surface residues are not always accounted in agricultural systems, this trash component at the soil surface can be significant (Figure 8.10).

Climatic factors, such as precipitation and temperature, appear to exert a great deal of control on the potential of conservation-tillage systems to sequester more soil organic C compared with conventional-tillage systems (Franzluebbers and Steiner, 2002). Potential soil organic C storage with no tillage compared with conventional tillage in North America was highest (~58 g m$^{-2}$ year$^{-1}$) in mesic, subhumid regions with mean annual precipitation-to-potential evapotranspiration ratios of 1.4 to 1.6 mm mm$^{-1}$ (Figure 8.18). Tillage comparisons in more extreme climates have often produced estimates of potential soil organic C storage with no tillage that are no different or less than those from under conventional tillage. For example, in the cold, semiarid climate in northern Alberta, soil organic C was not different between tillage systems in three of four soils (Franzluebbers and Arshad, 1996b). No tillage generally conserves surface soil moisture compared with conventional tillage. Shallow tillage in this semiarid environment incorporates residues near the soil surface, which dries frequently and more rapidly than under no tillage. Because soil is frozen for nearly 5 months of the year, with the remaining time devoted to crop production and utilization of available water, there are limited opportunities for decomposition to occur under either tillage system, resulting in little change in potential soil organic C storage with tillage.

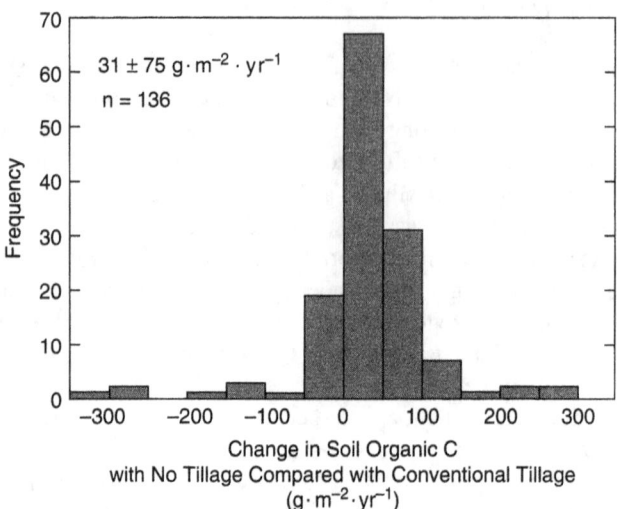

**FIGURE 8.17** Frequency distribution of 136 observations from North America on the change in soil organic C with no tillage compared with conventional tillage. Rate in upper left corner is mean and standard deviation from 136 observations. (Updated from Franzluebbers, A.J., and Steiner, J.L. 2002. In Kimble, J.M., Lal, R., and Follett, R.F. (Eds.), *Agriculture Practices and Policies for Carbon Sequestration in Soil*. CRC Press, Boca Raton, FL, pp. 71–86, with data compiled from Angers, D.A. et al. 1997. *Soil Tillage Res.* 41:191–201; Angers, D.A. et al. 1994. In *Proceedings of the 13th International Soil Tillage Research Organization*, Denmark, pp. 49–54; Beare, M.H. et al. 1994b. *Soil Sci. Soc. Am. J.* 58:777–786; Black, A.L., and Tanaka, D.L. 1997. In Paul, E.A., Paustian, K., Elliott, E.T., and Cole, C.V. (Eds.), *Soil Organic Matter in Temperate Agroecosystems*. CRC Press, Boca Raton, FL, pp. 335–342; Blevins, R.L. et al. 1977. *Agron. J.* 69:383–386; Cambardella, C.A., and Elliott, E.T. 1992. *Soil Sci. Soc. Am. J.* 56:777–783; Campbell, C.A. et al. 1995. *Can. J. Soil Sci.* 75:449–458; Campbell, C.A. et al. 1996. *Soil Tillage Res.* 37:3–14; Carter, M.R., and Rennie, D.A. 1982. *Can. J. Soil Sci.* 62:587–597; Carter, M.R. et al. 1988. *Soil Tillage Res.* 12:365–384; Carter, M.R. et al. 2002. *Soil Tillage Res.* 67:85–98; Clapp, C.E. et al. 2000. *Soil Tillage Res.* 55:127–142; Dick, W.A. et al. 1998. *Soil Tillage Res.* 47:235–244; Duiker, S.W., and Lal, R. 1999. *Soil Tillage Res.* 52:73–81; Edwards, J.H. et al. 1992. *Soil Sci. Soc. Am. J.* 56:1577–1582; Eghball, B. et al.1994. *J. Soil Water Conserv.* 49:201–205; Follett, R.F., and Peterson, G.A. 1988. *Soil Sci. Soc. Am. J.* 52:141–147; Franzluebbers, A.J., and Arshad, M.A. 1996c. *Can. J. Soil Sci.* 76:387–393; Franzluebbers, A.J. et al. 1994a. *Soil Sci. Soc. Am. J.* 58:1639–1645; Franzluebbers, A.J. et al. 1995b. *Soil Sci. Soc. Am. J.* 59:460–466; Franzluebbers, A.J. et al. 1998. *Soil Tillage Res.* 47:303–308; Franzluebbers, A.J. et al. 1999. *Soil Sci. Soc. Am. J.* 63:349–355; Halvorson, A.D. et al. 1997. In Paul, E.A., Paustian, K., Elliott, E.T., and Cole, C.V. (Eds.), *Soil Organic Matter in Temperate Agroecosystems*. CRC Press, Boca Raton, FL, pp. 361–370; Hendrix, P.F. et al. 1998. *Soil Tillage Res.* 47:245–251; Ismail, I. et al. *Soil Sci. Soc. Am. J.* 58:193–198; Karlen, D.L. et al. 1998. *Soil Tillage Res.* 48:155–165; Karlen, D.L. et al. 1994. *Soil Tillage Res.* 32:313–327; Lal, R. et al. 1994. *Soil Sci. Soc. Am. J.* 58:517–522; Lamb, J.A. et al. 1985. *Soil Sci. Soc. Am. J.* 49:352–356; Larney, F.J. et al. 1997. *Soil Tillage Res.* 42:229–240; McCarty, G.W. et al. 1998. *Soil Sci. Soc. Am. J.* 62:1564–1571; Mielke, L.N. et al. 1986. *Soil Tillage Res.* 5:355–366; Nyborg, M. et al. 1995. In Lal, R., Kimble, J., Levine, E., and Stewart, B.A. (Eds.), *Soil Management and Greenhouse Effect*. Lewis Publishers, CRC Press, Boca Raton, FL, pp. 93–99; Peterson, G.A. et al. 1998. *Soil Tillage Res.* 47:207–218; Pierce, F.J. et al. 1994. *Soil Sci. Soc. Am. J.* 58:1782–1787; Pikul, J.L., Jr., and Aase, J.K. 1995. *Agron. J.* 87:656–662; Potter, K.N. et al. 1997. *Soil Sci.* 162:140–147; Potter, K.N. et al. 1998. *Soil Tillage Res.* 47:309–321; Rhoton, F.E. et al. 2002. *Soil Tillage Res.* 66:1–11; Sainju, U.M. et al. 2002. *Soil Tillage Res.* 63:167–179; Salinas-Garcia, J.R. et al. 1997b. *Soil Tillage Res.* 42:79–93; Schomberg, H.H., and Jones, O.R. 1999. *Soil Sci. Soc. Am. J.* 63:1359–1366; Six, J. et al. 2000c. *Soil Sci. Soc. Am. J.* 64:681–689; Wander, M.M. et al. 1998. *Soil Sci. Soc. Am. J.* 62:1704–1711; Wanniarachchi, S.D. et al. A.F. 1999. *Can. J. Soil Sci.* 79:473–480; Yang, X.M., and Wander, M.M. 1999. *Soil Tillage Res.* 52:1–9; and Yang, X.M., and Kay, B.D. 2001. *Soil Tillage Res.* 59:107–114.)

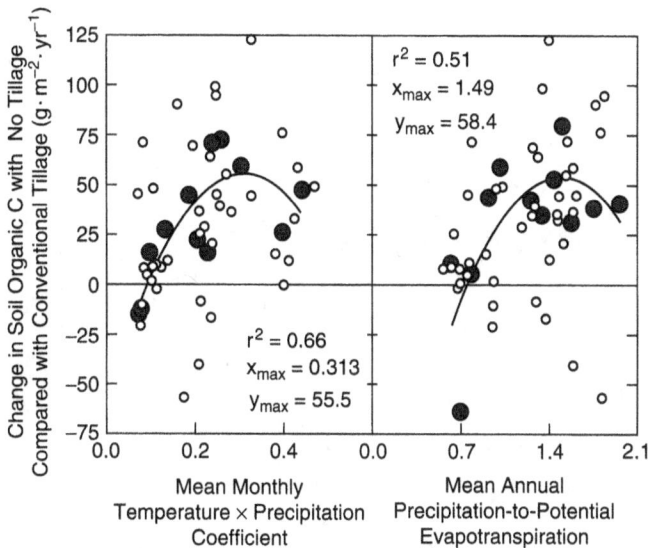

**FIGURE 8.18** Change in soil organic C with no tillage compared with conventional tillage in North America as a function of macroclimatic indices. Mean monthly temperature × precipitation coefficient was composed of a temperature coefficient (0 to 1; logarithmic relationship with maximum at 30°C) and a precipitation coefficient (0 to 1; linear-plateau relationship with maximum at 100 mm month⁻¹). Potential evapotranspiration was calculated by the Thornthwaite procedure. Small circles represent individual sites and large circles represent means of four consecutive sites in ranked climatic order. Regression parameters are based on means. (Updated from Franzluebbers, A.J., and Steiner, J.L. 2002. In Kimble, J.M., Lal, R., and Follett, R.F., Eds., *Agriculture Practices and Policies for Carbon Sequestration in Soil.* CRC Press, Boca Raton, FL, pp. 71–86. With permission.)

According to several published reports, the effect of tillage management on soil organic N content in the rooting zone suggests that no tillage leads to significantly higher soil organic N content than either plow or shallow tillage do (Table 8.5). Calculated on a yearly basis, soil organic N was $2.3 \pm 6.7$ g m⁻² year⁻¹ higher under no tillage than under plow tillage ($n = 24$). Soil organic N storage with no tillage compared with shallow tillage was $2.8 \pm 7.0$ g m⁻² year⁻¹ ($n = 26$). Although the mean change with adoption of no tillage compared with conventional tillage was positive among these studies, there was a great deal of variation. This variation suggests that much more work is needed to understand the mechanisms behind these differences. Detailed temporal analyses within several long-term studies would help separate random sampling variation from biogeochemical controls, including climate, mineralogy, soil texture, cropping system, and fertilization regime.

## PARTICULATE FRACTION OF ORGANIC MATTER

Particulate organic matter is defined as that portion of organic matter retained on a 50-μm screen following complete dispersion of soil. Particulate organic matter is considered to represent the slow pool of organic matter (Cambardella and Elliott, 1992), with an intermediate turnover time between the active and passive pools of organic matter (Parton et al., 1987). Particulate organic matter is derived from above- and belowground inputs of plant residues. Particulate organic C is often greater near the soil surface than at lower depths because of the dominant input from crop residues (Figure 8.19). Surface residue retention with no tillage can lead to higher particulate organic C near the soil surface than with inversion tillage systems (Figure 8.19). According to a compilation of studies in the literature, particulate organic C under no tillage is greater than under either plow or shallow tillage (Table 8.6). The effect of tillage system on particulate organic N content in the surface 15

**TABLE 8.5**
**Comparison of Total Soil N (g m$^{-2}$) among Tillage Systems**

| Soil Type | Texture | Location | Years | Depth (cm) | Plow | Shallow | Ridge | No. | Reference |
|---|---|---|---|---|---|---|---|---|---|
| Haplorthod | fSL | PEI, Canada | 8 | 60 | 688 | — | — | 607 | Angers et al. (1997) |
| Cryoboralf | L | PEI, Canada | 8 | 60 | 366 | 359 | — | — | Angers et al. (1997) |
| Humaquept | C | QC, Canada | 6 | 60 | 756 | — | — | 675 | Angers et al. (1997) |
| Humaquept | CL | QC, Canada | 4 | 60 | 520 | 483 | — | — | Angers et al. (1997) |
| Humaquept | SiC | QC, Canada | 3 | 60 | 815 | 715 | — | 823 | Angers et al. (1997) |
| Eutrochrept | SL | ON, Canada | 5 | 60 | 592 | — | — | 568 | Angers et al. (1997) |
| Haplaquoll | CL | ON, Canada | 11 | 60 | 1008 | — | — | 974 | Angers et al. (1997) |
| Aeric Haplaquept | SiL | QC, Canada | 11 | 24 | 472 | 488 | 472 | — | Angers et al. (1993b) |
| Paleudult | SCL | Brazil | 9 | 30 | 402 | — | — | 424 | Bayer et al. (2000b) |
| Rhodic Kanhapludult | SCL | Georgia | 13 | 15 | — | 235 | — | 259 | Beare et al. (1994b) |
| Typic Argiboroll | SiL | North Dakota | 6 | 30 | 570 | 566 | — | 565 | Black and Tanaka (1997) |
| Pachic Haplustoll | L | Nebraska | 20 | 20. | — | 325 | — | 326 | Cambardella and Elliott (1992) |
| Typic Haploboroll | SiL | SK, Canada | 4 | 15 | — | 304 | — | 313 | Campbell et al. (1999) |
| Typic Haploboroll | SiL | SK, Canada | 8 | 15 | — | 308 | — | 314 | Campbell et al. (1999) |
| Typic Haploboroll | SiL | SK, Canada | 12 | 15 | — | 298 | — | 312 | Campbell et al. (1999) |
| Typic Haploboroll | fSL | SK, Canada | 3 | 15 | — | 209 | — | 214 | Campbell et al. (1996) |
| Typic Haploboroll | fSL | SK, Canada | 7 | 15 | — | 187 | — | 194 | Campbell et al. (1996) |
| Typic Haploboroll | fSL | SK, Canada | 11 | 15 | — | 218 | — | 226 | Campbell et al. (1996) |
| Udic Pellustert | C | Qld, Australia | 13 | 30 | — | 429 | — | 461 | Dalal (1989) |
| Udic Pellustert | C | Qld, Australia | 20 | 10 | — | 141 | — | 141 | Dalal et al. (1991) |
| Typic Paleudalf | SiL | Kentucky | 11 | 15 | 377 | — | — | 479 | Doran (1987) |
| Aeric Ochraqualf | SiL | Illinois | 6 | 15 | 303 | — | — | 358 | Doran (1987) |
| Aquic Hapludoll | CL | Minnesota | 6 | 15 | 622 | — | — | 631 | Doran (1987) |
| Pachic Argiustoll | SiCL | Nebraska | 6 | 15 | 338 | — | — | 462 | Doran (1987) |
| Pachic Haplustoll | L | Nebraska | 13 | 15 | 259 | 277 | — | 296 | Doran (1987) |
| Aridic Argiustoll | SiL | Nebraska | 12 | 15 | 369 | 387 | — | 370 | Doran (1987) |
| Fluventic Ustochrept | SiCL | Texas | 9 | 20 | — | 258 | — | 333 | Franzluebbers et al. (1994a) |
| Typic Kanhapludult | SL | Georgia | 4 | 15 | — | 159 | — | 155 | Franzluebbers et al. (1999) |
| Aridic Paleustoll | SiL | Colorado | 15 | 20 | — | 216 | — | 223 | Halvorson et al. (1997) |

| Soil Type | Texture | Location | Years | Depth (cm) | Plow | Shallow | Ridge | No. | Reference |
|---|---|---|---|---|---|---|---|---|---|
| Typic Paleudalf | SiL | Kentucky | 20 | 30 | 621 | — | — | 645 | Ismail et al. (1994) |
| Typic Hapludalf | SiL | Wisconsin | 25 | 12 | 535 | 625 | — | 727 | Karlen et al. (1994) |
| Typic Haploboroll | SCL | AB, Canada | 16 | 15 | — | 298 | — | 303 | Larney et al. (1997) |
| Typic Haploboroll | SCL | AB, Canada | 8 | 15 | — | 325 | — | 343 | Larney et al. (1997) |
| Typic Haploboroll | SCL | AB, Canada | 9 | 15 | — | 374 | — | 376 | Larney et al. (1997) |
| Typic Hapludalf | SiL | Germany | 9 | 30 | 426 | 442 | — | — | Meyer et al. (1996) |
| Vertic Calcixeroll | C | Morocco | 11 | 20 | — | 328 | — | 352 | Mrabet et al. (2001) |
| Typic Cryoboralf | L | AB, Canada | 11 | 15 | — | 276 | — | 326 | Nyborg et al. (1995) |
| Typic Cryoboralf | L | AB, Canada | 11 | 15 | — | 776 | — | 785 | Nyborg et al. (1995) |
| Aeric Ochraqualf | L | Michigan | 11 | 20 | 333 | — | — | 379 | Pierce et al. (1994) |
| Typic Ochraqualf | SCL | Texas | 16 | 20 | 134 | 158 | — | 200 | Salinas-Garcia et al. (1997b) |
| Torrertic Paleustoll | CL | Texas | 8 | 12 | — | 88 | — | 98 | Schomberg and Jones (1999) |
| Pachic Haplustoll | L | Nebraska | 27 | 20 | 312 | — | — | 366 | Six et al. (2000c) |
| Typic Fragiudalf | SiL | Ohio | 34 | 20 | 294 | — | — | 346 | Six et al. (2000c) |
| Typic Hapludalf | SL | Michigan | 10 | 20 | 204 | — | — | 218 | Six et al. (2000c) |
| Typic Paleudalf | SiCL | Kentucky | 25 | 20 | 348 | — | — | 441 | Six et al. (2000c) |
| Typic Hapludalf | SiL | Germany | 20 | 30 | 480 | 490 | — | — | Stockfisch et al. (1999) |
| Aquic Argiudoll | SiL | Illinois | 10 | 30 | 466 | — | — | 532 | Wander et al. (1998) |
| Aquic Hapludoll | SiL | Illinois | 10 | 30 | 844 | — | — | 835 | Wander et al. (1998) |
| Typic Haplaquoll | SiCL | Illinois | 10 | 30 | 997 | — | — | 1066 | Wander et al. (1998) |
| 36 (Argiudoll-Argiaquoll) | SiCL | Illinois | >5 | 15 | 423 | — | — | 442 | Wander and Bollero (1999) |
| Plow vs. shallow tillage (Pr > F = 0.78) | | | | | 450 | 454 | — | | $n = 11$ |
| Plow vs. no tillage (Pr > F = 0.01) | | | | | 504 | — | — | 537 | $n = 25$ |
| Shallow vs. no tillage (Pr > F < 0.01) | | | | | — | 326 | — | 348 | $n = 26$ |

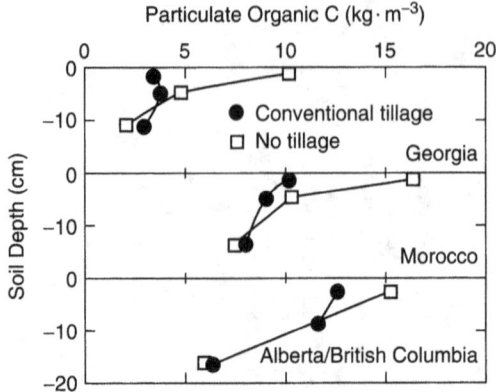

**FIGURE 8.19** Depth distribution of particulate organic C under conventional and no tillage in Georgia, near Settat Morocco, and in Alberta/British Columbia, Canada. (Data for Georgia from Franzluebbers, A.J. et al. 1999. *Soil Sci. Soc. Am. J.* 63:349–355; for Settat Morocco from Mrabet, R. et al. 2001. *Soil Tillage Res.* 57:225–235; and for Alberta/British Columbia, Canada from Franzluebbers, A.J., and Arshad, M.A. 1997a. *Soil Sci. Soc. Am. J.* 61:1382–1386.)

to 30 cm is less clear (Table 8.7). Paired *t*-tests of the effects of no tillage compared with other tillage systems on particulate organic N were not significant, although the trend was for numerically higher values under no-tillage compared with inversion-tillage systems, similar to that found for particulate organic C.

The decomposability of particulate organic matter does not appear to be affected by tillage management. In a set of four Cryoboralfs and Natriboralfs from Alberta and British Columbia, specific mineralization of particulate organic C was similar between shallow tillage and no tillage (Franzluebbers and Arshad, 1997a). However, the ratio of specific particulate organic C mineralization to specific whole-soil organic C mineralization was higher under no tillage (1.3) than under shallow tillage (1.0), suggesting that particulate organic C was of better quality (i.e., more mineralizable) under no tillage relative to other pools of soil organic C. Mineralizable whole-soil C under no tillage was significantly lower than under shallow tillage in two of the four soils (Franzluebbers and Arshad, 1996a, 1996b).

## DENSITY FRACTIONS OF ORGANIC MATTER

Soil organic matter can be separated by density to distinguish fractions along a decomposition gradient. Lightest fractions of organic matter represent recently deposited organic residues. Heaviest fractions of organic matter represent highly decomposed organic residues that become associated with mineral particles. Separation of light fractions from heavy fractions is typically in an unreactive salt solution with a density of 1.6 to 1.8 Mg m$^{-3}$. Light fractions float to the surface, whereas heavy fractions sink with sediment. Tillage effects on density fractions of organic matter have not been extensively investigated. Although not different between tillage systems to a depth of 20 cm, light-fraction C (<1.6 Mg m$^{-3}$) and medium-fraction C (1.6 to 2.0 Mg m$^{-3}$) were higher under no tillage than under plow tillage at a depth of 0–5 cm in a Typic Argiudoll in Argentina, but lower at a depth of 5–15 cm (Alvarez et al., 1998). Heavy-fraction C was unaffected by tillage management at all soil depths. In the surface 15 cm of Typic Haploborolls in Alberta, light-fraction C under no tillage for 8 to 16 years averaged 242 g m$^{-2}$ and 226 g m$^{-2}$ under shallow tillage (Larney et al., 1997). At a depth of 0 to 15 cm, light-fraction C was lower under no tillage than under blade cultivation in Alberta (Dormaar and Lindwall, 1989). At the end of 5 years on a Udic Dystrochrept in New Zealand, light-fraction C was not different at a depth of 0 to 5 cm under ryegrass that was either plowed or direct drilled annually (Haynes, 1999). Light-fraction C content to a depth of 20 cm was not statistically different in a Pachic Haplustoll under plow tillage (174 g m$^{-2}$) and no tillage (196 g m$^{-2}$) for 27 years (Six et al., 1998).

**TABLE 8.6**

**Comparison of Particulate Organic C (kg m$^{-2}$) among Tillage Systems**

| Soil Type | Texture | Location | Years | Depth (cm) | Plow | Shallow | Ridge | No. | Reference |
|---|---|---|---|---|---|---|---|---|---|
| Aeric Haplaquept | SiL | QC, Canada | 10 | 24 | 0.56 | 0.84 | 0.99 | — | Angers et al. (1994) |
| Aeric Haplaquept | SiL | QC, Canada | 11 | 8 | 0.38 | 0.46 | 0.43 | — | Angers et al. (1995) |
| Rhodic Kanhapludult | SCL | Georgia | 13 | 15 | — | 1.00 | — | 1.24 | Beare et al. (1994b) |
| Pachic Haplustoll | L | Nebraska | 20 | 20 | 0.56 | 0.68 | — | 0.93 | Cambardella and Elliott (1992) |
| Typic Cryoboralf | L | BC, Canada | 7 | 20 | — | 1.78 | — | 1.71 | Franzluebbers and Arshad (1997a) |
| Typic Cryoboralf | SiL | BC, Canada | 16 | 20 | — | 2.28 | — | 2.20 | Franzluebbers and Arshad (1997a) |
| Mollic Cryoboralf | CL | AB, Canada | 4 | 20 | — | 2.54 | — | 2.80 | Franzluebbers and Arshad (1997a) |
| Typic Natriboralf | C | AB, Canada | 6 | 20 | — | 1.29 | — | 1.58 | Franzluebbers and Arshad (1997a) |
| Typic Kanhapludult | SL | Georgia | 4 | 15 | — | 0.51 | — | 0.65 | Franzluebbers et al. (1999) |
| Typic Arigiboroll | SiL | North Dakota | 12 | 20a | — | 0.43 | — | 0.36 | Frey et al. (1999) |
| Pachic Haplustoll | L | Nebraska | 26 | 20a | 0.35 | — | — | 0.48 | Frey et al. (1999) |
| Aridic Argiustoll | L | Colorado | 11 | 20a | — | 0.38 | — | 0.30 | Frey et al. (1999) |
| Torrertic Paleustoll | CL | Texas | 15 | 20a | — | 0.18 | — | 0.16 | Frey et al. (1999) |
| Cumulic Haplustoll | SiL | Kansas | 22 | 20a | — | 0.36 | — | 0.41 | Frey et al. (1999) |
| Typic Paleudalf | SiCL | Kentucky | 26 | 20a | 0.44 | — | — | 0.52 | Frey et al. (1999) |
| Oxyaquic Fragiudalf | SiL | Illinois | 8 | 15a | 0.69 | 0.75 | — | 0.83 | Hussain et al. (1999) |
| Vertic Calcixeroll | C | Morocco | 11 | 20 | — | 1.70 | — | 1.84 | Mrabet et al. (2001) |
| 36 (Argiudoll-Argiaquoll) | SiCL | Illinois | >5 | 15a | 0.62 | — | — | 0.59 | Needelman et al. (1999) |
| Pachic Haplustoll | L | Nebraska | 27 | 20 | 0.37 | — | — | 0.53 | Six et al. (1999) |
| Typic Hapludalf | SL | Michigan | 10 | 20 | 0.47 | — | — | 0.45 | Six et al. (1999) |
| Typic Fragiudalf | SiL | Ohio | 34 | 20 | 0.52 | — | — | 0.79 | Six et al. (1999) |
| Typic Paleudalf | SiCL | Kentucky | 25 | 20 | 0.42 | — | — | 0.55 | Six et al. (1999) |
| Aquic Argiudoll | SiL | Illinois | 10 | 30 | 0.66 | — | — | 0.71 | Wander et al. (1998) |
| Aquic Hapludoll | SiL | Illinois | 10 | 30 | 1.21 | — | — | 1.45 | Wander et al. (1998) |
| Typic Haplaquoll | SiCL | Illinois | 10 | 30 | 2.61 | — | — | 2.44 | Wander et al. (1998) |
| 36 (Argiudoll-Argiaquoll) | SiCL | Illinois | >5 | 15b | 0.70 | — | — | 0.73 | Wander and Bollero (1999) |
| Plow vs. shallow tillage (Pr > F = 0.07) | | | | | 0.55 | 0.68 | — | — | n = 4 |
| Plow vs. ridge tillage (Pr > F = 0.43) | | | | | 0.47 | — | 0.71 | — | n = 2 |
| Plow vs. no tillage (Pr > F = 0.02) | | | | | 0.74 | — | — | 0.85 | n = 13 |
| Shallow vs. no tillage (Pr > F = 0.05) | | | | | — | 1.07 | — | 1.15 | n = 13 |

a Assumed a bulk density of 1.2 Mg m$^{-3}$ for soil at a depth of 0–5 cm and 1.4 Mg m$^{-3}$ for soil at a depth of either 5–15 or 5–20 cm.

b Assumed a bulk density of 1.33 Mg m$^{-3}$ for soil at a depth of 0–15 cm.

TABLE 8.7
Comparison of Particulate Organic N (g m$^{-2}$) among Tillage Systems

| Soil Type | Texture | Location | Years | Depth (cm) | Plow | Shallow | Ridge | No. | Reference |
|---|---|---|---|---|---|---|---|---|---|
| Rhodic Kanhapludult | SCL | Georgia | 13 | 15 | — | 75 | — | 99 | Beare et al. (1994b) |
| Pachic Haplustoll | L | Nebraska | 20 | 20 | 48 | 53 | — | 40 | Cambardella and Elliott (1992) |
| Typic Arigiboroll | SiL | North Dakota | 12 | 20[a] | — | 28 | — | 28 | Frey et al. (1999) |
| Pachic Haplustoll | L | Nebraska | 26 | 20[a] | 28 | — | — | 48 | Frey et al. (1999) |
| Aridic Argiustoll | L | Colorado | 11 | 20[a] | — | 26 | — | 22 | Frey et al. (1999) |
| Torrertic Paleustoll | CL | Texas | 15 | 20[a] | — | 14 | — | 14 | Frey et al. (1999) |
| Cumulic Haplustoll | SiL | Kansas | 22 | 20[a] | — | 27 | — | 32 | Frey et al. (1999) |
| Oxyaquic Fragiudalf | SiL | Illinois | 8 | 15[a] | 50 | 56 | — | 65 | Hussain et al. (1999) |
| Typic Paleudalf | SiCL | Kentucky | 26 | 20[a] | 33 | — | — | 36 | Frey et al. (1999) |
| Aquic Argiudoll | SiL | Illinois | 10 | 30 | 57 | — | — | 65 | Wander et al. (1998) |
| Aquic Hapludoll | SiL | Illinois | 10 | 30 | 94 | — | — | 123 | Wander et al. (1998) |
| Typic Haplaquoll | SiCL | Illinois | 10 | 30 | 235 | — | — | 217 | Wander et al. (1998) |
| 36 (Argiudoll-Argiaquoll) | SiCL | Illinois | >5 | 15[b] | 46 | — | — | 52 | Wander and Bollero (1999) |
| Plow vs. shallow tillage (Pr > F = 0.06) | | | | | 49 | 55 | — | — | n = 2 |
| Plow vs. no tillage (Pr > F = 0.24) | | | | | 74 | — | — | 81 | n = 8 |
| Shallow vs. no tillage (Pr > F = 0.52) | | | | | — | 40 | — | 43 | n = 7 |

[a] Assumed a bulk density of 1.2 Mg m$^{-3}$ for soil at a depth of 0–5 cm and 1.4 Mg m$^{-3}$ for soil at a depth of either 5–15 or 5–20 cm.

[b] Assumed a bulk density of 1.33 Mg m$^{-3}$ for soil at a depth of 0–15 cm.

## BIOLOGICALLY ACTIVE FRACTIONS OF ORGANIC MATTER

Biologically active fractions of soil organic matter are important in assessing nutrient cycling, decomposition potential, and biophysical manipulation of soil structure. Biologically active fractions of soil organic matter include microbial biomass, readily mineralizable C and N, and some chemical indices of labile organic substrates. All these fractions have a relatively short turnover time and would be part of the active pool of the active–slow–passive soil organic matter continuum (Parton et al., 1987). According to a compilation of studies from the literature, mineralizable C and N were generally higher under no-tillage than under inversion-tillage systems (Table 8.8 and Table 8.9). As with other pools of organic matter, differences in mineralizable C between tillage systems tend to be greatest nearest the soil surface (Figure 8.8). Data in Table 8.8 were compiled from the uppermost sampling depth reported and are therefore not necessarily representative of results that might occur summed to the surface 20 to 30 cm of soil. Mineralizable C represents potential activity under optimum temperature and moisture conditions. As such, it represents the lack of *in situ* mineralization that might occur in the field. Inversion tillage that stimulates microbial activity in the field leads to an exhaustion of substrates that contribute to this mineralizable C pool.

Lack of C input to feed the heterotrophic community of soil organisms will lead to a reduction in biologically active soil organic matter pools. When sorghum residues were removed for 6 years from an Entic Pellustert in Australia, mineralizable C of the surface 10 cm of soil declined by an average of 29% (Saffigna et al., 1989). Microbial biomass N in the surface 10 cm of soil was reduced by $16 \pm 8\%$ in an Udic Pellustert in Australia following 20 years of burning of wheat and barley residues compared with residue retention (Dalal et al., 1991). During the ninth and tenth years of a cropping system study on a Fluventic Ustochrept in Texas, mineralizable C increased linearly with additional C input from more intensive cropping systems under both conventional and no tillage (Figure 8.20). In this study, there was no evidence of an interaction between cropping intensity and tillage management on the response in mineralizable C, as slopes between tillage systems were essentially the same.

Seasonal differences in mineralizable C can occur as a result of pulses of C inputs from plant roots and aboveground residues. In a Fluventic Ustochrept in Texas, mineralizable C at wheat planting was 89% higher under no tillage than under conventional tillage at a depth of 0–5 cm, but 12% lower at a depth of 5–12.5 cm (Figure 8.21). At wheat flowering, mineralizable C increased in both tillage systems most notably toward the soil surface, but at all depths as a result of accumulation of roots and rhizodeposits, which provided readily decomposable substrates. Long-term tillage effects (i.e., 9 years) were maintained despite seasonal changes that occurred.

Seasonal changes in mineralizable N do not necessarily mirror seasonal changes in mineralizable C, because inputs of high levels of readily decomposable substrates can lead to net immobilization of N in the short term. In the long term, mineralizable C and N generally correspond more directly once steady-state levels of organic matter quality are reached. In a Fluventic Ustochrept in Texas, cyclical changes in mineralizable N were evident under both conventional and no tillage in a 2-year sorghum–wheat/soybean rotation (Figure 8.22). Differences in mineralizable N between · tillage systems occurred during approximately half of the rotation sequence, i.e., primarily during the soybean to sorghum phases of the rotation. Mineralizable N was suppressed to equal levels under both tillage systems during the wheat phase, probably because of the high level of rhizodeposition that occurs with the dense, fibrous root system of wheat. Roots and rhizosphere products can be low in N concentration and high in mineralizable C, leading to significant N immobilization (Mary et al., 1993).

Accumulation of residues at the soil surface with conservation tillage systems provides a habitat for a variety of soil fauna, which have important implications on the cycling of organic matter into biologically active pools (Kladivko, 2001). The most visible effect of conservation

**TABLE 8.8**
**Comparison of Mineralizable C in Surface Soil among Tillage Systems**

| Soil Type | Texture | Location | Years | Depth (cm) | Units | Plow | Shallow | Ridge | No. | Reference |
|---|---|---|---|---|---|---|---|---|---|---|
| Typic Argiudoll | SiL | Argentina | 15 | 5 | mg kg$^{-1}$ d$^{-1}$ | 14 | — | — | 51 | Alvarez et al. (1998) |
| Rhodic Kanhapludult | SCL | Georgia | 13 | 5 | mg kg$^{-1}$ d$^{-1}$ | — | 22 | — | 37 | Beare et al. (1994b) |
| Aridic Paleustoll | L | Colorado | 5 | 5 | kg ha$^{-1}$ d$^{-1}$ | — | 2.6 | — | 3.5 | Burke et al. (1995) |
| Ustollic Haplargid | SL | Colorado | 5 | 5 | kg ha$^{-1}$ d$^{-1}$ | — | 4.1 | — | 3.0 | Burke et al. (1995) |
| Typic Haploboroll | SiL | SK, Canada | 4 | 7.5 | kg ha$^{-1}$ d$^{-1}$ | — | 9.7 | — | 11.4 | Campbell et al. (1999) |
| Typic Haploboroll | SiL | SK, Canada | 8 | 7.5 | kg ha$^{-1}$ d$^{-1}$ | — | 5.8 | — | 5.8 | Campbell et al. (1999) |
| Typic Haploboroll | SiL | SK, Canada | 12 | 7.5 | kg ha$^{-1}$ d$^{-1}$ | — | 6.0 | — | 6.1 | Campbell et al. (1999) |
| Typic Haploboroll | L | SK, Canada | 6 | 7.5 | mg kg$^{-1}$ d$^{-1}$ | — | 12 | — | 21 | Campbell et al. (1989) |
| Typic Boroll | SiL | AB, Canada | 16 | 4 | mg kg$^{-1}$ d$^{-1}$ | — | 12 | — | 16 | Carter and Rennie (1984) |
| Udic Boroll | CL | SK, Canada | 12 | 5 | mg kg$^{-1}$ d$^{-1}$ | — | 15 | — | 20 | Carter and Rennie (1984) |
| Typic Boroll | CL | SK, Canada | 4 | 5 | mg kg$^{-1}$ d$^{-1}$ | — | 15 | — | 26 | Carter and Rennie (1984) |
| Typic Boroll | L | SK, Canada | 2 | 5 | mg kg$^{-1}$ d$^{-1}$ | — | 14 | — | 14 | Carter and Rennie (1984) |
| Typic Hapludalf | L | Michigan | 8 | 20 | mg kg$^{-1}$ d$^{-1}$ | 2.9 | — | — | 3.1 | Collins et al. (2000) |
| Mollic Ochraqualf | SiCL | Ohio | 30 | 20 | mg kg$^{-1}$ d$^{-1}$ | 4.0 | — | — | 7.9 | Collins et al. (2000) |
| Typic Fragiudalf | SiL | Ohio | 30 | 20 | mg kg$^{-1}$ d$^{-1}$ | 3.6 | — | — | 5.6 | Collins et al. (2000) |
| Pachic Haplustoll | L | Nebraska | 16 | 10 | mg kg$^{-1}$ d$^{-1}$ | 11 | 20 | — | 22 | Follett and Schimel (1989) |
| Mollic Cryoboralf | CL | AB, Canada | 4 | 5 | mg kg$^{-1}$ d$^{-1}$ | — | 63 | — | 42 | Franzluebbers and Arshad (1996b) |
| Typic Cryoboralf | SiL | BC, Canada | 16 | 5 | mg kg$^{-1}$ d$^{-1}$ | — | 49 | — | 50 | Franzluebbers and Arshad (1996b) |
| Typic Cryoboralf | L | BC, Canada | 7 | 5 | mg kg$^{-1}$ d$^{-1}$ | — | 50 | — | 34 | Franzluebbers and Arshad (1996b) |
| Typic Natriboralf | C | AB, Canada | 6 | 5 | mg kg$^{-1}$ d$^{-1}$ | — | 22 | — | 29 | Franzluebbers and Arshad (1996a) |
| Fluventic Ustochrept | SiCL | Texas | 9 | 5 | mg kg$^{-1}$ d$^{-1}$ | — | 11 | — | 21 | Franzluebbers et al. (1994a) |
| Typic Kanhapludult | SL | Georgia | 4 | 2.5 | mg kg$^{-1}$ d$^{-1}$ | — | 19 | — | 70 | Franzluebbers et al. (1999) |

| Soil Type | Texture | Location | Years | Depth (cm) | Units | Plow | Shallow | Ridge | No. | Reference |
|---|---|---|---|---|---|---|---|---|---|---|
| Udic Dystrochrept | SiL | New Zealand | 5 | 2.5 | mg kg$^{-1}$ d$^{-1}$ | 30 | — | — | 38 | Haynes (1999) |
| Typic Hapludalf | SiL | Wisconsin | 12 | 5 | mg kg$^{-1}$ d$^{-1}$ | 7 | 14 | — | 35 | Karlen et al. (1994) |
| Typic Haploboroll | SCL | AB, Canada | 16 | 15 | kg ha$^{-1}$ d$^{-1}$ | — | 11 | — | 11 | Larney et al. (1997) |
| Typic Haploboroll | SCL | AB, Canada | 8 | 15 | kg ha$^{-1}$ d$^{-1}$ | — | 6.9 | — | 8.6 | Larney et al. (1997) |
| Typic Haploboroll | SCL | AB, Canada | 9 | 15 | kg ha$^{-1}$ d$^{-1}$ | — | 18 | — | 20 | Larney et al. (1997) |
| Entic Pellustert | SC | Qld, Australia | 5 | 10 | mg kg$^{-1}$ d$^{-1}$ | — | 4.6 | — | 3.2 | Saffigna et al. (1989) |
| Typic Ochraqualf | SCL | Texas | 15 | 20 | mg kg$^{-1}$ d$^{-1}$ | 13 | 12 | — | 19 | Salinas-Garcia et al. (1997b) |
| Torrertic Paleustoll | CL | Texas | 12 | 4 | g m$^{-3-1}$ d$^{-1}$ | — | 16 | — | 21 | Schomberg and Jones (1999) |
| Pachic Haplustoll | L | Nebraska | 16 | 15 | kg ha$^{-1}$ d$^{-1}$ | 8 | — | — | 14 | Tracy et al. (1990) |
| 36 (Argiudoll-Argiaquoll) | SiCL | Illinois | >5 | 15 | kg ha$^{-1}$ d$^{-1}$ | 26 | — | — | 31 | Wander and Bollero (1999) |
| Typic Hapludult | fSL | Alabama | 10 | 5 | mg kg$^{-1}$ d$^{-1}$ | 7 | — | — | 14 | Wood and Edwards (1992) |
| Plow vs. no tillage (Pr > F = 0.01) | | | | | | 11.5 | — | — | 21.9 | $n = 11$ |
| Shallow vs. no tillage (Pr > F = 0.08) | | | | | | | 17.4 | — | 22.0 | $n = 25$ |

# TABLE 8.9
## Comparison of Mineralizable N in Surface Soil among Tillage Systems

| Soil Type | Texture | Location | Years | Depth (cm) | Units | Plow | Shallow | Ridge | No. | Reference |
|---|---|---|---|---|---|---|---|---|---|---|
| Rhodic Kanhapludult | SCL | Georgia | 13 | 5 | mg kg⁻¹ d⁻¹ | — | 1.9 | — | 2.7 | Beare et al. (1994b) |
| Aridic Paleustoll | L | Colorado | 5 | 5 | kg ha⁻¹ d⁻¹ | — | 0.31 | — | 0.36 | Burke et al. (1995) |
| Ustollic Haplargid | SL | Colorado | 5 | 5 | kg ha⁻¹ d⁻¹ | — | 0.22 | — | 0.20 | Burke et al. (1995) |
| Typic Haploboroll | SiL | SK, Canada | 4 | 7.5 | kg ha⁻¹ d⁻¹ | — | 0.81 | — | 0.92 | Campbell et al. (1999) |
| Typic Haploboroll | SiL | SK, Canada | 8 | 7.5 | kg ha⁻¹ d⁻¹ | — | 0.87 | — | 0.82 | Campbell et al. (1999) |
| Typic Haploboroll | SiL | SK, Canada | 12 | 7.5 | kg ha⁻¹ d⁻¹ | — | 0.75 | — | 0.73 | Campbell et al. (1999) |
| Typic Haploboroll | L | SK, Canada | 6 | 7.5 | mg kg⁻¹ d⁻¹ | — | 0.80 | — | 1.08 | Campbell et al. (1989) |
| Typic Boroll | SiL | AB, Canada | 16 | 4 | mg kg⁻¹ d⁻¹ | — | 1.02 | — | 1.45 | Carter and Rennie (1984) |
| Udic Boroll | CL | SK, Canada | 12 | 5 | mg kg⁻¹ d⁻¹ | — | 1.52 | — | 2.26 | Carter and Rennie (1984) |
| Typic Boroll | CL | SK, Canada | 4 | 5 | mg kg⁻¹ d⁻¹ | — | 1.21 | — | 1.98 | Carter and Rennie (1984) |
| Typic Boroll | L | SK, Canada | 2 | 5 | mg kg⁻¹ d⁻¹ | — | 1.19 | — | 1.19 | Carter and Rennie (1984) |
| Udic Pellustert | C | Qld, Australia | 13 | 10 | mg kg⁻¹ d⁻¹ | — | 4.0 | — | 5.0 | Dalal (1989) |
| Typic Paleudalf | SiL | Kentucky | 11 | 7.5 | kg ha⁻¹ d⁻¹ | 2.4 | — | — | 3.8 | Doran (1987) |
| Aeric Ochraqualf | SiL | Illinois | 6 | 7.5 | kg ha⁻¹ d⁻¹ | 1.5 | — | — | 2.2 | Doran (1987) |
| Aquic Hapludoll | CL | Minnesota | 6 | 7.5 | kg ha⁻¹ d⁻¹ | 1.6 | — | — | 2.1 | Doran (1987) |
| Pachic Argiustoll | SiCL | Nebraska | 6 | 7.5 | kg ha⁻¹ d⁻¹ | 0.9 | — | — | 1.4 | Doran (1987) |
| Pachic Haplustoll | L | Nebraska | 13 | 7.5 | kg ha⁻¹ d⁻¹ | — | 1.0 | — | 1.1 | Doran (1987) |
| Aridic Argiustoll | SiL | Nebraska | 12 | 7.5 | kg ha⁻¹ d⁻¹ | — | 0.9 | — | 1.0 | Doran (1987) |
| Pachic Haplustoll | L | Nebraska | 16 | 10 | mg kg⁻¹ d⁻¹ | 1.4 | 1.6 | — | 1.7 | Follett and Schimel (1989) |
| Mollic Cryoboralf | CL | AB, Canada | 4 | 5 | mg kg⁻¹ d⁻¹ | — | 1.2 | — | 1.5 | Franzluebbers and Arshad (1996b) |
| Typic Cryoboralf | SiL | BC, Canada | 16 | 5 | mg kg⁻¹ d⁻¹ | — | 1.4 | — | 2.5 | Franzluebbers and Arshad (1996b) |
| Typic Cryoboralf | L | BC, Canada | 7 | 5 | mg kg⁻¹ d⁻¹ | — | 0.5 | — | 1.5 | Franzluebbers and Arshad (1996b) |
| Typic Natriboralf | C | AB, Canada | 6 | 5 | mg kg⁻¹ d⁻¹ | — | 1.3 | — | 1.2 | Franzluebbers and Arshad (1996a) |
| Fluventic Ustochrept | SiCL | Texas | 9 | 5 | mg kg⁻¹ d⁻¹ | — | 1.0 | — | 2.1 | Franzluebbers et al. (1994a) |
| Typic Kanhapludult | SL | Georgia | 4 | 2.5 | mg kg⁻¹ d⁻¹ | — | 1.7 | — | 3.1 | Franzluebbers et al. (1999) |
| Solodic | SCL | Vic, Australia | 6 | 2.5 | mg kg⁻¹ d⁻¹ | — | 0.6 | — | 0.9 | Haines and Uren (1990) |
| Haplic Chernozem | fSL | Austria | 2 | 10 | mg kg⁻¹ d⁻¹ | 4.4 | 5.4 | — | 6.5 | Kandeler et al. (1999) |
| Haplic Chernozem | fSL | Austria | 6 | 10 | mg kg⁻¹ d⁻¹ | 2.9 | 7.2 | — | 7.4 | Kandeler et al. (1999) |
| Haplic Chernozem | fSL | Austria | 8 | 10 | mg kg⁻¹ d⁻¹ | 2.1 | 6.4 | — | 8.3 | Kandeler et al. (1999) |

| Soil Type | Texture | Location | Years | Depth (cm) | Units | Plow | Shallow | Ridge | No. | Reference |
|---|---|---|---|---|---|---|---|---|---|---|
| Typic Haploboroll | SCL | AB, Canada | 16 | 15 | mg kg$^{-1}$ d$^{-1}$ | — | 0.60 | — | 0.63 | Larney et al. (1997) |
| Typic Haploboroll | SCL | AB, Canada | 8 | 15 | kg ha$^{-1}$ d$^{-1}$ | — | 0.45 | — | 0.52 | Larney et al. (1997) |
| Typic Haploboroll | SCL | AB, Canada | 9 | 15 | kg ha$^{-1}$ d$^{-1}$ | — | 0.88 | — | 0.97 | Larney et al. (1997) |
| Aquic Hapludult | SiL | Maryland | 18 | 2.5 | mg kg$^{-1}$ d$^{-1}$ | 0.46 | — | — | 1.21 | McCarty and Meisinger (1997) |
| Aquic Hapludult | SiL | Maryland | 21 | 2.5 | mg kg$^{-1}$ d$^{-1}$ | 0.29 | — | — | 0.71 | McCarty and Meisinger (1997) |
| 36 (Argiudoll-Argiaquoll) | SiCL | Illinois | >5 | 5 | mg kg$^{-1}$ d$^{-1}$ | 3.0 | — | — | 4.5 | Needelman et al. (1999) |
| Entic Pellustert | SC | Qld, Australia | 5 | 10 | mg kg$^{-1}$ d$^{-1}$ | — | 0.36 | — | 0.53 | Saffigna et al. (1989) |
| Typic Ochraqualf | SCL | Texas | 15 | 20 | mg kg$^{-1}$ d$^{-1}$ | 0.95 | 0.90 | — | 1.32 | Salinas-Garcia et al. (1997b) |
| Torrertic Paleustoll | CL | Texas | 12 | 4 | g m$^{-3 \cdot 1}$ d$^{-1}$ | — | 0.72 | — | 0.64 | Schomberg and Jones (1999) |
| Humic Gleysol | L | QC, Canada | 3 | 7.5 | mg kg$^{-1}$ d$^{-1}$ | 1.0 | 1.4 | — | — | Simard et al. (1994) |
| Pachic Haplustoll | L | Nebraska | 16 | 15 | kg ha$^{-1}$ d$^{-1}$ | 0.9 | — | — | 1.5 | Tracy et al. (1990) |
| 36 (Argiudoll-Argiaquoll) | SiCL | Illinois | >5 | 15 | kg ha$^{-1}$ d$^{-1}$ | 8.3 | — | — | 8.5 | Wander and Bollero (1999) |
| Typic Hapludult | fSL | Alabama | 10 | 5 | mg kg$^{-1}$ d$^{-1}$ | 0.43 | — | — | 0.83 | Wood and Edwards (1992) |
| Plow vs. shallow tillage (Pr > F = 0.10) | | | | | | 2.1 | 3.8 | — | n = 6 | |
| Plow vs. no tillage (Pr > F = 0.01) | | | | | | 2.1 | 3.5 | — | n = 15 | |
| Shallow vs. no tillage (Pr > F < 0.01) | | | | | | 2.0 | 1.6 | — | n = 31 | |

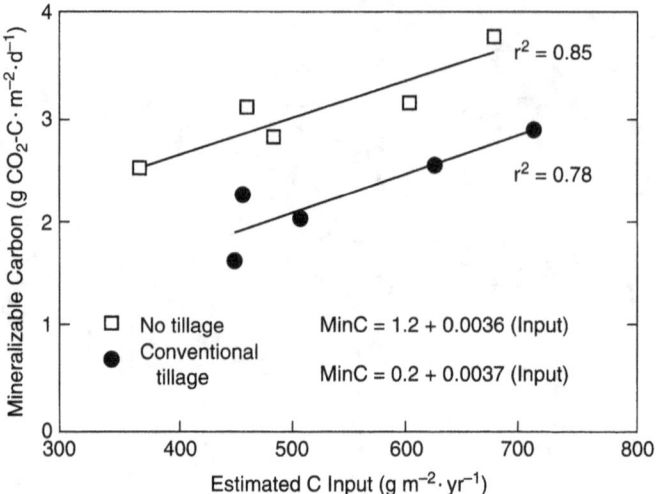

**FIGURE 8.20** Mineralizable C as a function of C input among five cropping systems under conventional and no tillage in southcentral Texas. (Data from Franzluebbers, A.J. et al. 1998. *Soil Tillage Res.* 47:303–308.)

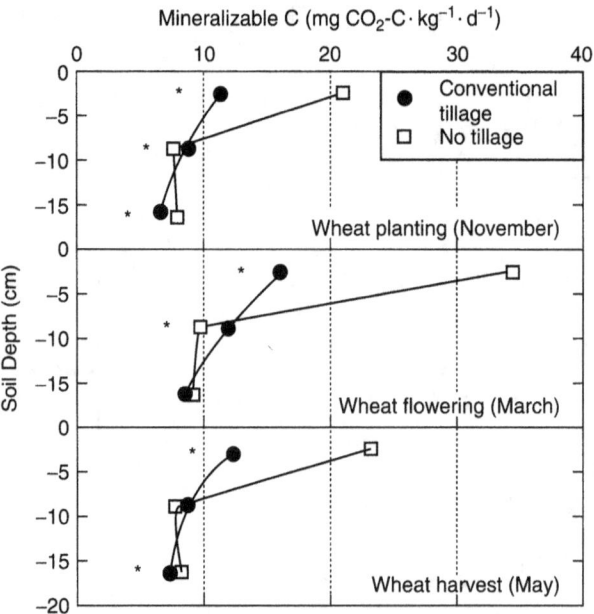

**FIGURE 8.21** Depth distribution of mineralizable C sampled at three growth stages of wheat under conventional and no tillage in southcentral Texas. * indicates significance between tillage systems at $p < 0.1$. (Data from Franzluebbers, A.J. et al. 1994b. *Soil Biol. Biochem.* 26:1469–1475.)

tillage is on earthworms. Earthworms require a moist environment with adequate organic debris, both provided by conservation tillage. In a Typic Rhodudult in Georgia, earthworms, microarthropods and various macroarthropods were two- to several-fold more numerous under no tillage than under conventional tillage as a result of the stratification of organic debris near the soil surface (House and Parmelee, 1985).

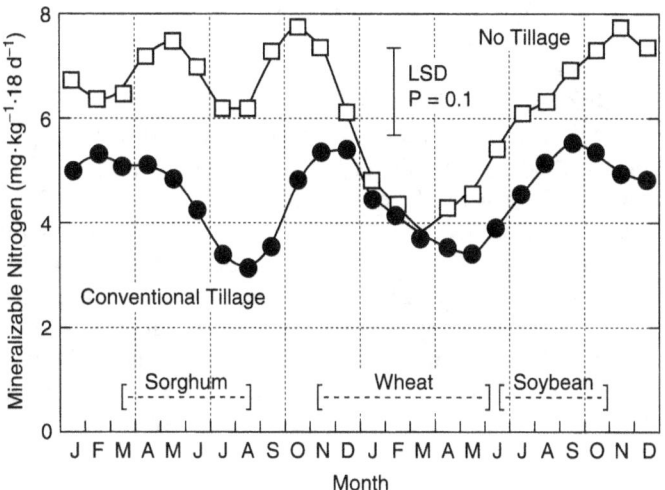

**FIGURE 8.22** Mineralizable N on a monthly basis during the ninth and tenth year under conventional and no tillage in southcentral Texas. (Data are from 3-month running averages reported in Franzluebbers, A.J., et al. 1996b. *Z. Pflanzenernähr. Bodenk.* 159:343–349.)

## SOIL ORGANIC MATTER AFFECTED BY INTERACTION OF TILLAGE WITH CROPPING INTENSITY

Sequestration of soil organic C is dependent on the net balance between C inputs and C outputs. Crop rotation and the intensity of cropping can affect the quantity and quality of organic inputs. The type of tillage management along with cropping intensity can also affect the decomposition environment, resulting in altered C output. Comparisons of continuous wheat with wheat–fallow rotations under shallow tillage and no tillage are most abundant in the literature. At the end of 12 years of tillage management on a Haploboroll in Saskatchewan, soil organic C at a depth of 0 to 15 cm was 0.05 kg m$^{-2}$ higher under no tillage than under shallow tillage in wheat–fallow and 0.14 kg m$^{-2}$ higher under no tillage in continuous wheat (Campbell et al., 1995). At the end of 11 years on a Typic Haploboroll in Saskatchewan, soil organic C at a depth of 0 to 15 cm was 0.06 kg m$^{-2}$ higher under no tillage than under shallow tillage in wheat–fallow and 0.18 kg m$^{-2}$ higher under no tillage in continuous wheat (Campbell et al., 1996). However, the difference in soil organic N between no tillage and shallow tillage was similar, whether the crop rotation was wheat–fallow ($\Delta$13 g m$^{-2}$) or continuous wheat ($\Delta$11 g m$^{-2}$). At the end of 9 years on a Typic Haploboroll in Alberta, no tillage had greater positive effects on mineralizable C and N in continuous wheat than in wheat–fallow (Larney et al., 1997). However, no tillage had greater positive effects on soil organic C and N and light-fraction C and N in wheat–fallow than in continuous wheat. At the end of 6 years on a Typic Argiboroll in North Dakota, soil organic C to a depth of 30 cm was 0.68 kg m$^{-2}$ lower under no tillage than under conventional tillage in wheat–fallow, but 0.64 kg m$^{-2}$ higher under no tillage in a wheat–wheat–sunflower rotation (Black and Tanaka, 1997). Soil organic N responded similarly to soil organic C: no tillage was 52 g m$^{-2}$ lower in wheat–fallow and 41 g m$^{-2}$ higher in wheat–wheat–sunflower. At the end of 7 years on a Torrertic Paleustoll in Texas, soil organic C under no tillage was 0.09 kg m$^{-2}$ higher than under stubble-mulch tillage in wheat–fallow and 0.13 kg m$^{-2}$ higher under no tillage in continuous wheat (Jones et al., 1997). Wheat–fallow has been utilized in the Great Plains region of North America, where precipitation is low and variable, to reduce risk of crop failure by filling the soil profile with water during the fallow period. However, precipitation use efficiency is improved with no-tillage management such that the fallow phase might not be economically productive compared with more intensive cropping (Peterson et al., 1996). In wheat–fallow, no tillage can keep the surface soil moister during the fallow phase such that organic matter decomposition is enhanced compared with the more extreme drying of the surface soil with tillage.

In warm, moist climates, more intensive cropping can make better use of environmental conditions by producing plant biomass throughout the year. Increased cropping intensity might increase the risk of a particular crop failure, but with extended time will likely capture more opportunities for enhanced C input via photosynthetic fixation. At the end of 9 years on a Fluventic Ustochrept in Texas, soil organic matter pools were always higher under no tillage than under conventional tillage, irrespective of cropping intensity (Figure 8.23). Absolute changes in soil organic matter pools with respect to tillage system were similar among all cropping intensities, suggesting no significant interaction between tillage system and cropping intensity on soil organic matter pools. However, the soil organic C sequestration rate per unit of estimated C input was significantly higher under no tillage than under conventional tillage at low cropping intensities (Figure 8.23).

## SOIL ORGANIC MATTER AFFECTED BY INTERACTION OF TILLAGE WITH SOIL TEXTURE

Soil texture might alter the response of soil organic matter pools to tillage management by altering plant productivity, soil moisture retention, and community structure and activity of soil organisms, all of which could have impacts on C inputs and outputs. Irrespective of tillage management, fine-textured soils, especially dominated by montmorillonitic clays, can store higher quantities of organic matter than can coarse-textured soils (Nichols, 1984; Hassink, 1994; Needelman et al., 1999). In a survey of 36 fields in Illinois, whole-soil and particulate organic C and N were higher under no tillage than under conventional tillage when sand content was <10% at a depth of 0–5 cm, but lower under no tillage when sand content was >10% at a depth of 5–15 cm (Needelman et al.,

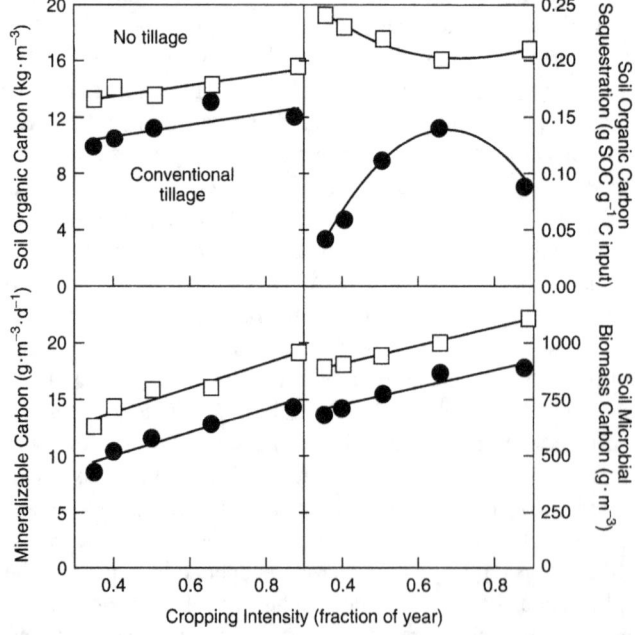

FIGURE 8.23 Soil organic matter pools at the end of 9 years of conventional and no tillage among five cropping systems that formed a cropping intensity gradient in southcentral Texas. (Data from Franzluebbers, A.J., Hons, F.M., and Zuberer, D.A. 1998. *Soil Tillage Res.* 47:303–308.)

1999). When the surface 15 cm was considered as a whole, soil and particulate organic C and N were not affected by the interaction of tillage and texture. From a set of four soils along a textural gradient in northern Alberta and British Columbia, tillage interacted with texture such that total, particulate, and microbial biomass C were not different between tillage systems in soils with low clay content, but were higher under no tillage than under conventional tillage in soils with high clay content.

According to a compilation of tillage studies on different soils, soil organic C storage with no tillage compared with conventional tillage was significantly higher in silty clay loams than in loams (Franzluebbers and Steiner, 2002). Overall, available data suggest that sequestration of soil organic C with no tillage compared with conventional tillage within the surface rooting zone might be higher in soils with finer texture.

## SOIL ORGANIC MATTER AFFECTED BY INTERACTION OF TILLAGE WITH CLIMATIC REGION

The climatic conditions of a region dictate to a large extent the type and sequence of crops grown. How the crops are managed can vary to some extent, such as crop variety selection, type and quantity of fertilization, type of pesticide applications, timing of planting, and type of tillage system employed. In some regions, forms of conservation tillage have been employed for many years, such as stubble-mulch tillage in the Great Plains of the U.S. or shallow blade or disk tillage in the western wheat region of Canada. These systems have become the convention rather than the exception.

According to a compilation of tillage studies from North America, the change in soil organic C with no tillage compared with conventional tillage was greatest when sites were located in the mesic subhumid region with a mean annual precipitation to potential evapotranspiration ratio of 1.4 to 1.6 mm mm$^{-1}$ (Figure 8.18). The relationship of the change in soil organic C with climate was not particularly strong, probably because the data were too limited to clearly separate cropping intensity, soil textural, and other management differences that might have interacted with climate. However, the derived shape of the response with climate is logical. Minimal benefit of no tillage on soil organic C storage compared with conventional tillage could be expected in dry, cold regions, because low precipitation would limit the potential of plants to fix C and limit decomposition even when crop residues are mixed with soil with tillage. In relatively wet, hot regions of North America, the benefit of no tillage on soil organic C storage compared with conventional tillage might also be limited, because abundant precipitation combined with warm temperature would allow surface-placed residues an ideal environment for rapid decomposition, similar to that with tillage. More long-term tillage studies under different soil and climatic conditions are clearly needed to accurately understand the dynamics of soil organic matter under the wide diversity of environments in the world.

## REFERENCES

Aase, J.K., and Pikul, Jr., J.L. 1995. Crop and soil response to long-term tillage practices in the northern Great Plains. *Agron. J.* 87:652–656.

Allmaras, R.R., Copeland, S.M., Copeland, P.J., and Oussible, M. 1996. Spatial relations between oat residue and ceramic spheres when incorporated sequentially by tillage. *Soil Sci. Soc. Am. J.* 60:1209–1216.

Allmaras, R.R., Schomberg, H.H., Douglas, C.L., Jr., and Dao, T.H. 2000. Soil organic carbon sequestration potential of adopting conservation tillage in U.S. croplands. *J. Soil Water Conserv.* 55:365–373.

Alvarez, R., Russo, M.E., Prystupa, P., Scheiner, J.D., and Blotta, L. 1998. Soil carbon pools under conventional and no-tillage systems in the Argentine Rolling Pampa. *Agron. J.* 90:138–143.

Angers, D.A., Bissonnette, N., Légère, A., and Samson, N. 1993a. Microbial and biochemical changes induced by rotation and tillage in a soil under barley production. *Can. J. Soil Sci.* 73:39–50.

Angers, D.A., Bolinder, M.A., Carter, M.R., Gregorich, E.G., Drury, C.F., Liang, B.C., Voroney, R.P., Simard, R.R., Donald, R.G., Beyaer, R.P., and Martel, J. 1997. Impact of tillage practices on organic carbon and nitrogen storage in cool, humid soils of eastern Canada. *Soil Tillage Res.* 41:191–201.

Angers, D.A., Légère, A., Avon, D., Ndayegamie, A., Côte, D., Samson, N., and Pageau, D. 1994. Effects of reduced tillage practices on soil quality in eastern Quebec. In *Proceedings of the 13th International Soil Tillage Research Organization*, Denmark, pp. 49–54.

Angers, D.A., Ndayegamie, A., and Côté, D. 1993b. Tillage-induced differences in organic matter of particle-size fractions and microbial biomass. *Soil Sci. Soc. Am. J.* 57:512–516.

Angers, D.A., Voroney, R.P., and Côté, D. 1995. Dynamics of soil organic matter and corn residues affected by tillage practices. *Soil Sci. Soc. Am. J.* 59:1311–1315.

Arshad, M.A., Franzluebbers, A.J., and Azooz, R.H. 1999a. Components of surface soil structure under conventional and no-tillage in northwestern Canada. *Soil Tillage Res.* 53:41–47.

Arshad, M.A., Franzluebbers, A.J., and Gill, K.S. 1999b. Improving barley yield on an acidic Boralf with crop rotation, lime, and zero tillage. *Soil Tillage Res.* 50:47–53.

Ball, B.C., Lang, R.W., O'Sullivan, M.F., and Franklin, M.F. 1989. Cultivation and nitrogen requirements for continuous winter barley on a gleysol and a camisol. *Soil Tillage Res.* 13:333–352.

Bandel, V.A., and Meisinger, J.J. 1993. In Steiner, R.A., and Herdt, R.W. (Eds.), *A Global Directory of Long-Term Agronomic Experiments*, Vol. 1. The Rockefeller Foundation, New York.

Bayer, C., Mielniczuk, J., Amado, T.J.C., Martin-Neto, L., and Fernandes, S.V. 2000b. Organic matter storage in a sandy clay loam Acrisol affected by tillage and cropping systems in southern Brazil. *Soil Tillage Res.* 54:101–109.

Beare, M.H., Cabrera, M.L., Hendrix, P.F., and Coleman, D.C. 1994a. Aggregate-protected and unprotected organic matter pools in conventional- and no-tillage soils. *Soil Sci. Soc. Am. J.* 58:787–795.

Beare, M.H., Hendrix, P.F., and Coleman, D.C. 1994b. Water-stable aggregates and organic matter fractions in conventional- and no-tillage soils. *Soil Sci. Soc. Am. J.* 58:777–786.

Black, A.L., and Tanaka, D.L. 1997. A conservation tillage-cropping systems study in the northern Great Plains of the United States. In Paul, E.A., Paustian, K., Elliott, E.T., and Cole, C.V. (Eds.), *Soil Organic Matter in Temperate Agroecosystems*. CRC Press, Boca Raton, FL, pp. 335–342.

Blevins, R.L., Thomas, G.W., and Cornelius, P.L. 1977. Influence of no-tillage and nitrogen fertilization on certain soil properties after 5 years of continuous corn. *Agron. J.* 69:383–386.

Brandt, S.A. 1992. Zero vs. conventional tillage and their effects on crop yield and soil moisture. *Can. J. Plant Sci.* 72:679–688.

Brown, P.L., and Dickey, D.D. 1970. Losses of wheat straw residue under simulated field conditions. *Soil Sci. Soc. Am. Proc.* 34:118–121.

Buckingham, F. 1976. *Fundamentals of Machine Operation: Tillage*. John Deere Service Publications, Moline, IL, 368 pp.

Burke, I.C., Elliott, E.T., and Cole, C.V. 1995. Influence of macroclimate, landscape position, and management on soil organic matter in agroecosystems. *Ecol. Appl.* 5:124–131.

Cambardella, C.A., and Elliott, E.T. 1992. Particulate soil organic-matter changes across a grassland cultivation sequence. *Soil Sci. Soc. Am. J.* 56:777–783.

Cambardella, C.A., and Elliott, E.T. 1993. Carbon and nitrogen distribution in aggregates from cultivated and native grassland soils. *Soil Sci. Soc. Am. J.* 57:1071–1076.

Campbell, C.A., Biederbeck, V.O., McConkey, B.G., Curtin, D., and Zentner, R.P. 1999. Soil quality — effect of tillage and fallow frequency: Soil organic matter quality as influenced by tillage and fallow frequency in a silt loam in southwestern Saskatchewan. *Soil Biol. Biochem.* 31:1–7.

Campbell, C.A., Biederbeck, V.O., Schnitzer, M., Selles, F., and Zentner, R.P. 1989. Effect of 6 years of zero tillage and N fertilizer management on changes in soil quality of an Orthic Brown Chernozem in southwestern Saskatchewan. *Soil Tillage Res.* 14:39–52.

Campbell, C.A., McConkey, B.G., Zentner, R.P., Dyck, F.B., Selles, F., and Curtin, D. 1995. Carbon sequestration in a Brown Chernozem as affected by tillage and rotation. *Can. J. Soil Sci.* 75:449–458.

Campbell, C.A., McConkey, B.G., Zentner, R.P., Selles, F., and Curtin, D. 1996. Tillage and crop rotation effects on soil organic C and N in a coarse-textured Typic Haploboroll in southwestern Saskatchewan. *Soil Tillage Res.* 37:3–14.

Cannell, R.Q., and Hawes, J.D. 1994. Trends in tillage practices in relation to sustainable crop production with special reference to temperate climates. *Soil Tillage Res.* 30:245–282.

Carter, M.R., Johnston, H.W., and Kimpinski, J. 1988. Direct drilling and soil loosening for spring cereals on a fine sandy loam in Atlantic Canada. *Soil Tillage Res.* 12:365–384.

Carter, M.R., and Rennie, D.A. 1982. Changes in soil quality under zero tillage farming systems: Distribution of microbial biomass and mineralizable C and N potentials. *Can. J. Soil Sci.* 62:587–597.

Carter, M.R., and Rennie, D.A. 1984. Nitrogen transformations under zero and shallow tillage. *Soil Sci. Soc. Am. J.* 48:1077–1081.

Carter, M.R., Sanderson, J.B., Ivany, J.A., and White, R.P. 2002. Influence of rotation and tillage on forage maize productivity, weed species, and soil quality of a fine sandy loam in the cool-humid climate of Atlantic Canada. *Soil Tillage Res.* 67:85–98.

Christensen, N.B., Lindemann, W.C., Salazar-Sosa, E., and Gill, L.R. 1994. Nitrogen and carbon dynamics in no-till and stubble mulch tillage systems. *Agron. J.* 86:298–303.

Clapp, C.E., Allmaras, R.R., Layese, M.F., Linden, D.R., Dowdy, R.H. 2000. Soil organic carbon and [13]C abundance as related to tillage, crop residue, and nitrogen fertilization under continuous corn management in Minnesota. *Soil Tillage Res.* 55:127–142.

Collins, H.P., Elliott, E.T., Paustian, K., Bundy, L.G., Dick, W.A., Huggins, D.R., Smucker, A.J.M., and Paul, E.A. 2000. Soil carbon pools and fluxes in long-term corn belt agroecosystems. *Soil Biol. Biochem.* 32:157–168.

Curtin, D., Wang, H., Selles, F., McConkey, B.G., and Campbell, C.A. 2000. Tillage effects on carbon fluxes in continuous wheat and fallow-wheat rotations. *Soil Sci. Soc. Am. J.* 64:2080–2086.

Dalal, R.C. 1989. Long-term effects of no-tillage, crop residue, and nitrogen application on properties of a vertisol. *Soil Sci. Soc. Am. J.* 53:1511–1515.

Dalal, R.C., Henderson, P.A., and Glasby, J.M. 1991. Organic matter and microbial biomass in a vertisol after 20 yr of zero-tillage. *Soil Biol. Biochem.* 23:435–441.

Dick, W.A., Blevins, R.L., Frye, W.W., Peters, S.E., Christenson, D.R., Pierce, F.J., and Vitosh, M.L. 1998. Impacts of agricultural management practices on C sequestration in forest-derived soils of the eastern Corn Belt. *Soil Tillage Res.* 47:235–244.

Dick, W.A., McCoy, E.L., Edwards, W.M., and Lal, R. 1991. Continuous application of no-tillage to Ohio soils. *Agron. J.* 83:65–73.

Doran, J.W. 1987. Microbial biomass and mineralizable nitrogen distributions in no-tillage and plowed soils. *Biol. Fertil. Soils* 5:68–75.

Dormaar, J.F., and Lindwall, C.W. 1989. Chemical differences in Dark Brown Chernozemic Ap horizons under various conservation tillage systems. *Can. J. Soil Sci.* 69:481–488.

Douglas, C.L., Jr., Allmaras, R.R., Rasmussen, P.E., Ramig, R.E., and Roager, N.C., Jr. 1980. Wheat straw composition and placement effects on decomposition in dryland agriculture of the Pacific Northwest. *Soil Sci. Soc. Am. J.* 44:833–837.

Duiker, S.W., and Lal, R. 1999. Crop residue and tillage effects on carbon sequestration in a Luvisol in central Ohio. *Soil Tillage Res.* 52:73–81.

Edwards, J.H., Thurlow, D.L., and Eason, J.T. 1988. Influence of tillage and crop rotation on yields of corn, soybean, and wheat. Agron. J. 80:76–80.

Edwards, J.H., Wood, C.W., Thurlow, D.L., and Ruf, M.E. 1992. Tillage and crop rotation effects on fertility status of a Hapludult soil. *Soil Sci. Soc. Am. J.* 56:1577–1582.

Eghball, B., Mielke, L.N., McCallister, D.L., and Doran, J.W. 1994. Distribution of organic carbon and inorganic nitrogen in a soil under various tillage and crop sequences. *J. Soil Water Conserv.* 49:201–205.

Endale, D.M., Schomberg, H.H., and Steiner, J.L. 2000. Long term sediment yield and mitigation in a small Southern Piedmont watershed. *Int. J. Sed. Res.* 14:60–68.

Follett, R.F., and Peterson, G.A. 1988. Surface soil nutrient distribution as affected by wheat-fallow tillage systems. *Soil Sci. Soc. Am. J.* 52:141–147.

Follett, R.F., and Schimel, D.S. 1989. Effect of tillage practices on microbial biomass dynamics. *Soil Sci. Soc. Am. J.* 53:1091–1096.

Franzluebbers, A.J. 2002a. Soil organic matter stratification ratio as an indicator of soil quality. *Soil Tillage Res.* 66:95–106.

Franzluebbers, A.J. 2002b. Water infiltration and soil structure related to organic matter and its stratification with depth. *Soil Tillage Res.* 66:197–205.

Franzluebbers, A.J., and Arshad, M.A. 1996a. Soil organic matter pools during early adoption of conservation tillage in northwestern Canada. *Soil Sci. Soc. Am. J.* 60:1422–1427.

Franzluebbers, A.J., and Arshad, M.A. 1996b. Soil organic matter pools with conventional and zero tillage in a cold, semiarid climate. *Soil Tillage Res.* 39:1–11.

Franzluebbers, A.J., and Arshad, M.A. 1996c. Water-stable aggregation and organic matter in four soils under conventional and zero tillage. *Can. J. Soil Sci.* 76:387–393.

Franzluebbers, A.J., and Steiner, J.L. 2002. Climatic influences on soil organic C storage with no tillage. In Kimble, J.M., Lal, R., and Follett, R.F. (Eds.), *Agriculture Practices and Policies for Carbon Sequestration in Soil.* CRC Press, Boca Raton, FL, pp. 71–86.

Franzluebbers, A.J., Hons, F.M., and Zuberer, D.A. 1994a. Long-term changes in soil carbon and nitrogen pools in wheat management systems. *Soil Sci. Soc. Am. J.* 58:1639–1645.

Franzluebbers, A.J., Hons, F.M., and Zuberer, D.A. 1994b. Seasonal changes in soil microbial biomass and mineralizable C and N in wheat management systems. *Soil Biol. Biochem.* 26:1469–1475.

Franzluebbers, K., Weaver, R.W., Juo, A.S.R., and Franzluebbers, A.J. 1994c. Carbon and nitrogen mineralization from cowpea plants part decomposing in moist and in repeatedly dried and wetted soil. *Soil Biol. Biochem.* 26:1379–1387.

Franzluebbers, A.J., Hons, F.M., and Saladino, V.A. 1995a. Sorghum, wheat and soybean production as affected by long-term tillage, crop sequence and N fertilization. *Plant Soil* 173:55–65.

Franzluebbers, A.J., Hons, F.M., and Zuberer, D.A. 1995b. Soil organic carbon, microbial biomass, and mineralizable carbon and nitrogen in sorghum. *Soil Sci. Soc. Am. J.* 59:460–466.

Franzluebbers, A.J., Arshad, M.A., and Ripmeester, J.A. 1996a. Alterations in canola residue composition during decomposition. *Soil Biol. Biochem.* 28:1289–1295.

Franzluebbers, A.J., Hons, F.M., and Zuberer, D.A. 1996b. Seasonal dynamics of active soil carbon and nitrogen pools under intensive cropping in conventional and no tillage. *Z. Pflanzenernähr. Bodenk.* 159:343–349.

Franzluebbers, A.J., and Arshad, M.A. 1997a. Particulate organic carbon content and potential mineralization as affected by tillage and texture. *Soil Sci. Soc. Am. J.* 61:1382–1386.

Franzluebbers, A.J., and Arshad, M.A. 1997b. Soil microbial biomass and mineralizable carbon of water-stable aggregates. *Soil Sci. Soc. Am. J.* 61:1090–1097.

Franzluebbers, A.J., Hons, F.M., and Zuberer, D.A. 1998. In situ and potential $CO_2$ evolution from a Fluventic Ustochrept in southcentral Texas as affected by tillage and cropping intensity. *Soil Tillage Res.* 47:303–308.

Franzluebbers, A.J., Langdale, G.W., and Schomberg, H.H. 1999. Soil carbon, nitrogen, and aggregation in response to type and frequency of tillage. *Soil Sci. Soc. Am. J.* 63:349–355.

Frey, S.D., Elliott, E.T., and Paustian, K. 1999. Bacterial and fungal abundance and biomass in conventional and no-tillage agroecosystems along two climatic gradients. *Soil Biol. Biochem.* 31:573–585.

Ghidey, F., and Alberts, E.E. 1993. Residue type and placement effects on decomposition: Field study and model evaluation. *Trans. ASAE* 36:1611–1617.

Griffith, D.R., Kladivko, E.J., Mannering, J.V., West, T.D., and Parsons, S.D. 1988. Long-term tillage and rotation effects on corn growth and yield on high and low organic matter, poorly drained soils. *Agron. J.* 80:599–605.

Groffman, P.M., Hendrix, P.F., and Crossley, D.A., Jr. 1987. Nitrogen dynamics in conventional and no-tillage agroecosystems with inorganic fertilizer or legume nitrogen inputs. *Plant Soil* 97:315–332.

Haines, P.J., and Uren, N.C. 1990. Effects of conservation tillage farming on soil microbial biomass, organic matter and earthworm populations, in north-eastern Victoria. *Aust. J. Exp. Agric.* 30:365–371.

Halvorson, A.D., Vigil, M.F., Peterson, G.A., and Elliott, E.T. 1997. Long-term tillage and crop residue management study at Akron, Colorado. In Paul, E.A., Paustian, K., Elliott, E.T., and Cole, C.V. (Eds.), *Soil Organic Matter in Temperate Agroecosystems.* CRC Press, Boca Raton, FL, pp. 361–370.

Hamblin, A.P. 1980. Changes in aggregate stability and associated organic matter properties after direct drilling and ploughing of some Australian soils. *Aust. J. Soil Res.* 18:27–36.

Hansmeyer, T.L., Linden, D.L., Allan, D.L., and Huggins, D.R. 1997. Determining carbon dynamics under no-till, ridge-till, chisel, and moldboard tillage systems within a corn and soybean cropping sequence. In Lal, R., Kimble, J.M., Follett, R.F., and Stewart, B.A. (Eds.), *Management of Carbon Sequestration in Soil*. CRC Press, Boca Raton, FL, pp. 93–97.

Hassink, J. 1994. Effects of soil texture and grassland management on soil organic C and N and rates of C and N mineralization. *Soil Biol. Biochem.* 26:1221–1231.

Haynes, R.J. 1999. Labile organic matter fractions and aggregate stability under short-term, grass-based leys. *Soil Biol. Biochem.* 31:1821–1830.

Hendrix, P.F., Franzluebbers, A.J., and McCracken, D.V. 1998. Management effects on C accumulation and loss in soils of the southern Appalachian Piedmont of Georgia. *Soil Tillage Res.* 47:245–251.

Hunt, P.G., Karlen, D.L., Matheny, T.A., and Quisenberry, V.L. 1996. Changes in carbon content of a Norfolk loamy sand after 14 years of conservation or conventional tillage. *J. Soil Water Conserv.* 51:255–258.

House, G.J., and Parmelee, R.W. 1985. Comparison of soil arthropods and earthworms from conventional and no-tillage agroecosystems. *Soil Tillage Res.* 5:351–360.

Hussain, I., Olson, K.R., and Ebelhar, S.A. 1999. Long-term tillage effects on soil chemical properties and organic matter fractions. *Soil Sci. Soc. Am. J.* 63:1335–1341.

Ismail, I., Blevins, R.L., and Frye, W.W. 1994. Long-term no-tillage effects on soil properties and continuous corn yields. *Soil Sci. Soc. Am. J.* 58:193–198.

Jones, O.R., Allen, R.R., Unger, P.W. 1990. Tillage systems and equipment for dryland farming. *Adv. Soil Sci.* 13:89–130.

Jones, O.R., Stewart, B.A., and Unger, P.W. 1997. Management of dry-farmed southern Great Plains soils for sustained productivity. In Paul, E.A., Paustian, K., Elliott, E.T., and Cole, C.V. (Eds.), *Soil Organic Matter in Temperate Agroecosystems: Long-Term Experiments in North America*. CRC Press, Boca Raton, FL, pp. 387–401.

Kandeler, E., Tscherko, D., and Spiegel, H. 1999. Long-term monitoring of microbial biomass, N minerali- sation and enzyme activities of a Chernozem under different tillage management. *Biol. Fertil. Soils* 28:343–351.

Kapkiyai, J.J., Karanja, N.K., Qureshi, J.N., Smithson, P.C., and Woomer, P.L. 1999. Soil organic matter and nutrient dynamics in a Kenyan nitisol under long-term fertilizer and organic input management. *Soil Biol. Biochem.* 31:1773–1782.

Karlen, D.L., Berry, E.C., Colvin, T.S., and Kanwar, R.S. 1991. Twelve-year tillage and crop rotation effects on yields and soil chemical properties in norheast Iowa. *Commun. Soil Sci. Plant Anal.* 22:1985–2003.

Karlen, D.L., Wollenhaupt, N.C., Erbach, D.C., Berry, E.C., Swan, J.B., Eash, N.S., and Jordahl, J.L. 1994. Long-term tillage effects on soil quality. *Soil Tillage Res.* 32:313–327.

Karlen, D.L., Kumar, A., Kanwar, R.S., Cambardella, C.A., and Colvin, T.S. 1998. Tillage system effects on 15-year carbon-based and simulated N budgets in a tile-drained Iowa field. *Soil Tillage Res.* 48:155–165.

Kern, J.S., and Johnson, M.G. 1993. Conservation tillage impacts on national soil and atmospheric carbon levels. *Soil Sci. Soc. Am. J.* 57:200–210.

Kladivko, E.J. 2001. Tillage systems and soil ecology. *Soil Tillage Res.* 61:61–76.

Knowles, T.C., Hipp, B.W., Graff, P.S., and Marshall, D.S. 1993. Nitrogen nutrition of rainfed winter wheat in tilled and no-till sorghum and wheat residues. *Agron. J.* 85:886–893.

Lal, R. 2001. Thematic evolution of ISTRO: Transition in scientific issues and research focus from 1955 to 2000. *Soil Tillage Res.* 61:3–12.

Lal, R., Logan, T.J., and Fausey, N.R. 1990. Long-term tillage effects on a Mollic Ochraqualf in north-west Ohio. III. Soil nutrient profile. *Soil Tillage Res.* 15:371–382.

Lal, R., Mahboubi, A.A., and Fausey, N.R. 1994. Long-term tillage and rotation effects on properties of a central Ohio soil. *Soil Sci. Soc. Am. J.* 58:517–522.

Lal, R., Kimble, J.M., Follett, R.F., and Cole, C.V. 1998. *The Potential of U.S. Cropland to Sequester Carbon and Mitigate the Greenhouse Effect*. Ann Arbor Press, Chelsea, MI.

Lamb, J.A., G.A. Peterson, and C.R. Fenster. 1985. Wheat-fallow tillage systems' effect on a newly cultivated grassland soils' nitrogen budget. *Soil Sci. Soc. Am. J.* 49:352–356.

Larney, F.J., Bremer, E., Janzen, H.H., Johnston, A.M., and Lindwall, C.W. 1997. Changes in total, mineral- izable and light fraction soil organic matter with cropping and tillage intensities in semiarid southern Alberta, Canada. *Soil Tillage Res.* 42:229–240.

Lascano, R.J., Baumhardt, R.L., Hicks, S.K., and Heilman, J.L. 1994. Soil and plant water evaporation from strip-tilled cotton: Measurement and simulation. *Agron. J.* 86:987–994.

Levin, A., Beegle, D.B., and Fox, R.H. 1987. Effect of tillage on residual nitrogen availability from alfalfa to succeeding corn crops. *Agron. J.* 79:34–38.

Mary, B., Fresneau, C., Morel, J.L., and Mariotti, A. 1993. C and N cycling during decomposition of root mucilage, roots and glucose in soil. *Soil Biol. Biochem.* 25:1005–1014.

Mataruka, D.F., Cox, W.J., Mt. Pleasant, J., van Es, H.M., Klausner, S.D., and Zobel, R.W. 1993. Tillage and nitrogen source effects on growth, yield, and quality of forage maize. *Crop Sci.* 33:1316–1321.

McCarty, G.W., Lyssenko, N.N., and Starr, J.L. 1998. Short-term changes in soil carbon and nitrogen pools during tillage management transition. *Soil Sci. Soc. Am. J.* 62:1564–1571.

McCarty, G.W., and Meisinger, J.J. 1997. Effects of N fertilizer treatments on biologically active N pools in soils under plow and no tillage. *Biol. Fertil. Soils* 24:406–412.

McFarland, M.L., Hons, F.M., and Saladino, V.A. 1991. Effects of furrow diking and tillage on corn grain yield and nitrogen accumulation. *Agron. J.* 83:382–386.

Meyer, K., Joergensen, R.G., and Meyer, B. 1996. The effects of reduced tillage on microbial biomass C and P in sandy loess soils. *Appl. Soil Ecol.* 5:71–79.

Mielke, L.N., Doran, J.W., and Richards, K.A. 1986. Physical environment near the surface of plowed and no-tilled surface soils. *Soil Tillage Res.* 5:355–366.

Mrabet, R., Saber, N., El-Brahli, A., Lahlou, S., and Bessam, F. 2001. Total, particulate organic matter and structural stability of a Calcixeroll soil under different wheat rotations and tillage systems in a semiarid area of Morocco. *Soil Tillage Res.* 57:225–235.

Needelman, B.A., Wander, M.M., Bollero, G.A., Boast, C.W., Sims, G.K., and Bullock, D.G. 1999. Interaction of tillage and soil texture: Biologically active soil organic matter in Illinois. *Soil Sci. Soc. Am. J.* 63:1326–1334.

Nichols, J.D. 1984. Relation of organic carbon to soil properties and climate in the southern Great Plains. *Soil Sci. Soc. Am. J.* 48:1382–1384.

Nyborg, M., Solberg, E.D., Malhi, S.S., and Izaurralde, R.C. 1995. Fertilizer N, crop residue, and tillage alter soil C and N content in a decade. In Lal, R., Kimble, J., Levine, E., and Stewart, B.A. (Eds.), *Soil Management and Greenhouse Effect*. Lewis Publishers, CRC Press, Boca Raton, FL, pp. 93–99.

Parton, W.J., Schimel, D.S., Cole, C.V., and Ojima, D.S. 1987. Analysis of factors controlling soil organic matter levels in Great Plains grasslands. *Soil Sci. Soc. Am. J.* 51:1173–1179.

Paustian, K., Collins, H.P., and Paul, E.A. 1997. Management controls on soil carbon. In Paul, E.A., Paustian, K., Elliott, E.T., and Cole, C.V. (Eds.), *Soil Organic Matter in Temperate Agroecosystems: Long-Term Experiments in North America*. CRC Press, Boca Raton, FL, pp. 15–49.

Peterson, G.A., Schlegel, A.J., Tanaka, D.L., and Jones, O.R. 1996. Precipitation use efficiency as impacted by cropping and tillage systems. *J. Prod. Agric.* 9:180–186.

Peterson, G.A., Halvorson, A.D., Havlin, J.L., Jones, O.R., Lyon, D.J., and Tanaka, D.L. 1998. Reduced tillage and increasing cropping intensity in the Great Plains conserves soil C. *Soil Tillage Res.* 47:207–218.

Pierce, F.J., Fortin, M.-C., and Staton, M.J. 1994. Periodic plowing effects on soil properties in a no-till farming system. *Soil Sci. Soc. Am. J.* 58:1782–1787.

Pikul, J.L., Jr., and Aase, J.K. 1995. Infiltration and soil properties as affected by annual cropping in the Northern Great Plains. *Agron. J.* 87:656–662.

Potter, K.N., Jones, O.R., Torbert, H.A., and Unger, P.W. 1997. Crop rotation and tillage effects on organic carbon sequestration in the semiarid southern Great Plains. *Soil Sci.* 162:140–147.

Potter, K.N., Torbert, H.A., Jones, O.R., Matocha, J.E., Morrison, J.E., and Unger, P.W. 1998. Distribution and amount of soil organic C in long-term management systems in Texas. *Soil Tillage Res.* 47:309–321.

Rasmussen, P.E., and Collins, H.P. 1991. Long-term impacts of tillage, fertilizer, and crop residue on soil organic matter in temperate semiarid regions. *Adv. Agron.* 45:93–134.

Rasse, D.P., and Smucker, A.J.M. 1999. Tillage effects on soil nitrogen and plant biomass in a corn-alfalfa rotation. *J. Environ. Qual.* 28:873–880.

Reicosky, D.C., Kemper, W.D., Langdale, G.W., Douglas, C.L., Jr., and Rasmussen, P.E. 1995. Soil organic matter changes resulting from tillage and biomass production. *Soil Water Conserv.* 50:253–261.

Rhoton, F.E., Shipitalo, M.J., and Lindbo, D.L. 2002. Runoff and soil loss from midwestern and southeastern U.S. silt loam soils as affected by tillage practice and soil organic matter content. *Soil Tillage Res.* 66:1–11.

Saffigna, P.G., Powlson, D.S., Brookes, P.C., and Thomas, G.A. 1989. Influence of sorghum residues and tillage on soil organic matter and soil microbial biomass in an Australian vertisol. *Soil Biol. Biochem.* 21:759–765.

Sainju, U.M., Singh, B.P., and Whitehead, W.F. 2002. Long-term effects of tillage, cover crops, and nitrogen fertilization on organic carbon and nitrogen concentrations in sandy loam soils in Georgia, USA. *Soil Tillage Res.* 63:167–179.

Salinas-Garcia, J.R., Hons, F.M., Matocha, J.E., and Zuberer, D.A. 1997a. Soil carbon and nitrogen dynamics as affected by long-term tillage and nitrogen fertilization. *Biol. Fertil. Soils* 25:182–188.

Salinas-Garcia, J.R., Matocha, J.E., and Hons, F.M. 1997b. Long-term tillage and nitrogen fertilization effects on soil properties of an Alfisol under dryland corn/cotton production. *Soil Tillage Res.* 42:79–93.

Sarrantonio, M., and Scott, T.W. 1988. Tillage effects on availability of nitrogen to corn following a winter green manure crop. *Soil Sci. Soc. Am. J.* 52:1661–1668.

Schomberg, H.H., and Jones, O.R. 1999. Carbon and nitrogen conservation in dryland tillage and cropping systems. *Soil Sci. Soc. Am. J.* 63:1359–1366.

Simard, R., Angers, D.A., and Lapierre, C. 1994. Soil organic matter quality as influenced by tillage, lime, and phosphorus. *Biol. Fertil. Soils* 18:13–18.

Singh, K.K., Colvin, T.S., Erbach, D.C., and Mughal, A.Q. 1992. Tilth index: An approach to quantifying soil tilth. *Trans. ASAE* 35:1777–1785.

Six, J., Elliott, E.T., Paustian, K., and Doran, J.W. 1998. Aggregation and soil organic matter accumulation in cultivated and native grassland soils. *Soil Sci. Soc. Am. J.* 62:1367–1377.

Six, J., Elliott, E.T., and Paustian, K. 1999. Aggregate and soil organic matter dynamics under conventional and no-tillage systems. *Soil Sci. Soc. Am. J.* 63:1350–1358.

Six, J., Elliott, E.T., and Paustian, K. 2000a. Soil structure and soil organic matter: II. A normalized stability index and the effect of mineralogy. *Soil Sci. Soc. Am. J.* 64:1042–1049.

Six, J., Elliott, E.T., and Paustian, K. 2000b. Soil macroaggregate turnover and microaggregate formation: A mechanism for C sequestration under no-tillage agriculture. *Soil Biol. Biochem.* 32:2099–2103.

Six, J., Paustian, K., Elliott, E.T., and Combrink, C. 2000c. Soil structure and organic matter. I. Distribution of aggregate-size classes and aggregate-associated carbon. *Soil Sci. Soc. Am. J.* 64:681–689.

Stockfisch, N., Forstreuter, T., and Ehlers, W. 1999. Ploughing effects on soil organic matter after twenty years of conservation tillage in Lower Saxony, Germany. *Soil Tillage Res.* 52:91–101.

Swinford, N. 1994. *Allis-Chalmers Farm Equipment 1914–1985.* American Society of Agricultural Engineering, St. Joseph, MI. 376 pp.

Tisdall, J.M., and Oades, J.M. 1982. Organic matter and water-stable aggregates in soils. *J. Soil Sci.* 33:141–163.

Tracy, P.W., Westfall, D.G., Elliott, E.T., Peterson, G.A., and Cole, C.V. 1990. Carbon, nitrogen, phosphorus, and sulfur mineralization in plow and no-till cultivation. *Soil Sci. Soc. Am. J.* 54:457–461.

Trimble, S.W. 1974. *Man-Induced Soil Erosion on the Southern Piedmont, 1700–1970.* Soil Conservation Society of America, Ankeny, IA.

Unger, P.W. 1984. Tillage and residue effects on wheat, sorghum, and sunflower grown in rotation. *Soil Sci. Soc. Am. J.* 48:885–891.

Varco, J.J., Frye, W.W., Smith, M.S., and MacKown, C.T. 1993. Tillage effects on legume decomposition and transformation of legume and fertilizer nitrogen-15. *Soil Sci. Soc. Am. J.* 57:750–756.

Vigil, M.F., and Kissel, D.E. 1991. Equations for estimating the amount of nitrogen mineralized from crop residues. *Soil Sci. Soc. Am. J.* 55:757–761.

Voorhees, W.B., and Lindstrom, M.J. 1984. Long-term effects of tillage method on soil tilth independent of wheel traffic compaction. *Soil Sci. Soc. Am. J.* 48:152–156.

Wagger, M.G., and Denton, H.P. 1992. Crop and tillage rotations: Grain yield, residue cover, and soil water. *Soil Sci. Soc. Am. J.* 56:1233–1237.

Wander, M.M., Bidart, M.G., and Aref, S. 1998. Tillage impacts on depth distribution of total and particulate organic matter in three Illinois soils. *Soil Sci. Soc. Am. J.* 62:1704–1711.

Wander, M.M., and Bollero, G.A. 1999. Soil quality assessment of tillage impacts in Illinois. *Soil Sci. Soc. Am. J.* 63:961–971.

Wanniarachchi, S.D., Voroney, R.P., Vyn, T.J., Beyaert, R.P., and MacKenzie, A.F. 1999. Tillage effects on the dynamics of total and corn-residue-derived soil organic matter in two southern Ontario soils. *Can. J. Soil Sci.* 79:473–480.

Wilson, D.O., and Hargrove, W.L. 1986.Release of nitrogen from crimson clover residue under two tillage systems. *Soil Sci. Soc. Am. J.* 50:1251–1254.

Wood, C.W., and Edwards, J.H. 1992. Agroecosystem management effects on soil carbon and nitrogen. *Agric. Ecosyst. Environ.* 39:123–138.

Yang, X.M., and Wander, M.M. 1999. Tillage effects on soil organic carbon distribution and storage in a silt loam soil in Illinois. *Soil Tillage Res.* 52:1–9.

Yang, X.M., and Kay, B.D. 2001. Rotation and tillage effects on soil organic carbon sequestration in a typic Hapludalf in southern Ontario. *Soil Tillage Res.* 59:107–114.

# 9 Strategies for Managing Soil Organic Matter to Supply Plant Nutrients

*Stefan Seiter and William R. Horwath*

## CONTENTS

## INTRODUCTION

Environmental and economic concerns have prompted agricultural producers and researchers to look for improved nutrient management strategies. Environmental and human health concerns about nutrient management are focused on nitrogen and phosphorus that are in excess of crop requirements and might escape from agroecosystems into ground and surface waters (Daniel et al., 1994). Economic considerations in nutrient management include efforts to reduce cost and increase the efficiency of agricultural inputs. Agricultural nutrient management thus aims to balance nutrient inputs with crop demand and to increase the degree of internal nutrient cycling. Management of soil organic matter (SOM) has emerged as a major strategy to help achieve these goals because of the central role SOM plays in storing and cycling nutrients.

    The two main objectives of organic matter management in agricultural systems are to (1) restore or maintain SOM to benefit soil quality and (2) supply crops with nutrients contained within or associated with SOM (Bruce et al., 1990). These two objectives are not always compatible (Bouldin, 1987) because mineralization that releases nutrients also destroys SOM. Conditions that favor SOM accumulation can also favor nutrient immobilization, which reduces the nutrients available for crop growth. Hendrix et al. (1992) noted that the two objectives of organic matter management are not necessarily mutually exclusive but might require different management approaches than those commonly used. Special considerations need to be given to the timing and intensity of management practices when trying to meet nutrient and organic matter management goals. The ultimate goal is to provide a continuous supply of nutrients while preventing loss of SOM. The chapter provides an overview of the general role of SOM in nutrient storage and nutrient availability and then discusses how various SOM management practices can contribute to sustainable nutrient management.

## NUTRIENTS IN SOIL ORGANIC MATTER

### COMPONENTS OF SOIL ORGANIC MATTER CONTROLLING NUTRIENT STORAGE

SOM provides a vast reservoir of nutrients for plants (Power, 1994; Brady and Weil, 1999). The mineralization of SOM is the primary source of available nitrogen, phosphorous, and sulfur in natural ecosystems. To a depth of 1 m a rich virgin soil can contain 17 t ha$^{-1}$ of N, not counting N in roots and surface litter (Jenny, 1985). Much of that N is contained in the SOM as a variety of compounds, ranging from amino acids to aromatic structures. Schulten and Leinweber (2000) developed molecular models of SOM. They calculated an elemental analysis of 54% C, 5.2% H, 4.7% N, 35.7% O, and 0.4% S for a total humic substance. However, the heterogeneity and dynamic nature of SOM result in a highly variable nutrient content. For example, the N content of SOM can range from less than 0.5% to more than 6%, depending on biotic and abiotic ecosystem properties such as climate, soil depth, annual input of organic materials, and soil mineralogy (Hassink, 1997). Nutrient content also varies widely between the different fractions of the SOM. For example, Paul and Clark (1996) found that fulvic acids contained 0.8% N and 0.3% S whereas humic acids contained 4.1% N and 1.1% S.

SOM is responsible for a large portion of the cation exchange capacity (CEC) in soil. Stevenson (1986) estimated that 20 to 70% of the whole soil CEC is because of humic substances, and the remainder can be attributed to silicate and nonsilicate mineral colloids. The relative contribution of SOM to total soil CEC in coarse-textured soils is usually greater than in fine-textured soils. Organic matter stabilization through association with clays means that SOM and the amount of CEC because of SOM increase as clay content increases. However, because of the increasing contribution of clay minerals to total soil CEC as clay content increases, the relative contribution of SOM to total CEC in fine-textured soils tends to be lower than in coarse-textured soils (see discussion in Chapter 1). The association of SOM with clay minerals provides physical protection from the mineralization activities of the soil organisms. The organomineral association, depending on its size, accounts for much of the potentially plant-available nutrients. Borogowski et al. (1976) showed that, depending on soil type, the organic–mineral complexes (<20 μm) can contain more than 90% of the exchangeable $Ca^{2+}$, $Mg^{2+}$, $K^+$, and $Na^+$. The availability of these nutrients is controlled by both equilibrium exchange into soil solution (which is analogous to mineral-associated CEC) and mineralization of SOM whereby nutrients are released from SOM as it degrades.

Conceptually, SOM is often divided into an active and a passive pool to describe the availability of nutrients from its complex assemblage of organic compounds and mineral interactions. The active pool provides many of the readily mineralizable nutrients and is composed of relatively recent plant residues, root exudates, and the microbial biomass (Tisdale and Oades, 1982). The passive pool is responsible for most of the CEC of the SOM and, in addition to exchangeable cations, contains nutrients that are tightly locked into complex organic–mineral assemblages (Stevenson, 1986). Intermediate pools of SOM are also involved in nutrient cycling along the continuum from active to passive SOM fractions. Jenkinson (1977) developed a model that described three SOM pools with different turnover times to describe C and N dynamics. Paustian et al. (1992) distinguished two litter and three SOM pools to describe SOM dynamics and nutrient cycling. The chemical extraction of SOM of classic humic fractions can also describe the fate of recently added N to soil. Bird et al. (2002) showed the importance of soil humic fractions, ranging from labile light fraction to resistant humin, in controlling the availability of recently added N fertilizer. The physical size separation of SOM-associated soil fractions has also shown promise in predicting available soil nutrients. Other approaches to determine the contribution of SOM to nutrient cycling include measuring the amounts of soil mineralizable C and N, microbial biomass, and enzymes (Gregorich et al., 1994). Despite extensive research by chemical, physical, biological and conceptual techniques, the accurate quantification of potentially available nutrients in SOM remains a challenge (Magid et al., 1996).

## PROCESSES AFFECTING NUTRIENT AVAILABILITY IN SOM

The availability of essentially all major nutrients is influenced by the presence of SOM (Magdoff and van Es, 2000). SOM supplies the available nutrient pool via mineralization and desorption and binds nutrients via immobilization and adsorption reactions (Figure 9.1). The fate of nutrients in the SOM is dependent on processes affecting organic matter decomposition and formation. The decomposition process is controlled predominantly by bacteria and fungi (Scow, 1997). Fauna that graze on microbes such as protozoa, nematodes, and earthworms also play a major role in nutrient cycling and are involved in the mineralization of nutrients previously thought to be attributed entirely to the microbial biomass (Clarholm, 1985; Coleman et al., 1984; Ruz-Jerez et al., 1992; Coleman and Crossley, 1997). Management strategies that target SOM accumulation for sustained nutrient availability must therefore provide a favorable environment to soil fauna and microflora because of their dominating role in mineralization–immobilization processes.

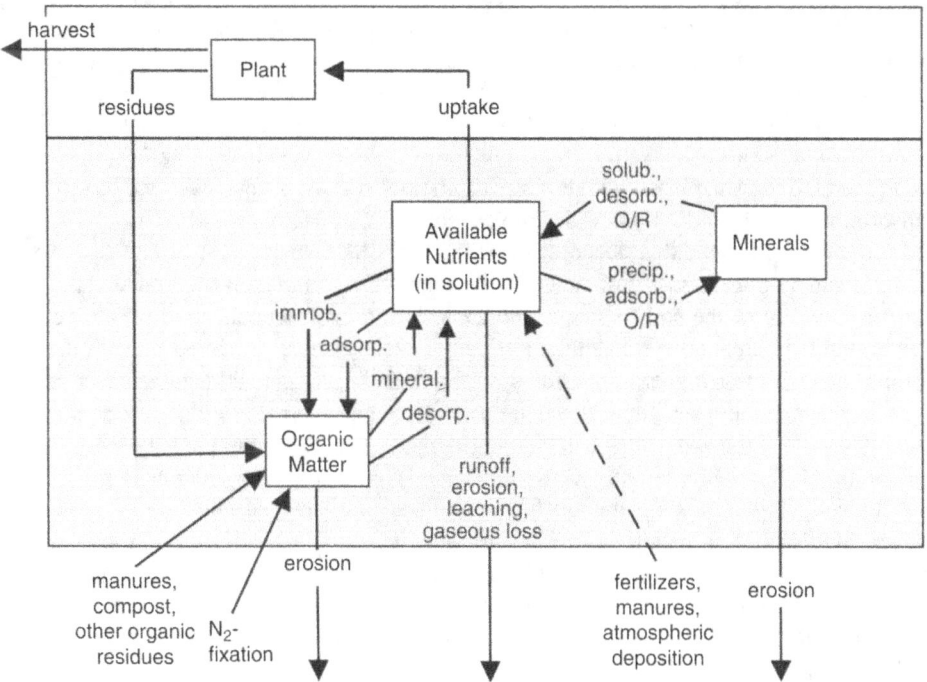

**FIGURE 9.1** Simplified nutrient flows and transformations in the soil-plant system. (Adapted from Magdoff, F. et al. 1997. *Adv. Agron.* 60:2–68. With permission.)

An often-cited goal of sustainable agroecosystem management is to accumulate and maintain SOM (Magdoff and van Es, 2000). The challenges in determining nutrient availability in cropping systems that are managed to accumulate SOM include assessing (1) the interaction of added nutrients (via organic residues and synthetic fertilizers) with soil organic nutrient pools and (2) the changes in SOM turnover dynamics due to management practices that slow the depletion of SOM (such as reduced tillage). Initially, in a low SOM situation the supply of plant-available nutrients is restricted because of inadequate active organic matter, specifically the particulate organic matter (POM) fraction and microbial biomass. As plant residues are added and SOM formation proceeds, soil microbial and fauna pools as well as POM increase (Hassink et al., 1994; Paul and Clark, 1996). These components of the active SOM are key to promoting nutrient mineralization in agroecosystems. Microbial and faunal biomass mediate the N mineralization, whereas POM contains much of the partially decomposed plant material that fuels mineralization (Hassink, 1995; Wilson et al., 2001).

In general, it is assumed that 1.5 to 3.5% of the SOM-N is mineralized annually in temperate climate agroecosystems (Brady and Weil, 1999). The actual rate at which nutrients are made available is highly variable and depends on a complex set of interacting factors, including vegetation type, SOM level, pH, soil texture, soil moisture, soil aeration, soil temperature, and management practices such as tillage and fertility amendments. Vegetation type and associated quality of residue inputs directly affect the availability of nutrients by influencing microbial C and nutrient use efficiency. Lower-quality plant residue having high C:N ratios, lignin, and polyphenol content can lead to immobilization or slow release of nutrients (Horwath et al., 2002). By contrast, very high-quality residues containing a low C:N ratio (and low lignin and polyphenols) can cause mineralization of nutrients far more than crop needs.

The quantity of N mineralized is not directly proportional to SOM. Magdoff (1991) showed that SOM level and soil texture interact to influence availability of N. At low soil N levels (and low SOM) in coarse-textured soils, mineralization rates are high but the low amounts of organic N mean that little N is made available. In fine-textured soils with high soil N (and high SOM), mineralization rates are low (probably because of stabilization of SOM in organomineral complexes), which also results in a relatively low N availability. SOM quantity and quality also affect the availability of nutrients other than N. Olk and Cassman (1995) found that SOM fractions rich in labile N could decrease K fixation by clay minerals in soils with high K-fixation potential. SOM can also increase P availability through mineralization of organic P sources as well as by reducing the adsorption onto Al and Fe oxides in tropical soils (Lopez-Hernandez, 1986).

Management practices that accumulate or maintain SOM usually also tend to have a high capacity to supply nutrients. Wilson et al. (2001) found that N mineralization potential (defined as the intrinsic ability of the soil to supply inorganic N through mineralization over time) (1) was higher in untilled perennial systems than in tilled annual systems, (2) increased with the addition of compost, and (3) was higher in rotation systems, including wheat and legume cover crops, than in corn–soybean rotations without cover crops. Management practices also affect how nutrient availability is synchronized with plant nutrient demand, which is important to reduce losses of soluble nutrients and increase nutrient use efficiency. Cover crops, for example, can buffer asynchrony by gradually releasing previously immobilized nutrients during time of peak demand. However, SOM and crop-residue mineralization might not supply sufficient nutrients during the peak demand times. Other management strategies can facilitate synchronization of nutrients from SOM. Hendrix et al. (1992) suggested cultivation to stimulate mineralization during plant growth, residue return to immobilize excess or residual inorganic nutrients, and continued organic inputs to replace nutrients removed from harvest.

Managing for SOM accumulation often produces improvements in soil quality, which can influence nutrient availability indirectly. Higher SOM levels generally increase porosity, which in turn promotes better root growth and distribution. This can lead to increased interception of nutrients and facilitate water-mediated nutrient movement to the roots. Nutrient availability is also influenced by the presence of chelating substances in the SOM. Chelators are substances of low molecular weight produced by soil microorganisms and present in SOM and organic amendments such as compost (Chen et al., 1998). Humic materials can also be important metal chelators in soils (Chapter 4). Chelating substances react with trace elements such as iron, zinc, copper, and manganese, forming bonds that protect these ions from precipitation reactions. In the absence of chelation, these nutrients would become insoluble and unavailable to plants at pH values commonly found in agricultural soils (Hodges, 1991).

The presence of growth-stimulating substances in the SOM can also contribute to enhanced nutrient capture and accumulation (Wiersum, 1974). These substances are biologically active metabolites of microbes produced during decomposition and formation of SOM, which stimulate plant root growth (Frankenberger and Arshad, 1995). Applications of humic material, such as humin, humate, and fulvic acids, have increased root growth and water uptake in agricultural crops (Russo and Berlyn, 1990) as well as sap flow in tree seedlings (Kelting et al., 1998a). Additions of these materials appear most effective in soils with low levels of humic substances (Mylonas and McCants, 1980), indicating that crops probably benefit from the higher levels of these substances present in soils well supplied with SOM (Kelting et al., 1998b). However, much of the effect of humic substances on plants can be their role as metal-chelating agents (Chapter 4).

## CROP MANAGEMENT STRATEGIES

### COVER CROPS

Cover crops are an important tool for integrated nutrient and SOM management. Cover crops can sustain nutrient cycling by adding N through fixation, retaining nutrients through SOM formation and nutrient uptake, and preventing nutrient leaching and runoff and erosion losses. One of the most important aspects of cover crops is the uptake of residual soil nutrients, which can significantly reduce nutrient movement off-site. Winter cover crops can significantly reduce soil nitrate in the soil profile (Figure 9.2). Cover crop management for nutrient retention and SOM accrual can be effective in a wide range of agricultural systems ranging from conventional to organic.

Vigorously growing legume cover crops can fix up to 300 kg ha$^{-1}$ of N, but 60 to 150 kg ha$^{-1}$ is the common range in temperate climate cropping systems, depending on cover crop species, plant density, and crop growth (Sarrantonio, 1994). Cover crops also contain significant amounts of phosphorous, potassium, and other nutrients. The need for externally supplied nutrient inputs can be reduced because nutrients in cover crop tissues become mineralized during decomposition and are potentially available to subsequent crops. Only a portion of the residue nutrients will be mineralized during the cropping season after killing of the cover crop. First-year legume N recovery rates range from 10 to more than 50% (Ladd et al., 1983; Hesterman et al., 1987; Bremer and van Kessel, 1992; Chung et al., 2000). Nutrient recovery from cover crop decomposition varies with differences in environmental factors (e.g., climate, soil conditions), type of management (e.g., shredding, mixing, soil incorporation), and tissue quality characteristics (e.g., content of C, N, cellulose, lignin, polyphenols).

For most cover crop species, the tissue C:N ratio increases with maturity. Therefore, the later a cover crop is killed, the less readily its N is mineralized. Herbaceous or vegetative-stage cover crops can rapidly decompose, providing N without increasing SOM (Kuo and Sainju, 1998). Under certain conditions, incorporating high-N cover crops can even result in a priming effect whereby the input of easily decomposable organic N can increase soil microbial activity, causing increased levels of SOM decomposition and associated nutrient release (Lovell and Hatch, 1998). In contrast, cover crops resistant to decomposition, such as mature small grains with high C:N ratios in the vegetative tissue, can increase SOM but provide only a small amount of readily available nutrients. Early studies reported that cover crops increased soil C in proportion to the quantity of C added (Pinck et al., 1948). However, the effects of cover cropping

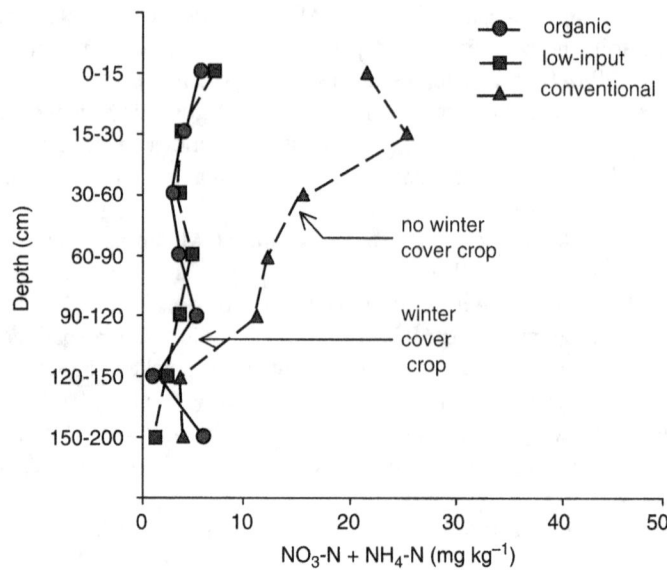

**FIGURE 9.2** Soil mineral N in the spring following tomato in the previous year at the Sustainable Agricultural Farming System Project at the University of California, Davis. A mixture of leguminous and grass cover crops was used in the organic and low-input agricultural systems. The conventional system was fallow during the winter. All systems received similar amounts of fertilizer as organic or inorganic sources. (From Poudel, D. D. et al. 2001. *Agric. Syst.* 68:253–268. With permission.)

on SOM levels vary widely. Ndiaye et al. (2000) found no consistent effect on total soil C when comparing a range of winter cover crops to winter fallow in rotation with summer vegetables for up to 7 years. When used at the same input rate of dry matter, cover crops usually increase SOM levels to a lesser degree than do animal manures or other more recalcitrant organic amendments such as peat (Gerzabek et al., 2001).

Various cover crop strategies have been developed to make nutrients readily available and at the same time build SOM. Drinkwater et al. (1998) showed that using high-N cover crops in combination with a diverse crop rotation can significantly increase soil C while meeting crop N needs. Another cover crop strategy involves growing mixtures of small grains and legumes, providing for a range of potential residue qualities in a single input (Kuo and Sainju, 1998). After incorporating the plant mixture, legume residues decompose quickly and provide nutrients whereas small grain residues (depending on the stage of development) decompose more slowly and contribute more to organic matter build-up. Incorporating cover crop residues with a range of residue C:N ratios can lead to the timely mineralization of available soil N for crop uptake. Collaa et al. (2000) showed that winter legume/grass cover crop mixtures maintained cash crop yields equivalent to those by using synthetic fertilizer (Table 9.1).

The concept of mixing residues with varying C:N ratios can be applied to a variety of organic amendments and cropping systems. For example, a mixture of slow and fast decomposing materials is used in alley cropping. In this system, crops are grown between rows of woody shrubs, which are cut periodically. Shredded prunings from the shrubs are then incorporated into the soil to serve as a green manure for the crops grown in the alleys (Kang et al., 1990). Decomposing leaves and small twigs provide readily available nutrients and decomposing pieces from woody branches provide slow-release nutrients and the raw material for organic matter build-up. Alternatively, use of cover crops in combination with manure applications can have similar effects (Poudel et al., 2001).

**TABLE 9.1**
**Two-Year Average Yield of Tomato in Conventional (Inorganic Fertilizer Only), Low-Input (One Half Conventional Fertilizer Rate Plus Winter Leguminous/Grass Cover Crop), and Organic (Manure and Winter Leguminous/Grass Cover Crop) Cropping Systems after 9 Years of Management**

| Nutrient Input | Marketable Yield (t ha$^{-1}$) | Unmarketable Yield (t ha$^{-1}$) | Total Yield (t ha$^{-1}$) |
|---|---|---|---|
| Full-rate inorganic fertilizer | 72.2 | 19.7 | 91.9 |
| Half-rate inorganic fertilizer + winter leguminous/grass cover crop | 72.6 | 25.4 | 98.0 |
| Manure + winter leguminous/grass cover crop | 69.0 | 26.9 | 95.9 |
| | N.S.[a] | N.S.[a] | N.S.[a] |

*Note:* All systems received similar amounts of N.
[a] Not significantly different. Determined by Duncan's multiple range test at the 0.05 probability level.
*Source:* Adapted from Collaa, G. et al. 2000. *Agron. J.* 92:924–932. With permission.

## CROP ROTATIONS

Carefully planned rotations can maintain or enlarge the active SOM pool to provide a steady supply of available nutrients for each crop in the sequence. Integration of perennial and high-residue crops into a crop rotation is a particularly important rotation strategy. Integrating crops with a diversity of life strategies (e.g., perennials vs. annuals), growth habit, and nutrient strategies ensures that over time residue materials of varying decomposition rates enter the soil to supply nutrients while maintaining SOM.

### Including Perennial Crops

Including perennial forage grasses or perennial legumes in a crop rotation that is otherwise composed of annual crops is an invaluable tool to increase SOM and nutrient supply for subsequent crops. During a long-term experiment in Germany, soil C content rose by 10% in plots with continuous grassland whereas soil C under continuous potato monocrop (fertilized with synthetic fertilizer) decreased by 50% over the course of 32 years (Haider et al., 1991). Tillage intensity and frequency are often found to be primary factors in determining C loss or accumulation (Wood et al., 1990). Weil et al. (1993) compared cropping systems of various management intensities and found SOM levels significantly higher in a grass system that was only mowed than in various annual cropping systems that included tillage (Table 9.2).

Active SOM fractions that provide potentially plant-available nutrients are markedly increased under perennial forages (Drury et al., 1991; Angers et al., 1993). The continuous contribution by the roots (rhizodeposition) is a major factor that promotes higher levels of these active SOM sources. Rhizodeposition in the form of fine root turnover and root excretion of organic compounds can account for more than 30% of the net primary production of perennial forages (Johansson, 1992; Beauchamp and Voroney, 1994). In cropping systems containing perennial legumes, decomposing nodules and excretion of organic compounds provide N-rich substrates for the soil microflora (Burity et al., 1989; Dubach and Russelle, 1994) and contribute significant quantities of available nutrients to succeeding crops. Weil et al. (1993) observed approximately twice as much metabolic activity in soil under grass compared with continuous tilled corn, indicating the value of including crops in rotation that contribute to the active SOM fractions.

**TABLE 9.2**
**Total Organic C and Extractable C in the 0- to 15-cm Soil Layer of Five Cropping Systems Established in 1985 as Determined by Wet Digestion in 1991**

| Cropping System | Total Organic C (g kg$^{-1}$) | Extractable C[a] (g kg$^{-1}$) |
|---|---|---|
| Continuous grass | 20.0 | 0.51 |
| Rotation of annual crops: no-till + chemicals | 11.0 | 0.32 |
| Rotation of annual crops: minimum till + reduced chemicals | 8.3 | 0.29 |
| Rotation of annual crops: reduced till, organic | 9.6 | 0.23 |
| Continuous corn: conventional till | 7.9 | 0.12 |
| | 4.8 (LSD$_{0.05}$) | 0.15 (LSD$_{0.05}$) |

[a] 0.5 $M$ NaHCO$_3$ extract.
*Source:* Adapted from Weil, R. R. et al. 1993. *Am. J. Altern. Agric.* 8:5–14.

In addition to increasing active SOM pools, perennial crops can also reduce the loss of total SOM. The permanent cover provided by perennial crops can protect the soil from water and wind erosion, which can carry away SOM-rich soil fractions (Tisdale and Oades, 1982). Furthermore, nutrient uptake and subsequent conversion into plant biomass in perennial crops occur over a longer time period in the cropping season compared with annual crops. The prolonged growth habit of perennials ensures scavenging of residual soil nutrients. Integrating perennials into a crop rotation thereby conserves nutrients in the soil ecosystem and is a positive step toward creating sustainable nutrient cycling from active SOM fractions.

## Including High-Residue Crops

In many agricultural systems worldwide, the decline in long-term soil fertility is often a direct consequence of partial or complete removal of aboveground biomass as food, feed, bedding, fuel, or building materials. Loss of near-surface nutrient-rich soil and active SOM fractions exacerbates this problem. Growing crops that return a large amount of residue to the soil is an important strategy to replace nutrient outputs, replenish SOM pools, and reduce the need for other inputs such as fertilizers, cover crops, or organic amendments. When corn (*Zea mays*) is harvested for silage, with the exception of short stubble, virtually all aboveground biomass is removed. At a yield of 45 Mg ha$^{-1}$ of 30% dry matter silage, the average nutrient removal amounts to 202 kg of N, 49 kg of P, and 205 kg of K (Jokela et al., 2000). Supplemental nutrients are needed to replenish the nutrient pool. Reliance on mineralization, desorption, and mineral dissolution for the supply of nutrients to succeeding crops when no organic amendments or cover crops are used leads to a long-term decline in nutrient reserves (Magdoff, 1991).

One of the most important factors determining the level of SOM is the size of the C inputs to the soil (Rasmussen and Collins, 1991; Park, 2001). Residue removal or prolonged fallow periods without considerable substrate additions via weeds or soil amendments can have a dramatic negative effect. Gerzabek et al. (2001) measured a 30% loss of the organic C from the topsoil layer over the course of 44 years in fallow plots of a long-term experiment in Sweden. Available nutrients are quickly depleted without continuous biomass input, because active SOM components are mineralized early in the decomposition process (Elliott and Papendick, 1986). Microbial activity decreases once labile components are depleted, which further limits the SOM to supply nutrients for plant growth. Conversely, nutrient turnover can quickly be restored when residues are added to soil.

Alvarez and Alvarez (2000) observed that the active microbial biomass highly correlated to the amount of plant residue during the first year after the introduction of different cropping systems. These studies indicate that a lack of C inputs reduces the active SOM fraction. At the same time, active SOM responds readily to C additions and the degree to which nutrient availability is restored depends also on the quality of the organic matter addition.

Hendrix et al. (1990) described hypothetical patterns of litter decomposition, nutrient availability, and plant nutrient uptake for four litter input scenarios to illustrate how residue quality influences the degree to which nutrient availability is synchronized with crop demand (Figure 9.3). High-quality residue (high N, low lignin, low polyphenol concentrations) can decompose quickly but not increase SOM even at a high dry matter input (Bruce et al., 1990). Conversely, returning large amounts of low-quality biomass can increase SOM and immobilize nutrients. For example, Powel and Hons (1991) investigated the effects of stover removal on SOM and extractable nutrients in a continuous sorghum cropping system. They found that when low-quality stover was returned to the soil over a 4-year period, sorghum yield and P uptake were lower, whereas complete stover removal resulted in decreased SOM levels but a net release of nutrients.

Residue return and tillage systems are interrelated in their effect on SOM. In cropping systems with a low residue return, such as silage corn, the type of tillage practice or its intensity might not affect total soil C significantly (Angers et al., 1993). On the other hand, intensive tillage can reduce or even negate the generally positive effect of high biomass additions on SOM levels by increasing SOM mineralization (Reicosky et al., 2002).

## Including a Diversity of Crops

At least since the 18th century, crops were classified as either soil enriching or soil depleting (Deane, 1790). This knowledge has led to the practice of alternating crops that differ in their effects on soil fertility. Although some of the specific beliefs from that era had to be adjusted (e.g., the belief that all root crops are soil enriching), the basic rotation principle is still valid and used at present. Several recent studies have found that diverse crop rotation can achieve both an increase in SOM and provide sufficient amounts of available nutrients. Topsoil organic C increased by 6.6 Mg ha$^{-1}$ over a 15-year period when legume cover crops were the sole supply of N for grain crops in a long-term study at the Rodale Institute in Pennsylvania (Drinkwater et al., 1998). Soil N levels paralleled the SOM increase. The authors attributed these results to the diversity of the cropping sequence and the associated qualitative differences in organic residues returned to the soil. Differences in tillage intensity might also have played a role in changes in SOM. Franco-Viscaino (1996) compared a wide range of high- and low-diversity cropping systems and showed that increased diversity of residues was associated with improved soil tilth, nutritional status (higher total N but not extractable N), and biological activity. Improvements were associated not with a single management practice but rather with diversity and frequency of residue sources coming from crop rotation, cover crops, or manure applications. The exact mechanism of how the diversity of residue inputs enhances SOM and nutrient levels is not well understood but probably relates to components of the active SOM and microbial processes. Diversity of inputs most likely leads to a complex microbial community structure (Neher, 1999; Chapter 7), which leads to a broad functional diversity, resulting in a greater substrate utilization efficiency and stress resistance (Kennedy, 1995).

A study by Sanchez et al. (2001) supports the link between substrate diversity and potentially available nutrients. Net N mineralization measured *in situ* at 70 d in a diverse cropping system was 70% higher compared to a corn monocrop system. Net N mineralization in the diverse system was enhanced by the incorporation of residues from legumes, grasses, and composted manure. However, they pointed out that the two systems did not differ in their ability to mineralize added substrate. There was no significant difference between both systems in the percentage of N mineralized from added legume residues or compost. Jenkinson (1977) also observed that the rate of mineralization of plant residues was independent of the rate of addition. These studies suggest that input history

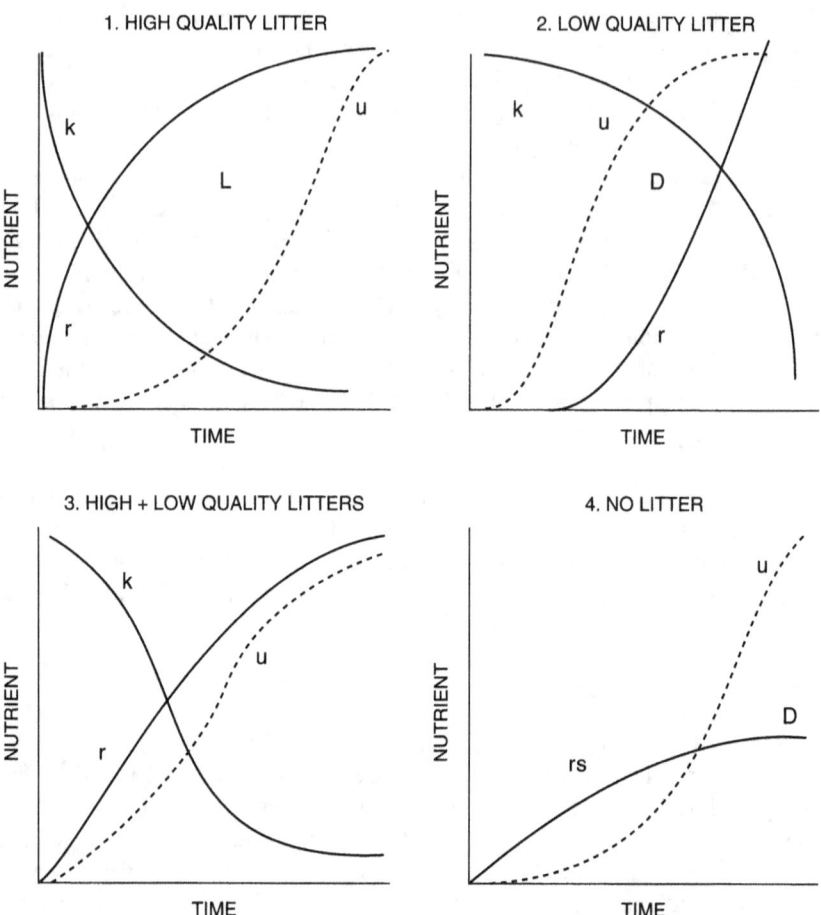

**FIGURE 9.3** Hypothesized pattern of decomposition (k), soil nutrient availability (r), and plant uptake of nutrients (u) in systems subject to inputs of various qualities; (rs) is release of nutrients from soil. Area L represents potential of nutrient loss by leaching; area D represents potential nutrient deficit (Swift, 1987). High-quality litter (Panel 1) decomposes quickly, releasing nutrients out of phase with plant uptake, resulting in high potential for loss. Low-quality residue (Panel 2) decomposes too slowly to provide nutrients for plant uptake and might stimulate microbial immobilization, resulting in a deficit for plants. Panel 3 represents ideal situation in which nutrient release is synchronized with plant demand. Nutrient release from soil organic and mineral pools (Panel 4) might also be synchronized with initial stages of plant growth but might not meet plant demand in many soils. (From Hendrix, P. F. et al. 1990. In Edwards, C. A., Lal, R., Madden, P., Miller, R. H., and House, G. (Eds.), *Sustainable Agricultural Systems*. Soil and Water Conservation Society, Ankeny, IA, 637–654. With permission.)

is of little importance in a cropping system's ability to mineralize nutrients and is more likely a result of the microbial biomass reacting to the input of easily decomposable compounds through the increase of its size and activity (Paul and Clark, 1996).

Alternating crop types with different growth habits also promotes the protection of SOM and nutrient availability. Rotating between deep- and shallow-rooted crops provides the obvious benefit of nutrient extraction from different soil layers. Including deep-rooted crops in the rotation also introduces new organic material into deeper soil layers by rhizodeposition. Park (2001) notes that continually growing shallow-rooting crops can lead to a gradual loss of organic material in deeper soil layers, the consequence of which could be a sharp decline in soil quality in that layer.

## TILLAGE

Tillage plays an important role in the management of soil nutrients through its influence on SOM dynamics. Tillage incorporates plant residues or living plants into the soil. The mixing action enhances aeration and the contact between soil and plant debris, resulting in favorable conditions for rapid mineralization of C and other organically bound nutrients (Parr and Papendick, 1978). In addition, the breaking apart of macroaggregates by tillage increases the availability of occluded SOM to soil organisms. Type, frequency, and intensity of tillage determine the degree to which mineralization processes occur. Higher intensity and frequency of tillage generally result in lower SOM (Blaesdent et al., 1990; Carter, 1992), nutrient retention, and nutrient cycling capacity (House et al., 1984). Gallaher and Ferrer (1987) showed that untilled soil contained 20 and 43% more Kjeldahl N than conventionally tilled soil in the 0- to 5-cm soil depth after 3 and 6 years, respectively. Staley et al. (1988) observed higher organic P levels in the top 0- to 15-cm soil depth at lower tillage intensity. Absence of tillage tends to increase N mineralization capacity (Weil et al., 1993). For example, Doran (1987) found potentially mineralizable N to be 37% higher in the surface layer of no-till soil compared with tilled soil.

Most changes in SOM fractions due to reduced tillage occur in the upper few centimeters of the soil (Chapter 8). Reduced-tillage systems tend to accumulate plant residues, fine roots, and microbial and microfaunal debris (Gregorich et al., 1994; Alvarez et al., 1998), thereby increasing the rapidly mineralized pool of organic matter. Intensive tillage, on the other hand, reduces labile SOM. Angers et al. (1993) found similar total soil C in reduced tillage and moldboard plowing treatments, but microbial biomass and labile sand-sized OM fractions (accounting for 1 to 20% of total soil C) were significantly larger under reduced tillage. Microbial biomass in their study accounted for 1.2–1.4% of the organic C in the moldboard plow treatment and 3.5–5.1% in the minimum-tillage treatment. Higher microbial biomass can result in higher immobilization of added fertilizer N in the 0- to 5-cm soil depth of no-till compared to conventional-till systems (Carter and Rennie, 1984).

The labile POM fraction is affected disproportionately by tillage and often adversely impacts C and N mineralization (Hassink, 1995; Hussain et al., 1999). POM is lost as tillage disrupts macroaggregates that provide physical protection from decomposing organisms (Elliott and Papendick, 1986). Conversely, no-till practices often increase the amount of organic C and N in POM (Wander and Bidart, 2000). Fine SOM fractions are less vulnerable to tillage disruption due to closer binding to soil minerals in the form of mineral–organic complexes. Over time, intensive tillage increases the percentage of the SOM associated with minerals as more free organic plant debris is lost (Tiessen et al., 1994). The reduced quantity of POM and fresh plant residues decreases the ability of the soil to supply available nutrients to agricultural crops because these SOM fractions are the main sources of mineralizable N in many soils (Wilson et al., 2001).

Contributing to the reduced mineralization potential in intensively tilled systems is the negative effect of tillage on the number of macro- and mesofauna, such as earthworms and arthropod species (Hendrix et al., 1986; Parmelee et al., 1990). Soil fauna contribute to soil C and N mineralization directly through their own C and N mineralization and indirectly through grazing on microbes and passing plant residues and soil through their digestive tracts. (Hassink et al., 1994; Coleman and Crossley, 1997). Without the grazing activities less N in the microbial biomass is mineralized and unavailable for plant uptake. Beare et al. (1992), for example, found a 25% higher N retention in plots in which fungivorous microarthropods were excluded from no-till surface litter compared with plots with microarthropods. The effects of soil fauna on nutrient availability are generally positive; however, there have been limited studies to assess their impact across a range of soil management practices.

The type of tillage system and the implements used determine how plant debris is distributed in the soil. Associated with the distribution of plant debris are SOM and nutrient levels. Plowing followed by harrowing, for example, distributes nutrients throughout the tillage depth (Hussain et al., 1999), whereas reduced-tillage systems tend to accumulate nutrients and decomposable organic matter in the soil surface (House et al., 1984; Karlen et al., 1994). Microbial activity levels mirror the organic matter distribution (Kandeler et al., 1999). Compared with plowed soil, higher microbial activity is consistently found in the surface layer of reduced-tillage soil, whereas less microbial activity is observed in deeper layers (Granastein et al., 1987; Angers et al., 1993; Friedel et al., 1996). Higher microbial biomass in no-till systems is observed usually only in the 0- to 7.5-cm layer, whereas biomass in the 7.5- to 15-cm layer is often more in the conventionally tilled soil (Staley et al., 1988). The stratification of fresh plant residue and decomposed organic matter under different tillage systems also affects the microbial community structure, leading to changes in soil C and N dynamics (Bossio et al., 1998; Chapter 10). Reduction in tillage intensity and retention of residues on the soil surface tend to favor fungi over bacteria. The higher C assimilation efficiency and slower turnover of fungi tend to promote slower nutrient cycling (Holland and Coleman, 1987; Hendrix et al., 1990; Beare et al., 1992).

Tillage-induced changes in SOM levels can directly affect nutrient storage and availability. For example, when intensive tillage lowers SOM levels, the CEC is also reduced. Various studies that compared different tillage systems have tracked CEC. Long-term plowing and harrowing on a clay soil caused a decline in SOM from 5.2 to 4.3% and a concurrent decline of CEC from 17.8 to 15.8 cmol kg$^{-1}$ (Magdoff and Amadon, 1980). Mahboubi et al. (1993) also found that after 28 years there was lower CEC with higher tillage intensity. Karlen et al. (1994), on the other hand, found no effect over 12 years. The mixed findings are likely a result of interacting environmental and management factors and the extended period of time, often 10 to 20 years, required to observe measurable changes in SOM. Hussain et al. (1999) noted that individual nutrients could be affected differently by various tillage systems because of tillage-induced changes in the soil matrix. In their study Ca, Mg, and Bray P-1 increased in surface soil of no-till plots compared with plowed soil, whereas increased leaching under no-till lowered exchangeable K levels.

## NUTRIENT APPLICATIONS

### INORGANIC FERTILIZER

Inorganic fertilizer applications affect SOM and nutrient management in many ways. First, as a nutrient source, inorganic fertilizers promote the production of plant biomass and therefore affect the potential amounts of crop residue that can be returned to the soil (Allison, 1973). Second, inorganic fertilizer nutrients such as P and N can become integrated into the soil matrix either directly into organic compounds of labile SOM fractions or as part of mineral–organic complexes (Polglase et al., 1992). Integration into the organic matter pools can be rapid. Balabane and Balesdent (1992), for example, recovered 26% of the $^{15}$N- labeled ammonium nitrate fertilizer in SOM 6 months after application. The microbial biomass is often a large part of the initial build-up of added N, which is then followed by their turnover into more stable fractions such as organomineral phases (Paul and Clark, 1996).

Inorganic fertilizer applications also affect decomposition rate of fresh residues and the integration of residues into the SOM (Parr and Papendick, 1978). Inorganic N in particular can drastically affect the microbially mediated breakdown of fresh plant residues because microbial activity is often limited by N (Jenkinson et al., 1985). To reduce the immobilization of N already present in the soil or to speed up the microbial decay process, farmers often use inorganic N applications when large quantities of organic materials with a high C:N ratio are incorporated into the soil. The amount and timing of the N application to achieve these goals depend on the organic material being decomposed and climate conditions that would promote decomposition.

Different views exist on the effect of inorganic N applications on the SOM levels. N fertilizer applications generally result in a decline of organic matter because the readily available N leads to rapid microbial decay of SOM in some soils. Green et al. (1995), for example, observed decreased

SOM in studies in which excessive amounts of nitrate fertilizer were added annually for multiple years. Based on laboratory incubation studies, another view on the effect of N fertilizer holds that there is only a short-term stimulation of the organic matter decay process but long-term SOM levels are not affected (Allison, 1973). Glendining and Powlson (1991) compared multiple long-term field studies and found that continuous inorganic N fertilization increases both the amounts of total soil N and readily mineralizable N. Increase in the proportion of mineralizable N over time indicated that the additional organic N returned to the soil as a result of using fertilizer N (i.e., via higher plant biomass or incorporation by the microbial biomass) is in a fraction that turns over more rapidly than the N in older organic matter.

Inorganic fertilizers generally enhance C inputs through increased biomass production, but excessive inorganic fertilizer applications can have negative effects on active SOM and soil N pools, such as microbial biomass (McCarty and Meisinger, 1997). Microbial biomass is usually lower in soils with long-term inorganic N applications than soils that have received organic amendments (Collins et al., 1992). Fauci and Dick (1994a) analyzed soils from a long-term study and found that of the crops that received additional N, microbial C and microbial N were lowest in plots with a 59-year history of inorganic N fertilizer, intermediate in plots fertilized with pea vine residues, and highest in manure plots. The differences were closely related to the quantity and quality (C:N ratios) of C substrate added.

## ANIMAL MANURE

The application of animal manure is an important tool for an integrated nutrient management strategy because applications can simultaneously increase SOM levels and supply nutrients for crop growth. The mix of feces, urine, and bedding material present in many types of animal manure generally provides a combination of recalcitrant and labile organic materials. For example, annual application of 34 Mg ha$^{-1}$ of fresh dairy manure provided all necessary nutrients to crops and tripled SOM levels over the course of 120 years at the Rothamstead experiments (Jenkinson, 1991). The rate of SOM accumulation is usually highest in the first 10 years of manure application and slows thereafter (Sommerfeldt et al., 1988). Increases are generally lower when initial SOM levels are already high. Magdoff and Amadon (1980) showed that yearly applications of 66 Mg ha$^{-1}$ of fresh dairy manure were needed to increase SOM from 5.2 to 5.5% over the course of 11 years on a land on which continuous silage corn was produced by using conventional tillage (Table 9.3). Yearly applications of 44 Mg ha$^{-1}$ were needed to maintain SOM at the original level.

**TABLE 9.3**
**Effect of 11 Years of Solid Dairy Manure Additions on Properties of a Panton Clay Soil (Typic Ochraqualf) in Vermont**

|  |  | Yearly Application Rate (Mg ha$^{-1}$ Wet Weight) | | | |
|---|---|---|---|---|---|
|  | Original | None | 22 | 44 | 66 |
| Organic matter (%) | 5.2 | 4.3 | 4.8 | 5.2 | 5.5 |
| CEC (me 100g$^{-1}$) | 17.8 | 15.8 | 17.0 | 17.8 | 18.9 |
| pH | 6.4 | 6.0 | 6.2 | 6.3 | 6.4 |
| P[a] (mg kg$^{-1}$) | 4 | 6.0 | 7.0 | 14.0 | 17.0 |
| K[a] (mg kg$^{-1}$) | 129 | 121.0 | 159.0 | 191.0 | 232.0 |
| Total pore space (%) | N.A.[b] | 44.0 | 45.0 | 47.0 | 50.0 |

[a] Determined by modified Morgan extraction.
[b] Not available.
*Source:* From Magdoff, F., and van Es, H. 2000. *Building Soils for Better Crops.* USDA Sustainable Agriculture Network (www.sare.org), Beltsville, MD. With permission.

After manure is applied to the soil, organic nutrients become either part of the vegetation or the SOM, or they are lost from the field. For N, losses occur via runoff, erosion, volatilization, and denitrification. Of the N that becomes part of the SOM, some fractions are mineralized rapidly and available for plant uptake in a short time whereas other fractions are mineralized more slowly. The rate of manure N mineralization has been reported to range from 13 to 86% (Chae and Tabatabai, 1986). This rate is affected by a multitude of factors, including manure composition (manure N concentration, C:N ratio, stability of C and N), animal type, environmental factors (soil temperature, soil moisture, aeration), soil texture, and various factors that affect microbial activity (pH, soluble salt content, toxic chemicals, heavy metals).

Animal and manure management have a strong influence on whether manure nutrients are integrated into the stable SOM or the labile SOM pools. For example, the type and quantity of bedding can significantly alter the C:N ratio and thus the balance between immobilization and mineralization. Magdoff and van Es (2000) noted that dairy and beef manures that contain a high amount of lignified substances, such as bedding or undigested parts of forage, contribute to SOM maintenance much more than layer chicken manure, which commonly does not contain bedding material. The manure storage system in part determines manure dry matter content, which also plays a significant role in organic nutrient fluxes. In a 4-year study in California, Pratt et al. (1976) found that dry manure gave higher increases in SOM compared with liquid manure (19% vs. 4%), whereas liquid manure had larger gaseous losses (24% vs. 2%) by ammonia volatilization and denitrification. As an approximate guide for nutrient management planners, Klausner (1997) estimated availability of different forms of N in dairy manure for northeastern U.S., based on time of application, dry matter content, and residual organic manure N (Figure 9.4).

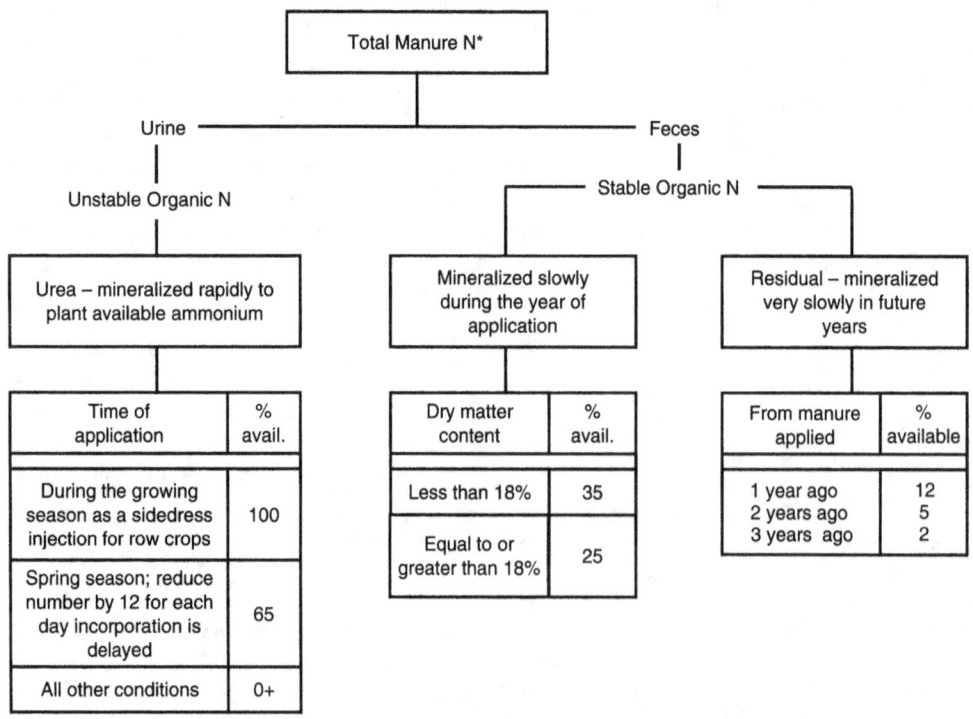

∗ *In solid dairy manure approximately 50% of total manure N is in the urine and 50% in the feces.

**FIGURE 9.4** Estimated availability of different forms of N in dairy manure in northeastern U.S. (Adapted from NRAES, 1997. *Nutrient Management: Crop Production and Water Quality*. Publication 101, Northeast Regional Agricultural Engineering Service (NRAES), Ithaca, NY. With permission.)

Organic manure N not mineralized becomes integrated into the SOM and is an important residual nutrient source for future crops. Pratt et al. (1973) developed the concept of a decay series that described the amount of inorganic N that becomes available with time. Klausner et al. (1994) developed a decay series for dairy manure. They suggested that under the climate conditions of northeastern U.S., ca. 21% of the initial total N mineralizes in the first year, 9% of the N remaining after the first year mineralizes in the second year, and 3% of N remaining after the second year mineralizes in the third year. Residual organic N can supply the entire N need for crops in combination with the rapidly available N fraction from a current manure application. Klausner (1997) reported that without residual N or supplemental inorganic fertilizer, an application of 168 Mg fresh weight of solid dairy manure per hectare was required to provide sufficient available nutrients to achieve optimal corn silage yield in trials in northeastern U.S. In a separate trial, only 52 Mg fresh weight of solid dairy manure per hectare was required to achieve optimal corn grain yield when applied continuously to create large residual nutrient pools.

The gradual release of inorganic N from soil that is continuously supplied with manure is often well synchronized with the nutrient demand of long-season agronomic crops such as corn. Ma et al. (1999) compared N mineralization, N uptake, and yield in corn to which solid dairy manure was applied in the fall with corn that received $NH_4NO_3$ fertilizer in the spring. They found that available N in the manured system was better synchronized with plant demand than in the inorganic fertilizer system. Yield and plant N uptake were higher in the manured crops, whereas mineral N losses from the rooting zone were higher in the system that received inorganic fertilizer. In contrast, Kirchmann and Bergström (2001) noted that the nitrate-leaching potential from manures is high when organic N is mineralized in the soil and no N uptake by the crop occurs. In temperate-climate agricultural systems, this usually occurs in the spring when the cash crops have not been planted or are still very small and also occurs in the fall when the cash crop is already harvested and no cover crops were planted.

## COMPOST

Compost applications can form the foundation of a successful whole-farm nutrient management strategy. Composts are a major source of nutrients in organic farming systems (Lampkin, 1990). Biodynamic and biointensive farming also rely heavily on compost applications for crop nutrient supply (Pia, 1998; Carpenter Boggs et al., 2000). Nutrient release rates vary widely with type of raw materials, method of composting, and maturity (Sims, 1995). Nutrients in composts are usually less available than in fresh manure (Hadas et al., 1996) because of stabilization by microbial assimilation and humification during the composting process (Castellanos and Pratt, 1981). Within one growing season, Eghball (2000) measured 11 and 21% organic N mineralization from a field applied composted and noncomposted beef cattle feedlot manure, respectively. Haddas et al. (1996) found high N mineralization rates of composted cattle manure under favorable conditions. In their study up to 26% of the total N was mineralized in only 33 weeks. A flush of net N mineralization in the first week was followed by net N immobilization between 2 and 4 weeks after application. They noted that percent recovery of compost N as inorganic N was independent of soil history or rate of application, but largely reflected the properties of the compost.

Depending on the total amount of compost applied and the residual from previous applications, the supply of mineralized nutrients from a compost application, particularly N, might not be adequate to achieve optimum yield (Eriksen et al., 1999; Chung et al., 2000). Inorganic fertilizers are sometimes applied along with compost if there is concern about low N availability. This practice might result in a priming effect in which N mineralization from compost is increased. Application of organic sources of labile N, such as clover residues, in combination with compost has the same effect (Sanchez et al., 2001). It appears that the addition of labile N sources mainly stimulates mineralization of compost and less of SOM (Sikora and Yakavchenko, 1996). However, the practice can result in high nitrate leaching rates if plant nutrient uptake is not synchronized with mineralization of the added compost (Gerke et al., 1999). C and nutrient losses during the composting process are significant and need to be considered when assessing the value of compost applications to long-term soil fertility.

Depending on raw materials and composting method, 20 to 90% of the C present in the raw materials can be lost (Eghball et al., 1997; Shellinger and Breitenbeck, 1998). In a comparison of windrow composting of nine different raw materials, Shellinger and Breitenbeck (1998) observed N, P, and K losses of 0–62%, 9–70%, and 1–79%, respectively. They noted that greatest losses of these nutrients occurred in silage, bark, and cotton gin trash, all raw materials that contain large amounts of readily degraded organic substrate and low amounts of mineral matter, such as clay or other soil contaminants. Eghball et al. (1997) measured 40% loss of total N where ammonia volatilization accounted for 92% of the N loss. Compost storage can play a significant role in the extent of nutrient losses. Sommer (2001) found that compacting and covering with a porous tarpaulin reduced N leaching losses from 28% to 12–18%. The same treatments also reduced K leaching.

Nutrient and C losses during composting challenge the notion that composts are an efficient way to increase the capacity of the soil to meet long-term nutrient needs. Nevertheless, high compost applications can effectively increase SOM and nutrients. In addition, compost is less likely to act as a priming agent (Smith, 1979), whereas applying less stable materials such as fresh animal manures can lead to mineralization of SOM (Glendining et al., 1996; Ma et al., 1999). In the long term, however, the amount of organic matter applied can be more important to SOM accumulation than the type of organic amendments used (Delschen, 1999).

The increase of SOM in soil continually supplied with composts can result in improvement of soil quality indicators that facilitate nutrient availability and uptake (Roe, 1998). For example, Pinamonti (1998) showed that improved porosity and water retention as well as reduced temperature fluctuations in a vineyard supplied with compost were more important in improving nutrient uptake than increased availability of soil nutrients. Biological soil quality indicators, such as biomass C and basal respiration, also improve with compost applications (Pascual et al., 1997). Application of composts that contain high levels of heavy metals, however, can inhibit biological soil quality indicators, such as enzyme activity (Moreno et al., 1998) and microbial biomass (Moreno et al., 1999) and might impede nutrient mineralization processes.

## EXCESS NUTRIENT LOADING ASSOCIATED WITH ORGANIC AMENDMENTS

The need to reduce input cost and the heightened awareness of the potential soil quality benefits have contributed to the increased popularity of manure, compost, and other organic amendments in agricultural production. At the same time, improper storage and excessive land application rates of organic amendments (especially animal manures) have contributed to some of the most serious environmental issues facing agriculture at present. These issues have led to an intense debate on the appropriate management of animal-derived nutrients, particularly in regions that have a high concentration of farm animals and limited land area to apply manure.

Throughout this chapter we have argued that adding plant and animal residues to the soil has positive effects on soil and environmental quality. There is a widely held belief that compost applications in particular are environmentally benign because nutrients are released relatively slowly. However, there is increasing evidence that animal manure and compost applications applied at rates used by some farmers can result in dramatic overloading of available nutrients in soil. Soil and tissue testing show that excessive soil nitrate concentrations during and beyond the growing season are commonplace in agricultural systems that use animal manures and compost (Maynard, 1994; Leclerc et al., 1995). Excessive P levels are frequently created by basing manure and compost application rates on the N need of the crop. The combination of low N:P ratios in many organic amendments and low crop P removal rates leave much of the applied P unused in the soil. In addition, several studies have found that manure P is more mobile than synthetic fertilizer P (Eghball et al., 1996; Parham et al., 2002), enhancing its potential for off-site movement.

Potential consequences of overloading the soil with nutrients and off-site movement include leaching of nitrate into groundwater (Sommerfeldt et al., 1988) and the accumulation of P in the soil (Gartley and Sims, 1994). This accumulation increases the chance of harmful P concentrations leaving the field through surface runoff and subsurface flow (Daniel et al., 1994). An example of the potential problems associated with organic amendments is a recent study at the Rodale Research Institute in Pennsylvania, which compared nutrient budgets and soil C levels following four composts, raw dairy manure, and mineral fertilizer applications (Reider et al., 2000). Application rates were chosen to provide similar amounts of available N to all treatments, but total nutrient inputs differed widely (Table 9.4). After 3 years, three (dairy manure leaf compost, yard clipping compost, and controlled microbial compost) of the four compost treatments created a significant increase in SOM, whereas mineral fertilizer and raw manure did not. The compost treatments created a significant nutrient surplus and at the same time elevated levels of extractable soil potassium. Extractable P and total N were less affected, suggesting that the vast surplus of these nutrients were fixed in the soil matrix or were lost from the topsoil layer through runoff, erosion, leaching, or volatilization. The researchers recommended that compost applications should be reduced and substituted with alternative nutrient sources.

The striking difference in residual nutrient content between organic additions and nonorganic fertilizers requires distinctly different management strategies. Inorganic fertilizers are usually managed on a seasonal basis, with emphasis on meeting crop demand during the growth period. On the other hand, the use of organic amendments probably means higher total nutrient application rates during the initial transition from inorganic-fertilizer-based systems. Mineralization of organic residues from the previous year's application, as well as earlier applications, add to the pool of available nutrients. Thus, if the annual application rate of an organic amendment is constant, over time the ratio of the quantity of nutrients made available by mineralization relative to total nutrient input from organic amendments will increase. For example, Suzuki et al (1990) found that when applying compost annually at a rate chosen to supply the entire N needed in the first year, the amount of N mineralized each year increased from about the equivalent of 70% of the total N applied annually after 20 years to ca. 90% after 50 years. Because of the increased availability of nutrients from residual sources, a gradual reduction of the annual input is needed to reduce nutrient overloading (Figure 9.5 and Figure 2.2). Quantification of the appropriate reduced rates over time under various soil and climate conditions has so far not received adequate attention in long-term research studies.

Integrating legume cover crops to supply N as part of the fertility management regime will also reduce the potential of nutrient overloading and reduce possible problems with excess salt build-up. Another strategy involves a combination of organic wastes and inorganic fertilizers to take full advantage of all the potential nutrient sources and reduce losses to the environment. Fauci and Dick (1994b) found that, under greenhouse conditions, a decreasing rate of synthetic fertilizer over the course of several cropping cycles in combination with poultry manure or pea vine residues was an effective means of maintaining crop productivity during the transition from inorganic to organic fertility sources. Gerke et al. (1999) used a combination of kitchen and garden waste compost at a rate of 10 Mg ha$^{-1}$ year$^{-1}$ and synthetic N additions of ca. 20 kg ha$^{-1}$ year$^{-1}$ in a crop rotation dominated by winter grain to achieve relatively high yields and acceptably low nitrate concentrations. They further suggested that similar results can be achieved with a combination of matured and nonmatured compost.

## CONCLUSIONS

Nutrient management goals at the field level require such strategies as greater return of organic materials, use of perennial and cover crops, and reduced tillage. These strategies contribute to the accumulation of the active fraction of the organic matter, which is crucial for a sustainable management of nutrient cycles in agroecosystems. Management practices that promote SOM

**TABLE 9.4**
**Average Dry Matter Application Rate and Cumulative N, P, K Budget of Organic Amendments in a Three-Year Study in Pennsylvania**

| Organic Amendment | DM Input[a] (Mg ha$^{-1}$) | Nutrient Surplus[b] | | | Changes in Soil Properties[c] | | | |
|---|---|---|---|---|---|---|---|---|
| | | N | P | K | Total N[d] | Extractable P[e] | Extractable K[e] | Total C[d] |
| Inorganic fertilizer | | 96 | 11 | 141 | -356 **** | -4 ns | +32 ns | -1808 |
| Raw dairy manure | 7.5 | 505 | 146 | 337 | -170 ns | +21 ns | +49 * | -19 |
| Dairy manure and leaf compost | 27 | 865 | 299 | 384 | +425 * | +55 ns | +56 ** | +7963 |
| Broiler litter and leaf compost | 14 | 614 | 351 | 326 | +41 ns | +79 ns | +79 *** | +2174 |
| Controlled microbial compost[f] | 47.5 | 795 | 445 | 972 | +231 ns | +46 ns | +223 **** | +5421 |
| Yard clipping compost | 14.5 | 485 | 60 | 127 | +4 ns | 0 ns | +23 ns | +3504 |

*Note:* ANOVAs were done by replication and treatment with year and replication as main effect. Significant change from 1992 to 1995 noted as follows: ns, not significant; * $p < 0.5$; ** $p < 0.01$; *** $p < 0.001$, **** $p < 0.0001$.

[a] Average yearly application rate applied for 3 years.

[b] Surplus equals the total inputs minus harvested outputs.

[c] Temporal changes in N, P, K from October 1992 to November 1995.

[d] Calculated using bulk density measurements of the plow layer (0- to 20-cm depth) and discounting weight of rock fragments >2 mm.

[e] Calculated based on the assumption that the plow layer contains 2,240,000 kg ha$^{-1}$.

[f] Mix of farm animal manure and bedding, clay loam, rock powder, microbial inoculant, and finished compost from a previous batch.

*Source:* Adapted from Reider, C. et al. 2000. *Compost Sci. Util.* 8:328–339.

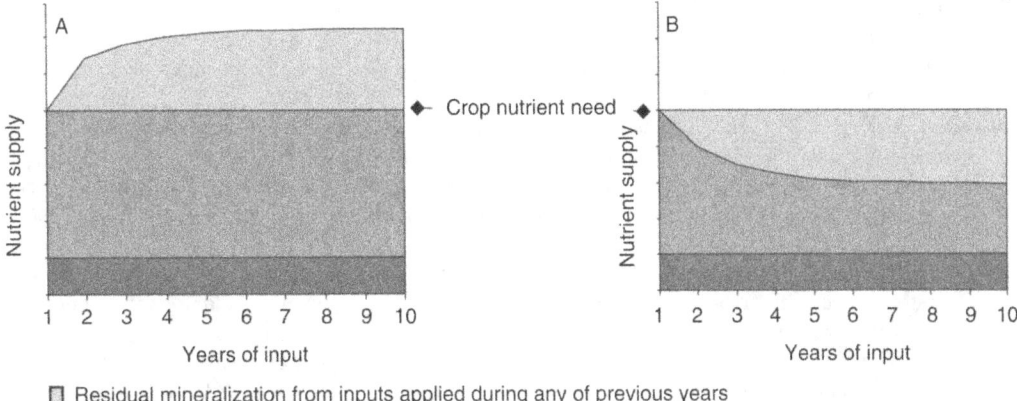

FIGURE 9.5 Two hypothetical scenarios of nutrient availability from long-term annual organic amendment inputs: (A) constant annual input rates leading to excess nutrient loading and (B) decreasing annual input rates leading to a constant nutrient supply.

accumulation also address farm- and regional-level nutrient management goals. Increased SOM, for example, might diminish the need for transport of nutrients on farm and between farms, thereby saving economic resources and reducing the risk of environmental impact. Higher SOM levels might similarly benefit regional agroecosystems in which the dependence of large quantities of imported nutrients has negatively impacted water resources and increased economic vulnerability (Burkart et al., 1995).

The shift toward environmentally sound agricultural systems incorporates the objective to increase SOM for sustained nutrient availability. To what level SOM should be increased can depend on the management objective. For the purpose of soil fertility, Domsch (1985) argues that the highest possible humus level for a specific location is desirable because it represents a steadily flowing nutrient and energy source for the plant and soil organism communities. However, as the preceding discussion showed, managing for an increase in SOM is generally beneficial, but doing so without attention to potential problems can lead to a reduction of nutrient use efficiency or excess loading of soil nutrients. The variability in soil, climate, and crops produces unique management scenarios for regional and local agroecosystems, which must be understood when managing for SOM and its associated benefits.

## REFERENCES

Allison, F. E. 1973. *Soil Organic Matter and its Role in Crop Production.* Elsevier, Amsterdam.

Alvarez, C. R., and Alvarez, R. 2000. Short term effects of tillage systems on active soil microbial biomass. *Biol. Fertil. Soils* 31:157–161.

Alvarez, C. R., Alvarez, R., Grigera, M. S., and Lavado, R. S. 1998. Associations between organic matter and fractions of the active microbial biomass. *Soil Biol. Biochem.* 30:767–773.

Angers, D. A., Bisonette, N., Legere, A., and Smason, N. 1993. Microbial and biochemical changes induced by rotation and tillage in a soil under barley production. *Can. J. Soil Sci.* 73:39–50.

Balabane, M., and Balesdent, J. 1992. Input of fertilizer-derived labeled N to soil organic matter during a growing season of maize in the field. *Soil Biol. Biochem.* 24:89–96.

Beare, M. H., Parmelee, R. W., Hendrix, P. F., and Cheng, W. 1992. Microbial and faunal interactions and effects on litter nitrogen and decomposition in agroecosystems. *Ecol. Monogr.* 62:569–591.

Beauchamp, E.G., and Voroney, R. P. 1994. Crop carbon contribution to the soil with different cropping and livestock systems. *J. Soil Water Conserv.* 49:205–209.

Bird, J. A., van Kessel, C., and Horwath, W. R. 2002. Nitrogen dynamics in humic fractions under alternative straw management in temperate rice. *Soil Sci. Soc. Am. J.* 66:478–488.

Blaesdent, J., Mariotti, A., and Boisgontier, D. 1990. Effect of tillage on soil organic carbon mineralization estimated from $^{13}C$ abundance in maize fields. *J. Soil Sci.* 41:587–596.

Borogowski, Z., Dorzanski, B., Kusinka, A., and Zembrzycka, K. 1976. An attempt to diagnose the genetic horizons of soils on the basic content of exchangeable metal cations in mechanical fractions. *Pol. J. Soil Sci.* 9:115–122.

Bossio, D. A., Scow, K. M., Gunupala, N., and Graham, K. J. 1998. Determinants of soil microbial communities: Effects of agricultural management, season and soil type on phospholipid fatty acid profiles. *Microb. Ecol.* 36:1–12.

Bouldin, 1987. Effect of green manure on soil organic matter content and nitrogen availability. In *Proceedings Symposium on Sustainable Agriculture: The Role of Green Manure Crops in Rice Farming Systems,* International Rice Research Institute, Los Baños, Philippines, pp. 151–163.

Brady, N. C., and Weil, R. R. 1999. *The Nature and Properties of Soils.* Prentice-Hall, Upper Saddle River, NJ.

Bremer, E., and van Kessel, C. 1992. Plant-available nitrogen from lentil and wheat residues during a subsequent growing season. *Soil Sci. Soc. Am. J.* 56:1155–1160.

Bruce, R. R., Langdale, G. W., and West, L. T. 1990. Modification of soil characteristics of degraded soil surfaces by biomass input and tillage affecting soil water regime. *Trans. 14th Conf. Soil Sci.*6:4–9.

Burity, H. A., Ta, T. C., Farris, M. A., and Coulman, B. E. 1989. Estimation of nitrogen fixation and transfer from alfalfa to associated grasses and mixed swards under field conditions. *Plant Soil* 144:249–255.

Burkart, M. R., James, D. E., and Oberle, S. L. 1995. Exploring diversity within regional agroecosystems. In Olson, R., Francis, C., and Kaffka, S. (Eds.), *Exploring the Role of Diversity in Sustainable Agricultural Systems,* American Society of Agronomy, Crop Science Society of America, and Soil Science Society of America, Madison, WI, pp.195–224.

Carpenter-Boggs, L., Kennedy, A. C., and Reganold, J. P. 2000. Organic and biodynamic management: effects on soil biology. *Soil Sci. Soc. Am. J.* 64:1651–1659.

Carter, M. R. 1992. Influence of reduced tillage systems on organic matter, microbial biomass, macroaggregate distribution and structural ability of the surface soil in a humid climate. *Soil Tillage Res.* 23:361–372.

Carter, M. R., and Rennie, D. A. 1984. Dynamics of soil microbial biomass N under zero and shallow tillage for spring wheat, using $^{15}$urea. *Plant Soil* 76:157–164.

Castellanos, J. Z., and Pratt, P. F. 1981. Mineralization of manure nitrogen: Correlation with laboratory indexes. *Soil Sci. Soc. Am. J.* 45:354–357.

Chae, Y. M., and Tabatabai, M. A. 1986. Mineralization of nitrogen in soils amended with organic wastes. *J. Environ. Qual.* 15:193–198.

Chen, L., Dick, W. A., Streeter, J. G., and Hoitnik, H. A. J. 1998. Fe chelates from compost microorganisms improve Fe nutrition of soybean and oat. *Plant Soil* 200:139–147.

Chung, R. S., Wang, C. H., Wang, C. W., and Wang, Y. P. 2000. Influence of organic matter and inorganic fertilizer on the growth and nitrogen accumulation of corn plants. *J. Plant Nutr.* 23:297–311.

Clarholm, M. 1985. Interactions of bacteria, protozoa and plants leading to mineralization of soil nitrogen. *Soil Biol. Biochem.* 17:181–187.

Coleman, D. C., Cole, C. V., and Elliott, E. T. 1984. Decomposition, organic matter turnover and nutrient dynamics in agroecosystems. In Lowrance, R., Stiner, B. R. and House, G. J. (Eds.), *Agricultural Ecosystems — Unifying Concepts.* Wiley-Interscience, New York, pp. 83–104.

Coleman, D. C., and Crossley, D. A. 1997. *Fundamentals of Soil Ecology.* Academic Press, San Diego, CA.

Collaa, G., Mitchell, J. P., Joyce, B. A., Huyck, L. M., Wallender, W. W., Temple, S. R., Hsiao, T. C., and Poudel, D. 2000. Soil physical properties and tomato yield and quality in alternative cropping systems. *Agron. J.* 92:924–932.

Collins, H. P., Rasmussen, P. E., and Douglas, C. L. J. 1992. Crop rotation and residue management effects on soil carbon and microbial dynamics. *Soil Sci. Soc. Am. J.* 56:783–788.

Daniel, T. C., Sharpley, A. N., Edwards, D. R., Wedepoh, L. R., and Lemunyon, J. L. 1994. Minimizing surface water eutrophication from agriculture by phosphorus management. *J. Water Soil Conserv.* 49:30–38.

Deane, S. D. D. 1790. *New England Farmer.* Press of Isaiah Thomas, Worcester, MA.

Delschen, T. 1999. Impacts of long-term application of organic fertilizers on soil quality parameters in reclaimed loess of the Rhineland lignite mining area. *Plant Soil* 213:43–54.

Domsch, K.-H. 1985. *Funktionen und Belastbarkeit des Bodens aus der Sicht der Bodenmikrobiologie*. Verlag W. Kohlhammer, Stuttgart.

Doran, J. W. 1987. Microbial biomass and mineralizable nitrogen distribution in no-tillage on plowed soils. *Biol. Fertil. Soils* 5:68–75.

Drinkwater, L. E., Wagoner, P., and Sarrantonio, M. 1998. Legume-based cropping systems have reduced carbon and nitrogen losses. *Nature* 396:262–264.

Drury, C. F., Stone, J. A., and Findlay, W. I. 1991. Microbial biomass and soil structure associated with corn, grasses and legumes. *Soil Sci. Soc. Am. J.* 55:805–811.

Dubach, M., and Russelle, M. P. 1994. Forage legume roots and nodules and their role in nitrogen transfer. *Agron. J.* 86:259–266.

Eghball, B., 2000. Nitrogen mineralization from field-applied beef cattle feedlot manure or compost. *Soil Sci. Soc. Am. J.* 64:2024–2030

Eghball, B., Binford, G. D., and Baltensperger, D. D. 1996. Phosphorous movement and adsorption in a soil receiving long term manure and fertilizer application. *J. Environ. Qual.* 25:1339–1343.

Eghball, B., Power, J. F., Gilley, J. E., and Doran, J. W. 1997. Nutrient, carbon and mass loss during composting of beef cattle feedlot manure. *J. Environ. Qual.* 26:189–193.

Elliott, L. F., and Papendick, R. I. 1986. Crop residue management for improved soil productivity. *Biol. Agric. Hortic.* 3:131–142.

Eriksen, G. N., Coale, F. J., and Bollero, G. A. 1999. Soil nitrogen dynamics and maize production in municipal solid waste amended soil. *Agron. J.* 91:1009–1016.

Fauci, M. F., and Dick, R. P. 1994a. Soil microbial dynamics: Short and long-term effects of inorganic and organic nitrogen. *Soil Sci. Soc. Am. J.* 58:801–806.

Fauci, M. F., and Dick, R. P. 1994b. Plant response to organic amendments and decreasing inorganic nitrogen rates in soils from long-term experiments. *Soil Sci. Soc. Am. J.* 58:134–138.

Franco-Viscaino, E. 1996. Soil quality in central Michigan: Rotations with high and low diversity of crops and manure. In Doran, J. W., and Jones, A. J. (Eds.), *Methods of Assessing Soil Quality*. Soil Science Society of America, Madison, WI, pp. 327–336.

Frankenberger, W. T., and Arshad, M. 1995. *Phytohormones in Soils: Microbial Production and Function*. Marcel Dekker, New York.

Friedel, J. K., Munch, J. C., and Fischer, W. R. 1996. Soil microbial properties and the assessment of available soil organic matter in a haplic luvisol after several years of different cultivation and crop rotation. *Soil Biol. Biochem.* 28:479–488.

Gallaher, R. N., and Ferrer, M. B. 1987. Effect of no-tillage versus conventional tillage on soil organic matter and nitrogen content. *Commun. Soil Sci. Plant Anal.* 18:1061–1076.

Gartley, K. L., and Sims, J. T. 1994. Phosphorus soil testing: environmental uses and implications. *Commun. Soil Sci. Plant Anal.* 25: 1565–1582.

Gerke, H. H., Arning, M., and Stoppler-Zimmer, H. 1999. Modeling long-term compost application effects on nitrate leaching. *Plant Soil* 213:75–92.

Gerzabek, M. H., Haberhauer, G., and Kirchmann, H. 2001. Soil organic matter and carbon 13 natural abundance in particle size fractions of a long-term agricultural field experiment receiving organic amendments. *Soil Sci. Am. J.* 62:352–358.

Glendining, M. J., and Powlson, D. S. 1991. The effect of long-term application of inorganic nitrogen fertilization on soil organic nitrogen. In Wilson, W. S. (Ed.), *Advances in Soil Organic Matter Research: The Impact on Agriculture and the Environment*. The Royal Society of Chemistry, Cambridge, U.K., pp. 329–338.

Glendining, M. J., Powlson, D. S., Poulton, P. R., Bradbury, N. J., Palazzo, D., and Li, X. 1996. The effect of long-term applications of inorganic fertilizer on soil nitrogen in the Broadbalk wheat experiment. *J. Agric. Sci. (Camb.)* 127:347–363.

Granastein, D. M., Bezdicek, D. F., Cochran, V. L., Elliot, L. F., and Hammel, J. 1987. Long-term tillage and rotation effects on soil microbial biomass, carbon and nitrogen. *Biol. Fertil. Soils* 5:265–270.

Green, C. J., Blackmer, A. M., and Horton, R. 1995. Nitrogen effects on conservation of carbon during corn residue decomposition in soil. *Soil Sci. Soc. Am. J.* 59:453–459.

Gregorich, E. G., Carter, M. R., Angers, A. D., Monreal, C. M., and Ellert, B. H. 1994. Towards a minimum data set to assess soil organic matter quality in agricultural soils. *Can. J. Soil Sci.* 74:367–385.

Hadas, A., Kautsky, L., and Portnoy, R. 1996. Mineralization of composted manure and microbial dynamics in soil as affected by long-term nitrogen management. *Soil Biol. Biochem.* 28:733–738.

Haider, K., Groeblinghoff, F. F., Beck, T., Schulten, H. R., Hempfling, R., and Luedman, H. D. 1991. Influence of soil management practices on the soil organic matter structure and biochemical turnover of plant residue. In Wilson, W. S. (Ed.), *Advances in Soil Organic Matter Research: The Impact on Agriculture and the Environment*. The Royal Society of Chemistry, Cambridge, U.K., pp. 79–92.

Hassink, J. 1995. Decomposition rate constant of size and density fractions of soil organic matter. *Soil Sci. Am. Soc. Am. J.* 59:1631–1635.

Hassink, J. 1997. The capacity of soil to preserve organic C and N by their association with clay and silt particles. *Plant Soil* 191:77–87.

Hassink, J., Neutel, A. M., and De Ruiter, P.C. 1994. C and N mineralization in sandy and loamy grassland soils: the role of microbes and microfauna. *Soil Biol. Biochem.* 26:1565–1571.

Hendrix, P.F., Coleman, D.C., and Crossley D. A., Jr. 1992. Using knowledge of soil nutrient cycling processes to design sustainable agriculture. *J. Sust. Agric.* 2:63–79.

Hendrix, P. F., Crossley D. A., Jr., Blair, J. M., and Coleman, D. C. 1990. Soil biota as components of sustainable agroecosystems. In Edwards, C. A., Lal, R., Madden, P., Miller, R. H., and House, G. (Eds.), *Sustainable Agricultural Systems*. Soil and Water Conservation Society, Ankeny, IA, 637–654.

Hendrix, P. F., Parmelee, R. W., D.A. Crossley, J., Coleman, D. C., Odum, E. P. and Groffman, P. M. 1986. Detritus foodwebs in conventional and no-tillage agroecosystems. *Bioscience* 36:374–380.

Hesterman, O. B., Russelle, M. P., Sheaffer, C. C., and Heichel, G. H. 1987. Nitrogen utilization from fertilizer and legume residues in legume-corn rotations. *Agron. J.* 79:726–731.

Hodges, R. D. 1991. Soil organic matter: its central position in organic farming. In Wilson, W. S. (Ed.), *Advances in Soil Organic Matter Research: The Impact on Agriculture and the Environment*. The Royal Society of Chemistry, Cambridge, U.K., pp. 355–365.

Holland, E. A., and Coleman, D. C. 1987. Litter placement effects on microbial communities and organic matter dynamics. *Ecology* 68:425–433.

Horwath, W. R., Devevre, O. C., Doane, T. A., Kramer, A. W., and van Kessel, C. 2002. Soil C sequestration management effects on N cycling and availability. In Kimble, J., Lal, R., and Follett, R. (Eds.), *Agricultural Practices and Policies for Carbon Sequestration in Soil*. Lewis Publishers, New York, pp. 155–164.

House, G. J., Stinner, B. R., and Crossley D. A., Jr. 1984. Nitrogen cycling in conventional and no-tillage agroecosystems: analysis of pathways and processes. *J. Appl. Ecol.* 21:991–1012.

Hussain, I., Olson, K. R., and Eblehar, S. A. 1999. Long-term tillage effects on soil chemical properties and organic matter fractions. *Soil Sci. Soc. Am. J.* 63:1335–1341.

Jenkinson, D. S. 1977. Studies on decomposition of plant material in soil. 4. Effect of rate of addition. *J. Soil Sci.* 28:417–423.

Jenkinson, D. S. 1991. The Rothamsted long-term experiments: Are they still of use. *Agron. J.* 83:2–10.

Jenkinson, D. S., Fox, R. H., and Rayner, J. H. 1985. Interactions between fertilizer nitrogen and soil nitrogen-the so-called "priming" effect. *J. Soil Sci.* 36:425–444.

Jenny, H. 1985. The making and unmaking of a fertile soil. In Jackson, W., Berry, B., and Coleman, B. (Eds.), *Meeting the Expectations of the Land*. North Point Press, San Francisco, 42–55.

Johansson, G. 1992. Release of organic C from growing roots of meadow fescue. *Soil Biol. Biochem.* 24:427–433.

Jokela, B., Magdoff, F., Bartlett, R., Bosworth, S., and Ross, D. 2000. *Nutrient Recommendations for Field Crops in Vermont, BR 1390*. University of Vermont, Burlington, VT.

Kandeler, E., Tscherko, D., and Spiegel, H. 1999. Long-term monitoring of microbial biomass, N mineralization and enzyme activities of a chernozem under different tillage management. *Biol. Fertil. Soil* 28:343–351.

Kang, B. T., Reynolds, L., and Atta-Krah, A. N. 1990. Alley farming. *Adv. Agron.* 43:315–359.

Karlen, D. L., Wollenhaupt, N. C., Erbach, D. C., Swan, J. B., Eash, N. S., and Jordahl, J. L. 1994. Long-term tillage effects on soil quality. *Soil Tillage Res.* 32:313–327.

Kelting, M., Harris, R. J., Fanelli, J., and Appleton, B. 1998a. Humate-based biostimulants affect early post-transplant root growth and sapflow of balled and burlapped Red Maple. *HortSci.* 33:342–344.

Kelting, M., Harris, R. J., Fanelli, J., and Appleton, B. 1998b. Biostimulants and soil amendments affect two-year post transplant growth of red maple and Washington hawthorn. *HortSci.* 33:819–822.

Kennedy, A. C. 1995. Soil microbial diversity in agricultural systems. In Olson, R., Francis, C., and Kaffka, S. (Eds.), *Exploring the Role of Diversity in Sustainable Agricultural Systems.* American Society of Agronomy, Crop Science Society of America, and Soil Science Society of America, Madison, WI, pp. 35–54.

Kirchmann, H., and Bergström, L. 2001. Do organic farming practices reduce nitrate leaching? *Commun. Soil Sci. Plant Anal.* 32:997–1028.

Klausner, S. D., Kanneganti, V. R., and Bouldin, D. R. 1994. An approach for estimating a decay series for organic N in animal manure. *Agron. J.* 86:897–903.

Kuo, S., and Sainju, U. M. 1998. Nitrogen mineralization and availability of mixed leguminous and non-leguminous cover crop residues in soil. *Biol. Fertil. Soils* 26:346–353.

Ladd, J. N., Amato, M., Jackson, R. B., and Butler, J. H 1983. Utilization by wheat crops of nitrogen from legume residues decomposing in soils in the field. *Soil Biol. Biochem.* 15:213–238.

Lampkin, N. 1990. *Organic Farming.* Farming Press, Ipswich, U.K.

Leclerc, B., Georges, P., Cauwel, B., and Lairon, D. 1995. A five year study on nitrate leaching under crops fertilized with mineral and organic fertilizers in lysimeters. *Biol. Agric. Hortic.* 11:301–308.

Lopez-Hernandez, D., Siegert, S., and Rodrigues, J. V. 1986. Competitive adsorption of phosphate with malate and oxalate by tropical soils. *Soil Sci. Soc. Am. J.* 50:1460–1462.

Lovell, R. D., and Hatch, D. J. 1998. Stimulation of microbial activity following spring applications of nitrogen. *Biol. Fertil. Soils* 26:28–30.

Ma, B. L., Dywer, L. M., and Gregorich, E. G. 1999. Soil nitrogen amendment effects on seasonal nitrogen mineralization and nitrogen cycling in maize production. *Agron. J.* 91:1003–1009.

Magdoff, F. R. 1991. Field nitrogen dynamics. *Commun. Soil Sci. Plant Anal.* 22:1507–1517.

Magdoff, F. R., and Amadon, J. F. 1980. Yield trends and soil chemical changes resulting from N and manure application to continuous corn. *Agron. J.* 72:629–632.

Magdoff, F., Lanyon, L., and Liebhardt, B. 1997. Nutrient cycling, transformations, and flows: Implications for a more sustainable agriculture. *Adv. Agron.* 60:2–68.

Magdoff, F., and van Es, H. 2000. *Building Soils for Better Crops.* USDA Sustainable Agriculture Network (www.sare.org), Beltsville, MD.

Magid, J., Gorrissen, A., and Giller, K. E. 1996. In search of the elusive "active" fraction of the soil organic matter: Three size-density fractioning methods for tracing the fate of homogeneously $^{14}C$-labelled plant materials. *Soil Biol. Biochem.* 28:89–99.

Mahboubi, A. A., Lal, R., and Faussey, N. R. 1993. Twenty-eight years of tillage effects on two soils in Ohio. *Soil Sci. Soc. Am. J.* 57:506–512.

Maynard, A. A. 1994. Effect of annual amendments of compost on nitrate leaching in nursery stock. *Compost Sci. Util.* 2:54–55.

McCarty, G. W., and Meisinger, J. J. 1997. Effects of N fertilizer treatment on biologically active N pools in soils under plow and no tillage. *Soil Biol. Biochem.* 24:406–412.

Moreno, J. L., Garcia, C., Hernandez, T., and Garcia, C. 1998. Changes in organic matter and enzymatic activity of an agricultural soil amended with metal contaminated sewage sludge compost. *Commun. Soil Sci. Plant Anal.* 28:230–237.

Moreno, J. L., Hernandez, T., and Garcia, C. 1999. Effects of cadmium contaminated sewage sludge compost on dynamics of organic matter and microbial activity in an arid soil. *Biol. Fertil. Soils* 28:230–237.

Mylonas, V. A., and McCants, C. B. 1980. Effects of humic and fulvic acids on growth of tobacco. *Plant Soil* 54:485–490.

Ndiaye, E. L., Sandeno, J. M., McGrath, D., and Dick, R. P. 2000. Integrative biological indicators for detecting change in soil quality. *Am. J. Altern. Agric.* 15:26–36.

Neher, D. A. 1999. Soil community composition and ecosystem processes: Comparing agricultural ecosystems with natural ecosystems. *Agrofor. Syst.* 45:159–185.

NRAES, 1997. *Nutrient Management: Crop Production and Water Quality.* Publication 101, Northeast Regional Agricultural Engineering Service (NRAES), Ithaca, NY.

Olk, D. C., and Cassman, K. G. 1995. Reduction of potassium fixation by two humic acid fractions in vermiculitic soils. *Soil Sci. Soc. Am J.* 59:1250–1258.

Parham, J. A., Deng, S P. Raun, W. R., and Johnson, G. V. 2002. Long term cattle manure application in the soil. I. Effect on soil phosphorous levels, microbial biomass C, and dehydrogenase and phosphatase activities. *Biol. Fertil. Soils* 35:328–337.

Park, J. 2001. Changing soil biological health in agroecosystems. In Shiyomi, M., and Koizumi, H. (Eds.), *Structure and Function in Agroecosystem Design and Management.* CRC Press, Boca Raton, FL, pp. 335–350.

Parmelee, R. W., Beare, M., Cheng, W., Hendrix, P. F., Rider, S. J., Crossley, D. A. J., and Coleman, D. C. 1990. Earthworms and enchytraeids in conventional and no-tillage agroecosystems: a biocide approach to assess their role in organic matter breakdown. *Biol. Fertil. Soils* 10:1–10.

Parr, J. F., and Papendick, R. I. 1978. Factors affecting the decomposition of crop residues by microorganisms. In Oschwald, W. R. (Ed.), *Crop Residue Management Systems.* American Society of Agronomy, Madison, WI, pp. 101–129.

Pascual, J. A., Garcia, C., Hernandez, T., and Ayuso, M. 1997. Changes in microbial activity in an arid soil amended with urban organic wastes. *Biol. Fertil. Soils* 24:429–434.

Paul, E. A., and Clark, F. E. 1996. *Soil Microbiology and Biochemistry.* Academic Press, San Diego, CA.

Paustian, K., Parton, J. W., and Perrson, J. 1992. Modeling soil organic matter in organic-amended and nitrogen fertilized long-term plots. *Soil Sci. Soc. Am. J.* 56:476–488.

Pia, J. F. 1998. CIESA Project: Biointensive mini-farming in Patagonia. In Foguelman, D., and Lockeretz, W. (Eds.), *Organic Agriculture — The Credible Solution for the 21st Century.* Proceedings of the 12th International IFOAM Scientific Conference, Mar del Plata, Argentina International Federation of Organic Agricultural Movements (IFOAM), Tholey-Theley, Germany, pp. 64–67.

Pinamonti, F. 1998. Compost mulch effects on soil fertility, nutritional status and performance of grapevine. *Nutr. Cycl. Agroecosyst.* 51:239–248.

Pinck, L. A., Allison, F. E., and Gaddy, V. L. 1948. The effect of green manure crops of varying carbon-nitrogen ratios upon nitrogen availability and soil organic matter content. *J. Am. Soc. Agron.* 40:237–248.

Polglase, P. J., Comerford, N. B., and Jokela, E. J. 1992. Mineralization of nitrogen and phosphorous form soil organic matter in a southern pine plantation. *Soil Sci. Soc. Am. J.* 56:921–927.

Poudel, D. D., Horwath, W. R., Mitchell, J. P., and Temple, S. R. 2001. Impacts of cropping systems on soil nitrogen storage and loss. *Agric. Syst.* 68:253–268.

Powel, J. M., and Hons, F. M. 1991. Sorghum stover removal effects on soil organic matter content, extractable nutrients and crop yield. *J. Sustain. Agric.* 2:25–39.

Power, J. F. 1994. Understanding the nutrient cycling process. *J. Water Soil Conserv.* 49:16–23.

Pratt, P. F., Broadbent, F. E., and Martin, J. P. 1973. Using organic wastes as nitrogen fertilizers. *Calf. Agric.* 27:10–13.

Pratt, P. F., Davis, S., and Sharpless, R. G. 1976. A four year trial with manures. *Hilgardia* 44:99–125.

Rasmussen, P. E., and Collins, H. P. 1991. Long-term impacts of tillage, fertilizer and crop residue on soil organic matter in temperate semi-arid regions. *Adv. Agron.* 45:93.

Reicosky, D. C., Evans, S. D., Cambardella, C. A., Allmaras, R. R., Wilts, A. R., and Huggins, D. R. 2002. Continuous corn with moldboard tillage: residue and fertility effects on soil carbon. *J. Soil Water Conserv.* 57:277–284.

Reider, C., Herdman, W., Drinkwater, L. E., and Janke, R. 2000. Yields and nutrient budgets under compost, raw dairy manure and mineral fertilizer. *Compost Sci. Util.* 8:328–339.

Roe, N. 1998. Compost utilization for vegetable and fruit crops. *HortSci.* 33:934–937.

Russo, R. O., and Berlyn, G. P. 1990. The use of organic biostimulants to help low input sustainable agriculture. *J. Sustain. Agric.* 1:19–42.

Ruz-Jerez, B. E., Ball, R., and Tillman, R. W. 1992. Laboratory assessment of the nutrient release from pasture soil receiving grass or clover residues in the presence and absence of *Lumbricus rubellus* or *Eisenia fetida. Soil Biol. Biochem.* 24:1529–1534.

Sanchez, J. E., Willson, T. C., Kizilkaya, K., Parker, E., and Harwood, R. R. 2001. Enhancing the mineralizable nitrogen pool through substrate diversity in long term cropping systems. *Soil Sci. Soc. Am. J.* 65:1442–1447.

Sarrantonio, M. 1994. *Northeast Cover Crop Handbook.* Rodale Institute, Emmaus, PA.

Schulten, H. R., and Leinweber, P. 2000. New insights into organic-mineral particles: Composition, properties and models of molecular structure. *Biol. Fertil. Soils* 30:399–432.

Scow, K. M. 1997. Soil microbial communities and carbon flow in agroecosystems. In Jackson, L. E. (Ed.), *Ecology in Agriculture*. Academic Press, CA, San Diego, pp. 367–413.

Shellinger, D. A., and Breitenbeck, G. 1998. Quantifying losses of plant nutrients and elements during windrow composting of various feedstocks. In Das, K. C., and Graves, E. F. (Eds.), *Composting in the Southeast*. University of Georgia, Athens, GA, pp. 41–48.

Sikora, L. J., and Yakavchenko, V. 1996. Soil organic matter mineralization after compost amendment. *Soil Sci. Am. J.* 60:1401–1404.

Sims, J.T. 1995. Organic wastes as alternative nitrogen sources. In P.E. Bacon (Ed.), *Nitrogen Fertilization in the Environment*. Marcel Dekker, New York, pp. 487–536.

Smith, O. L. 1979. Application of a model of decomposition of organic matter. *Soil Biol. Biochem.* 11:607–618.

Sommer, S. G. 2001. Effect of composting on nutrient loss and nitrogen availability. *Eur. J. Agron.* 14:123–133.

Sommerfeldt, T. G., Chang, C., and Entz, T. 1988. Long-term annual manure applications increase soil organic matter and nitrogen and decrease carbon to nitrogen. *Soil Sci. Soc. Am. J.* 52:1668–1672.

Staley, T. E., Edwards, W. M., Scott, C. L., and Owens, L. B. 1988. Soil microbial biomass and organic component alteration in no-tillage chronosequences. *Soil Sci. Soc. Am J.* 52:998–1005.

Stevenson, F. J. 1986. *Cycles of the Soil*. Wiley-Interscience, New York.

Suzuki, M., Kamekawa, K., Sekiya, S., and Shiga, H. 1990. Effect of continuous application of organic or inorganic fertilizer for sixty years on soil fertility and rice yield in paddy field. *Trans. 14th Int. Congr. Soil Sci.* 4:14–19.

Tiessen, H., Cuevas, E., and Chacon, P. 1994. The role of soil organic matter in sustaining soil fertility. *Nature* 371:783–785.

Tisdale, J. M., and Oades, J. M. 1982. Organic matter and water-stable aggregates in soils. *J. Soil Sci.* 33:141–163.

Wander, M. M., and Bidart, M. G. 2000. Tillage influence on the physical protection, bioavailability, and composition of particulate organic matter. *Biol. Fertil. Soils* 32:360–367.

Weil, R. R., Lowell, K. A., and Shade, H. M. 1993. Effects of intensity of agronomic practices on a soil ecosystem. *Am. J. Altern. Agric.* 8:5–14.

Wiersum, L. K. 1974. The activity of specific growth stimulating substances in the soil in relation to the application of organic matter. *Trans. 10th Int. Congr. Soil Sci.* 3:123–129.

Wilson, T., Paul, E. A., and Harwood, R. R. 2001. Biologically active soil organic matter fractions in sustainable cropping systems. *Appl. Soil Ecol.* 16:63–76.

Wood, C. W., Westfal, D. G., Peterson, G. A., and Burke, I. C. 1990. Impacts of cropping intensity on carbon and nitrogen mineralization under no-till dryland agroecosystems. *Agron. J.* 82:1115–1120.

# 10 Soil and Crop Management Effects on Soil Microbiology

*Ann C. Kennedy, Tami L. Stubbs, and William F. Schillinger*

## CONTENTS

## INTRODUCTION

Life in soil is responsible for a multitude of processes vital to soil function. Microorganisms can have a profound effect on plant growth, soil organic matter (SOM) accumulation, and soil condition or soil quality. For more than 3.5 billion years, microorganisms have been a life force on earth, establishing communities well before any other life forms. Since the beginning, natural selection has ever increased the microbial diversity in soils. All life is dependent on microbial processes (Price, 1988), and SOM transformations are due to microbial processes (Altieri, 1999). In turn, SOM sustains that life and is crucial to soil function. Strategies that increase SOM tend to enhance soil biological processes and vice versa. Understanding these processes and implementing strategies to enhance SOM, improve soil quality, and maintain biological diversity will help attain sustainable agriculture.

## Soil Quality

Soil quality is defined as the capacity of a soil to function within ecosystem boundaries to sustain biological productivity, maintain environmental quality, and promote plant and animal health (Doran and Parkin, 1994). It is easy to visualize a healthy, rich soil and to remember its smell. Descriptive and analytical measurements of the physical, chemical, and biological properties are sometimes used to characterize soil quality. Indicators of soil quality are needed to measure changes in soil function that occur because of alteration in management. Total organic matter can be an indicator; however, changes in total SOM usually respond very slowly to changes in management and thus lack sensitivity. Soil organisms contribute to the maintenance of soil quality because they control many key processes. Soil microorganisms and their communities are continually changing and adapting to changes in their environment. A high-quality soil is biologically active and contains a balanced population of microorganisms. The dynamic nature of soil microorganisms makes them a sensitive indicator to assess changes in soil quality due to management (Kennedy and Papendick, 1995).

This chapter explores microbiological changes occurring with soil and crop management in farming systems. Our discussion of community structure includes microbial survival strategies and delineation of groups of organisms, such as bacteria and fungi, nutritional-based groups or species, and functional determinations. Our goal is to describe changes in the soil biota with management to help identify soil microbial parameters useful in assessing management practices for conserving and enhancing SOM, soil quality, and crop production.

## SOIL MICROBIAL COMMUNITIES

The number of microbial species on earth is estimated to exceed 100,000 and may be more than a million (Hawksworth, 1991b; American Society for Microbiology, 1994). Unfortunately, only 3 to 10% of the earth's microbial species have been identified or studied in any detail (Hawksworth, 1991a). The full potential of these groups of organisms has not been explored. The diversity of microorganisms is thought to exceed that of any other life form (Torsvik et al., 1990; Ward et al., 1992). It is estimated that several thousand genomes are present in each gram of soil (Torsvik et al., 1990).

Soil microorganisms are responsible for many soil processes, such as SOM turnover, soil humus formation, cycling of nutrients, and building soil tilth and structure (Table 10.1; Lynch, 1983; Wood,

---

**TABLE 10.1**
**Beneficial Functions of Soil Microorganisms in Agricultural Systems**

- Release plant nutrients from insoluble inorganic forms
- Decompose organic residues and release nutrients
- Form beneficial soil humus by decomposing organic residues and through synthesis of new compounds
- Produce plant growth-promoting compounds
- Improve plant nutrition through symbiotic relationships
- Transform atmospheric nitrogen into plant-available N
- Improve soil aggregation, aeration, and water infiltration
- Have antagonistic action against insects, plant pathogens, and weeds (biological control)
- Help in pesticide degradation

---

1991). These functions are performed by many different genera and species. Beneficial soil bacteria enhance plant performance by increasing solubility of minerals (Okon, 1982), $N_2$ fixation (Albrecht et al., 1981), producing plant hormones (Brown, 1972; Arshad and Frankenberger, 1998), and suppressing harmful pathogens (Chang and Kommendahl, 1968). Beneficial mycorrhizal fungi can enhance plant growth by increasing nutrient (Fitter, 1977; Hall, 1978; Rovira, 1978; Ocampo, 1986) and water (Tinker, 1976) uptake and soil structure by enhancing aggregate formation and stability (Wright and Upahyaya, 1998; Chapter 6). Conversely, plant-suppressive bacteria impair seed germination and delay plant development by producing phytotoxic substances (Woltz, 1978; Suslow and Schroth, 1982; Alstrom, 1987; Schippers et al., 1987). Pathogenic fungi greatly reduce the survival, growth, and reproduction of plants (Shipton, 1977; Bruehl, 1987; Burdon, 1987). Another example of the importance of microorganisms to agriculture is the production of antibiotics by strains of fluorescent *Pseudomonas* bacteria that suppress the root disease take-all (*Gaeumannomyces graminis* var. *tritici*) in continuous winter wheat (*Triticum aestivum* L.) cropping systems (Thomashow and Weller, 1988).

Specific microorganisms can be manipulated to produce beneficial effects for agriculture and the environment (Lynch, 1983), e.g., rhizobia to increase plant available N (Sprent, 1979), mycorrhizal associations to assist nutrient and water uptake (Sylvia, 1998; Mohammad et al., 1995), or biological control of plant pests to reduce chemical inputs (Cook and Baker, 1983; Kennedy et al., 1991). Bacterial or fungal inoculants can be added to soil to aid in the bioremediation of harmful substances such as petroleum hydrocarbons (Rhykerd et al., 1999; Mohn and Stewart, 2000), polycyclic aromatic hydrocarbons (Allen et al., 1999), and a wide range of environmental pollutants (Cameron et al., 2000).

The presence of a large and diverse soil microbial community is crucial to the productivity of any agroecosystem. This diversity is influenced by almost all crop and soil management practices, including the type of crops grown. Plants and their exudates influence soil microorganisms and the soil microbial community found near roots (Duineveld et al., 1998; Ibekwe and Kennedy, 1998; Ohtonen et al., 1999). In turn, the composition of the microbial community influences the rate of residue decomposition and nutrient cycling in agroecosystems (Beare et al., 1993). The basic groups of microorganisms in soil are bacteria (including actinomycetes), fungi, algae, and protozoa. Bacteria and fungi are decomposers involved in nutrient cycling and SOM processes and are critical in the functioning of the soil food web. Ninety-five percent of plant nutrients must pass through these organisms to higher trophic levels (Moore, 1994).

Bacteria are diverse metabolically and perform numerous functions. Bacteria convert SOM into carbon (energy sources) used by others in the soil food web, break down pesticides and pollutants, and immobilize and maintain valuable nutrients such as N in the root zone. Bacteria readily colonize the substrate-rich rhizosphere (Figure 10.1). Actinomycetes are a specialized group of soil bacteria that degrade plant materials such as cellulose. Actinomycetes are important in mineralization of nutrients and some can produce antibiotics. Actinomycetes can tolerate low soil water potential better than other bacteria, but are not tolerant of low soil pH (Alexander, 1998).

Fungi, like bacteria, are vital members of the food web. Fungi are especially important at lower pH, because many bacteria are adversely affected by acid soils. Fungi are able to withstand unfavorable conditions, such as water stress and extreme temperatures, better than other microorganisms (Papendick and Campbell, 1975). They are critical for residue decomposition and accumulation of stable SOM fractions through breakdown of more complex carbon sources such as cellulose, lignin, and other organic materials. These decomposition products are then available for use by other organisms. Fungal mycelia bind soil particles together to form aggregates that increase water-holding capacity and infiltration and reduce erosion. Fungi can be saprophytes on detrital material or in associations with plant roots (Swift and Boddy, 1984). The more recalcitrant material left from decomposition then accumulates as SOM. Hyphae of arbuscular mycorrhizal (AM) fungi produce the protein glomalin, which improves soil structure (Chapter 6).

**FIGURE 10.1** Scanning electron micrograph of soil bacteria from a Palouse silt loam.

Algae occur in soil at populations of $10^3$ to $10^4$ $g^{-1}$ soil, far fewer than bacteria and fungi. The greatest populations of algae are found in moist soil, but their numbers decrease with increasing soil depth. Some algal species are nitrogen fixers and produce mucigel, which can stabilize soil aggregates. Algae are susceptible to soil disturbance and can be good indicators of soil quality. Their populations increase in agricultural systems with reduced disturbance where the surface soil and residue maintain a higher moisture regime for longer periods (Harris et al., 1995), and as a result foster algal growth on the soil surface.

Protozoa are found at populations of $10^3$ to $>10^5$ $g^{-1}$ soil. These single-celled organisms prey on bacteria and other microorganisms, and thus regulate bacterial populations (Opperman et al., 1989) and influence SOM decomposition by regulating decomposer populations. Protozoa are crucial to the functioning of soil and other ecosystems because of their role in nutrient cycling and in providing energy for other microorganisms, plants, and animals (Foissner, 1999). Fluctuations in microbial populations with tillage affect protozoan populations because protozoa feed on these organisms. Protozoa can be useful indicators of changes in soils because their populations react rapidly to changes in the environment (Foissner, 1999).

## MICROBIAL DIVERSITY

There are two primary ways that diversity can be evaluated: species diversity and functional diversity. Functional diversity can be a better parameter than species diversity to learn about soil processes and stable SOM fraction formation (Mikola and Setälä, 1998). However, it is often difficult to obtain actual measurements of functional diversity, whereas evaluating species diversity, when specific species can be assessed, is easier. The number of organisms in various microbial groups might not be sufficient to illustrate the breadth of diversity found in the soil. Although an increase in microbial products, such as SOM or $CO_2$, can be an indicator of increased functioning, it might not necessarily be due to higher functional diversity. One of the earliest studies involving soil diversity and soil respiration (Salonius, 1981) established differences in bacterial and fungal diversity by inoculating soil with varying soil suspensions. Respiration rate was reduced with the lower dilution or the assumed lower microbial diversity. The true extent or dimension of the diversity of soil microorganisms is unknown, although molecular investigations suggest that culturing techniques underestimate population numbers (Holben and Tiedje, 1988; Torsvik et al., 1990). The functioning of a group of organisms is as important as the number of species in regulating ecosystem processes (Grime, 1997; Wardle et al., 1997; Bardgett and Shine, 1999). How much diversity is required to ensure sustainable and efficient SOM turnover, as well as other important functions? Greater use of diversity indices is limited by absence of detailed information on the composition of microbial species in soil (Torsvik et al., 1990). Diverse systems are thought to have higher agricultural productivity, resilience to stress, and be more sustainable and provide risk protection (Giller et al., 1997; Wolters, 1997). A diverse system has a wider range of function with more interactions among microorganisms that influence each other to varying degrees. A higher number of different types of organisms present in a system means there are more to perform various processes and fill a niche that might not be filled if a particular group is inhibited by stress (Andren et al., 1995).

Substrate-utilization patterns have been used to obtain fingerprints of community structure (Garland, 1996; Bossio and Scow, 1995; Haack et al., 1995; Wunsche et al., 1995; Zak et al., 1994). These measures can also indicate functional diversity, metabolic potential (Degens, 1999; Haack et al., 1995; Wunsche et al., 1995), and nutritional strategies (Zak et al., 1994). Soil microbial communities as indicated by whole-soil fatty acid methyl ester (FAME) analysis can be differentiated by geographic region (Kennedy and Busacca, 1995) and cropping pattern (Cavigelli et al., 1995). The living microbiological component of soil can be estimated by phospholipid fatty acid (PLFA) analyses (Zelles et al., 1994). Another method for measuring microbial diversity is the DNA hybridization technique, which uses similarity indices. This technique illustrated that extracted bacteria and whole-community DNA had 75% similarity (Griffiths et al., 1996). The DNA microarray technology can be used to rapidly analyze microbial communities based on phylogenetic groupings and increases the ease of molecular analyses (Guschin et al., 1997). These analyses can help further understand the changes occurring among soil communities with various management practices.

Microbial diversity can be linked to susceptibility and resiliency of soil to stress, and thus might affect some soil functions such as SOM decomposition. Partial fumigation of grassland soils produces differing degrees of diversity, with longer fumigation times producing soils with less diversity. There is no direct correlation between the progressive fumigation to reduce diversity and measures of soil function, such as soil microbial biomass, soil respiration, and N mineralization. However, soils with lower diversity initially have more ability to decompose added grass residue (Griffiths et al., 2000). There is greater susceptibility to copper toxicity with decreasing diversity. Soils that contained the most diverse populations showed the greatest resilience to copper-induced stress by quickly rebounding, as shown by an increase in grass residue decomposition rates. In a similar study, no differences were seen in decomposition of *Medicago* residues even though the residues were added to both organic and conventionally farmed soils with different SOM levels (Gunapala et al., 1998). Organically farmed soils initially contained a more abundant microbial population as measured by microbial biomass C and N. When organic amendments were added, soil from the conventionally farmed system increased in microbial biomass C to a level that was comparable to the soil in the organic system. The biotic community in the conventionally farmed soil was sufficient and could respond to added substrate as well as the organic soils did. The microbial communities in this study functioned adequately whether from conventional or organic farming systems (Gunapala et al., 1998).

A reduction in functional diversity does not necessarily impede a soil's ability to decompose residue. Degens (1998) used fumigation to alter functional diversity in a grassland and measured *in situ* catabolic potential (Degens and Harris, 1997) to characterize the ability of the soil community to metabolize C substrates, with substrate added to the soil directly. The functional indices were different among fumigated, unfumigated, and fumigated and inoculated with untreated soil. There was no relationship between functional diversity and decomposition of wheat straw added into these systems. Water potential might have been the overriding factor controlling decomposition rate, because soils with reduced functional diversity continued decomposing the wheat straw under optimum moisture conditions.

Diversity of soil microorganisms can impact antagonists of pathogens and pathogen load, thus influencing their impact on plant growth. Decreased diversity of actinomycetes, some of which are antagonists of pathogens, correlated with an increase in pathogens of tomato (Workneh and van

Bruggen, 1995). *Cochliobolus sativus,* a pathogen causing a serious disease in wheat, was found in higher numbers, and individual isolates exhibited greater pathogenicity in a continuous wheat rotation than in wheat in a 3-year rotation. This increased pathogenicity was attributed to a reduction in microbial diversity (El Nashaar and Stack, 1989). Take-all decline of wheat occurs after several years of monoculture and is correlated with the appearance of several different types of organisms and alterations in microbial populations in the rhizosphere (McSpadden-Gardener and Weller, 2001). The impact of the microbial community on pathogen load and pathogenicity is complex and changes with the make-up and diversity of the community.

Assuming all functional groups are present, more microbial diversity might not necessarily be crucial to ecosystem functioning. Soil biodiversity and nutrient cycling were not linked in a study of Nigerian tropical soils (Swift et al., 1998). A study comparing native bush soils with those under cultivation showed greater abundance and diversity of soil fauna in the former, but little difference in decomposition of surface residues. Although variation in species richness might not be discernible in many environments, differences can be important in stressed systems or when conditions are altered (Yachi and Loreau, 1999). Organic matter accumulation and rate of decomposition can be important, although slowly changing indicators of ecosystem functioning in less-stressed systems.

The quality and quantity of substrate can affect community structure. Griffiths et al. (1999) used synthetic root exudates to study community structure. Microbial community changes occurred with continual substrate loading increases, and fungi dominated over bacteria in high-substrate conditions. Different organisms have the ability to be a dominant portion of the community when changes in efficiency occur because of changes in optimal growth factors, substrate quality, or substrate concentration. This knowledge is important when considering additions of organic amendments to agricultural soils.

## NUTRITIONAL STRATEGIES

The concept of *r*- and *K*-strategies is an ecological classification system based on the ability of an organism to survive in different environments (MacArthur and Wilson, 1967). To indicate two contrasting methods of selection in animals, *K* refers to the carrying capacity and *r* to the maximum intrinsic rate of natural increase ($r_{max}$). Although most microorganisms are considered r-strategists and plants and animals K-strategists, there are differences in growth strategies among microorganisms (Andrews and Harris, 1986; Table 10.2). *K*-strategists favor competition at carrying capacity, whereas *r*-strategists take advantage of easily available substrates with fast growth rates to facilitate colonization of new habitats in response to a flush of nutrients or other fluxes. Organisms can be both *r*- and *K*-strategists, depending on circumstances. An organism can exhibit an *r*-strategy when faced with fresh resources and an unstable environment, i.e., when organic amendments are applied, but become a *K*-strategist after resources are depleted and only more recalcitrant substrate is available. Age of plant roots and plant type can also influence the dominant strategy. *K*-strategists were found in higher numbers on older wheat roots than in younger roots (De Leij, 1993). The root surface of ryegrass had more *K*-strategists than that of white clover (Sarathchandra et al., 1997). Spore formation is a tactic of *r*-strategists to survive during low nutrient availability. Although the initial colonists of a residue might be *r*-strategists, organisms involved in humus degradation or lignin and cellulose degradation are *K*-strategists. Most soil bacteria are generally considered *r*-strategists, whereas fungi and actinomycetes are usually *K*-strategists (Bottomley, 1998).

The type of strategy used and various processes influence soil and plant functioning. For example, when root exudates were added to soils contaminated with heavy metals, certain bacterial populations increased, the dominance of various strategy organisms depending on availability of substrate and soil conditions (Kozdroj and van Elsas, 2000). Exudates added to these polluted soils decreased the overall diversity in favor of *r*-strategists, whereas *K*-strategists dominated soils not amended with exudates. In another study, organisms with the same community structure exhibited

**TABLE 10.2**
**Characteristics of *r*- and *K*-Strategists in Ecological Classification**

| Characteristic | *r*-Strategist | *K*-Strategist |
|---|---|---|
| General | Rapid reproductive rate, extreme fluctuation | Adapt to environment, stable and permanent |
| Growth rate | Rapid | Moderate |
| Substrate-utilization efficiency | Low efficiency | Higher efficiency |
| Diversity of substrates utilized | Simple, readily available, not resource limited | Complex, diverse, may be resource limited |
| Phenotype | Polymorphic to monomorphic | Monomorphic |
| Morphology | Smaller cells, mycelium not highly differentiated | Larger cells, well-developed mycelium |
| Reproduction | Simple genetic exchange, rapid rate | Complex genetic exchange, slow rate |
| Population dynamics | Explosive, density-independent nonequilibrium, below carrying capacity, recolonization, high migration | Stable, density dependent by competition or grazing, equilibrium dynamics at or near carrying capacity, low migration |
| Tolerance to niche overlap | High tolerance | Low tolerance |
| Residue colonists | Early | Late |
| Competitive adaptations | Few | Many |
| Microbial types | Cyanobacteria, dinoflagellates blooms; *Aspergillus, Penicillium, Pseudomonas, Bacillus*; heterotrophs, spore formers | Humus, lignin and cellulose degraders, spirilla, vibrious, *Agrobacterium, Corynebacterium* and basidiomycetes |

*Source:* Modified from Andrews, J. H. 1984. In M. G. Klug and C. A. Reddy (Eds.), *Current Perspectives in Microbial Ecology.* American Society for Microbiology, Washington, D.C., pp. 1–7.

different catabolic response profiles when grown in different soil environments, illustrating the effect of management on the community's functional diversity (Degens, 1999).

In addition to *r*- and *K*-strategies, oligotrophic response can be used to characterize organisms in an ecosystem. Organisms are grouped based on their nutritional strategies. Oligotrophs are organisms that grow under low nutrient supply and subsist on more resistant SOM, whereas copiotrophs flourish in nutrient-rich environments. Bacteria with enhanced growth under high nutrient concentrations are described as copiotrophs. Oligotrophs are more prevalent than copiotrophs in low-substrate concentrations. The proportion of copiotrophs to oligotrophs varies over time; the ratio of copiotrophs to oligotrophs increased immediately after cover crop residue incorporation but decreased 26 d later when readily available C declined (Hu et al., 1999). High quantities of readily available C early in the experiment might have inhibited oligotroph growth (Hu et al., 1999). Crop selection, region of the root system, and proximity to plant roots influence the number of oligotrophs and copiotrophs as well as their ratio (Maloney et al., 1997). It is important to understand the response of the microbial community to varying levels of C inputs to better manage for residue decomposition, competition with crop pathogens, and to improve the survival of introduced microorganisms (Hu and van Bruggen, 1997). Analysis of microbial community survival and nutritional strategies can aid in investigations of changes with management.

## MANAGEMENT EFFECTS ON SOIL MICROBIAL COMMUNITIES

Throughout each season, crop management, resource additions, or soil disturbance influence the microbial community (Figure 10.2). Each crop or soil management practice affects the microbial community and formation or degradation of SOM.

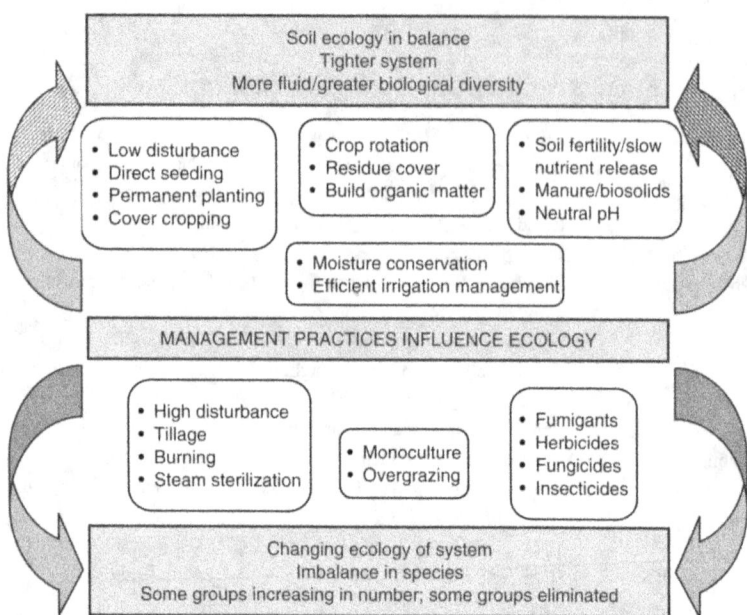

**FIGURE 10.2** Management effects on soil biology. Practices that favor build-up of soil organic matter can lead to higher biological diversity, whereas practices that involve high disturbance and reliance on chemical additives can result in limited microbial diversity or elimination of some biological groups.

## PLANT INFLUENCES

### Roots and Rhizosphere

The rhizosphere is a dynamic zone of soil under the influence of plant roots (Bowen and Rovira, 1999; Pinton et al., 2001) and has high microbial numbers (Grayston et al., 1998), activity, and diversity (Kennedy, 1998). The rhizosphere is a region of intense microbial activity because of its proximity to plant root exudates, making rhizosphere microbial communities distinct from those of bulk soil (Curl and Truelove, 1986; Whipps and Lynch, 1986). Nutrients exuded by the root or germinating seed stimulate microbial activity (Rouatt and Katznelson, 1961). Interactions between plants and rhizosphere microorganisms can play a critical role in plant competition. Competitive interaction among plants can also be important to develop rhizosphere soil communities. Free-living bacteria and fungi from rhizospheres of different pairs of plant species in two fields utilized different substrates and grew differently in the presence of antibiotics, osmotic stresses, and zinc (Westover et al., 1997). Results from these two fields suggest that adjacent plant species influence populations of rhizosphere bacteria and fungi, creating local microscale heterogeneity in rhizosphere soil (Westover et al., 1997). Similar results have been obtained for AM communities associated with certain grass species (McGonigle and Fitter, 1990; Johnson et al., 1992), rhizosphere bacterial populations associated with particular wheat genotypes (Neal et al., 1973), and root bacterial communities following bacterial inoculation (Gilbert et al., 1993).

Composition of plant species can influence the microbial community because of differences in chemical composition of root exudates (Christensen, 1989). Peas and oats exude different amounts of amino acids (Rovira, 1956). Environmental factors regulating plant growth can affect root exudation, including temperature (Rovira, 1959; Vancura, 1967; Martin and Kemp, 1980), light (Rovira, 1956), and soil water (Martin, 1977). Plants significantly influence the make-up of their own rhizosphere microbial communities (Miller et al., 1989). This is the result of different plant species and cultivars transporting varying amounts of C to the rhizosphere (Liljeroth et al., 1990) as well as different compositions of exudates. Ibekwe and Kennedy (1999) showed that wheat

(*Triticum aestivum* L.), barley (*Hordeum vulgare* L.), pea (*Pisum sativum* L.), jointed goatgrass (*Aegilops cylindrica* L.), and downy brome (*Bromus tectorum* L.) grown in two soil types had different rhizosphere microbial communities. Barley cultivars differed in the abundance of fungi and bacteria present in their rhizoplanes and rhizospheres, and these differences were sustained over different stages of plant growth (Liljeroth and Bååth, 1988). Two corn (*Zea mays*) cultivars (*Fusarium* susceptible and resistant) and grass (*Poa pratense*) lines (disease susceptible and resistant) differed in the numbers of rhizosphere bacteria, with susceptible lines having the highest numbers (Miller et al., 1989). These results were obtained even with no known presence of the pathogen. The rhizosphere microbial communities as determined by Biolog (Biolog® GN microtiter plates, Hayward, CA) differed with plant species of wheat, ryegrass (*Lolium perenne*), bentgrass (*Agrostis capillaris*), and clover (*Trifolium repens*). Plant species affected C-utilization profiles of the rhizosphere microbial communities of wheat, ryegrass, clover, and bentgrass. Microorganisms in the rhizospheres of wheat, ryegrass, and clover had higher utilization of C sources than in the bentgrass rhizosphere. Soil type, however, did not affect the nonrhizosphere soil microbial community profiles (Grayston et al., 1998). In natural plant communities, different plant combinations exhibited unique rhizosphere populations of free-living bacteria and fungi with differing abilities to utilize C substrates and withstand stresses (Westover et al., 1997). Unique C-source utilization patterns among rhizosphere communities of hydroponically grown wheat, white potato (*Solanum tuberosum*), soybean (*Glycine max*), and sweet potato (*Ipomoea batatas*) were found by using Biolog plates (Garland, 1996). C-source utilization patterns could distinguish among soils from six plant communities (Zak et al., 1994).

Substrate-utilization patterns have been used successfully to differentiate bacteria associated with different cropping and management practices (Garland, 1996; Zak et al., 1994). Crop effects can be associated with plant exudates as a result of the enhanced utilization or inhibition of substrates, similar to the organic content of root hairs, mucilage, or root cell lysates of the particular crop (Garland, 1996). Bossio and Scow (1995) found pattern differences associated with rice straw treatments and flooding. These systems are highly reactive to changes in their environment and can thus serve as easily attained, reliable fingerprints of community shifts as a function of substrate use.

## Plant Competition

Competitive interactions of the plant can influence plant productivity and are affected by soil microorganisms, such as mycorrhizal fungi (Crowell and Boerner, 1988; Hetrick and Wilson, 1989; Allen and Allen, 1990) and *Rhizobium* (Turkington et al., 1988; Turkington and Klein, 1991; Chanway and Holl, 1993). Evidence suggests that soilborne pathogens affect plant competitiveness and plant succession (Van der Putten and Peters, 1997). A pathogen-resistant species, sand fescue (*Festuca rubra* ssp. *arenaria*), outcompeted the susceptible species, marram grass (*Ammophila arenaria*), when both coastal grasses were exposed to pathogens (Van der Putten and Peters, 1997).

## Plant Diversity/Crop Rotation

Plant species and numbers can drive the make-up of the microbial community and the diversity of rhizosphere microbial populations. The above- and belowground plant community can influence microbial spatial heterogeneity in soil. Aboveground shoot material contributes organic material to the surface layers of soil. Decaying root systems also function as a source of nutrients for the surrounding microorganisms (Swinnen et al., 1995). Compared with monocropping, crop rotation can improve conditions for diversity in soil organisms because of variability in type and amount of organic inputs, and allow for time periods, or breaks, when there is no host available for a particular pest (Altieri, 1999). Diversity in crop rotation can allow for higher C inputs and diversity of plant material added to soils, depending on the residue level and carbon quality of the crops in rotation. Crop rotation enhances beneficial microorganisms, increases microbial diversity, interrupts the cycle of pathogens, and reduces weed and insect populations. Legumes in a crop rotation supply symbiotically fixed nitrogen to the system, use less water than many other crops, and reduce pathogen load. Studies have long shown the positive effects of crop rotation on crop growth, attributing these to changes in composition of microbial community (Shipton, 1977; Cook, 1981; Johnson et al., 1992).

Crop rotation and plant cover affected soil microbial biomass C and N of long-term field experiments in Iowa, with the highest values found in the longer rotations (4 years vs. 2 years) and multicropping systems, and the lowest in the continuous corn–soybean system. The varied diversity and quality of crop residues, amount of readily decomposable organic material, and root density led to increased soil microbial biomass under crop rotation. N fertilization did not affect microbial biomass in these studies (Moore et al., 2000).

Allelopathic interactions can occur between crops and weeds, between two crops, from decomposing crop and weed residues, and from crop and weed exudates (Anaya, 1999). Nonpathogenic allelopathic bacteria can produce plant-inhibiting compounds (Barazani and Friedman, 1999). Crop rotation can be used to alleviate the allelopathic or autopathic effects a crop plant might have on itself. Monocropping encourages proliferation of allelopathic bacteria (Barazini and Friedman, 1999).

By rotating crops, it is possible to lessen the negative effects a crop might have on itself and on subsequent crops (Rice, 1995). The populations and aggressiveness of pathogens can be altered with crop rotation, illustrating changes in microbial diversity and function due to management (El Nashaar and Stack, 1989). In a long-term study, *Cochliobolus sativus*, a pathogen of spring wheat, was found in higher numbers and individual isolates exhibited greater aggressiveness or ability to cause severe disease in continuous wheat, when compared with wheat in a 3-year rotation. Continuous monocropping led to changes in the soil community, which increased pathogen load and reduced barley growth compared with that by grains in multiple-crop rotation (Olsson and Gerhardson, 1992). Continuous monocropping of wheat, however, can lead to suppression of the take-all pathogen or take-all decline. This natural defense occurs in soils in the presence of fluorescent pseudomonad bacteria that produce the antibiotics phenazine and phloroglucinol (Mazzola et al., 1995). Barley plants produce compounds that can help protect it from fungus (*Drechslera teres*) and armyworm (*Mythimna convecta*) larvae (Lovett and Hoult, 1995).

Crop rotation can influence root colonization by mycorrhizae. In years following spinach (*Spinacea oleraceae*) and bell pepper (*Capsicum annuum*), spore populations of most species of AM were depressed and had lower infectivity compared with that in years following wheat, rice, or corn (Douds et al., 1997). Cover crops, such as autumn-sown cereals or vetches, increase the AM inoculum potential for subsequent crops (Boswell et al., 1998; Galvez et al., 1995). Cover crops aid in maintaining a viable mycelial network. A cover crop of winter wheat inoculated with AM increased AM infection rate, and in turn increased the growth and yield of a subsequent corn crop (Boswell et al., 1998). Soil from no-till, low-input fields with a hairy vetch cover crop maintained higher levels of colonization by indigenous AM than soils that had been tilled or received high-input management (Galvez et al., 1995). Use of cover crops can maintain AM when inoculum levels might otherwise be low and enhance infection of the subsequent crop.

## Crop Residue

Additions of crop residue are critical to maintain or increase SOM levels in agricultural soils (Figure 10.3). Cropping systems vary in residue quality and quantity, the microbial community supported, contributions to SOM, and ability to withstand the effects of disturbance. The residue decomposition process depends on the organisms present, type of SOM, and environmental conditions (Martin, 1933). Residue decomposition can also be affected by availability of carbon for microbial growth, physical separation because of landscape position, soil horizonation, or encapsulation of SOM in

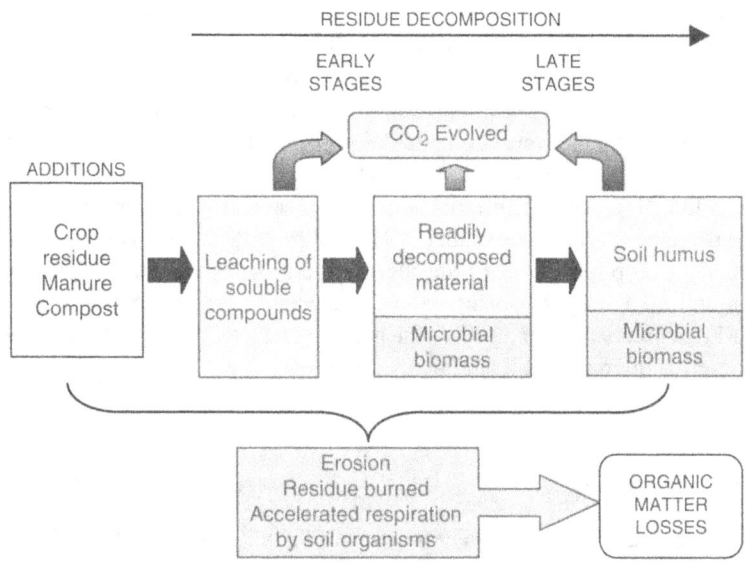

**FIGURE 10.3** Fate of organic amendments added to croplands. As residue decomposes, a portion is lost to the atmosphere through $CO_2$ evolution. The remainder can be utilized by soil microorganisms, eventually increasing soil organic matter content. Rapid removal of organic residues through processes such as erosion, burning, and intensive tillage can slow the formation of, or over time deplete, soil organic matter.

soil aggregates and clay peds, and low N availability (Paul, 1984). Water, temperature, soil pH, aeration and oxygen supply, nutrient availability, crop residue composition, C:N ratio of crop residue, and microflora are critical factors in residue degradation (Parr and Papendick, 1978).

Residue quality and management influence the composition of the microbial community. Higher numbers of culturable bacteria, including actinomycetes, were observed from decomposing soybean residues in buried bags than from wheat or corn. Fungal populations were highest on corn and lowest on wheat (Broder and Wagner, 1988). Sorghum (*Sorghum bicolor* (L.) Moench) residues buried by conventional tillage contained greater fungal hyphal length but fewer actual fungal propagules than with no-till, whereas no-till mineral soil had greater fungal hyphal length but no difference in propagule counts compared with conventionally tilled soils (Beare et al., 1993). In their study of sorghum residues, Beare et al. (1993) identified genera of fungi that were specialized for surface residue, whereas buried residues contained no specialized fungal community.

Crop species (Cookson et al., 1998) and cultivar residues (Chalaux et al., 1995) vary in their decomposition rate as well as the microbial populations they support. The amount of C mineralized from crop residues depends on the type of residue and residue composition (Henriksen and Breland, 1999). Amino acids and simple sugars, which are metabolized most rapidly in the residue-decomposition process, support populations of *r*-strategists. More complex compounds such as cellulose and lignin are broken down by *K*-strategists or oligotrophs. Lupine (*Lupinus albus*) residue decomposes more rapidly and supports higher populations of bacteria and fungi than does wheat or barley residue (Cookson et al., 1998).

Surface management (undisturbed on surface, incorporation, burning, or mechanical removal) of wheat or barley stubble also affects decomposition and microbial populations. However, wheat straw incorporated into soil had a higher decomposition rate, mass lost, and substrate-induced respiration than where stubble was burned or removed (Cookson et al., 1998). Residue management did not affect residue decomposition or microbial activity of barley or lupine (Cookson, et al., 1998). The authors hypothesized that the higher lignin:N ratio of wheat caused the response to incorporation.

Decomposition of residue by microorganisms is dependent on the presence of mineral N supply or high N residues (low C:N ratios; Wagner and Wolf, 1998). Decay rate is better correlated with initial N content of residue than with lignin content or low soil N. To adequately meet the needs of the microorganisms involved in decomposition without requiring either added fertilizer or mineral N sources from the soil, residues must contain at least 1.5 to 1.7% N, corresponding to a C:N ratio of ca. 25 or 30 (Parr and Papendick, 1978). The effects of adding inorganic N fertilizer to hasten decomposition of low N residue occur quickly, and after several months the effects of added N on decay cannot be detected (Parr and Papendick, 1978). Knowledge of changes in C and N availability is required to manage crop fertility needs throughout the growing season. Active C and N pools in SOM in agricultural fields vary seasonally, and are dependent on crop rotation, tillage depth, and N fertilization (Franzleubbers et al., 1994). Each of these factors affects the type and amount of substrate available for microbial utilization.

## RESOURCES

### Nutrient Status/Cycling

Nutrient availability and the role of microorganisms in nutrient immobilization are important concerns in agriculture. Manipulations of food webs to maintain plant nutrition while minimizing N losses are worthwhile goals of sustainable agricultural systems (Altieri, 1999).

Bacterial and fungal abundance in the rhizosphere is influenced by the nutrient status of both plant and soil. The percent mycorrhizal cover on roots of *Plantago lanceolata* was positively correlated with leaf N and P, whereas root colonization by bacteria and other fungi was negatively correlated with plant P (Newman et al., 1981). It might be difficult to separate the effects of soil nutrients on rhizosphere populations from effects involved with increased or altered root exudation of organic compounds. Grasses grown in monoculture can modify N availability (Wedin and Tilman, 1990), and it has been hypothesized that changes in soil N availability influenced by plant species affect composition of AM fungal communities (Johnson et al., 1992). Microbial population changes occur with added fertilizer and tillage. Nitrogen fertilization increased numbers of fungi and Gram-negative bacteria in rhizosphere of rice (Emmimath and Rangaswami, 1971). Kirchner et al. (1993) found that in no-till treatments receiving N fertilizer, fungal populations were higher than under no-till conditions with no fertilizer added. Higher fungal populations in the fertilized treatment were due to increased corn crop growth and higher amounts of residue to serve as substrate for microbial populations, as well as increased root growth and higher amounts of root exudates, which, in turn, increase microbial biomass. Soils that were conventionally tilled, planted with a crimson clover cover crop, and rotated with corn had more actinomycetes, *Bacillus* spp., and total culturable bacteria than corn grown under conventional tillage with fertilizer added, whereas fungi and Gram-negative bacteria were not different.

### Plant Growth-Regulating Compounds

Plant growth-regulating compounds are substances produced by plants and microorganisms in the rhizosphere that enhance seed germination and plant growth (Arshad and Frankenberger, 1998). Soil microorganisms synthesize plant growth regulators, such as auxins, abscissic acid, cytokinins, ethylene, and gibberellins (Frankenberger and Arshad, 1995). These compounds and the organisms that produce them can protect against plant pathogens and stimulate biofertilization (fixation of atmospheric $N_2$ or solubilization of nutrients) and plant growth (Figure 10.4). The mechanisms of action are often not readily apparent. Initially, it was thought that $N_2$ fixation by *Azotobacter* and *Azospirillum* was the major reason for plant growth promotion; however, other substances, such as auxins, cytokinins and gibberellins, can stimulate growth (Hussain and Vancura, 1970; Barbieri et al., 1993; Janzen et al., 1992). *Bacillus* and *Rhizobium* also produce plant growth-stimulating compounds (Frankenberger and Arshad, 1995). Plant growth promotion can also be an indirect

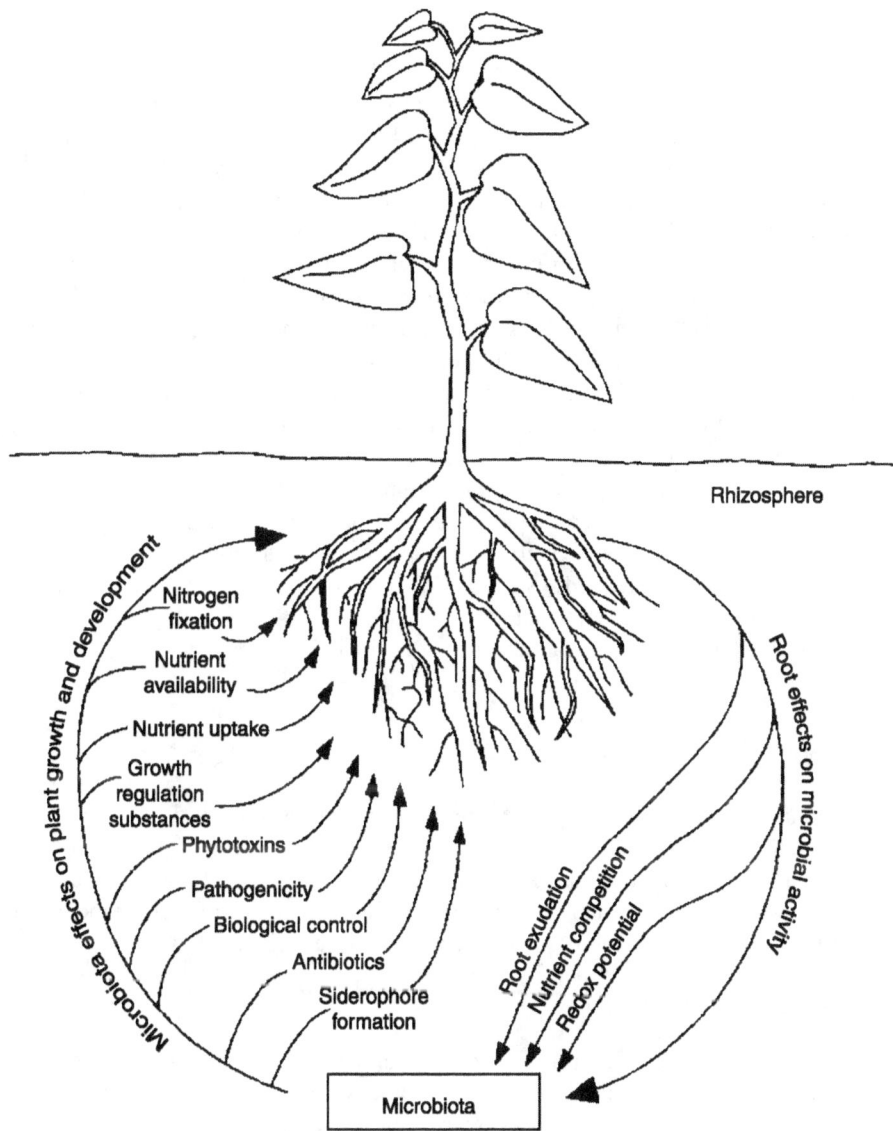

**FIGURE 10.4** Plant–microbe interactions affecting plant growth. (From Frankenberger, W.T., Jr., and M. Arshad. 1995. *Phytohormones in Soils: Microbial Production and Function.* Marcel Dekker, New York, 503 pp. With permission.)

effect of siderophore or antibiotic production, which leads to the reduction in pathogen colonization and infection of the seed and root (Kloepper et al., 1989; Glick 1995).

Microbially derived auxins, ethylene, and other compounds have also been implicated in plant growth inhibition. Plant growth inhibition can be correlated with elevated indole acetic acid levels produced by rhizobacteria in sugarbeet (*Beta vulgaris*; Loper and Schroth, 1986), sour cherry (*Prunus corasus* L.; Dubeikovsky et al., 1993), maize (Sarwar and Frankenberger, 1994), lettuce (*Latuca sativa*; Barazani and Friedman, 1999), and several weed species (Sarwar and Kremer, 1995). Ethylene, produced by plants, soil fungi, yeasts, and bacteria, can affect plant development from seed germination to senescence (Beyer et al., 1984). Microbial synthesis of ethylene can be affected by the availability of organic substrates and crop residues (Goodlass and Smith, 1978; Lynch and Harper, 1980; Arshad and Frankenberger, 1990).

## Amendments

Throughout the history of agriculture, farmers have added amendments to soil to increase crop yield. These additions have the potential to increase SOM accumulation and plant productivity. Soil amendments can also cause alterations to the soil microbial community. These changes can be quantified by microbial population structure and soil enzyme techniques. Soil microbial populations and activity increase, in general, with manure additions (Altieri, 1999). Manure also increases populations of Collembola and earthworms (Altieri, 1999).

Biosolids (sewage sludge) is a material that is sometimes applied for agricultural benefit. Long-term application of biosolids with low and high metal contents added (Cd, Cu, Ni, Zn) showed few differences in the effects on culturable bacteria, but had a dramatic effect on the whole bacterial community, where numbers of the $\alpha$ subdivision of Proteobacteria increased with high metal concentration (Sandaa et al., 2001). Caution must be exercised before applying large amounts of biosolids to agricultural lands, and biosolids should be tested for their heavy-metal content before application. Heavy-metal accumulations from biosolids application can negatively affect microbial communities. Zinc-contaminated agricultural soils (from biosolids) were tested for microbial diversity and catabolic versatility (Wenderoth and Reber, 1999). Microbial diversity was reduced, and the microbial community experienced a shift to less Gram-positive bacteria and more Gram-negative bacteria compared with the nonstressed system. The diversity of the Gram-negative bacteria declined under high zinc stress. Stress or heavy-metal contamination can affect overall populations, specific groups, and also the diversity within various groups, as the individual species have different ways to adapt to stress.

## Agromicrobials

Numerous agromicrobial products have been touted to increase soil fertility, microbial diversity, and crop yields. Microbial inoculants such as effective microorganisms (EM) containing yeasts, fungi, bacteria and actinomycetes increase yields of onion (*Allium cepa* L.) and pea, and increase cob weights of sweet corn (Daly and Stewart, 1999). The consistency of plant response to these types of products has not yet been demonstrated, and further critical study is needed.

### Arbuscular Mycorrhiza (AM)

Mycorrhiza are nonpathogenic fungi that form symbiotic associations with plant roots (Chapter 6). Mycorrhiza are involved in the nutrient cycling process, especially in stressed environments (e.g., P-deficient soils) and can play an active role in SOM accumulation by increasing plant growth by solubilization of nutrients and by producing recalcitrant compounds (glomalin). Fungal hyphal threads allow roots to expand the volume of soil that can be explored for nutrients and water that otherwise might be inaccessible to the plant. Mycorrhizal associations enhance nutrient uptake in the rhizosphere and expand the volume of soil the root can explore (Sylvia, 1998). This relationship is especially beneficial under moisture-limiting conditions. Wheat plants inoculated with AM and subjected to water stress at three different times had higher grain yield and biomass than plants that were not inoculated with AM (Ellis et al., 1985). In the Palouse region of eastern Washington, mycorrhizal fungi lessened the severity of water stress in winter wheat (Mohammed et al., 1995). AM species, abundance, and spore distribution are affected by tillage and crop inputs (Douds et al., 1995). *Glomus occultum* numbers were higher under no-till, whereas other Glomus species were more abundant under conventional tillage in a corn–soybean rotation (Douds et al., 1995).

The interactions of AM and other microorganisms often benefit plants, although the relationships are not always readily evident (Edwards et al., 1998). Presence of AM can enhance relationships with introduced organisms in the rhizosphere of crops. Edwards et al. (1998) found that biological control agents for *P. fluorescens* did not affect AM function in the rhizospheres of tomato (*Lycopersicon esculentum*) and leek (*Allium porrum*). AM plants had higher shoot weights than non-AM plants, and *P. fluorescens* populations were higher in the presence of AM.

## Biological Control

Biological control is the use of pathogens, parasites, or other predators to reduce the population or activity of pest organisms (DeBach, 1964). Another broader definition includes all forms of intervention, such as genes and gene products, to reduce the impact of pests on crops and beneficial organisms (Cook, 1987). The three major strategies for biological control are classical, inundative, and integrated management (DeBach, 1964; TeBeest, 1991). The classical approach involves the importation of exotics or the use of natural enemies for release, dissemination, and self-perpetuation on target pests. The addition of a virulent strain to suppress pests is the inundative approach. The biocontrol agent is not self-sustaining and must be applied to the target host every season. Integrated management is a broad approach that involves management practices to conserve or enhance native enemies of pests. Biological control is an alternative to pesticides and is part of sustainable agriculture management.

Biological control agents have been investigated for their control of diseases, such as take-all root disease in wheat. The phenomenon known as take-all decline (the reduction in severity of take-all disease) is attributed to naturally occurring strains of fluorescent pseudomonads that produce the antibiotics phenazine and phloroglucinol in annually monocropped wheat (Thomashow and Weller, 1988). Deleterious rhizobacteria have been shown to inhibit such weeds as downy brome (*Bromus tectorum* L.; Kennedy et al., 1991), jointed goatgrass (*Aegilops cylindrica* Host.; Kennedy et al., 1992), and velvetleaf (*Abutilon theophrasti* Medik.; Kremer, 1987). Several fungal isolates have been investigated for use in weed biological control, such as *Fusarium* spp. against leafy spurge (*Euphorbia* spp.) in the rangelands of the U.S. and Canada (Caesar et al., 1999), *Exserohilum monoceras* for grass weeds (*Echinochloa* spp.) in rice production (Zhang and Watson, 1997), and *Colletotrichum gloeosporioides* f. sp. aeschynomene to control northern jointvetch (*Aeschynomene virginica*; Luo and TeBeest, 1999). Several conditions must be met before a biological control agent can be widely used in a crop or rangeland situation. The agent must have adequate shelf life (Cross and Polonenko, 1996), the ability to be mass-produced (Oleskevich et al., 1998), the ability to survive and compete in a field situation, and a simple method of application of the organism and subsequent delivery of the plant-inhibitory compound (Kremer and Kennedy, 1996).

## Organic/Low-Input Farming

Organic farming does not allow use of synthetic pesticides or fertilizers and is intended to reduce the detrimental effects of agriculture on soils, animals, food, and the environment. Organic matter and microbial biomass are higher in organic farming systems than in conventional systems (Fließbach and Mäder, 1997; Reganold et al., 1993; Murata and Goh, 1997; Wander et al., 1994). AM fungi were 30 to 60% higher in roots of plants from low-input practices in a long-term field trial that compared organic and conventional systems (Mader et al., 2000). In this study, AM was highest in the control soils, lowest in the conventional system, and intermediate in the organic system. The control soils were not fertilized, whereas the pesticide use, disturbance, and high fertility in the conventional systems reduced AM infection. Soils under animal-based organic management had higher levels of the light fraction of particulate SOM than crop-based organic systems or conventional systems (Wander et al., 1994). This might be the result of a more biologically active substrate pool due to a lower C:N ratio and higher respiration rate, higher amounts of organic residue added, and less soil disturbance in the animal-based system. Microbial biomass C and dissolved organic C increased as organic inputs increased, and microbial communities as determined by PLFA were different in organic farms and conventional farms (Lundquist et al., 1999). Organic management systems that employed animal manure and legumes for N supply were equally profitable as higher-input conventional systems after 15 years in a study in Pennsylvania (Drinkwater et al., 1998). The organic management systems had lower leaching losses of N and higher levels of soil organic C and N.

Biodynamic agriculture is an organic farming system that uses specific fermented preparations as either field sprays or compost inoculants (Koepf et al., 1976). Soil quality parameters were not different among biodynamic, organic, and conventional management systems, but differed with fertilization level (Penfold et al., 1995). Field-applied biodynamic sprays and compost did not alter soil microbial characteristics compared with conventional practices in a cereal–legume cropping system in the state of Washington (Carpenter-Boggs et al., 2000a). In other studies in the state of Washington, however, biodynamic management resulted in higher microbial biomass, respiration, and SOM than organically managed or conventionally managed systems (Goldstein, 1986). Bio-dynamic preparations for compost development altered compost microbial community and increased compost temperature (Carpenter-Boggs et al., 2000b).

## Genetically Modified Organisms (GMOs)

The impact of the addition of genetically modified organisms (GMOs) on soil populations and plant productivity is of interest as more GMOs are introduced into agricultural systems. In a microcosm study of soils from Canada and the U.S., assessment of nontarget effects of two GMOs indicated that there were functional and community differences as long as GMOs persisted in soil; however, effects differed with the GMOs used (Gagliardi et al., 2001). Inoculation of transgenic potatoes with two bacterial biological control agents did not reduce survival of bacterial biological control agents compared with nontransgenic potatoes, nor was the indigenous bacterial community impacted by the introduced bacteria (Lottman et al., 2000). When a GMO and wild-type *Pseudomonas fluorescens* were inoculated into the rhizosphere of wheat, both bacterial strains caused shifts in the native microbial populations in the rhizosphere and phylloplane of wheat; however, there were no changes in nonrhizosphere soil and no negative effects on plant health (De Leij et al., 1995). Addition of a genetically modified *P. fluorescens* in the rhizosphere of pea affected soil enzymes activities and microbial communities (Naseby and Lynch, 1998). Differences are evident with the introduction of some GMOs, but the impact of these differences on soil microbial community, plant productivity, soil quality, and SOM accumulation is case specific, and long-term impacts are not clear.

### DISTURBANCE

Agroecosystem function and SOM dynamics are greatly influenced by anthropogenic activities. Soil erosion caused by excessive tillage is the most visual example of humankind's influence on soil function. Microorganisms are highly sensitive to physical soil disturbance (Elliott and Lynch, 1994), and their population dynamics can serve as early warning indicators of changes in soil quality. Fluxes in microbial diversity and functional diversity can contribute greatly to the understanding of soil quality and the development of sustainable agroecosystems (Thomas and Kevan, 1993; di Castri and Younes, 1990; Hawksworth, 1991a). Soil organisms are useful in classifying disturbed or contaminated systems, because diversity can be affected by minute changes in the ecosystem. Severe disturbances, such as those caused by heavy tillage with a moldboard plow (which completely inverts the surface soil), overgrazing, and pollutants, can reduce aboveground plant diversity and growth. This reduction in plant biomass and lack of a varied carbon source decreased microbial growth and functioning (Christensen, 1989; Zak, 1992).

## 100 Years of Dryland Farming in the Inland Pacific Northwest, U.S.A.

Winter wheat–fallow, in which only one crop is grown every 2 years, is the dominant cropping system in the low-precipitation dryland cropping region of the inland Pacific Northwest. Early settlers grew a wheat crop every year after they first broke the land out of native bunch grass and sagebrush in the 1880s. Grain yields were frequently low; however, it soon became apparent that farmers could better stabilize yields by having a fallow year to store soil moisture between wheat crops. During the fallow year, essential nutrients, mainly N, were released through mineralization of SOM. The native grass prairie provided a reserve of readily decomposable SOM that supplied nitrogen for crop use for many years. A rapid reduction in SOM content occurred when prairie soils were brought under cultivation, especially when alternated with fallow. In some soils, more than 25% of the organic matter was lost in the first 20 years of farming. In undisturbed native soil, SOM in the top 10 cm of soil is 4% at Pullman, WA (530 mm precipitation), 3% at Pendleton, OR (410 mm precipitation), and 1.5% at Lind, WA (240 mm precipitation). Distinct decreases in SOM have been observed on farmland compared to undisturbed native soil in all three areas. Organic matter content in the top 10 cm of cropland soil at Lind, for example, is at present less than 1%. Data are available on the long-term fate of SOM in a continuous winter wheat–fallow rotation from an ongoing 70-year-old study at Pendleton, OR (Rasmussen et al., 1989; Figure 10.5). Since 1931, SOM has continually declined under all residue management methods except when 22 mt (fresh weight) ha$^{-1}$ of cattle manure was applied every other year. SOM decline has been highest when stubble was burned in the fall and when no nitrogen was applied. Maintaining an adequate nitrogen supply and returning all residue to the soil has reduced, but not arrested, SOM decline (Figure 10.5).

## Tillage

Up to 50% of the SOM in some soils has been lost after years of intensive tillage, clearing vegetation, and draining wetlands for farming (Cambardella and Elliott, 1993). Carbon is sequestered in soils through "humification of organic residues, building of organomineral complexes to form aggregates, positioning of SOM beneath the tillage layer, use of deep rooting crops, and calcification" (Bruce et al., 1999). Emissions of $CO_2$ from agricultural soils can be reduced by minimizing or eliminating tillage and by growing perennial crops (Lal et al., 1999). In the U.S., many farmers have 10-year contracts under the Conservation Reserve Program (CRP) to grow perennial grasses and shrubs for environmental and soil-conserving benefits (USDA-FSA, 2000). Improved farming techniques, higher productivity from farmland, and government programs that pay landowners to plant permanent vegetation on highly erodible lands have combined to increase levels of SOM in many soils (Lal et al., 1999). No-till farming results in less $CO_2$ released to the atmosphere than do intensive tillage (moldboard plowing) and minimum tillage (disk harrowing; Reicosky and Lindstrom, 1993). Incorporated straw emits more $CO_2$ than does surface straw or soil with no straw applied (Curtin et al., 1998). Gale and Cambardella (2000) compared contributions of root and surface residues to soil organic C in no-till soils, and found that the greatest increases were due to maintenance of root-derived C, illustrating the importance of leaving root biomass intact (no disturbance) for maximum SOM accumulation.

Conservation-tillage systems maintain at least 30% of the crop residue on the soil surface (Stroo et al., 1989) and help prevent soil erosion by wind and water (Figure 10.6; Papendick, 1996). Positive attributes of retaining crop residue on the soil surface are improved soil quality (Karlen et al., 1994; Dalal et al., 1991; Duran et al., 1996) through increased biological activity, leading to improved soil aggregation and more SOM content (Elliott and Papendick, 1986).

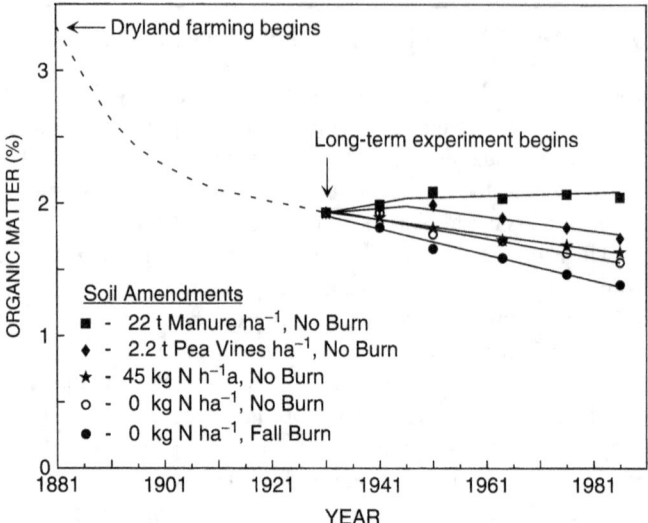

**FIGURE 10.5** Soil organic matter decline in a winter wheat–summer fallow rotation at Pendleton, OR. Rapid decline in SOM occurred after the onset of dryland farming in the early 1880s. A long-term experiment was initiated in 1931 to test effects of soil amendments (cattle manure and pea vines), N fertility (0 and 45 kg ha[-1]), and burning of residue. SOM steadily declined in all treatments except on addition of 22 mt manure ha[-1] every other year with no burning of residue. (Modified from Rasmussen, P. E., H.P. Collins, and R.W. Smiley. 1989. Long-term Management Effects on Soil Productivity and Crop Yields in Semi-arid Regions of Eastern Oregon. Oregon State University Bulletin 675, Corvallis, OR.)

**FIGURE 10.6** No-till planting in the dryland wheat production region of the inland Pacific Northwest. No-till preserves plant residue on the soil surface for erosion control, promotes microbial populations, and provides other environmental benefits.

Conservation tillage, increased cropping intensity (e.g., reduction in fallow), crop rotation, and use of cover crops improve soil quality (Karlen et al., 1992). Crop residues on the soil surface, however, can negatively affect crop yield by impairing seedling emergence, serving as hosts for pathogens, or nutrient immobilization (Elliott and Papendick, 1986). Even distribution of crop residue at harvest and selection of a no-till planter for specific soil and residue conditions reduce the possibility of yield loss.

No-till increases microbial biomass in surface soils (0 to 15 cm; Drijber et al., 2000), increases the ratio of fungi to bacteria, and provides for a more diverse population of residue decomposers and a slower release of nutrients than does conventional tillage (Altieri, 1999). The changes in the physical and chemical properties of soil resulting from tillage greatly alter the matrix supporting growth of the microbial population. Within a given soil, there is considerable variation in the composition of the microbial community and diversity with depth in the profile. In agricultural systems, microbial activities differ drastically with depth, with the highest microbial activity occurring near the surface in no till, and more evenly distributed activity throughout the plow layer of tilled soil (Doran, 1980).

Composition of the microbial community influences the rate of residue decomposition and nutrient cycling in both no-till and tillage-based systems (Beare et al., 1993). Fungi dominate decomposition in a no-till system, whereas the bacterial component is responsible for a greater portion of the decomposition of residue with tillage. In a study of the diversity of native prairie and cultivated soils, diversity indices were higher with tillage than with grassland (Kennedy and Smith, 1995). With the substrate exposed by tillage, more surface area was available for colonization and more activity occurred. Increase in diversity seen early on with disturbance indicates a change in the microbial community to one that exhibits a greater range of substrate utilization and stress resistance. When comparing microbial numbers among burned, tilled, or no-tilled fields of double-cropped wheat and soybean in Georgia, there were no changes in total bacteria or nitrifiers with burning or tillage. Plots that were not burned or tilled initially had higher numbers of algae, actinomycetes, and fungi; however, there were no treatment differences later in the growing season (Harris et al., 1995). Preemergent herbicides had no effect on microbial numbers in that study. Buried sorghum residue under conventional tillage contained more fungal hyphae and CFU than surface residue in no-till soil did; however, in bulk soil there were no differences in fungal CFU between treatments, and higher numbers of fungal hyphae were found in no-till soil (Beare et al., 1993). These studies illustrate the alteration of the make-up of microbial communities and possibly the diversity of basic microbial groups with disturbance.

Studies have shown varying results with regards to N immobilization in reduced-tillage systems. When farmers first convert to minimum- or no-till cropping, they often encounter lower N availability for the first several years because of reduced mineralizable N. SOM (and N) accumulates under no-till, however, and a new equilibrium is established in which mineralized N and microbial biomass C are higher than under intensive tillage (Simard et al., 1994). Higher tillage intensity under conventional tillage decreased the amount of N mineralized per unit of biomass C, which could lead to a decline in SOM quality. In a long-term (11-year) study in Canada, soil C and mineralizable N were highest with no till compared with conventional tillage, regardless of cropping sequence or cropping frequency (Campbell et al., 1996). Conversely, decomposing surface residues in some no-till systems can immobilize enough N to cause N deficiency in succeeding crops (Knowles et al., 1993), and increased tillage intensity reduces the potential for N immobilization (Follett and Schimel, 1989). In these instances, cropping intensity and rotation (Kolberg et al., 1999; Knowles et al., 1993) and fertilizer placement (Knowles et al., 1993) must be managed to ensure success of no-till farming systems. Carbon gains in the soil are a function of both residue input and clay content of the soil. After 20 years of wheat and barley residue maintenance by using no-till and high N fertilization in Australia, organic C, total N, and microbial biomass were higher and pH was lower at the soil surface than under conventional tillage in which residue was burned and lower levels of N were applied (Dalal et al., 1991).

SOM increases when crop residue is retained on the soil surface (as in no-till systems), when erosion is reduced, and crops are adequately fertilized (Campbell and Zentner, 1993). In the Canadian prairies, potential exists for C sequestration under a long-term (12- to 15-year) no-till continuous wheat farming system compared with a system using conventional tillage wheat–fallow because of lower $CO_2$ flux from the soil and more organic matter accumulation with reduced residue disturbance and continuous cropping (Curtin et al. 2000). Four years of no till increased SOM in the top 2.5 cm of soil compared with conventional tillage in a Mississippi study with grain sorghum–corn, cotton and soybean–wheat rotations (Rhoton, 2000). No-till in the sorghum–corn rotation showed the maximum accumulations of soil organic content, especially in the top 2 cm of soil, compared with conventional tillage and rotations that included soybean in the Great Plains (McCallister and Chien, 2000).

Although numerous studies indicate higher SOM under no till vs. conventional tillage, some studies show little or no difference in SOM between the two systems, especially in low-precipitation regions where residue production from crops is minimal (A.C. Kennedy, unpublished data). Macroorganic matter (>50 um) and microbial biomass-C can be good indicators of changes in residue management; however, effects of tillage might be limited to vertical distribution without influencing SOM turnover (Angers et al. 1995). Needelman et al. (1999) found that no-till fields in a corn–soybean rotation in Illinois had higher SOM in the top 0 to 5 cm of soil than that in conventionally tilled fields, but SOM was not different between the two tillage treatments when the entire sampling depth (0 to 30 cm) was considered. SOM levels did not differ from conventional tillage levels in the top 15 cm of soil after 30 years of a no-till wheat–sorghum–fallow rotation in Kansas (Thompson and Whitney, 2000). Maintaining or increasing SOM is critical to crop production to improve soil water-holding capacity and aeration, provide nutrients for plants and microbes, and maintain soil physical properties such as friability and low bulk density. The amounts of SOM that accumulate in different systems vary greatly with geographic location (soil type, precipitation, and climate), length of time for which a particular management scheme is used, tillage intensity, and crop residue inputs.

Population and diversity of genomic patterns of the $N_2$-fixing bacteria *Bradyrhizobium* increased with no-till compared to conventional tillage in southern Brazil (Ferreira et al. 2000), even though the field was last inoculated 15 years before the study. Along with no-till, crop rotations containing soybean increased populations of *Bradyrhizobium*. Treatments that did not include soybean in rotation for 17 years and were in conventional tillage contained the least amount of *Bradyrhizobium*.

Wardle et al. (1999) studied three methods of controlling weeds (mulching, herbicides, tillage) and found that mulching (adding residue) increased soil C, microbial biomass, and activity in surface soils (1 to 10 cm) over the course of a 7-year study in New Zealand; however, some immobilization of N might have occurred late in the study. Herbicide application did not adversely affect microbial biomass and activity. Where less weed biomass was present, microbial respiration was reduced, probably because of more decomposition of weeds than crop plants. Tillage for weed control was not detrimental to substrate-induced respiration, $CO_2$-C released from chloroform fumigation, or soil organic C in the study. The authors emphasized the need for long-term studies, as many of their results were not apparent until after 6 years.

Although most of the effects of reduced tillage are positive, there are some instances wherein more physical soil disturbance is advantageous. Direct seeding wheat into cereal or grass residue increases the risk of infection by pathogens causing the diseases take-all, Rhizoctonia (*Rhizoctonia solani*) root rot, Cephalosporium (*Cephalosporium gramineum*) stripe, and *Pythium* root rot (*Pythium* spp.; Cook and Haglund, 1991). Crop residue can serve as a host for the pathogens (Cook, 1986). Crop rotation and tillage are suggested to alleviate disease pressure in wheat (Cook and Haglund, 1991). Abawi and Widmer (2000) cite numerous examples in which yield of bean was increased because of less disease with intensive tillage compared with reduced tillage or no-till. The increase in yield with tillage was attributed to reduced compaction, improved drainage, and higher soil temperature, which led to improved bean root competition against pathogens.

## Grazing

Careful management of grazing lands is needed to protect soils from the negative effects of overgrazing and to maintain benefits of permanent plant cover. Livestock grazing is thought to be less damaging to soil quality than is conventional crop management; however, soil quality can be impacted by compaction and continual removal of plant cover (Southorn and Cattle, 2000). Cattle grazing can also affect the biomass and biodiversity of plants by causing patches that differ in size and plant species (Cid and Brizuela, 1998). Also, because of overgrazing, species composition in grazing lands can shift from perennial species to annual grasses (DiTomaso, 2000). This land then becomes more susceptible to invasion by broadleaf weed species, which degrades soil productivity (DiTomaso, 2000). Although many weed species have deep taproots, they produce less aboveground biomass than do most crop plants, often leaving surface soil vulnerable to erosion. A high infestation of spotted knapweed (*Centaurea maculose*) reduced water infiltration rates (DiTomaso, 2000). Additionally, overgrazing and recolonization with weed species led to less soil moisture available to grass species and less contribution to SOM than did fibrous root systems of grass species (DiTomaso, 2000). Abril and Bucher (1999) showed the negative effects of overgrazing of range-lands in Argentina, where the overgrazed site had the lowest SOM and microbial activity. In a comparison of restored, partially restored, and overgrazed rangelands, they found that soil water, SOM, N content, and microbial activity were highest at the restored site. In another study on the soil quality of grasslands in New Mexico, Liu et al. (2000) found no negative effects on microbial diversity (substrate utilization) or microbial activity (enzyme analysis) from intense grazing; however, burning reduced microbial diversity.

Integrated systems combining crop production and livestock production with perennials are suggested as a means to improve soil quality and combat the decline in organic C of soil from the Great Plains of the U.S. after decades of cultivation (Krall and Schuman, 1996). Well-managed grasslands used for livestock grazing adjacent to streams can protect or enhance soil quality by stabilizing stream banks and reducing erosion (Lyons et al. 2000). These managed riparian areas can reduce the impact of livestock grazing and help restore degraded stream banks.

## STRATEGIES FOR MANAGING MICROORGANISMS

Although the technology for managing microorganisms for sustainable agricultural production systems has not yet been developed, several strategies have been used for centuries to optimize soil life (see Chapter 2 for a discussion on SOM management). First, management practices that increase SOM should be used, especially in SOM-depleted systems. Organic matter is responsible for providing substrate for microorganisms, but also improves microbial habitat. Organic matter, in various forms of decay, improves soil physical properties, increases water-holding capacity and nutrient availability, and acts as a cementing agent for holding soil particles together. SOM can be maintained or increased by incorporating crop residues, crop rotation, cover crops, permanent plantings, maintaining soil fertility, and adding animal manures or biosolids. Addition of plentiful amounts of organic residues helps ensure a productive soil and stimulates plant growth by providing food for microorganisms. The movement toward sustainable farming systems, with diverse, healthy soil microbial communities that closely imitate the processes of native, undisturbed systems, can be realized by using these practices or adopting a combination of several practices.

A second strategy is to ensure a diverse plant community through crop rotation or grazing management. Minimizing fallow or increasing root growth in soil will provide substrate additions and adequate nutrition for a healthy soil and large, diverse populations of microorganisms. Tillage and burning of crop residues often negatively and dramatically affect the chemical and physical properties of soil, which alter growth of microorganisms and processes for which they are responsible. Minimum tillage or no till helps prevent erosion of valuable topsoil.

### Options for Farmers

The following management principles will help maximize soil quality in low-precipitation areas, such as in the inland Pacific Northwest:

- Minimize tillage to the degree feasible to leave as much residue as possible on or near the soil surface.
- Maintain adequate nitrogen inputs. Because of the linkage between soil N and organic matter, adequate (but not excessive) nitrogen inputs are a requisite for optimum crop growth and residue return.
- Minimize the use of summer fallow, if possible. Consider recropping to spring wheat or barley after wet winters. Use a no-till drill, if feasible, to plant seed and fertilize in one pass through the standing residue of the previous crop.
- Emphasize wind and water erosion control, because any loss of topsoil increases loss of SOM.

Applying large quantities of organic materials, such as cattle manure, can increase SOM. This, however, is not a realistic option for most farmers because of the large quantities of manure required, as the size of the average farm exceeds 1000 ha.

## CONCLUSIONS

Microorganisms are responsible for a multitude of soil processes, such as SOM dynamics, nutrient cycling, and changes in soil structure. In agroecosystems, microorganisms can affect all levels within the ecosystem through functions such as N and C cycling, plant growth promotion and inhibition, and natural biological control. Microorganisms have more diversity than does any other group of organisms on earth, but our knowledge of these organisms is still limited. We need to increase our understanding of microbial communities and their functioning in agroecosystems. Several strategies have been suggested to optimize soil life. The most critical to sustainability is to use management practices that increase SOM, reduce disturbance, and maintain a diverse plant community. There is a wealth of genetic potential in the soil, but we do not presently have the means or understanding to use the full potential of the earth's oldest inhabitants. With a better understanding of soil ecology and changes in soil biota with management, best management practices can be developed to conserve and enhance SOM, soil quality, and crop production for sustainable agricultural systems.

## References

Abawi, G. S., and T. L. Widmer. 2000. Impact of soil health management practices on soil borne pathogens, nematodes, and root diseases of vegetable crops. *Appl. Soil Ecol.* 15:37–47.

Abril, A., and E. H. Bucher. 1999. The effects of overgrazing on soil microbial community and fertility in the Chaco dry savannas of Argentina. *Appl. Soil Ecol.* 12:159–167.

Albrecht, S. L., Y. Okon, J. Lonnquist, and R. H. Burris. 1981. Nitrogen fixation by corn-*Azospirillum* associations in a temperate climate. *Crop Sci.* 21:301–306.

Alexander, D. B. 1998. Bacteria and archaea. In Sylvia, D. M., J. J. Fuhrmann, P. G. Hartel, and D. A. Zuberer (Eds.), *Principles and Applications of Soil Microbiology.* Prentice-Hall, New York, pp. 44–71.

Allen, C. C. R., D. R. Boyd, F. Hempenstall, M. J. Larkin, and N. D. Sharma. 1999. Contrasting effects of a nonionic surfactant on the biotransformation of polycyclic aromatic hydrocarbons to *cis*-dihydrodiols by soil bacteria. *Appl. Environ. Microbiol.* 65:1335–1339.

Allen, E. B., and M. F. Allen. 1990. The mediation of competition by mycorrhizae in successional and patchy environments. In Grace, J. B. and D.A. Tilman (Eds.), *Perspectives on Plant Competition*. Academic Press, San Diego, CA, pp. 367–389.

Alstrom, S. 1987. Factors associated with detrimental effects of rhizobacteria on plant growth. *Plant Soil* 102: 3–9.

Altieri, M. A. 1999. The ecological role of biodiversity in agroecosystems. *Agric. Ecosyst. Environ.* 74:19–31.

American Society for Microbiology. 1994. *Microbial Diversity Research Priorities*. American Society for Microbiology, Washington, D.C., 7 pp.

Anaya, A. L. 1999. Allelopathy as a tool in the management of biotic resources in agroecosystems. *Crit. Rev. Plant Sci.* 18:697–739.

Andren, O., J. Bengtsson, and M. Clarholm. 1995. Biodiversity and species redundancy among litter decomposers. In Collins, H.P., G.P. Robertson, and M. J. Klug (Eds.), *The Significance and Regulation of Soil Biodiversity*. Kluwer, Dordrecht, pp. 141–151.

Andrews, J. H. 1984. Relevance of r- and K-theory to the ecology of plant pathogens. In Klug, M. G. and C. A. Reddy (Eds.), *Current Perspectives in Microbial Ecology*. American Society for Microbiology, Washington, D.C., pp. 1–7.

Andrews, J., and H. R. F. Harris. 1986. r- And K-selection in microbial ecology. *Adv. Microb. Ecol.* 9:99–147.

Angers, D. A., R. P. Voroney, and D. Cote. 1995. Dynamics of soil organic matter and corn residues affected by tillage practices. *Soil Sci. Soc. Am. J.* 59:1311–1315.

Arshad, M., and W. T. Frankenberger, Jr. 1990. Ethylene accumulation in soil in response to organic amendments. *Soil Sci. Soc. Am. J.* 54:1026–1031.

Arshad, M., and W. T. Frankenberger. 1998. Plant growth-regulating substances in the rhizosphere: Microbial production and functions. *Adv. Agron.* 62:45–151.

Barazani, O., and J. Friedman. 1999. Allelopathic bacteria and their impact on higher plants. *Crit. Rev. Plant Sci.* 18:741–755.

Barbieri, P., and E. Galli. 1993. Effect on wheat root development of inoculation with an *Azospirillum brasilense* mutant with altered indole-3-acetic acid production. *Res. Microbiol.* 144:69–75.

Bardgett, R. D., and A. Shine. 1999. Linkages between plant litter diversity, soil microbial biomass and ecosystem function in temperate grasslands. *Soil Biol. Biochem.* 31:317–321.

Beare, M. H., B. R. Pohlad, D. H. Wright, and D. C. Coleman. 1993. Residue placement and fungicide effects of fungal communities in conventional and no-tillage soils. *Soil Sci. Soc. Am. J.* 57:392–399.

Beyer, E. M., Jr., P. W. Morgan, and S. F. Yang. 1984. Ethylene. In Wilkins, M. B. (Ed.), *Advanced Plant Physiology*. Pitman Publishing, London, pp. 111–126.

Bossio D. A., and K. M. Scow. 1995. Impacts of carbon and flooding on the metabolic diversity of microbial communities in soils. *Appl. Environ. Microbiol.* 61: 4043–4050.

Boswell, E. P., R. T. Koide, D. L. Shumway, and H. D. Addy. 1998. Winter wheat cover cropping, VA mycorrhizal fungi and maize growth and yield. *Agric. Ecosyst. Environ.* 67:55–65.

Bottomley, P. J. 1998. Microbial ecology. In Sylvia, D. M., J. J. Fuhrmann, P. G. Hartel, and D. A. Zuberer (Eds.), *Principles and Applications of Soil Microbiology*. Prentice-Hall, New York, pp. 149–167.

Bowen, G. D., and A. D. Rovira. 1999. The rhizosphere and its management to improve plant growth. *Adv. Agron.* 66: 1–102.

Broder, M. W., and G. H. Wagner. 1988. Microbial colonization and decomposition of corn, wheat, and soybean residue. *Soil Sci. Soc. Am. J.* 52:112–117.

Brown, M. E. 1972. Plant growth substances produced by microorganisms of soil and rhizosphere. *J. Appl. Bacteriol.* 35: 443–451.

Bruce, J. P., M. Frome, E. Haites, H. Janzen, R. Lal, and K. Paustian. 1999. Carbon sequestration in soils. *J. Soil Water Conserv.* 54:382–389.

Bruehl, G. W. 1987. *Soilborne Plant Pathogens*. MacMillan, New York.

Burdon, J. J. 1987. *Diseases and Plant Population Biology*. Cambridge University Press, New York.

Caesar, A. J., G. Campobasso, and G. Terragitti. 1999. Effects of European and U.S. strains of *Fusarium* spp. pathogenic to leafy spurge on North American grasses and cultivated species. *Biol. Control* 15:130–136.

Cambardella, C. A., and E. T. Elliott. 1993. Carbon and nitrogen distribution in aggregates from cultivated and native grassland soils. *Soil Sci. Soc. Am. J.* 57:1071–1076.

Cameron, M. D., S. Timofeevski, and S. D. Aust. 2000. Enzymology of *Phanerochaete chrysosporium* with respect to the degradation of recalcitrant compounds and xenobiotics. *Appl. Microbiol. Biotechnol.* 54:751–758.

Campbell, C. A., B. G. McConkey, R. P. Zentner, F. Selles, and D. Curtin. 1996. Long-term effects of tillage and crop rotations on soil organic C and total N in a clay soil in southwestern Saskatchewan. *Can. J. Soil Sci.* 76:395–401.

Campbell, C. A., and R. P. Zentner. 1993. Soil organic matter as influenced by crop rotations and fertilization. *Soil Sci. Soc. Am. J.* 57:1034–1040.

Carpenter-Boggs, L., A. C. Kennedy, and J. P. Reganold. 2000a. Organic and biodynamic management: Effects on soil biology. *Soil Sci. Soc. Am. J.* 64:1651–1659.

Carpenter-Boggs, L., J. P. Reganold, and A. C. Kennedy. 2000b. Effects of biodynamic preparations on compost development. *Biol. Agric. Hortic.* 17: 313–328.

Cavigelli, M. A., G. P. Robertson, and M. J. Klug. 1995. Fatty acid methyl ester (FAME) profiles as measures of soil microbial community structure. In Collins, H. P., G. P. Robertson, and M. J. Klug (Eds.), *The Significance and Regulation of Soil Biodiversity.* Kluwer, Dordrecht, pp. 99–113.

Chalaux, N., S. Libmond, and J.-M. Savoie. 1995. A practical enzymatic method to estimate wheat straw quality as raw material for mushroom cultivation. *Bioresour. Technol.* 53:277–281.

Chang, I. P., and T. Kommendahl. 1968. Biological control of seedling blight of corn by coating kernels with antagonistic microorganisms. *Phytopathology* 58: 1395–1401.

Chanway, C. P., and F. B. Holl. 1993. First year field performance of spruce seedlings inoculated with plant growth promoting rhizobacteria. *Can. J. Microbiol.* 39:1084–1088.

Christensen, M. 1989. A view of fungal ecology. *Mycologia* 81:1–19.

Cid, M. S, and M. A. Brizuela. 1998. Heterogeneity in tall fescue pastures created and sustained by cattle grazing. *J. Range Manage.* 51:644–649.

Cook, R. J. 1981. The influence of rotation crops on take-all decline phenomena. *Phytopathology* 71:189–192.

Cook, R. J. 1986. Plant health and the sustainability of agriculture, with special reference to disease control by beneficial microorganisms. *Biol. Agric. Hortic.* 3:211–232.

Cook, R. J. 1987. Research Briefing Panel on Biological Control in Managed Ecosystems. Committee on Science, Engineering, and Public Policy, National Academy of Sciences, National Academy of Engineering and Institute of Medicine. National Academy Press, Washington, D.C., 12 pp.

Cook, R. J., and K. F. Baker. 1983. *The Nature and Practice of Biological Control of Plant Pathogens.* American Phytopathological Society, St. Paul, MN, 539 pp.

Cook, R. J., and W. A. Haglund. 1991. Wheat yield depression associated with conservation tillage caused by root pathogens in the soil not phytotoxins from the straw. *Soil Biol. Biochem.* 23:1125–1132.

Cookson, W. R., M. H. Beare, and P. E. Wilson. 1998. Effects of prior crop residue management on microbial properties and crop residue decomposition. *Appl. Soil Ecol.* 7:179–188.

Cross, J. V., and D. R. Polonenko. 1996. An industry perspective on registration and commercialization of biocontrol agents in Canada. *Can. J. Plant Pathol.* 18:446–454.

Crowell, H. F., and R. E. Boerner. 1988. Influences of mycorrhizae and phosphorus on belowground competition between two old-field annuals. *Environ. Exp. Bot.* 28:381–392.

Curl, E., and B. Truelove. 1986. *The Rhizosphere.* Springer-Verlag, Berlin, 288 pp.

Curtin, D., F. Selles, H. Wang, C. A. Campbell, and V. O. Biederbeck. 1998. Carbon dioxide emissions and transformation of soil carbon and nitrogen during wheat straw decomposition. *Soil Sci. Soc. Am. J.* 62:1035–1041.

Curtin, D., H. Wang, F. Selles, B. G. McConkey, and C. A. Campbell. 2000. Tillage effects on carbon fluxes in continuous wheat and fallow-wheat rotations. *Soil Sci. Soc. Am. J.* 64:2080–2086.

Dalal, R. C., P. A. Henderson, and J. M. Glasby. 1991. Organic matter and microbial biomass in a Vertisol after 20 yr. of zero-tillage. *Soil Biol. Biochem.* 23:435–451.

Daly, M. J., and D. P. C. Stewart. 1999. Influence of "effective microorganisms" (EM) on vegetable production and carbon mineralization: A preliminary investigation. *Am. J. Sust. Agric.* 14:15–25.

De Leij, F. A. A. M., E. J. Sutton, J. M. Whipps, J. S. Fenlon, and J. M. Lynch. 1995. Impact of field release of genetically modified *Pseudomonas fluorescens* on indigenous microbial populations of wheat. *Appl. Environ. Microbiol.* 61:3443–3453.

De Leij, F. A. A. M., J. M. Whipps, and J. M. Lynch. 1993. The use of colony development for the characterization of bacterial communities in soil and roots. *Microb. Ecol.* 27:81–97.

DeBach, P. 1964. *Biological Control of Insects, Pests and Weeds*. Reinhold, New York, 844 pp.

Degens, B. P. 1998. Decreases in microbial functional diversity do not result in corresponding changes in decomposition under different moisture conditions. *Soil Biol. Biochem.* 30:1989–2000.

Degens, B. P. 1999. Catabolic response profiles differ between microorganisms grown in soils. *Soil Biol. Biochem.* 31: 475–477.

Degens, B. P., and J. A. Harris. 1997. Development of physiological approach to measuring the catabolic diversity of soil microbial communities. *Soil Biol. Biochem.* 29:1309–1320.

di Castri, F., and T. Younes. 1990. Ecosystem function of biological diversity. *Biol. Int.* 22 (Special Issue):1–20.

DiTomaso, J. M. 2000. Invasive weeds in rangelands: species, impacts and management. *Weed Sci.* 48:255–265.

Doran, J. W. 1980. Soil microbial and biochemical changes associated with reduced tillage. *Soil Sci. Soc. Am. J.* 44: 765–771.

Doran, J. W., and T. B. Parkin. 1994. In Doran, J. W., D. C. Coleman, D. F. Bezdicek, and B. A. Stewart (Eds.), *Defining Soil Quality for a Sustainable Environment*. Special Publication No. 35, American Society of Agronomy, Madison, WI, pp. 1–45.

Doran, J. W., M. Sarrantonio, and M. A. Liebig. 1996. Soil health and sustainability. In D.L. Sparks (Ed.), *Advances in Agronomy*. Academic Press, New York, pp. 1–54.

Douds, D. D., Jr., L. Galvez, M. Franke-Snyder, C. Reider, and L. E. Drinkwater. 1997. Effect of compost addition and crop rotation point upon VAM fungi. *Agric. Ecosyst. Environ.* 65:257–266.

Douds, D. D., Jr., L. Galvez, R. R. Janke, and P. Wagoner. 1995. Effect of tillage and farming system upon populations and distribution of vesicular-arbuscular mycorrhizal fungi. *Agric. Ecosyst. Environ.* 52:111–118.

Drijber, R. A., J. W. Doran, A. M. Parkhurst, and D. J. Lyon. 2000. Changes in soil microbial community structure with tillage under long-term wheat-fallow management. *Soil Biol. Biochem.* 32:1419–1430.

Drinkwater, L. E., P. Wagoner, and M. Sarrantonio. 1998. Legume-based cropping systems have reduced carbon and nitrogen losses. *Nature* 396:262–265.

Dubeikovsky, A. N., E. A. Mordukhova, V. V. Kochetkov, F. Y. Polikarpova, and A. M. Boronin. 1993. Growth promotion of blackcurrant softwood cuttings by recombinant strain *Pseudomonas fluorescens* BSP53a synthesizing an increased amount of indole-3-acetic-acid. *Soil Biol. Biochem.* 25:1277–1281.

Duineveld, B. M., A. S. Rosado, J. D. vanElsas, and J. A. vanVeen. 1998. Analysis of the dynamics of bacterial communities in the rhizosphere of the chrysanthemum via denaturing gradient gel electrophoresis and substrate utilization patterns. *Appl. Environ. Microbiol.* 64:4950–4957.

Edwards, S. G., J. Peter, W. Young, and A. H. Fitter. 1998. Interactions between *Pseudomonas fluorescens* biocontrol agents and *Glomus mosseae*, an arbuscular mycorrhizal fungus, within the rhizosphere. *FEMS Microbiol. Lett.* 166:297–303.

El Nashaar, H. M., and R. W. Stack. 1989. Effect of long-term continuous cropping of spring wheat on aggressiveness of *Cochliobolus sativus*. *Can. J. Plant Sci.* 69:395–400.

Elliott, L. F., and J. M. Lynch. 1994. Biodiversity and soil resilience. In Greenland, D. J. and I. Szabolcs (Eds.), *Soil Resilience and Sustainable Land Use*. CAB International, Wallingford, U.K., pp. 353–364.

Elliott, L. F., and R. I. Papendick. 1986. Crop residue management for improved soil productivity. *Biol. Agric. Hortic.* 3:131–142.

Ellis, J. R., H. J. Larsen, and M. G. Boosalis. 1985. Drought resistance of wheat plants inoculated with vesicular-arbuscular mycorrhizae. *Plant Soil* 86:369–378.

Emmimath, V. S., and G. Rangaswami. 1971. Studies of the effect of heavy doses of nitrogenous fertilizer on the soil and rhizosphere microflora of rice. *Mysore J. Agric. Sci.* 5:39–58.

Ferreira, M. C., D. D. Andrade, L.M.D. Chueire, S. M. Takemura, and M. Hungria. 2000. Tillage method and crop rotation effects on the population sizes and diversity of *Bradyrhizobia* nodulating soybean. *Soil Biol. Biochem.* 32:627–637.

Fitter, A. H. 1977. Influence of mycorrhizal infection on competition for phosphorus and potassium by two grasses. *New Phytol.* 79:119–125.

Fließbach, A., and Mäder, P. 1997. Carbon source utilization by microbial communities in soils under organic and conventional farming practice. In Insam, H. and A. Rangger (Eds.), *Microbial Communities — Functional versus Structural Approaches*. Springer-Verlag, Berlin, pp. 109–120.

Foissner, W. 1999. Soil protozoa as bioindicators: Pros and cons, methods, diversity, representative examples. *Agric. Ecosyst. Environ.* 74:95–112.

Follett, R. F., and D. S. Schimel. 1989. Effect of tillage practices on microbial dynamics. *Soil Sci. Soc. Am. J.* 53:1091–1096.

Frankenberger, W. T., Jr., and M. Arshad. 1995. *Phytohormones in Soils: Microbial Production and Function.* Marcel Dekker, New York, 503 pp.

Franzluebbers, A. J., F. M. Hons, and D. A. Zuberer. 1994. Seasonal changes in soil microbial biomass and mineralizable C and N in wheat management systems. *Soil Biol. Biochem.* 26:1467–1475.

Gagliardi, J. V., J. S. Buyer, J. S. Angle, and E. Russek-Cohen. 2001. Structural and functional analysis of whole-soil microbial communities for risk and efficacy testing following microbial inoculation of wheat roots in diverse soils. *Soil Biol. Biochem.* 33:25–40.

Gale, W. J., and C. A. Cambardella. 2000. Carbon dynamics of surface residue- and root-derived organic matter under simulated no-till. *Soil Sci. Soc. Am. J.* 64:190–195.

Galvez, L., D. D. Douds, Jr., P. Wagoner, L. R. Longnecker, L. E. Drinkwater, and R. R. Janke. 1995. An overwintering cover crop increases inoculum of VAM fungi in agricultural soil. *Am. J. Altern. Agric.* 10:152–156.

Garland, J. L. 1996. Patterns of potential C source utilization by rhizosphere communities. *Soil Biol. Biochem.* 28:223–230.

Gilbert, G. S., J. L. Parke, M. K. Clayton, and J. Handelsman. 1993. Effects of an introduced bacterium on bacterial communities on roots. *Ecology* 74:840–854.

Giller, K. E., M. H. Beare, P. Lavelle, A. N. Izac, and M. J. Swift. 1997. Agricultural intensification, soil biodiversity and agroecosystem function. *Appl. Soil Ecol.* 6:3–16.

Glick, B. R. 1995. The enhancement of plant growth by free-living bacteria. *Can. J. Microbiol.* 41:109–117.

Goldstein, W. Q. 1986. Alternative crops, rotations and management systems for the Palouse. Ph.D. Dissertation, Washington State University, Pullman, WA (Dissertation Abstract 87–11988).

Goodlass, G., and K. A. Smith. 1978. Effects of organic amendments on evolution of ethylene and other hydrocarbons from soil. *Soil. Biol. Biochem.* 10:201–205.

Grayston, S. J., S. Wang, C. D. Campbell, and A. C. Edwards. 1998. Selective influence of plant species on microbial diversity in the rhizosphere. *Soil Biol. Biochem.* 30:369–378.

Griffiths, B. S., K. Ritz, R. D. Bardgett, R. Cook, S. Christensen, F. Ekelund, S. Sorensen, E. Bååth, J. Bloem, P. C. de Ruiter, J. Dolfing, and B. Nicolardot. 2000. Ecosystem response of pasture soil communities to fumigation-induced microbial diversity reductions: An examination of the biodiversity-ecosystem function relationship. *Oikos* 90:279–294.

Griffiths, B. S., K. Ritz, N. Ebblewhite, and G. Dobson. 1999. Soil microbial community structure: Effects of substrate loading rates. *Soil Biol. Biochem.* 31:145–153.

Griffiths, B. S., K. Ritz, and L. A. Glover. 1996. Broad-scale approaches to the determination of soil microbial community structure: Application of the community DNA hybridization technique. *Microb. Ecol.* 31:269–280.

Grime, J. P. 1997. Biodiversity and ecosystem function: The debate deepens. *Science* 277:1260–1261.

Gunapala, N., R. C. Venette, H. Ferris, and K. M. Scow. 1998. Effects of soil management history on the rate of organic matter decomposition. *Soil Biol. Biochem.* 30:1917–1927.

Guschin, D. Y., B. K. Mobarry, D. Proudnikov, D. A. Stahl, B. Rittman, and A. D. Mirzabekov. 1997. Oligonucleotide microchips as genosensors for determinative and environmental studies in microbiology. *Appl. Environ. Microb.* 63:2397–2402.

Haack, S. K., H. Garchow, M. J. Klug, and L. J. Forney. 1995. Analysis of factors affecting the accuracy, reproducibility, and interpretation of microbial community carbon source utilization patterns. *Appl. Environ. Microbiol.* 61:1458–1468.

Hall, J. R. 1978. Effects of endomycorrhizas on the competitive ability of white clover. *N.Z. J. Agric. Res.* 21:509–515.

Harris, P. A., H. H. Schomberg, P. A. Banks, and J. Giddens. 1995. Burning, tillage and herbicide effects on the soil microflora in a wheat-soybean double-crop system. *Soil Biol. Biochem.* 27:153–156.

Hawksworth, D. L. 1991a. *The Biodiversity of Microorganisms and Invertebrates: Its Role in Sustainable Agriculture.* CAB International, Redwood Press, Melksham, U.K., 302 pp.

Hawksworth, D. L. 1991b. The fungal dimension of biodiversity: Magnitude, significance, and conservation. *Mycol. Res.* 95:641–655.

Henriksen, T. M., and T. A. Breland. 1999. Evaluation of criteria for describing crop residue degradability in a model of carbon and nitrogen turnover in soil. *Soil Biol. Biochem.* 31:1135–1149.

Hetrick, B. A. D., and G. W. T. Wilson. 1989. Suppression of mycorrhizal fungus spore germination in nonsterile soil: Relationship to mycorrhizal growth response in big bluestem. *Mycologia* 81:382–390.

Holben, W. E., and J. M. Tiedje. 1988. Applications of nucleic acid hybridization in microbial ecology. Ecology 69:561–568.

Hussain, A., and A. Vancura. 1970. Formation of biologically active substances by rhizosphere bacteria and their effect on plant growth. *Folia Microbiologia* 19:468–478.

Hu, S. J., A. H. C. van Bruggen, and N. J. Grünwald. 1999. Dynamics of bacterial populations in relation to carbon availability in a residue-amended soil. *Appl. Soil Ecol.* 13:21–30.

Hu, S., and A. H. C. van Bruggen. 1997. Microbial dynamics associated with multiphasic decomposition of $^{14}$C labeled cellulose in soil. *Microb. Ecol.* 33:134–143.

Janzen, R. A., S. B. Rood, J. F. Dormaar, and W. B. McGill. 1992. *Azospirillum brasilense* produces gibberellin in pure culture and on chemically defined medium in co-culture on straw. *Soil Biol. Biochem.* 24:1061–1064.

Ibekwe, A. M., and A. C. Kennedy. 1998. Phospholipid fatty acid profiles and carbon utilization patterns for analysis of microbial community structure under field and greenhouse conditions. *FEMS Microbiol. Ecol.* 26:151–163.

Ibekwe, A. M., and A. C. Kennedy. 1999. Fatty acid methyl ester (FAME) profiles as a tool to investigate community structure of two agricultural soils. *Plant Soil* 209:151–161.

Johnson, N. C., P. J. Copeland, B. K. Crookston, and F. L. Pfleger. 1992. Mycorrhizae: Possible explanation for yield decline with continuous corn and soybean. *Agron. J.* 84:387–390.

Karlen, D. L., N. S. Eash, and P. W. Unger. 1992. Soil and crop management effects on soil quality indicators. *Am. J. Altern. Agric.* 7:48–55.

Karlen, D. L., N. C. Wollenhaupt, D. C. Erbach, E. C. Berry, J. B. Swan, N. S. Eash, and J. L. Jordahl. 1994. Crop residue effects on soil quality following 10 years of no-till corn. *Soil Tillage Res.* 31:149–167.

Kennedy, A. C. 1998. The rhizosphere and spermosphere. In Sylvia, D. M., J. J. Fuhrman, R. G. Hartel, and D. A. Zuberer (Eds.), *Principles and Applications of Soil Microbiology*. Prentice-Hall, New York, pp. 389–407.

Kennedy, A. C., and A. J. Busacca. 1995. Microbial analysis to identify source of PM-10 material. In *Proceedings of the Air and Waste Management Association Specialty Conference on Particulate Matter: Health and Regulatory Issues.* April 4–6, 1995, Pittsburgh, PA, pp. 670–675.

Kennedy, A. C., L. F. Elliott, F. L. Young, and C. L. Douglas. 1991. Rhizobacteria suppressive to the weed downy brome. *Soil Sci. Soc. Am. J.* 55:722–727.

Kennedy, A. C., A. G. Ogg, Jr., and F. L. Young. 1992. Biocontrol of jointed goatgrass. United States Patent 5,163,991, November 17, 1992.

Kennedy, A. C., and R. I. Papendick. 1995. Microbial characteristics of soil quality. *J. Soil Water Conserv.* 50:243–248.

Kennedy, A. C., and K. L. Smith. 1995. Soil microbial diversity and the sustainability of agricultural soils. *Plant Soil* 170:75–86.

Kirchner, M. J., A. G. Wollum II, and L. D. King. 1993. Soil microbial populations and activities in reduced chemical input agroecosystems. *Soil Sci. Soc. Am. J.* 57:1289–1295.

Kloepper, J. W., F. M. Scher, M. Laliberte, and B. Tipping. 1989. Free-living bacterial inocula for enhancing crop productivity. *Trends Biotechnol.* 7:39–43.

Knowles, T. C., B. W. Hipp, P. S. Graff, and D. S. Marshall. 1993. Nitrogen nutrition of rainfed winter wheat in tilled and no-till sorghum and wheat residues. *Agron. J.* 85:886–893.

Koepf, H. H., B. D. Pettersson, and W. Schaumann. 1976. *Bio-Dynamic Agriculture: An Introduction*. Anthroposophic Press, Spring Valley, NY.

Kolberg, R. L., D. G. Westfall, and G. A. Peterson. 1999. Influence of cropping intensity and nitrogen fertilizer rates on *in situ* nitrogen mineralization. *Soil Sci. Soc. Am. J.* 63:129–134.

Kozdroj, J., and J. D. van Elsas. 2000. Response of the bacterial community to root exudates in soil polluted with heavy metals assessed by molecular and cultural approaches. *Soil Biol. Biochem.* 32:1405–1417.

Krall, J. M., and G. E. Schuman. 2000. Integrated dryland crop and livestock production systems on the Great Plains: Extent and outlook. *J. Crop Prod.* 9:187–191.

Kremer, R. J. 1987. Identity and properties of bacteria inhabiting seeds of selected broadleaf weed species. *Microb. Ecol.* 14:29–37.

Kremer, R. J., and A. C. Kennedy. 1996. Rhizobacteria as biocontrol agents of weeds. *Weed Technol.* 10:601–609.

Lal, R., R. F. Follett, J. Kimble, and C. V. Cole. 1999. Managing U.S. cropland to sequester carbon in soil. *J. Soil Water Conserv.* 54:374–381.

Liljeroth, E., and E. Bååth. 1988. Bacteria and fungi on roots of different barley varieties (*Hordeum vulgare* L.). *Biol. Fertil. Soils* 7:53–57.

Liljeroth, E., J. A. VanVeen, and H. J. Miller. 1990. Assimilate translocation to the rhizosphere of two wheat lines and subsequent utilization by rhizosphere microorganisms at two soil nitrogen concentrations. *Soil Biol. Biochem.* 22:1015–1021.

Liu, X., W. C. Lindemann, W. G. Whitford, and R. L. Steiner. 2000. Microbial diversity and activity of disturbed soil in the northern Chihuahuan Desert. *Biol. Fertil. Soils* 32:243–249.

Loper, J. E., and M. N. Schroth. 1986. Influence of bacterial sources of indole-3-acetic acid on root elongation of sugar beet. *Phytopathology* 76:386–389.

Lottmann, J., H. Heuer, J. de Vries, A. Mahn, K. During, W. Wackernagel, K. Smalla, and G. Berg. 2000. Establishment of introduced antagonistic bacteria in the rhizosphere of transgenic potatoes and their effect on the bacterial community. *FEMS Microbiol. Ecol.* 33:41–49.

Lovett, J. V., and A. H. C. Hoult. 1995. Allelopathy and self-defense in barley. In Inderjit, K. M., M. Dakshini, and F. A. Einhellig (Eds.), *Allelopathy: Organisms, Processes, and Applications.* American Chemical Society, Washington, D.C., pp. 170–183.

Lundquist, E. J., K. M. Scow, L. E. Jackson, S. L. Uesugi, and C. R. Johnson. 1999. Rapid response of soil microbial communities from conventional, low input and organic farming systems to a wet/dry cycle. *Soil Biol. Biochem.* 31:1661–1675.

Luo, Y., and D. O. TeBeest. 1999. Effect of temperature and dew period on infection of northern jointvetch by wild-type and mutant strains of *Colletotrichum gloeosporioides* f. sp. *aeschynomene. Biol. Control* 14:1–6.

Lynch, J. M. 1983. *Soil Biotechnology: Microbiological Factors in Crop Productivity.* Blackwell Scientific Publications, Oxford, 191 pp.

Lynch, J. M., and S. H. T. Harper. 1980. Role of substrates and anoxia in the accumulation of soil ethylene. *Soil. Biol. Biochem.* 12:363–368.

Lyons, J., S. W. Trimble, and L. K. Paine. 2000. Grass versus trees: Managing riparian areas to benefit streams of central North America. *J. Am. Water Resour. Assoc.* 36:919–930.

MacArthur, R. H., and E. O. Wilson. 1967. *The Theory of Island Biogeography.* Princeton University Press, Princeton, NJ, 203 pp.

Mader, P., S. Edenhoffer, T. Boller, A. Wiemken, and U. Niggli. 2000. Arbuscular mycorrhizae in a long-term field trial comparing low-input (organic, biological) and high-input (conventional) farming systems in a crop rotation. *Biol. Fertil. Soils* 31:150–156.

Maloney, P. E., A. H. C. van Bruggen, and S. Hu. 1997. Bacterial community structure in relation to the carbon environments in lettuce and tomato rhizospheres and in bulk soil. *Microb. Ecol.* 34:109–117.

Martin, J. K. 1977. Effect of soil moisture on the release of organic carbon from wheat roots. *Soil Biol. Biochem.* 9:303–304.

Martin, J. K., and J. R. Kemp. 1980. Carbon loss from roots of wheat cultivars. *Soil Biol. Biochem.* 12:551–554.

Martin, T. L. 1933. Influence of chemical composition of organic matter on rate of decomposition. *J. Am. Soc. Agron.* 25:341–346.

Mazzola, M., D. K. Fujimoto, L. S. Thomashow, and R. J. Cook. 1995. Variation in sensitivity of *Gaeumannomyces graminis* to antibiotics produced by fluorescent *Pseudomonas* spp., and effect on biological control of take-all of wheat. *Appl. Environ. Microbiol.* 61:2554–2559.

McCallister, D. L., and W. L. Chien. 2000. Organic carbon quantity and forms as influenced by tillage and cropping sequence. *Commun. Soil Sci. Plant Anal.* 31:465–479.

McSpadden-Gardener, B. B., and D. M. Weller. 2001. Changes in populations of rhizosphere bacteria associated with take-all disease of wheat. *Appl. Environ. Microbiol.* 67: 4414–4425.

McGonigle, T. P., and A. H. Fitter. 1990. Ecological specificity of vesicular-arbuscular mycorrhizal associations. *Mycol. Res.* 94:120–122.

Mikola, J., and H. Setälä. 1998. Relating species diversity to ecosystem functioning: Mechanistic backgrounds and experimental approach with a decomposer food web. *Oikos* 83:180–194.

Miller, H. J., G. Henken, and J. A. VanVeen. 1989. Variation and composition of bacterial populations in the rhizospheres of maize, wheat, and grass cultivars. *Can. J. Microbiol.* 35:656–660.

Mohammed, M. J., W. L. Pan, and A. C. Kennedy. 1995. Wheat responses to vesicular-arbuscular mycorrhizal fungal inoculation of soils from eroded toposequence. *Soil Sci. Soc. Am. J.* 59:1086–1090.

Mohn, W. W., and G. R. Stewart. 2000. Limiting factors for hydrocarbon biodegradation at low temperature in Arctic soils. *Soil Biol. Biochem.* 32:1161–1172.

Moore, J. C. 1994. Impact of agricultural practices on soil food web structure: Theory and application. *Agric. Ecosyst. Environ.* 51:239–247.

Moore, J. M., S. Klose, and M. A. Tabatabai. 2000. Soil microbial biomass carbon and nitrogen as affected by cropping systems. *Biol. Fertil. Soils* 31:200–210.

Murata, T., and K. M. Goh. 1997. Effects of cropping systems on soil organic matter in a pair of conventional and biodynamic mixed cropping farms in Canterbury, New Zealand. *Biol. Fertil. Soils* 25:372–381.

Naseby, D. C., and J. M. Lynch. 1998. Establishment and impact of *Pseudomonas fluorescens* genetically modified for lactose utilization and kanamycin resistance in the rhizosphere of pea. *J. Appl. Microbiol.* 84:169–175.

Neal, J. L., R. I. Larson, and T. G. Atkinson. 1973. Changes in rhizosphere populations of selected physiological groups of bacteria related to substitution of specific pairs of chromosomes in spring wheat. *Plant Soil* 39:209–212.

Needelman, B. A., M. M. Wander, G. A. Bollero, C. W. Boast, G. K. Sims, and D. G. Bullock. 1999. Interaction of tillage and soil texture: biologically active soil organic matter in Illinois. *Soil Sci. Soc. Am. J.* 63:1326–1334.

Newman, E. I., A. J. Heap, and R. A. Lawley. 1981. Abundance of mycorrhizas and root-surface microorganisms of *Plantago lanceolata* in relation to soil and vegetation: A multi-variate approach. *New Phytol.* 89:95–108.

Ocampo, J. A. 1986. Vesicular-arbuscular mycorrhizal infection of "host" and "non-host" plants: Effect on the growth responses of the plants and the competition between them. *Soil Biol. Biochem.* 18: 607–610.

Ohtonen, R., H. Fritze, T. Pennanen, A. Jumpponen, and J. Trappe. 1999. Ecosystem properties and microbial community changes in primary succession on a glacier forefront. *Oecologia* 119:239–246.

Okon, Y. 1982. Azospirillum: Physiological properties, modes of association with roots and its application for the benefit of cereal and forage grass crops. *Isr. J. Bot.* 31:214–220.

Oleskevich, C., S. F. Shamoun, R. F. Vesonder, and Z. K. Punja. 1998. Evaluation of *Fusarium avenaceum* and other fungi for potential as biological control agents of invasive *Rubus* species in British Columbia. *Can. J. Plant Pathol.* 20:12–18.

Olsson, S., and B. Gerhardson. 1992. Effects of long-term barley monoculture on plant-affecting soil microbiota. *Plant Soil* 143:99–108.

Opperman, M. H., M. Wood, and P. J. Harris. 1989. Changes in microbial populations following the application of cattle slurry to soil at two temperatures. *Soil Biol. Biochem.* 21:263–268.

Papendick, R. I. 1996. Farming systems and conservation needs in the Northwest wheat region. *Am. J. Altern. Agric.* 11:52–57.

Papendick, R. I., and G. S. Campbell. 1975. Water potential in the rhizosphere and plant and methods of measurement and experimental control. In Bruehl, G. W. (Ed.), *Biology and Control of Soil-Borne Pathogens.* American Phytopathological Society, St. Paul, MN, pp. 39–49.

Parr, J. F., and R. I. Papendick. 1978. Factors affecting the decomposition of crop residues by microorganisms. In W. R. Oschwald (Ed.), *Crop Residue Management Systems.* Madison, WI. ASA Special Publication Number 31, American Society of Agronomy, Crop Science Society of America, and Soil Science Society of America, pp. 101–129.

Paul, E. A. 1984. Dynamics of organic matter in soils. *Plant Soil* 76:275-285.

Penfold, C. M., M. S. Miyan, T. G. Reeves, and I. T. Grierson. 1995. Biological farming for sustainable agricultural production. *Aust. J. Exp. Agric.* 35:849–856.

Pinton, R., Z. Varanini, and P. Nannipieri. 2001. *The Rhizosphere.* Marcel Dekker, New York, p. 424.

Price, W. P. 1988. An overview of organismal interactions in ecosystems in evolutionary and ecological time. *Agric. Ecosyst. Environ.* 2:369–377.

Rasmussen, P., H. P. Collins, and R. W. Smiley. 1989. Long-term management effects on soil productivity and crop yields in semi-arid regions of eastern Oregon. Oregon State University Bulletin 675, Corvallis, OR.

Reganold, J. P., A. S. Palmer, J. C. Lockhart, and A. N. Macgregor. 1993. Soil quality and financial performance of biodynamic and conventional farms in New Zealand. *Science* 260:344–349.

Reicosky, D. C., and M. J. Lindstrom. 1993. Fall tillage method: Effect on short-term carbon dioxide flux from soil. *Agron. J.* 85:1237–1243.

Rhoton, F. E. 2000. Influence of time on soil response to no-till practices. *Soil Sci. Soc. Am. J.* 64:700–709.

Rhykerd, R. L., B. Crews, K. J. McInnes, and R. W. Weaver. 1999. Impact of bulking agents, forced aeration, and tillage on remediation of oil-contaminated soil. *Bioresour. Technol.* 67:279–285.

Rice, E. L. 1995. *Biological Control of Weeds and Plant Diseases: Advances in Applied Allelopathy.* University of Oklahoma Press, Norman, OK, 439 pp.

Rouatt, J. W., and H. Katznelson. 1961. A study of the bacteria on the root surface and in the rhizosphere soil of crop plants. *J. Appl. Bacteriol.* 24:164–171.

Rovira, A. D. 1956. Plant root excretions in relation to the rhizosphere effect: I. The nature of root exudates from oats and peas. *Plant Soil* 7: 178–194.

Rovira, A. D. 1959. Root excretions in relation to the rhizosphere effect. IV. Influence of plant species, age of plant, light, temperature, and calcium nutrition on exudation. *Plant Soil* 11:53–64.

Rovira, A. D. 1978. Microbiology of pasture soils and some effects of microorganisms on pasture plants. In Wilson, J. R. (Ed.), *Plant Relations in Pastures.* CSIRO, East Melbourne, Australia, pp. 95–110.

Salonius, P. O. 1981. Metabolic capabilities of forest soil microbial populations with reduced species diversity. *Soil Biol. Biochem.* 13:1–10.

Sandaa, R.-A., V. Torsvik, and O. Enger. 2001. Influence of long-term heavy-metal contamination on microbial communities in soil. *Soil Biol. Biochem.* 33:287–295.

Sarathchandra, S.U., G. Burch, and N. R. Cox. 1997. Growth patterns of bacterial communities in the rhizoplane and rhizosphere in white clover (*Trifolium repens* L.) and perennial ryegrass (*Lolium perenne* L.) in long-term pasture. *Appl. Soil Ecol.* 6:293–299.

Sarwar, M., and W. T. Frankenberger, Jr. 1994. Influence of L-tryptophan and auxins applied to the rhizosphere on the vegetative growth of *Zea mays* L. *Plant Soil* 160:97–104.

Sarwar, M., and R. J. Kremer. 1995. Enhanced suppression of plant growth through the production of L-tryptophan derived compounds by deleterious rhizobacteria. *Plant Soil* 172:261–269.

Schippers, B., A. W. Bakker, and P. A. Bakker. 1987. Interactions of deleterious and beneficial rhizosphere microorganisms and the effect of cropping practices. *Annu. Rev. Phytopathol.* 25:339–358.

Shipton, P.J. 1977. Monoculture and soilborne plant pathogens. *Annu. Rev. Phytopathol.* 15:387–407.

Simard, R. R., D. A. Angers, and C. Lapierre. 1994. Soil organic matter quality as influenced by tillage, lime, and phosphorus. *Biol. Fertil. Soils* 18:13–18.

Southorn, N., and S. Cattle. 2000. Monitoring soil quality for central tablelands grazing systems. *Commun. Soil Sci. Plant Anal.* 31:2211–2229.

Sprent, J. L. 1979. *The Biology of Nitrogen-Fixing Organisms.* McGraw-Hill, London.

Stroo, H. F., K. L. Bristow, L. F. Elliott, R. I. Papendick, and G. S. Campbell. 1989. Predicting rates of wheat residue decomposition. *Soil Sci. Soc. Am. J.* 53:91–99.

Suslow T. V., and M. N. Schroth. 1982. Role of deleterious rhizosphere bacteria as minor pathogens in reducing crop growth. *Phytopathology* 72:111–115.

Swift, M. J., and L. Boddy. 1984. Animal-microbial interactions in wood decomposition. In Anderson, J. M., A. D. M. Rayner, and W. H. Walton (Eds.), *Invertebrate–Microbial Interactions.* Cambridge University Press, Cambridge, U.K. pp. 89–131.

Swift, M. J., O. Andren, L. Brussaard, M. Briones, M. M. Couteaux, K. Ekschmitt, A. Kjoller, P. Loiseau, and P. Smith. 1998. Global change, soil biodiversity, and nitrogen cycling in terrestrial ecosystems: three case studies. *Glob. Change Biol.* 4:729–743.

Swinnen, J., J. A. Van Veen, and R. Merckx. 1995. Root decay and turnover of rhizodeposits in field-grown winter wheat and spring barley estimated by $^{14}C$ pulse-labelling. *Soil Biol. Biochem.* 27:211–217.

Sylvia, D. M. 1998. Mycorrhizal symbioses. In Sylvia, D. M., J. J. Fuhrman, R. G. Hartel, and D. A. Zuberer (Eds.), *Principles and Applications of Soil Microbiology.* Prentice-Hall, New York, pp. 408–426.

TeBeest, D. O., 1991. *Microbial Control of Weeds.* Chapman and Hall, New York, 284 pp.

Thomas, V. G., and P. G. Kevan. 1993. Basic principles of agroecology and sustainable agriculture. *J. Agric. Environ. Ethics* 5:1–18.

Thomashow, L. S., and D. M. Weller. 1988. Role of a phenazine antibiotic from *Pseudomonas fluorescens* in biological control of *Gaeumannomyces graminis* var. *tritici. J. Bacteriol.* 170:3499–3508.

Thompson, C. A., and D. A. Whitney. 2000. Effects of 30 years of cropping and tillage systems on surface soil test changes. *Commun. Soil Sci. Plant Anal.* 31:241–257.

Tinker, P. B. 1976. Roots and water. Transport of water to plant roots in soil. *Philos. Trans. R. Soc. Lond., Ser. B: Biol. Sci.* 273:445–461.

Torsvik, V., J. Goksoyr, and F. L. Daae. 1990. High diversity in DNA of soil bacteria. *Appl. Environ. Microbiol.* 56:782–787.

Turkington, R., F. B. Holl, C. P. Chanway, and J. D. Thompson. 1988. The influence of microorganisms, particularly *Rhizobium*, on plant competition in grass-legume communities. *Symp. Brit. Ecol. Soc.* 28:343–366.

Turkington, R., and E. Klein. 1991. Competitive outcome among four pasture species in sterilized and unsterilized soils. *Soil Biol. Biochem.* 23:837–843.

USDA-FSA. 2000. The conservation reserve program 20th signup. U.S. Department of Agriculture, Farm Service Agency, Washington, D.C.

Van der Putten, W. H., and B. A. M. Peters. 1997. How soil-borne pathogens may affect plant competition. *Ecology* 78:1785–1795.

Vancura, V. 1967. Root exudates of plants: III. Effects of temperature and "cold shock" on the exudation of various compounds from seeds and seedlings of maize and cucumber. *Plant Soil* 27:319–328.

Wagner, G. H., and D. C. Wolf. 1998. Carbon transformations and soil organic matter formation. pp. 218–258. In Sylvia, D. M., J. J. Fuhrman, R. G. Hartel, and D. A. Zuberer (Eds.), *Principles and Applications of Soil Microbiology*. Prentice-Hall, New York.

Wander, M. M., S. J. Traina, B. R. Stinner, and S. E. Peters. 1994. Organic and conventional management effects on biologically active soil organic matter pools. *Soil Sci. Soc. Am. J.* 58:1130–1139.

Ward, D., M. Bateson, R. Weller, and A. Ruff-Roberts. 1992. Ribosomal RNA analysis of microorganisms as they occur in nature. *Adv. Microb. Ecol.* 12:219–286.

Wardle, D. A., G. W. Yeates, K. S. Nicholson, K. I. Bonner, and R. N. Watson. 1999. Response of soil microbial biomass dynamics, activity and plant litter decomposition to agricultural intensification over a seven-year period. *Soil Biol. Biochem.* 31:1707–1720.

Wardle, D. A., O. Zachrisson, G. Hornberg, and C. Gallet. 1997. The influence of island area on ecosystem properties. *Science* 277:1296–1299.

Wedin, D. A., and D. Tilman. 1990. Special effects of nitrogen cycling: A test with perennial grasses. *Oecologia* 84:433–441.

Wenderoth, D. F., and H. H. Reber. 1999. Correlation between structural diversity and catabolic versatility of metal-affected prototrophic bacteria in soil. *Soil Biol. Biochem.* 31:345–352.

Westover, K. M., A. C. Kennedy, and S. E. Kelley. 1997. Patterns of rhizosphere microbial community structure associated with co-occurring plant species. *J. Ecol.* 85:863–873.

Whipps, J. M., and J. M. Lynch. 1986. The influence of rhizosphere on crop productivity. *Adv. Microb. Ecol.* 9:187–244.

Wolters, V. 1997. *Functional Implications of Biodiversity in Soil*. Office for Official Publications of the European Communities, Luxembourg.

Woltz, S. S. 1978. Nonparasitic plant pathogens. *Annu. Rev. Phytopathol.* 16:403–430.

Workneh, F., and A. H. C. van Bruggen. 1995. Bacterial density, composition and diversity in organically and conventionally managed rhizosphere soil in relation to suppression of corky root of tomatoes. *Appl. Soil Ecol.* 1:219–230.

Wood, M. 1991. Biological aspects of soil protection. *Soil Use Manage.* 7:130–136.

Wright, S. F., and A. Upadhyaya. 1998. A survey of soils for aggregate stability and glomalin, a glycoprotein produced by hyphae of arbuscular mycorrhizal fungi. *Plant Soil* 198:97–107.

Wunsche, L., L. Bruggemann, and W. Babel. 1995. Determination of substrate utilization patterns of soil microbial communities: An approach to assess population changes after hydrocarbon pollution. *FEMS Microbiol. Ecol.* 17:295–306.

Yachi, S., and M. Loreau. 1999. Biodiversity and ecosystem productivity in a fluctuating environment. The insurance hypothesis. *Proc. Natl. Acad. Sci. USA* 96:1463–1468.

Zak, J. C. 1992. Response of soil fungal communities to disturbance. In G. C. Carroll and D. T. Wicklow (Eds.), *The Fungal Community: Its Organization and Role in the Ecosystem*. Marcel Dekker, New York, pp. 403–426.

Zak, J. C., M. R. Willig, D. L. Moorhead, and H. G. Wildman. 1994. Functional diversity of microbial communities: A quantitative approach. *Soil Biol. Biochem.* 26:1101–1108.

Zelles, L., Q. Y. Gai, R. X. Ma, R. Rackwitz, K. Winter, and F. Beese. 1994. Microbial biomass, metabolic activity and nutritional status determined from fatty acid patterns and poly-hydrozybutyrate in agriculturally managed soils. *Soil Biol. Biochem.* 26:439–446.

Zhang, W., and A. K. Watson. 1997. Efficacy of *Exserohilum monoceras* for the control of *Echinochloa* species in rice (*Oryza sativa*). *Weed Sci.* 45:144–150.

# 11 Interactions among Organic Matter, Earthworms, and Microorganisms in Promoting Plant Growth

*Clive A. Edwards and Norman Q. Arancon*

## CONTENTS

0-8493-1294-9/04/$0.00+$1.50
© 2004 by CRC Press LLC

# INTRODUCTION

The importance of soil biota in soil pedogenesis and in maintaining soil structure, organic matter breakdown, recycling of nutrients, and fertility is not always fully appreciated by physical and chemical soil scientists. Earthworms are probably the most important component of the soil fauna in terms of soil formation, nutrient cycling, and global distribution. Although they are not numerically dominant, their size makes them one of the major contributors to invertebrate biomass and their activities are extremely important in maintaining soil fertility in many ways (Edwards and Bohlen, 1996; Edwards, 1998).

Aristotle first drew attention to the earthworm's role in turning over the soil and aptly called them "intestines of the earth." Charles Darwin (1881) in his definitive work *The Formation of Vegetable Mould through the Action of Worms* first pointed out the importance of earthworms in breakdown of dead plant and animal matter; release of nutrients; and maintenance of soil structure, aeration, drainage, and fertility. Before this, earthworms were commonly regarded as pests, until Darwin's views were supported, expanded, and validated by other contemporary scientists such as Muller (1878) and Urquhart (1887).

Earthworms belong to the order Oligochaeta, which contains ca. 3000 species, although many of these are aquatic in habitat and considerable controversy surrounds their systematics. They are found in most parts of the world, except those with extreme climates, such as deserts and areas under constant snow and ice. Some species of earthworms, particularly those belonging to Lumbricidae, are widely distributed (peregrine) and often when introduced to new areas become dominant over the endemic species. This situation probably applies to large areas of the northern U.S. and Canada, where lumbricid earthworms were eliminated by glaciation in the Ice Age. Evidence for this is their spread from major waterways used by colonists (Reynolds, 1998).

Although all species of earthworms contribute to the breakdown of plant-derived organic matter, they differ in the ways by which they degrade organic matter. Their activities can be of three kinds, each associated with a different group of earthworm species. Some species are limited mainly to the plant litter layer on the soil surface, composed of decaying organic matter or wood, and seldom penetrate soil more than superficially. The main role of these species seems to be shredding of the organic matter into fine particles, which facilitates increased microbial activity. Other species live just below the soil surface for most of the year, except when it is very cold or very dry; these do not have permanent burrows and ingest both organic matter and inorganic materials in soils. These species produce organically enriched soil materials in the form of casts, which they deposit either randomly in the surface layers of the soil or as distinct casts on the soil surface. The truly soil-inhabiting species have permanent burrows that penetrate deep into the soil. These species feed primarily on organic matter, but also ingest considerable quantities of inorganic materials and mix these thoroughly through the soil profile. These species are of primary importance in pedogenesis. Finally, some species are almost exclusively limited to living in organic materials and cannot survive long in soil; these species are commonly used in vermiculture and vermicomposting. All earthworm species depend on consuming organic matter in some form, and they play an important role, mainly by promoting microbial activity in various stages of organic matter decomposition, which eventually includes humification into complex and stable amorphous colloids containing phenolic materials.

There is little doubt that in many habitats, earthworms are the key invertebrate organisms in the breakdown of plant organic matter. Populations of earthworms usually increase with the availability of organic matter, and in many temperate and even tropical forests, earthworms have the capacity to consume the total annual litter fall. Such a total turnover of organic litter fall has been calculated for an English mixed woodland (Satchell, 1967), an English apple orchard (Raw, 1962), a Nigerian tropical forest (Madge, 1969), and a Japanese oak forest (Sugi and Tanaka, 1978). Similar calculations could have been made for other sites (Edwards and Bohlen, 1996).

During feeding by earthworms, the C:N ratio in the organic matter falls progressively, and the residual N is converted mainly into the ammonium or nitrate forms, which can be readily taken up by plants. At the same time, other nutrients such as P and K are converted into forms more available to plants. Forested soils that have poor populations of earthworms often develop a mor structure, with a mat of undecomposed organic matter at the surface (Kubiena, 1955). This can also occur in grasslands and is common on poor upland grasslands in temperate countries and in countries such as New Zealand in areas where earthworms have only recently been introduced and where introduction of earthworms into pasture is a common agricultural practice (Stockdill, 1966).

Earthworm fecal material takes the form of casts, which can vary greatly in size and form, and these are deposited on the soil surface, in the lining of earthworm burrows, or in spaces and cavities below the soil surface, thereby playing a major role in the development of soil horizons. Casts tend to be much more microbially active than the surrounding soil, and the plant nutrients in them are converted into forms that can be utilized readily by plants. By facilitating these various interactions, earthworms are key organisms in the overall breakdown of organic matter and the transformation and cycling of macro- and micronutrients, processes central to maintaining soil fertility and promoting plant growth.

In recent years, interactions of earthworms with microorganisms in degrading organic matter have been used commercially in systems designed to dispose of agricultural and urban organic wastes and convert these materials into valuable soil amendments for crop production. Commercial enterprises processing wastes in this way are expanding worldwide and diverting organic wastes from more expensive and environmentally harmful ways of disposal, such as incinerators and landfills (Edwards and Neuhauser, 1988; Edwards, 1998).

## BREAKDOWN OF ORGANIC MATTER AND NUTRIENT CYCLING IN THE FIELD

### ORGANIC MATTER BREAKDOWN

Plant and animal organic material that reaches the soil is subject to the action of many agents, including microorganisms and invertebrates, that promote decomposition. Some plant and animal residues are decomposed rapidly by microorganisms; however, much of the organic matter, particularly the tougher plant leaves, stems, and root material, breaks down much more readily after being fragmented by soil-inhabiting invertebrates, which facilitates microbial and enzymatic activity in the invertebrates' intestines. In many soils, earthworms are probably the most important macroinvertebrates involved in the initial stages of recycling of organic matter and release of nutrients for plant growth.

Early evidence of this importance was provided by Edwards and Heath (1963), who placed 5-cm diameter disks, cut from freshly fallen oak and beech leaves, in nylon bags of four different mesh sizes and buried them in woodland or old pasture soils. Only bags with the largest mesh (7 mm) allowed the entry of earthworms. After 1 year, none of the 50 oak disks originally placed in each of the 7-mm mesh bags remained intact, and 92% of the total oak-leaf material and 70% of the beech had been removed. Much less had disappeared from disks in bags that allowed access to only micro and mesoarthropods. Earthworms consumed not only the softer parts of the leaves but also veins and ribs (Edwards and Heath, 1963). Curry and Byrne (1992) in a similar experiment in which wheat litter was confined by meshes of different sizes in a winter wheat field in Ireland reported that decomposition rates of straw accessible to the earthworms increased by 26 to 47% compared with straw from which earthworms were excluded. MacKay and Kladivko (1985) placed maize and soybean residues on the soil surface in pots with and without earthworms in a greenhouse. After 36 d, pots with no earthworms retained 60% of the soybean residues and 85% of the maize residues, whereas pots with earthworms had only 34% of the original soybean residues and 52% the original maize residues.

Organic matter that passes through the earthworm gut and is excreted in their casts is broken down into much finer particles by their grinding gizzards, thereby exposing a much greater surface area of the organic matter to microbial decomposition. Martin (1991) reported that casts of the tropical earthworm *Pheretima anomala* had much less coarse organic matter than the surrounding soil, indicating that the larger particles of organic matter were fragmented during passage through the earthworm gut. Parmelee et al. (1990), who used the vermicidal insecticide carbofuran to decrease earthworm populations in no-tillage agroecosystems by more than 90%, reported that after 292 d, the amounts of fine, coarse, and total particulate organic matter in the treated plots increased by 43%, 30%, and 32%, respectively, compared with those in the control plots. Such commonly reported increases in particulate organic matter resulting from decreased earthworm populations illustrate the importance of earthworms in the fragmentation and breakdown of organic matter and the release of nutrients.

Feeding habits of different earthworm species can influence their effects on litter fragmentation and incorporation into soil. Bouché (1971) separated lumbricid earthworms into three major eco-logical groups: (1) anecic earthworm species, such as *Lumbricus terrestris* L., live in deep burrows and feed at the soil surface, incorporate large amounts of organic matter into soil, and can break down and feed on large litter fragments by stripping off smaller particles with their mouthparts; (2) epigeic earthworm species, such as *L. rubellus* and *Dendrobaena octaedra*, reside mainly in the surface organic litter, consume large amounts of organic materials, but do not incorporate much of it into the mineral soil layers; and (3) endogeic earthworm species, such as *Allolobophora caliginosa*, reside close to the soil surface, and feed mainly on fragmented organic matter, mixing it thoroughly with mineral soil.

Ferriere (1980) examined the gut contents of 10 species of lumbricid earthworms in a pasture and observed distinct differences in the types of food consumed by the various species. Epigeic species fed primarily on relatively undecomposed fragments of leaves and roots, anecic species fed on partially decomposed but identifiable fragments of aboveground plant litter, and endogeic species fed mainly on unidentifiable organic matter together with roots and leaves that were in a more advanced stage of decomposition. Anecic and endogeic species of earthworms occur together in many soils and probably have a synergistic effect on the redistribution of organic matter throughout the soil profile. Shaw and Pawluk (1986) reported that when the anecic species *L. terrestris* and the endogeic species *Octolasion cyaneum* were kept together in soil microcosms, they distributed the crop residues from the soil surface much more evenly throughout the soil matrix than when either species was present alone.

Earthworm species such as *L. terrestris* are responsible for a large proportion of the overall fragmentation and incorporation of litter in many woodlands of the temperate zone and are primarily responsible for the formation of mull soils, which are forest soils in which the surface litter and organic layers are mixed thoroughly with the mineral soil (Muller, 1878; Scheu and Wolters, 1991a). Soils with small or no earthworm populations often have a well-developed layer of undecomposed litter and organic matter on the soil surface, separated from the underlying mineral soil by a sharp boundary. These are termed mor soils, which represent the opposite extreme to mull soils, along a continuum of different forest soil types (Edwards and Bohlen, 1996).

Earthworms can convert mor soils to mulls 3 to 4 years after they colonize a site that previously lacked earthworms. Mixing and fragmentation of forest litter by earthworms were identified as being of fundamental importance to the renewal of spruce forest ecosystems in the French Alps (Bernier and Ponge, 1994). Anecic species, such as *L. terrestris,* play a particularly important role in mixing the surface humus horizons with mineral soil in these ecosystems, forming a favorable environment for the germination and growth of spruce seedlings. Elimination of earthworms from forest soils, such as by changes in food quality or a decrease in soil pH from factors such as acid precipitation, results in a decreased litter bioturbation, a slowing of organic matter decomposition, and development of distinct litter and organic layers. Beyer et al. (1991) reported such changes in oak forests in Germany, which they attributed to a steady decline in earthworm populations resulting from decreased soil pH due to air pollution and acid precipitation.

The effectiveness of *L. terrestris* in initiating the fragmentation and incorporation of fallen leaves in an apple orchard was demonstrated by Raw (1962), who compared the soil profile and structure of an orchard with a high *L. terrestris* population with one in which earthworms were almost totally absent, because of frequent and heavy spraying with a copper-based fungicide. The orchard with few earthworms had an accumulated surface mat, 1- to 4-cm thick, made up of leaf material decomposing at a very slow rate, demarcated sharply from the underlying soil, which had a poor crumb structure.

Earthworms in agricultural grassland ecosystems also play an important role in incorporating surface organic matter into soil. In New South Wales, pastures containing no earthworms normally accumulated surface mats or thatches up to 4 cm thick, but these disappeared progressively after earthworms were introduced experimentally, which is at present a common agricultural practice (Barley and Kleinig, 1964; Stockdill, 1966). Potter et al. (1990) reported that the rates of thatch breakdown in plots of Kentucky blue grass (*Poa pratense* L.) in the U.S. was much slower in plots from which earthworms had been eliminated with insecticides. Clements et al. (1991) studied plots of perennial ryegrass (*Lolium perenne*) from which earthworms had been absent for 20 years, because of regular application of the insecticide phorate. After this 20-year period, they reported a dramatic increase in the depth of the leaf litter layer and a great reduction in the soil organic matter content in plots from which earthworm populations had been eliminated.

Many kinds of organic litter that first fall on to the soil surface are not acceptable to earthworms. Some kinds of litter require a period of weathering before they become palatable to earthworms, and we suggest that this weathering leaches water-soluble polyphenols and other unpalatable substances from the leaves (Edwards and Heath, 1963). For instance, Zicsi (1983) offered four litter-feeding species of earthworms, including *L. terrestris,* litter from five tree species. Earthworms began feeding immediately on the rapidly decomposing higher-quality litter, such as maple (*Acer platanoides*), but did not feed on the lower-quality litter of beech (*Fagus sylvatica* L.) and oak (*Quercus* spp.) until it had been weathered for several months. The type of organic litter affects its rate of breakdown; for example, beech leaves disappeared much more slowly than oak leaves (Edwards and Heath, 1963), which in turn were more resistant to attack by earthworms than were apple leaves (Raw, 1962). Elm, lime, and birch disappear more rapidly than beech (Heath et al., 1966). Earthworms are much more attracted to moist than to dry litter (Edwards and Heath, 1963). Haimi and Huhta (1990) showed that *L. rubellus* increased the mass loss of coniferous forest humus by a factor of 1.4 in a 48-week laboratory incubation. Earthworms can also accelerate the decomposition of pine litter. Earthworms apparently do not influence the primary stages of decomposition of pine needles, but have a progressively important role during their later stages of decomposition (Ponge, 1991).

The final stage in the degradation of plant organic matter is known as humification, which is basically the breaking down of large particles of organic matter into complex amorphous colloids containing phenolic groups. Only ca. 25% of the total fresh organic matter reacting in soil gets converted to humus this way. Much of the humification process is due to soil microorganisms, although it is accentuated by activities of small soil-inhabiting invertebrates such as mites (Acarina), springtails (Collembola), and other arthropods. Rates of humification accelerate considerably by the passage of the organic material through the guts of earthworms. Some of the final stages of humification are probably due to the diverse intestinal microflora in the earthworms' guts (Edwards and Fletcher, 1988), because most of the evidence reported indicates that the chemical processes of humification are facilitated mainly by the microflora. Earthworms accelerated the rates of straw humification in pot experiments by 17–24% and in a field experiment by 15–42% (Atlavinyte, 1975). Neuhauser and Hartenstein (1978) suggested that earthworms enhance the polymerization of aromatic organic compounds, possibly facilitating the formation of humus as an end product. The guts of earthworms have a high, specific peroxidase activity, which is a key enzyme in the polymerization reactions (Hartenstein, 1982). There is considerable evidence that humification is accelerated greatly by vermicomposting (Edwards, 1998).

## AMOUNTS OF ORGANIC MATTER CONSUMED BY EARTHWORMS

Earthworms can ingest very large amounts of plant litter, and the amounts they consume seem to depend more on the quantities of available suitable organic matter than on other factors. If the physical soil conditions of moisture and temperature are suitable, the numbers of earthworms usually increase until food becomes a limiting factor. Many researchers have calculated the amounts of leaf litter of different plant species consumed by different species of earthworms, and there is considerable variability in these calculations. For instance, the consumption of beech litter during laboratory incubations lasting 24 weeks was estimated to be 19 mg per gram wet weight of earthworms per day for *Lumbricus rubellus* and 26 mg per g wet weight per day for *Denbrobaena octaedra* (Haimi and Huhta, 1990). *Lumbricus terrestris* consumed 10 to 15 mg litter per gram fresh weight per day in reclaimed peat soils in Ireland (Curry and Bolger, 1984). Kaushal et al. (1994) fed a variety of leaves (corn, wheat, and mixed grasses) to the tropical earthworm *Amynthas alexandri* and reported food consumption rates ranging from 36 to 69 mg per gram live worm per day. Daniel (1991) showed that rates of leaf litter consumption by juvenile *L. terrestris* could be described by a nonlinear function based on three main factors: soil temperature, soil water potential, and food availability. These three factors probably govern the amounts and rates of food consumed by most litter-feeding earthworm species.

Earthworms can consume a large portion of the entire annual litter fall in some ecosystems. In an apple orchard, *L. terrestris* consumed the equivalent of 2000 kg/ha of leaf litter between leaf fall and the end of February in the U.K. (98.6% of the total leaf fall; Raw, 1962). Based on an estimate of litter consumption of 27 mg dry litter per gram wet weight of earthworms per day, Satchell (1967) estimated that a population of *L. terrestris* in a mixed forest in England could consume the entire annual leaf fall of 300 $g/m^2$ in ca. 3 months. Nielson and Hole (1964) reported that earthworm populations in mixed forests in Wisconsin could consume the entire annual leaf fall of a forest. Knollenberg et al. (1985) suggested that a population of *L. terrestris* in a woodland flood plain in Michigan could consume 94% of the annual leaf fall in 4 weeks during spring. Sugi and Tanaka (1978) calculated that a population of earthworms, composed of six species of *Pheretima* and one species of *Allolobophora,* could ingest 1071 g litter/$m^2$/year from the soil surface in evergreen oak forests in Japan, which is 1.4 times the annual litter fall in these forests, suggesting that the earthworms could only obtain adequate food by reingesting their casts or feeding on other fractions of organic matter in the soil. At a site with lower earthworm populations, Sugi and Tanaka (1978) estimated that earthworms consumed ca. 56% of the total annual leaf fall. Lavelle (1978), working in the Lamto region of Ivory Coast, calculated that a mixed population of eudrilid and megascolecid earthworms annually ingested ca. 30% of the litter decomposed in a grass savanna and 27% of that decomposed in a shrub savanna. The consumption of dung produced by dairy cattle (675 t/ha) is only 25% of the amount that a typical earthworm population over the same area could consume (Satchell, 1967). Hendriksen (1991) estimated that a field population of *L. festivus* and *L. castaneus* in a pasture in Denmark could consume 10 to 15 t manure/ha in 180 d. This corresponds to the amounts of manure produced by two or three dairy cows, which is slightly above the normal stocking rate per hectare.

Even when suitable organic material such as litter or animal manure is freely available to earthworms, many species also ingest large quantities of mineral soil. When individuals of *A. caliginasa* had unlimited quantities of litter available, they still ingested 200 to 300 mg of soil per gram body weight per day, and the ingested mineral soil passed through the gut in ca. 20 h (Barley, 1961). Scheu (1987) estimated that a population of *Allolobophora caliginosa* in a beechwood in Germany consumed up to 6 $kg/m^2$ of soil per year. James (1991) studied rates of organic matter processing by a mixed earthworm community containing several species of the native North American genus *Diplocardia* and the European lumbricids *A. caliginosa* and *Octolasion cyaneum.* He estimated that the earthworms annually consumed 4 to 10% of the soil and 10% of the total organic matter in the top 15 cm of soil.

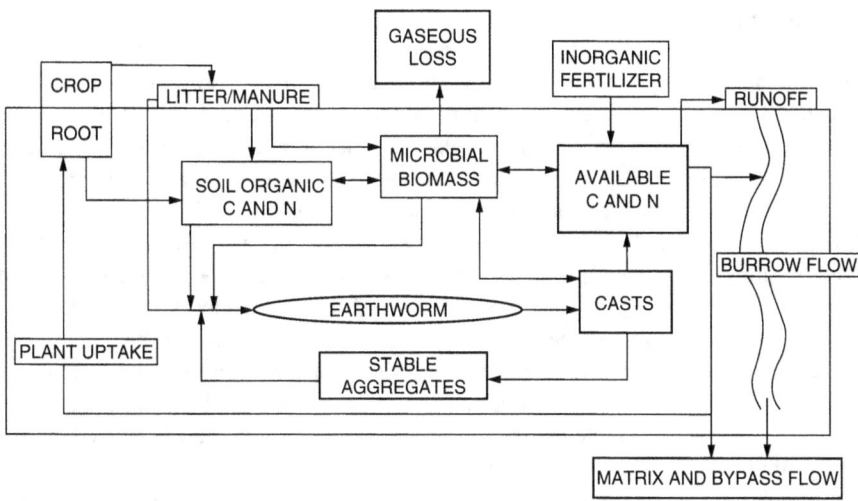

**FIGURE 11.1** Ecosystem budget model to examine pools and fluxes of C and N in the presence of earthworms. Bold boxes indicate pools and fluxes where earthworms are predicted to have a particularly significant impact. (From Parmelee et al. 1995. With permission.)

## NUTRIENT CYCLING

Earthworms have major influences on the soil nutrient cycling processes in many ecosystems. By ingesting and turning over large amounts of soil and organic matter, they increase the rates of mineralization of organic matter, converting organic forms of nutrients into inorganic forms that can be taken up more readily by plants (Figure 11.1).

Earthworms influence nutrient cycles in four ways: (1) during transit of litter through the earthworm gut, (2) in freshly deposited earthworm casts, (3) in aging casts, and (4) during the long-term genesis of the whole soil profile (Lavelle and Martin, 1992). Earthworm effects at all these scales are influenced by soil type, climate, vegetation, and availability and quality of organic matter. Integrating across these scales and understanding the interrelationships among multiple factors are essential to assessing the overall influence of earthworms on nutrient cycling processes. Many of the influences of earthworms on nutrient cycling and mineralization are mediated by the interactions between earthworms and microorganisms. (See the section on interactions between earthworms and microorganisms.)

Although earthworms consume and turn over large amounts of organic matter, their contribution to total heterotrophic soil respiration is relatively small, accounting usually for only 5 to 6% of the total energy flow in terrestrial ecosystems (Edwards and Bohlen, 1996). For a population of the species *Allolabophora caliginosa* in Australia, earthworms were responsible for only 4% of total C consumption (Barley and Kleinig, 1964), and in two English woodlands, *L. terrestris* was responsible for only 8% of the total C consumption (Satchell, 1967). The researchers assumed that the consumption of 22.9 l $O_2/m^2$ was equivalent to a C consumption of 118.6 kg/ha, and that 3000 kg of litter that was 50% C fell on to the soil surface per hectare.

The small contribution of earthworms to overall $CO_2$ output from ecosystems is probably due to their relatively low assimilation efficiencies. C assimilation efficiencies of 2 to 18% have been reported for several species of endogeic earthworms (Bolton and Phillipson, 1976; Barois et al., 1987; Scheu, 1991; Martin et al., 1992). Assimilation efficiencies of litter-feeding earthworms tend to be higher than those of endogeic species. For example, Dickschen and Topp (1987) reported assimilation efficiencies of 30 to 70% for *L. rubellus,* depending on the quality of the litter ingested by the earthworms and the temperature at which they were incubated. Daniel (1991) reported assimilation efficiencies of 43 to 55% for *L. terrestris* that fed on fresh dandelion leaves, although under natural conditions actual field assimilation efficiencies for *L. terrestris* feeding on decaying plant litter are probably much lower. However, earthworms can make substantial contributions to total soil respiration when populations are large and active. Hendrix et al. (1987) estimated that earthworms were responsible for ca. 30% of the total heterotrophic soil respiration during late winter and early spring in a no-tillage agroecosystem in the southeastern U.S. Earthworm population densities at their site reached a maximum of nearly 1000 individuals/m$^2$.

Earthworms can assimilate C from recently deposited fractions of soil organic matter, which is composed of more readily decomposable substances. Martin et al. (1992) incubated earthworms in soils where recent changes in vegetation had led to distinctive patterns of $^{13}C$:$^{12}C$ ratio in the pool of recently deposited organic matter. The $^{13}C$:$^{12}C$ ratios of the earthworms matched those of the recently deposited organic matter in the soil, indicating that the worms assimilated C primarily from recent organic matter pools than from older, much more humified and recalcitrant pools.

Large amounts of water-soluble organic compounds are converted to mucus materials as food passes through the earthworm gut (Barois and Lavelle, 1986). These high-energy mucous materials stimulate microbial activity in the earthworm gut and enable the intestinal microflora to digest some of the more complex organic compounds of the soil. Although a large proportion of these high-energy water-soluble compounds are resorbed in the posterior portion of the gut, some are excreted in earthworm casts (Scheu, 1991), where they continue to serve as energy substrates for microorganisms.

## Carbon

The forms and amounts of C in earthworm casts differ from those of the surrounding soil. There are considerable increases in the polysaccharide contents of casts relative to those in uningested soil (Parle, 1963b; Bhandari et al., 1967). Shaw and Pawluk (1986) reported higher amounts of clay associated with clay in earthworm casts than in surrounding soil, which they suggested promote the stabilization of soil C through binding with clays. The C contents of casts usually tend to be higher than in the surrounding soil, in part due to the addition of intestinal mucus but also because earthworms might consume selectively soil fractions enriched in organic compounds (Lee, 1985; Blair et al., 1994). The turnover of C by earthworms is quite rapid. Ferriere and Bouché (1985) labeled the earthworm *Nicodrilus longus* by feeding it algae labeled with $^{14}C$ and $^{15}N$. They reported that the entire C content of the earthworm tissues could turnover in 40 d, and a considerable portion of this turnover was due to mucus excretion. Scheu (1991) reported that secretion of mucus in casts and from the body wall accounted for 63% of total C losses (mucus excretion plus respiration) from the geophagous earthworm *Octolasion lacteum*, and that this corresponded to a daily loss of 0.7% of total C for this species. Respiration, by contrast, accounted for only 37% of total C losses due to earthworms. Lavelle (1988) estimated that populations of *Pontoscolex corethrurus* in tropical pastures of Mexico can secrete up to 50 Mg mucus/ha in a single year, which equates to 20% of the total C in the soil.

A fundamental unanswered question regarding the influences of earthworms on the cycling of soil C is whether the net effect of earthworms is to increase or decrease the overall storage of organic C (Blair et al., 1994). Earthworms can increase the amounts of C stored by increasing rates of plant growth, but most research suggests that earthworms increase the rates of loss of C from soil by stimulating the mineralization of organic matter. O'Brien and Stout (1978) estimated that the annual flux of C from a New Zealand pasture might have increased by 300 to 1000 kg/ha after earthworms were introduced and the mean residence time of organic C decreased from 180 to 67 years. However, more recent research suggests that stabilization of organic matter in earthworm casts can lead to increased C storage and decreased mineralization of organic matter in the long term. Martin (1991) reported that fresh earthworm casts from *Pheretima anomala* contained 2% less total C than the surrounding soil did, demonstrating a short-term increase in the rates of mineralization of organic matter. However, in longer-term incubations of 1 year, C mineralization in the casts (3%/year) was much lower than in the noningested soil (11%/year). Lavelle and Martin (1992) claimed that the stabilization of organic matter in earthworm casts can be an important mechanism to stabilize organic matter in tropical soils, and this method of organic matter stabilization is probably important in temperate soils as well.

## Nitrogen

Significant amounts of N pass directly through the earthworm biomass in terrestrial ecosystems. Satchell (1963) estimated that 60–70 kg N/ha/year was returned to the soil in the dead tissues of *L. terrestris* in a woodland in England and an additional 30–40 kg N/ha/year was returned in urine and mucus deposited by this species. Keogh (1979) estimated that *A. caliginosa* contributed ca. 109 to 147 kg N/ha/year to mineral N pools in a New Zealand pasture, or ca. 20% of the total amount of N mineralized in the pasture. These amounts are usually higher than those of N turnover in earthworm casts (up to 100 kg/ha; Lavelle et al., 1992). Nowak (1975) estimated that the turnover of N through earthworm tissues in a pasture in Poland equaled 3 to 17% of the total N input from plant litter. Rosswall and Paustian (1984) calculated that 10 kg N/ha/year flowed through an earthworm population that contained a mean annual standing stock of 3.0 kg N/ha. The direct flux of N through earthworm biomass in a no-till agroecosystem in Georgia was 63 kg N/ha/year, or nearly 38% of the total N uptake by the crop (Parmelee and Crossley, 1988). Christensen (1988) reported that dead earthworm tissues contributed 20 to 42 kg N/ha to the soil during the autumn in three arable systems in Denmark.

Dead earthworms decompose very rapidly, and the N in earthworm tissues is mineralized quickly. Satchell (1967) reported that nearly 70% of the N in dead earthworm tissue was mineralized in 10 to 20 d. Ferriere and Bouché (1985) reported that the entire N (and C) content of the earthworms could turn over within 40 d. Barois et al. (1987) labeled individuals of *Pontoscolex corethrurus* with $^{15}$N and reported that 14% of the incorporated label was lost within 5 d and 30% was lost after 30 d. Hameed et al. (1994) also labeled *L. terrestris* with $^{15}$N and reported that the earthworms lost 80% of a $^{15}$N label after 48 d in the field and calculated that the N flow through the earthworms was 16% of their total body N/d.

Earthworms consume large amounts of plant organic matter that contains considerable quantities of N, and much of the N that they assimilate into their own tissues is eventually returned to the soil in their excretions. The presence of earthworms in well-aerated moist soil can increase the rates of $O_2$ consumed and the accumulation of ammonium and nitrate during the early stages of degradation. These excretions, which include mucoproteins secreted by gland cells in the epidermis, and ammonia, urea, and possibly uric acid and allantoin in fluid urine excreted from the nephridiopores, contribute additions of a significant amount of readily assimilable N to soils. Lee (1983) estimated an annual N excretion rate of 18 to 50 kg N/ha for a typical population of lumbricid earthworms. There are no reliable estimates of the N assimilation efficiencies of earthworms, and this represents a considerable void in our understanding of basic earthworm biology (Blair et al., 1994).

The concentrations of inorganic N in fresh earthworm casts and around the lining of their burrows are usually much higher than in bulk soil, with ammonium and nitrates usually being the dominant forms of inorganic N in the casts (Lavelle and Martin, 1992). Overall increases in inorganic N in earthworm casts are probably due to excretory products and mucus from the earthworm as well as through increased rates of mineralization of organic N by microorganisms in the casts. The rates of nitrification in casts can be high, and several authors have noted simultaneous increases in nitrate and decrease in ammonium as casts age (Lavelle et al., 1992).

A key question is whether the total amounts of available N deposited in earthworm casts can significantly contribute to the total amounts of N available in soil for plant growth. Lee (1985) calculated that *A. caiiginosa* casts contribute only 22 to 28 g N/ha/year to soils in the Adelaide region of Australia. Lee (1985) calculated the additional input of available N because of earthworm casts to 35 to 50 g /ha/year. Others have reported significant turnover of N in earthworm casts. For example, James (1991) used earthworm population estimates, soil climate data, and cast production–temperature relationships to estimate that the total amount of mineral N produced in earthworm casts (5 to 5.5 kg N/ha/year) was 10 to 12% of the total N taken up by plants in the N-limited tallgrass prairie in Kansas. It is clear that earthworms can make a substantial contribution to the overall turnover of available forms of mineral N, especially when the amounts produced in earthworm casts as well as those produced in mucus secretions and from the decaying tissues of dead earthworms are considered.

Earthworms increase the rates of mineralization of N, but surprisingly there are few estimates of the influence of earthworms on the overall net mineralization of N in bulk soils. The enhanced mineralization of N caused by earthworm activity is linked to the enhanced mineralization of C, suggesting that certain fractions of organic matter protected physically from mineralization become mobilized during passage through the earthworm gut (Scheu, 1994). Anderson et al. (1983) measured rates of N mineralization in forest soils incubated with oak litter with or without the earthworm *L. rubellus*. The earthworms increased the mobilization of nitrate-N by 10 times and that of ammonium-N by 80 times relative to that in soil without earthworms. Ruz-Jerez et al. (1992) reported that mineral N concentrations were ca. 50% higher in soils with earthworms than in soils without earthworms in laboratory incubation of grassland soil with different plant residues added. Scheu (1987) observed a direct relationship between the biomass of *A. caliginosa* and increased rate of N mineralization in laboratory incubations. He used this relationship, combined with laboratory-derived data on interactions between temperature and N mineralization, to calculate that a field population of *A. caliginosa* could cause an additional mineralization of 4.23 kg N/ha/year in a beechwood site on limestone soil. Obviously, earthworms can mobilize significant amounts of N, but much more research is needed in a variety of ecosystems to reinforce our relatively sparse understanding of their net effects on N mineralization in the field.

Earthworms can increase rates of loss of N by increasing the rates of denitrification and the leaching of nitrate and other mobile N compounds. Fresh earthworm casts usually have higher denitrification rates than the surrounding soil (Svensson et al., 1986; Elliot et al., 1990). Knight et al. (1992) estimated that earthworm casts on the soil surface in English pastures could account for 12% of the total denitrification losses from an unfertilized pasture and 26% of the losses from a fertilized pasture. They also reported that earthworms tripled the amounts of nitrate in leachates from these pastures. The degree to which earthworms increased the losses of N depended on the amounts and types of fertilizer added, losses being higher when large amounts of inorganic fertilizer were added to the soil (Blair et al., 1995).

The C:N ratio in organic matter added to soil is important because net mineralization does not occur unless the C:N ratio is 20:1 or lower. The C:N ratio of freshly fallen leaf litter is usually much higher than this: 25:1 for elm, 28:1 for ash, 38:1 for lime, 42:1 for oak, 44:1 for birch, 54:1 for rowan, and 91:1 for Scots pine (Wittich, 1953). Succulent leaf material often has much lower C:N ratios, whereas tougher tree leaves with a high percentage of resistant constituents, such as cellulose and lignin, that are unpalatable to earthworms and other litter animals often have high C:N ratios (Witkamp, 1966). During the process of leaf litter breakdown and decomposition, the C:N ratio of the litter decreases progressively, because of respiratory losses, until the ratio falls to ca. 20:1, after which net mineralization of N begins and the mineralized N can be taken up directly by plants (Edwards et al., 1995; Edwards and Bohlen, 1996). Earthworms can also lower the C:N ratio by C combustion during respiration.

Earthworms can alter the C:N ratio of the material that passes through their digestive tracts, and several authors have reported that earthworm casts have C:N ratios higher than those of the surrounding soil (Wasawo and Visser, 1959; Graff, 1971; Czerwinski et al., 1974; Aldag and Graff, 1975). This could occur either if earthworms ingest material enriched in C selectively or if they have higher assimilation efficiencies for N than for C. However, a few researchers have reported higher C:N ratios in earthworm casts than in the surrounding soil (Lavelle, 1978; Syers et al., 1979). However, it seems that in most instances earthworms decrease the C:N ratio in soil significantly, because they increase rates of combustion of C by enhancing total soil respiration.

## INTERACTIONS BETWEEN EARTHWORMS
## AND MICROORGANISMS

Earthworms have many complex interrelationships with microorganisms (Figure 11.2). They depend on particular groups of microorganisms as their major source of nutrients, promote microbial activity and diversity in decaying organic matter by fragmenting it and inoculating it with microorganisms, and disperse microorganisms widely through soils.

### MICROORGANISMS IN THE INTESTINES OF EARTHWORMS

There is a great increase in the total numbers of bacteria, fungi, and actinomycetes occurring in the earthworm gut compared with those in the surrounding soil; microbial population increase exponentially from the anterior to the posterior portions of the earthworm gut (Parle, 1959, 1963). Usually, microbial populations are also greatly increased during passage through earthworm casts compared with those of the surrounding soil (Blair et al., 1995). The large increases in the numbers of microorganisms in the earthworm gut can be due partially to the considerable amounts of water and mucus that earthworms secrete into their guts. Barois and Lavelle (1986) showed that the intestinal mucus produced by the earthworm *Pontoscolex corethrurus* contained large amounts of water-soluble, low-molecular-weight organic compounds that could be assimilated easily by the rapidly multiplying microbial communities in the gut. Most of the species of microorganisms that commonly occur in the alimentary canals of earthworms are the same as those in the soils in which the earthworms live. In an early work, Bassalik (1913) isolated more than 50 species of bacteria from the alimentary canal of *L. terrestris* and reported none that differed from those in the soil from which the worms had been taken. This was confirmed for three other species of earthworms by Parle (1963), who reported that most of the cellulose and chitinase enzymes that occur in the alimentary canals of earthworms are secreted by the earthworms and not by symbiotic microorganisms, as they are in some arthropods.

Several researchers have shown that particular groups of microorganisms are stimulated selectively during passage of organic matter through the earthworm gut. These include the actinomycetes *Nocardia, Oerskovia,* and *Streptomyces* spp. and the bacteria *Vibrio* spp. (Mariaglieti, 1979; Contreras, 1980; Szabo et al., 1990; Krištfek et al., 1993). Of the microbes isolated from the gut of

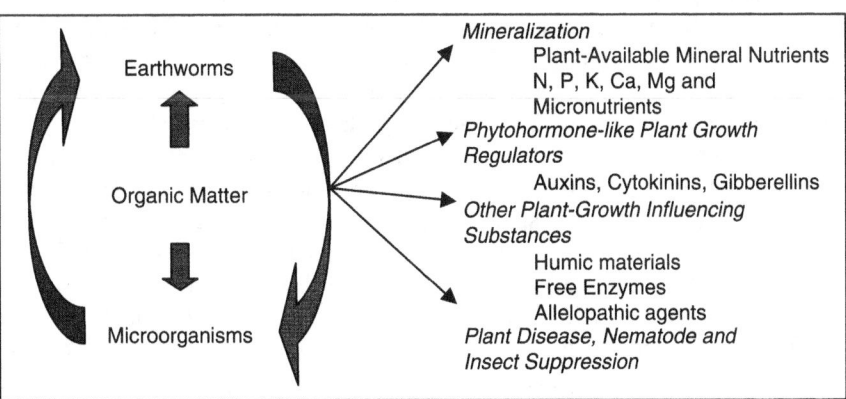

**FIGURE 11.2** Effects of interactions between earthworms and microorganisms on the availability of nutrients and production of plant growth-influencing substances and plant disease antagonists.

*Eisenia lucens, Vibrio* spp. accounted for 73% of the total bacteria and *Streptomyces lipmanii* accounted for 90% of the actinomycetes, although these species were of relatively low abundance in the wood substrate where the earthworms were living (Contreras, 1980). The species composition and relative abundance of actinomycetes in the hindgut differed among different species of earthworm, at different times of the year, and with different types of food ingested (Ravasz et al., 1987; Krištkef et al., 1990, 1993), and this is probably true for species and populations of other bacteria and fungi. Striganova et al. (1989) reported that certain species of fungi, namely *Aspergillus fumigatus* and *Penicillium roqueforti,* were abundant in the digestive tracts of *Nicodrilus caliginosa,* but were absent from the surrounding soil in turf-podzolic soils in Russia. These fungal species might be suppressed in the soil but can be obligate inhabitants of the earthworm intestine.

Satchell (1967) concluded that it was unlikely that earthworms have an indigenous gut flora, but there is still considerable controversy over this. The controversy stems from the general difficulty in culturing all the microorganisms that live either in soil or in the intestines of earthworms. Certain groups of microorganisms are clearly more abundant in earthworm guts than in the surrounding soil, but the extent to which this selective stimulation of particular microbial species constitutes a true mutualistic association remains to be demonstrated. Jolly et al. (1993) used scanning and transmission electron microscopy to examine the hindgut epithelium of *Octolasion lacteum* and *Lumbricus terrestris* to provide evidence of a physical link between bacterial cells and the epithelium of the hindgut. The electron micrographs revealed segmented filamentous bacteria that were connected to the hindgut via a socket-like structure, as well as cocci and bacilli attached to the gut wall via a mucopolysaccharide-like material. These physical links do not necessarily prove the existence of truly indigenous microbial strains, but they do indicate that some microbial strains are highly adapted to living in the alimentary tract of earthworms.

Although many species of microorganisms can survive passage through the earthworm gut, not all emerge in a viable form. Aichberger (1914) reported that the crops, gizzards, and intestines of earthworms contained few live organisms that did not possess a firm outer coat and found no diatoms, desmids, blue-green algae, rhizopods, or live yeasts. Dawson (1947) reported that the number of species of bacteria in soil that passed through the gut of an earthworm were lesser than those in the organic matter consumed whereas those of fungi seemed unaffected. Day (1950) stated that when soil with a large inoculum of *Bacillus cereus* var. *mycoides* passed through the gut of *L. terrestris,* the numbers of these bacilli decreased greatly, although a few survived passage through the gut, suggesting that vegetative cells rather than spores were destroyed. Two other species of bacteria, *Serratia marcescens* (Day, 1950) and *Escherichia coli* (Brusewitz, 1959), which had been introduced into soil by inoculation, were killed after the soil had been ingested by *L. terrestris.* Khambata and Bhatt (1957) reported that the bacillus *E. coli* was usually absent from the intestines of species of *Pheretima,* although these earthworms often live in soil that is regularly manured with human excreta, and they suggested that secretions in the intestine of the earthworms possibly prevented the growth of this and other human pathogens. Dash et al. (1979) examined the number of species of fungi in the soil of a tropical pasture in India and in the gut and casts of the earthworm *Drawida caleb* living in the pasture. These authors isolated 19 species of fungi from the soil, 16 species from the anterior portion of the earthworm's gut, and 8 from the posterior portion of the gut. All the microfungal species found in the earthworm gut were also found in the surrounding soil. The reductions in the number of species that occurred during passage from the anterior to the posterior portion of the earthworm gut indicate that at least half of the ingested microfungal species were killed during passage through the gut, probably because of selective digestion of fungal mycelia and spores. However, Krištkef et al. (1992) reported that overall populations of some fungi increased during passage through the gut of *L. rubellus,* indicating that the viability and potential for multiplication of some fungal species might be enhanced during passage through the earthworm gut. Several researchers have shown that spores of some species of fungi can survive passage through the alimentary canal of earthworms (Harinikumar et al., 1991; Reddell and Spain, 1991; Harinikumar and Bagyaraj, 1994). Various fungal spores that have thick walled or wrinkled coats

(Dash et al., 1979) and the spores and mycelia of certain dark-colored fungi are resistant to breakdown by the intestinal enzymes of earthworms (Striganova et al., 1989).

Antibiotic substances produced by actinomycetes in the intestines of earthworms can inhibit the growth of fungi and Gram-positive bacteria and can explain why some actinomycetes and antibiotic-resistant Gram-negative bacteria predominate in the gut (Ravasz et al., 1986; Krištkef et al., 1993). Others have suggested that earthworms might produce antibiotic substances. For instance, it was shown that the growth of certain fungi on soil in a petri dish ceased whenever an earthworm was introduced (van der Bruel, 1964). Ghabbour (1966) reported that when earthworms were placed in dilute glucose or glycine solutions, fungi did not grow until the earthworms died.

## POPULATIONS OF MICROORGANISMS IN EARTHWORM CASTS AND BURROWS

Many researchers have observed that there are much larger populations of fungal, bacterial, actinomycete, microorganisms and higher enzymatic activity in earthworm casts than in bulk soil (Edwards and Bohlen, 1996; Ponomareva, 1962; Zrazhevskii, 1957; Went, 1963; Shaw and Pawluk, 1986; Tiwari et al., 1989; Tiwari and Mishra, 1993). Daniel and Anderson (1992) kept *L. rubellus* in four different soils containing different amounts of particulate organic matter and observed much higher bacterial plate counts in the earthworm casts than in the surrounding soils. Differences in the size of microbial populations of casts compared with the surrounding soil might result from changes in numbers of microorganisms occurring in the earthworm's intestine or because the selection of food material ingested by the earthworm forms a richer substrate for microbial activity. It is not usually easy to determine which of these is the major factor involved.

Tiwari and Mishra (1993) collected samples of earthworm casts and adjacent soil from 30 sites in India and reported that the casts usually contained larger fungal populations and more fungal species than the associated soil. Some researchers have shown that earthworm casts contain higher numbers of cellulolytic, hemicellulolytic, nitrifying, and denitrifying bacteria than the surrounding soil does (Bhatnagar, 1975; Loquet et al., 1977). Shaw and Pawluk (1986) examined the microorganisms in the casts of the anecic earthworm species *L. terrestris* and two endogeic species *O. tyraeum* and *A. turgida* after keeping the earthworms for 1 year under controlled conditions in three different types of calcareous soil (sandy loam, clay loam, and silty clay loam). Casts of both the anecic and endogeic earthworms had densities of bacteria and actinomycetes one to three orders of magnitude higher than those in soils kept without earthworms. Such results demonstrate that the influence of earthworms on the density of microorganisms can vary with different types of soil and different species of earthworm.

It has been suggested that microbial activity in earthworm casts might have an important effect on soil crumb structure by increasing the stability of the worm-cast soil relative to that of surrounding soil. Many researchers have shown that earthworm casts contain more water-stable aggregates than the surrounding soil does, and part of this might be due to polysaccharide gums produced by the bacteria in the earthworm intestine (Satchell, 1958) and by the proliferation of fungal hyphae on the surface of casts (Marinissen and Dexter, 1990). The walls of earthworm burrows can also be enriched in microorganisms compared with the surrounding soil. Bhatnagar (1975) analyzed the microorganisms associated with earthworm burrows in a grassland in France and reported that 42% of soil aerobic nitrogen-fixing bacteria, 13% of anaerobic nitrogen-fixing bacteria, and 16% of denitrifying bacteria were associated with earthworm burrows. The numbers of ammonifying, denitrifying, nitrogen-fixing, and proteolytic bacteria were much higher in the burrow walls than in the surrounding soil.

Newly deposited casts are usually rich in ammonium-N and partially digested organic matter and thus provide a good substrate for growth of microorganisms. However, as the casts age, much of the N is converted into the nitrate form (Blair et al., 1995; Lavelle et al., 1992; Figure 11.3). Some of the intestinal mucus secreted during passage through the earthworm gut is excreted with the casts, where it continues to stimulate microbial activity and growth (Barois and Lavelle, 1986;

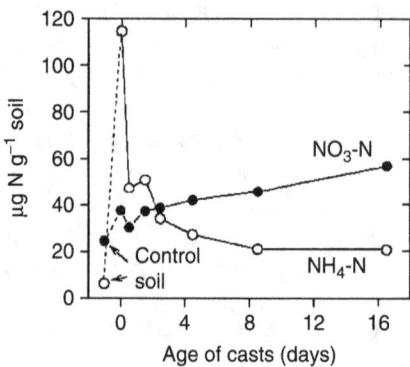

FIGURE 11.3 Temporal changes in concentrations of ammonium and nitrate nitrogen in aging casts of *Pontoscolex corethrurus* fed on an Amazonian Ultisol. (From Lavelle, P. et al. 1992. *Biol. Fertil. Soils* 14:49–53.)

Scheu, 1991). Parle (1963b) reported that yeasts and fungi, which occurred in the soil as spores, germinated as soon as they were in worm casts, and most hyphae were formed in 15-d-old casts. Microbial activity, as indicated by $O_2$ consumption, declined from the time casts were produced (Parle, 1963b). The simultaneous decline in $O_2$ consumption and increase in microbial populations is probably because as the casts age, an increasing proportion of the microorganisms pass into resting stages.

There is more recent evidence that microbial communities and activity associated with contents of fresh earthworm casts begin to change soon after the casts are deposited by the earthworms. Scheu (1987) observed that within 4 h of being deposited, fresh casts of *A. caliginosa* contained 130% more microbial biomass than surrounding soil, but this decreased to ca. 90% of that in soil after 2 weeks. However, microbial respiration in casts was much higher than in soils throughout the 30-d incubation period, suggesting that although the microbial biomass of casts might be less than that in soil, it is usually more metabolically active. Lavelle et al. (1992) reported a six- to sevenfold increase in the amount of microbial biomass N in fresh casts relative to that in uningested soil, but within 12 h, the amount of microbial biomass N in the casts decreased to slightly more than twice that in surrounding soil and declined only slightly during the remainder of the 16-d incubation. The higher relative stimulation of microbial biomass observed by Lavelle et al. (1992) compared with that by Scheu (1987) might have occurred because the soil used by Lavelle contained much less organic matter and microbial biomass than that used by Scheu. This hypothesis is supported by some of the results of Lavelle et al. (1993), who reported that casts of earthworms in soil containing low concentrations of soluble organic matter had higher microbial activity than the uningested soil, whereas the casts of earthworms provided with soil containing high concentrations of soluble organic matter had less microbial activity than noningested soil.

There is evidence that earthworms can decrease the total biomass of microorganisms in soil by causing a change to a smaller but more metabolically active microbial community. Wolters and Joergensen (1992) kept earthworms in six different soil types for 21 d and observed that although the microbial biomass was lower than that in control soils without earthworms in five of the six soils, earthworms increased the metabolic activity per unit of microbial biomass in all six soils. In the same experiment, Wolters and Joergensen (1992) removed earthworms from soil after 21 d and incubated the soil for a further 21 d to determine the longer-term effects of earthworms on microbial biomass following cessation of earthworm activity. Following this additional time of incubation, the microbial biomass responded differently between soil types and was higher in soils with earthworms than in those without. Bohlen and Edwards (1996) reported that earthworms caused microbial biomass to decrease in a laboratory incubation of silty loam soils with organic or inorganic nutrient inputs. After 112 d of incubation, soil with earthworms had less microbial biomass N than soil without earthworms, particularly in soil treated with an inorganic fertilizer. Based on this, Bohlen and Edwards (1995) speculated that earthworms fed on microbial biomass, releasing the nutrients bound in microbial tissues (Figure 11.2). This process was overshadowed in soils that received organic inputs because of the stimulatory effects of these inputs on the microbial community. The results from the various studies cited underscore the complexity of earthworm–microbial interactions and emphasize that interactions between earthworms and the microbial community are often specific to soil type, organic matter resources, and the time scale considered. Differences among these various factors can explain some of the conflicting results reported in the literature.

## IMPORTANCE OF MICROORGANISMS AS FOOD FOR EARTHWORMS

Microorganisms constitute the main nutritional components of the earthworm diet. Edwards and Fletcher (1988) summarized their experimental evidence that microorganisms provide a source of nutrients for earthworms and from axenic cultures they concluded that bacteria are of minor importance, algae are of moderate importance, and fungi, and to a lesser extent protozoa and possibly nematodes, are major sources of nutrients. Moreover, they emphasized that earthworms cannot grow on pure cultures of microorganisms and need mixed groups or species of microorganisms to develop satisfactorily. Feeding-preference studies have shown that earthworms prefer materials inoculated with particular groups of microorganisms. For instance, Cooke and Luxton (1980) and Cooke (1983) showed that *L. terrestris* preferred to feed on paper disks inoculated with particular species of fungi such as *Fusarium oxysporum, Alternaria solani*, and *Trichoderma viride* and rejected, or were not stimulated by, disks inoculated with other species such as *Cladosporium cladosporoides, Poronia piliformis*, and *Chaetonia globosum*. Large numbers of fungal hyphae can be observed in the intestines of earthworms, and many of these hyphae are digested as they pass through the earthworm gut (Dash et al., 1979, 1984; Spiers et al., 1986). Domsche and Banse (1972) reported that fungal hyphae were digested completely during passage through the earthworm gut, although most others have reported that some of the fungi in the gut remain undigested. Dash et al. (1984) calculated that earthworms digested and assimilated up to 54% of the fungal material they ingested. Most of the digestion of fungal hyphae occurs in the anterior portion of the gut. By examining the gut contents of field-collected earthworms, Piearce (1978) concluded that fungi and algae composed a significant component of the food of six different lumbricid species. Atlavinyte and Pociene (1973) reported that earthworms grew best in soil containing green and blue-green algae, indicating the importance of algae to the earthworm diet.

The best experimental evidence for the importance of microorganisms in the diet of earthworms comes from studies on *Eisenia fetida*. Miles (1963) introduced this species into soils inoculated with fungi and bacteria and showed that the earthworms were unable to reach sexual maturity unless protozoans were also added to the cultures. Protozoa are normally abundant in the habitat of this earthworm, which lives in compost and manure heaps, and it has also been suggested that protozoa are essential components of the diet of *E. fetida*. Other researchers have shown that ciliates and amoebae exposed to the digestive juices of *E. fetida* and *L. terrestris* are killed and digested (Piearce and Phillips, 1980; Rouelle, 1983). Neuhauser et al. (1980) reported that *E. fetida* increased in weight in the presence, but not in the absence, of seven species of microorganisms (two bacteria, two protozoa, and three fungi). The rates of growth of earthworms on certain specific microorganisms was not significantly different when either dead or live microorganisms were offered as food. Flack and Hartenstein (1984) showed that earthworms grew well when provided with many species of protozoa and bacteria, although the growth rates were 20% higher in the presence of protozoa than in the presence of bacteria alone. Hand and Hayes (1988) provided 18 individual species of bacteria and 22 different species of fungi to *E. fetida* and showed that earthworm growth improved in the presence of some microorganisms but was unaffected by others. Some species of bacteria, such as *Flavobacterium lutescens, Pseudomonas fluorescens, P. putida*, and fungi of *Streptomyces* spp., had a toxic effect on earthworms. In general, fungi had a much higher nutritional value than bacteria, although one bacterial species, *Acinetobacter lwoffi*, produced significant weight gains in the earthworms. The relative importance of particular groups of microorganisms as food for earthworms probably differs between different earthworm species, particularly those having markedly different feeding habits. However, data on this subject are limited to studies on only a few earthworm species and the importance of different microorganisms in the diet of most earthworm species has yet to be established.

## Dispersal of Microorganisms by Earthworms

Earthworms can enhance the dispersal of microorganisms by ingesting them at one location from a particular food source and excreting them elsewhere, or by transporting microbes that adhere to their body surface. Many of the microorganisms transported by earthworms are those involved in the decomposition of organic materials, but earthworms also consume and transport other beneficial microbial groups, such as plant-associated mycorrhizae (Cavender and Atiyeh, 2000) and other root symbionts, biocontrol agents, and microbial antagonists of plant pathogens. Thick and thin-walled spores tend to lose little of their viability during passage through the intestines of earthworms (Hoffman and Purdy, 1964; Dash et al., 1979; Striganova et al., 1989), and the spores of dwarf bunt (*Tilletia controversa*) lost none of their viability during passage through the earthworm gut. It has also been suggested that earthworms can disperse spores of harmful fungi such as *Pythium* (Baweja, 1939) and *Fusarium* (Khambata and Bhatt, 1957). Hutchinson and Kamel (1956) inoculated sterilized soil with several different species of fungi and reported that the rate of spread of the fungi through the soil was much higher when earthworms were present than when they were absent. Huss (1989) isolated 11 species of slime molds from the guts of *A. caliginosa* and *O. tyrteum* collected in northeastern Kansas. He force-fed adult *L. terrestris* with separate suspensions of spores and myxamoebae of *Dictyostelium mucoroides* and found that spores survived passage through the gut well, suggesting that earthworms play an important role in the short-range dispersal of slime mold propagules.

Earthworms can also have adverse effects on the spread of fungi. For example, the ascospores of *Ventura inaequalis* (Coole) Wint, which causes apple scab, are released from perithecia on overwintering dead leaves lying on the soil surface in the spring, and these infest the new growth (Hirst et al., 1955). However, a large population of *L. terrestris* can remove most of these leaves from the soil surface during the winter, thus preventing at least a proportion of the ascospores from being able to infect trees. There are probably many other important relationships between earthworms and plant pathogens (see later), but much more work is required to assess them (Arancon et al., 2002b).

Earthworms have been shown to have a significant influence on the dispersal of vesicular-arbuscular mycorrhizae (VAM) fungi, which form an important mutualistic association with plant roots (see Chapter 6). Rabatin and Stinner (1989) reported that 25% of earthworms in conventional corn, 83.3% from no-tillage corn, and 50% from pastures contained propagules of VAM fungi in their guts. Reddell and Spain (1991) surveyed the casts of 13 earthworm species from 60 sites in Australia and found intact spores of VAM fungi in all but one collection. They also found VAM root fragments in the casts, and the diversity of VAM spores in casts was similar to that of surrounding soil, but the numbers of VAM spores were highest in casts of *P. corethrurus* and *Diplotrema heteropora*. In greenhouse experiments, earthworms have also been shown to enhance the spread of VAM on roots of soybean (McIlveen and Cole, 1976) and seedlings of tropical fruit trees. Propagules of VAM can survive for several months in air-dried casts of *E. eugeniae* (Harinikumar et al., 1991) and for at least 12 months in those of *L. terrestris* (Harinikumar and Bagyaraj, 1994). Gange (1993) found that earthworms had a significant impact on the distribution of VAM propagules in early (1 and 3 years) and later (5, 8, and 11 years) successional plant communities. Earthworm casts contained nearly twice as many spores as the surrounding soil in most communities. The influence of earthworms on infective VAM propagules was even greater, particularly at the later successional sites, in which casts contained up to 10 times as many infective VAM propagules as did surrounding soil. Cavender et al. (2003) reported that vermicomposts produced by earthworms stimulated mycorrhizal colonization of roots of *Sorghum bicolor* and affected shoot and root dry weights. The high casting rates of earthworms in the early successional sites and the abundance of VAM spores in later successional sites indicate the considerable potential of earthworms to affect the establishment and competitive ability of mycorrhizal plants in these communities.

Dispersal of nitrogen-fixing bacteria that form mutualistic associations with plant roots can also be enhanced by earthworm activity. Reddell and Spain (1991) investigated the ability of the earthworm *Pheretima corethrurus* to transfer infective propagules of *Frankia,* an endophytic actinomycete that fixes nitrogen in association with roots of certain nonleguminous plants. They inoculated seedlings of *Casuarina equisetifolia* with either a crushed nodule suspension of *Frankia* or casts of *P. corethrurus,* raised in sterilized soil in which crushed *Frankia* nodules had been thoroughly mixed. The shoot and nodule dry weights of seedlings treated with casts from eleven species of earthworms were similar to those of seedlings inoculated with crushed nodules. Another very important group of beneficial nitrogen-fixing microorganisms influenced by earthworms are the *Rhizobium* bacteria, which fix nitrogen in nodules formed on the roots of leguminous plants. For instance, *L. rubellus* has been shown to enhance the translocation of *Bradyrhizobium japonicum* to greater soil depths (Madsen and Alexander, 1982). Rouelle (1983) reported that *L. terrestris* increased the spread of *Rhizobium japonicum* and the formation of nodules on soybean roots. Stephens et al. (1993) reported that *A. trapezoids* increased the rates of dispersal of *Rhizobium meliloti* and levels of root nodulation in infected alfalfa plants. Doube et al. (1994) showed that *Allolobophora trapezoides* increased the number of nodules on roots of *Trifolium subterraneum* in pot experiments in which sheep manure inoculated with *Rhizobium trifolii* was applied to the soil surface in pots with or without earthworms. Thompson et al. (1993) observed that *L. terrestris* and *Aporrectodea* spp. increased root nodulation on *Trifolium dubium* by up to 100 times and increased the proportion of *Trifolium* threefold in simple plant communities grown in controlled environmental chambers.

Thus, there is good overall evidence that earthworms are very important in inoculating soils with microorganisms and that their casts are foci for dissemination of many species of soil microorganisms (Ghilarov, 1963). Earthworms can enhance the multiplication and dispersal of several important groups of beneficial microorganisms, sometimes by several orders of magnitude. The literature on this subject is relatively sparse, which suggests that there is an excellent opportunity for important developments through future research. Knowledge gained in this area can provide the basis for a new technology to introduce and disperse beneficial microorganisms in soil (Doube et al., 1994).

## STIMULATION OF MICROBIAL DECOMPOSITION BY EARTHWORMS

The rates of decomposition of organic material can be accelerated when simple nitrogenous compounds are added to soil (Harmsen and van Schreven, 1955), particularly if the organic material is poor in N. Because excreted cast material from earthworms is usually rich in nitrogenous compounds, large populations of earthworms can not only help decompose organic material in the soil by ingestion, disintegration, and transport, but their waste products can also stimulate other microbial decomposition processes in the soil.

Many microorganisms in the soil are in a dormant stage, with low metabolic activity, awaiting suitable conditions to become active (Lavelle et al., 1992). The earthworm gut provides suitable conditions for the vigorous multiplication of particular microorganisms, which are stimulated to decompose ingested organic matter. Earthworms secrete large amounts of water-soluble organic compounds, which can be assimilated readily by microorganisms in the earthworm gut (Barois and

Lavelle, 1986). Addition of these compounds might help prime the microorganisms in the earthworm gut to break down more complex organic compounds in the ingested soil (Barois and Lavelle, 1986; Lavelle et al., 1993). This process by which the microorganisms benefit from the mucus secretions of the earthworm and the earthworm benefits from the enhanced microbial decomposition of ingested organic matter has been described as a mutualistic digestive system (Barois, 1992; Trigo and Lavelle, 1993).

Because microbial activity remains higher in earthworm casts than in surrounding soil, the enhanced rates of decomposition of organic matter, which begins in the earthworm gut, are maintained for some time after the gut contents are egested. Kozlovskaya and Zhdannikova (1961) reported that the decomposition of organic matter was much faster and more intensive in earthworm casts than in the surrounding soil. Parle (1959) showed that $O_2$ consumption, which is an indicator of microbial activity, was considerably higher in cast soil than in the surrounding soil even 50 d after being excreted. Daniel and Anderson (1992) showed that rates of $CO_2$ production were higher in earthworm casts than in the soil ingested by the earthworms. Increased respiration rates in earthworm casts were accompanied by an increase in the numbers of bacteria and soluble organic C in the casts. After a preliminary stage of high respiratory activity in fresh casts, microbial activity tended to decrease, gradually returning to the same level as that of the surrounding soil.

As earthworm casts age, there is a reorganization of mineral and organic components of the casts, which results in lower rates of decomposition in casts than that of the surrounding soil, because of the physical protection of organic matter in the compact structure of the casts (Martin, 1991; Lavelle and Martin, 1992). This process might be more important in poorly aggregated soils where climate and soil texture favor rapid mineralization of soil C. The physical changes that occur in aging casts are accompanied by biological changes, in which slower-growing soil fungi and dormant microbial stages begin to predominate in older casts. Thus, although the short-term effect of earthworms is to stimulate microbial decomposition of organic matter, ultimately the long-term effect might be to decrease rates of microbial decomposition to some extent by increasing the physical protection of organic matter in the casts. The consequences of this for the long-term net storage or loss of organic matter in soil remain unknown.

Another way in which earthworms might affect the microbial decomposition of soil organic matter is by influencing the ratio of fungi to bacteria in the soil (Blair et al., 1994). Earthworms can change the fungal to bacterial ratios in soil by increasing the amounts of soluble organic C that can be mineralized rapidly by bacteria and other microbial groups and also by feeding preferentially on fungi. Changes in the ratio of fungi to bacteria are important because of the differences between fungi and bacteria in the efficiency to assimilate C. Bacteria tend to be less efficient than fungi at assimilating C and thus respire more C as $CO_2$ for each unit of C consumed (Adu and Oades, 1978). Furthermore, fungal hyphae contain C compounds resistant to degradation. Evidence from short-term incubations and investigations of the microflora in earthworm casts indicates the potential for earthworms to stimulate soil bacteria preferentially. There is a need to link earthworm-induced changes in the ratio of fungi to bacteria to C cycling processes under natural field conditions. It is clearly that earthworms stimulate microbial activities in soils considerably and accelerate rates of organic matter decomposition.

## THE POTENTIAL OF VERMICOMPOSTING IN PROCESSING AND UPGRADING ORGANIC WASTES AS PLANT GROWTH MEDIA

### INTRODUCTION

Thermophilic composting is being increasingly used to process a wide range of organic wastes. However, the various methods of composting do not always produce high-quality products that have good potential for soil and land improvement and plant growth. Over the past 20 years, interest has increased progressively about the potential of a related process, termed vermicomposting, which involves the use of earthworms to promote microbial activity in organic waste and break them down into materials that can be used in crop production. In recent years, the potential of earthworms to be used in various systems of breaking down organic wastes has been explored in much more depth. The basic research, which began at the State University of New York, Syracuse, in the 1970s under the leadership of Dr. Roy Hartenstein focused mainly on the use of earthworms for processing sewage solids. This laboratory research was expanded in the early 1980s to development of field-scale practical methods for disposing of poultry, pig, and cattle wastes in an interdisciplinary research program at the Rothamsted Experimental Station, U.K., which involved nearly 50 scientists, including biologists, agricultural engineers, economists, and representatives from various commercial enterprises (Edwards and Neuhauser, 1998). These studies, which have since been complemented by more recent research in the U.S., France, Germany, Italy, Spain, the Philippines, and Australia, have demonstrated a very considerable economic potential of using earthworms to convert efficiently a wide range of animal, plant, and industrial organic wastes into effective and valuable plant growth media with great economic potential for exploitation in horticulture and agriculture (Edwards, 1998).

Populations of organic-waste-degrading earthworms can increase rapidly, fragmenting organic wastes and increasing microbial activity in them dramatically. The main difference between composting and vermicomposting is that composting is a thermophilic process reaching temperatures of 60 to 70°C for several days and requiring turning or forced ventilation for aeration, and vermicomposting systems must be maintained at temperatures below 35°C and require much less manipulation. Exposure of most species of earthworms to temperatures above this, even for short periods, will kill them; therefore, careful management of the rates of addition of wastes to vermicompost systems is required to avoid overheating. Earthworms actively consume organic wastes in only a relatively narrow aerobic layer of 15 to 25 cm below the surface of a windrow, bed, or other container, the layers above this too dry and those below too warm for earthworm activity. The key to successful vermicomposting lies in adding organic wastes in successive thin layers to the waste at frequent intervals, so that little thermophilic heating occurs and temperatures can be maintained in the optimum range (20 to 25°C) for earthworm growth and activity.

Almost any agricultural, urban, or industrial organic waste can be used for vermicomposting, but some might need some form of preprocessing such as washing, precomposting, macerating, or mixing to make them acceptable to earthworms. Food and paper industry wastes, sewage solids, yard wastes, garden and food wastes, and sewage biosolids are particularly suitable for vermicomposting. Often, mixtures of several different organic wastes can be processed more effectively than individual wastes, because they are easier to maintain in terms of relatively constant aerobicity, moisture content, and temperature.

An extensive but small-scale cottage industry in the U.S. and elsewhere grows earthworms for fish bait in a variety of organic wastes. These use, almost exclusively, outdoor ground beds or windrows. Such systems require relatively large areas of land for large-scale production and are relatively labor intensive even when machinery is used for adding wastes to the beds. More importantly, windrow systems process wastes relatively slowly, taking 6 to 18 months to process a 4-cm-deep layer. There is good evidence (Edwards, 1998) that a large proportion of the relatively soluble essential plant nutrients are either washed out or volatilized during this long processing period. Such nutrient losses can contribute to groundwater pollution and also result in a product low in nutrients with relatively poor potential as a plant growth medium.

## SCIENTIFIC BASIS FOR VERMICOMPOSTING ORGANIC MATTER

Only a few species of earthworms are specific to organic wastes, and these can consume organic materials very rapidly and fragment them into much finer particles by passing them through a grinding gizzard in the mouth. Earthworms obtain their nourishment not from organic wastes but from fungi and bacteria and other microorganisms that grow on them; at the same time they promote microbial activity in the wastes, so that the casts, or vermicomposts, that they produce are much more fragmented and microbially active than the parent organic wastes. During this process, the important plant nutrients such as N, P, K, and Ca are converted into materials much more soluble and available to plants than those in the original wastes. The retention time of the waste in the earthworms' body is much less than 24 h, and very large quantities of organic matter can pass through an average population of earthworms in a short time. This can be compared with thermophilic aerobic composting, in which organic wastes have to be turned regularly or artificially aerated to maintain aerobic conditions in the waste. This might often involve extensive engineering, machinery, and technology to process the wastes rapidly. By contrast, in vermicomposting, earthworms, which can survive only under aerobic conditions, both turn the waste and maintain it in an aerobic condition, thereby minimizing any need for expensive engineering. Indeed, earthworms have been called ecological engineers (Lavelle et al., 1993).

The major constraint to vermicomposting is that vermicomposting systems must be maintained at temperatures above freezing and below 35°C. The processing of organic wastes by earthworms occurs most rapidly between 15°C and 25°C and at moisture contents of 70 to 90% (g water/dry weight solids; Edwards, 1988). Outside these limits, earthworm activity and productivity and rates of organic waste processing can fall. For maximum efficiency, the wastes should be maintained under cover and as close to these environmental limits as possible. Earthworms are also sensitive to certain chemical conditions in the wastes. In particular, earthworms are very sensitive to ammonia and salts and certain other substances. For instance, they die quite quickly if exposed to wastes containing more than 0.5 mg of ammonium radical/g organic waste and more than 5 mg/g salts (Edwards, 1988). However, both salts and ammonia can be washed out of organic wastes readily or dispersed quite rapidly by precomposting. Contrary to common belief, earthworms do not have many serious natural enemies, diseases, or predators and can survive exposure to many adverse conditions, provided they are not exposed to extremes of temperature and moisture (Edwards and Bohlen, 1996).

## Vermicomposting Technologies Available

A number of species or earthworms specific to the breakdown of organic wastes and not surviving long in soils are used in vermicomposting. The temperate species most commonly used are *Eisenia fetida* (the tiger or brandling worm) and *Denbroboena veneta*. Another suitable temperate species is *Lumbricus rubellus* (the red worm), and two tropical species, *Eudrilus eugeniae* (the African night-crawler), and *Perionyx excavatus*, an Asian species; the latter two species are very productive, but unable to withstand temperatures below 5°C for extended periods. Each species has particular environmental requirements, and it is important to choose the best species for processing in particular climates and for types of waste (Edwards, 1998).

Traditionally, vermiculture has been based on outdoor beds or windrows on the ground 1 to 5 m wide containing organic wastes up to 45-cm deep, but these have many technical drawbacks, particularly in terms of land and labor requirements. Moreover, when the vermicompost is collected it is necessary to use sloping mechanically driven rotating mesh cylinders or trommels or other mechanical means of separating earthworms from the processed organic materials.

It is also possible to process organic wastes with earthworms by using batch vermicomposting systems involving bins, crates, or larger containers often stacked one above the other in racks. However, such container systems need considerable handling and lifting machinery, and if stacked in racks there are problems in adding water to maintain constant moisture contents and also in adding additional layers of waste to the surface at frequent intervals. However, small-scale container systems for household use have been used extensively to process organic domestic and institutional food wastes. They range from simple raised containers (Appelhof, 1981) to more sophisticated commercially produced stacking systems that collect the vermicompost at the bottom.

In recent years, more sophisticated and efficient systems of vermicomposting have been developed (Edwards, 1998). These use large containers raised on legs above the ground. This allows organic wastes to be added in thin layers at the top from mobile gantries, and the vermicomposts are collected mechanically through mesh floors at the bottom by using manual power or driven breaker bars, which travel up and down the length of the system. Such reactors range from relatively low technology systems, using manual loading and waste collection systems, to large (40 m long × 2.8 m wide × 1 m deep or 1-m-long legs), completely automated and hydraulically driven continuous flow reactors, which have operated successfully in the U.K. and U.S. for several years (Edwards, 1998). Earthworm populations in such reactors tend to reach an equilibrium biomass of ca. 9 kg/m$^2$. Such reactors can fully process the entire 1 m depth of suitable organic wastes in 30 to 45 d (Edwards, 1998). Economic studies have shown such reactors to have much greater economic potential to produce high-grade plant growth media very quickly and efficiently than do windrows or ground beds.

## EFFECTS OF VERMICOMPOSTS ON PLANT GROWTH

### INTRODUCTION

Vermicomposts are materials derived from the accelerated biological degradation of organic wastes by interactions between earthworms and microorganisms. Earthworms ingest and fragment organic wastes into much finer particles by passing them through a grinding gizzard and derive their nourishment from microorganisms that grow on the organic matter. This process accelerates decomposition of the organic matter, changes and improves the physical and chemical properties of the material that influence plant growth, and accelerates humification to well-oxidized and stabilized products (Albanell et al., 1988; Orozco et al., 1996) termed vermicomposts (Edwards and Neuhauser, 1988; Edwards, 1998).

Vermicomposts are finely divided peat-like materials with high porosity, aeration, drainage, and water-holding capacity (Edwards and Burrows, 1988). Their specific surface area greatly exceeds that of either the parent organic matter or traditional thermophilic composts, providing more microsites for microbial decomposition and strong adsorption and retention of nutrients (Shiwei and Fu-zhen, 1991). Albanell et al. (1988) reported that vermicomposts tended to have pH values near neutrality, which might be due to the production of $CO_2$ and organic acids during microbial metabolism. They also reported that if moisture was not added, the moisture content of the organic matter was reduced progressively during vermicomposting to 45 to 60%, the ideal moisture contents for land-applied composts (Edwards, 1983). However, earthworm activity is highest at 75 to 85% moisture content, and therefore water is usually added.

Businelli et al. (1994) documented the rates of humification of a range of animal manures and municipal waste by *Lumbricus rubellus*. Elvira et al. (1996) reported increased humification rates of paper-pulp mill sludge as it was worked by *Eisenia andrei*. Studies of transformations into humic compounds during passage through the earthworm gut revealed that rates of humification of ingested organic matter intensified during transit through the earthworm gut (Kretzschmar, 1984). Orlov and Biryukova (1996) reported that vermicomposts contained 17–36% humic acid and 13–30% fulvic acid. Senesi et al. (1992) compared the quality of humic acids present in vermicomposts with those found in natural soils, using spectroscopic analysis procedures. They demonstrated that the organometallic complexes containing iron and copper present in vermicomposts were similar to the humic acids common in soils.

Edwards and Burrows (1988) reported that vermicomposts, especially those from animal manures, usually contained higher amounts of mineral elements than commercially produced plant growth media, and N, P, K, Ca, and Mg were changed to forms more readily taken up by plants. This is one reason why vermicomposts are better soil amendments than traditional composts, which have comparatively little N in plant-available forms. Orozco et al. (1996) reported that processing of coffee pulp by *Eisenia fetida* increased the availability of nutrients such as P, Ca, and Mg. Available P was 64% higher in vermicomposts than in the parent organic material, probably because of increased phosphatase activity from the direct action of gut enzymes and indirectly by the stimulation of microbial activity. Werner and Cuevas (1996) reported that most vermicomposts contained adequate amounts of macronutrients and trace elements of various kinds, but the types and amounts were dependent on the sources of the earthworm feedstock. Businelli et al. (1984) reported similar differences in the chemical compositions of vermicomposts, based on the nature of the parent substrate used. In their experiments, the highest elemental values were recorded in vermicomposts from a cattle and horse manure mixture with 38.8% organic C, 2.7% total N, and 1080 mg/kg $NO_3$-N. The lowest elemental concentrations were recorded in municipal waste compost, which had only 9.5% organic C, 1.0% total N, and 503 mg/kg $NO_3$-N. Edwards (1988) reported higher amounts of available nutrients in a range of earthworm-processed animal wastes than in commercial plant-growth media. The wastes he investigated were separated cattle solids, separated pig solids, cattle solids on straw, pig solids on straw, duck solids on straw, and chicken solids on shavings. These materials contained (% dry weight) 2.2–3.0 N, 0.4–2.9 P, 1.7–2.5 K, and 1.2–9.5 Ca compared with those of a commercial plant growth medium (Levington compost), which had only 1.80 N, 0.21 P, 0.48 K, and 0.94 Ca. The changed amounts and form of the nutrients in vermicomposts can be explained by the accelerated mineralization of organic matter, breakdown of polysaccharides, and high rates of humification occurring during vermicomposting compared with rates of humification in soils (Elvira et al., 1996; Albanell et al., 1988). In studies of the bioconversion of solid paper-pulp mill sludge by earthworms, it was reported that the total carbohydrate content decreased whereas total extractable C, nonhumified fraction, and humification rates increased significantly by the end of the experiment.

Vermicomposts have many valuable biological properties that make them useful in production of crops. They are rich in bacteria, actinomycetes, fungi (Edwards, 1983; Tomati et al., 1983, 1987; Werner and Cuevas, 1996), and cellulose-degrading bacteria (Werner and Cuevas, 1996). Vermicomposts had much larger populations of bacteria ($5.7 \times 10^7$), fungi ($22.7 \times 10^4$), and actinomycetes ($17.7 \times 10^6$) compared with those in conventional composts made from the same starting material. The physicochemical and biological properties of vermicomposts make them excellent materials as additives to greenhouse container media, organic fertilizers, or soil amendments for various field horticultural crops.

## EFFECTS OF VERMICOMPOSTS ON GROWTH OF GREENHOUSE CROPS

Greenhouse experiments have clearly demonstrated that vermicomposts can have consistently positive effects on plant germination growth and yields. Edwards and Burrows (1988) reported that vermicomposts increased emergence of a range of ornamental and vegetable seedlings compared with germination in control commercial plant growth media. They used a wide range of test plants, such as peas, lettuce, wheat, cabbages, tomatoes, and radishes. In other similar investigations, they showed that ornamental shrubs such as *Eleagnus pungens, Cotoneaster conspicua, Pyracantha, Viburnum bodnantense, Chaemaecyparis lawsonia, Cupressocyparis leylandii*, and *Juniperus communis* usually grew much better in vermicompost-supplemented mixtures than in a commercial plant growth medium when transplanted into larger pots or grown outdoors. Plants even responded significantly in growth to a substitution of 5% of a mixture of pig and cattle manure vermicomposts into 95% of a commercial plant growth medium (Scott, 1988). Buckerfield et al. (1999), using 0 to 100% mixtures of vermicompost and 100 to 0% sand, reported similar increased growth trends in plant growth trials in the greenhouse. Although the germination of radish decreased with increasing vermicompost concentrations, radish harvest weights increased proportionally to the application rates of vermicomposts, with yields of plants in 100% vermicompost up to 10 times more than in 10% vermicompost, but these experiments depended on the vermicompost to supply nutrients. Research at the Soil Ecology Laboratory, The Ohio State

University (OSU), has shown consistent acceleration of the germination of a wide range of greenhouse and field crops by vermicomposts. Buckerfield et al. (1999), using water extracts from vermicomposts, reported that the first applications inhibited germination, but subsequently weekly applications of the diluted extracts improved plant growth and increased radish yields significantly by up to 20%. Growths of tomatoes, lettuces, and peppers were best after substitution of vermicomposts into soils at 8 to 10%, 8%, and 6%, respectively, using duck waste vermicompost and peat mixtures (Wilson and Carlile, 1989). At higher vermicompost substitutions, the inhibition of growth that commonly occurred was attributed to higher salt content and electicial conductivity, combined with excessive nutrient levels in these materials. Subler et al. (1998) reported increased growth of tomatoes and marigolds in a commercial soilless medium, Metro-Mix 360 (MM360), with a range of levels of different vermicomposts substituted into the MM360 in comparison with growth in MM360 only, or traditional composts made from biosolids or yard wastes. MM360 is a soilless growth medium prepared from vermiculite, Canadian sphagnum peat moss, ark ash, and sand, and contains a starter nutrient fertilizer in its formulation. In Subler's experiments in 1998, increases in chlorophyll contents in response to vermicomposts were observed at early stages of marigold growth. Significant increases in tomato seedling weights after substitution of 10 and 20% vermicompost into 90 and 80% MM360 were also reported (Atiyeh et al., 2000b). Raspberries grown in commercial media substituted with 20% pig manure vermicompost produced much larger shoot dry weights than did plants that received a complete inorganic fertilizer. In investigations by Scott (1988), who used hardy nursery stocks of *Juniperus, Chamaecyparis,* and *Pyracantha,* 20 to 50% substitution of vermicomposts produced from cattle manure, pig manure, and duck waste into a commercial plant growth medium, with regular application rates of nutrients, produced better growth than plants grown in a peat–sand mixture used as a control, which was also treated regularly with nutrients.

Chan and Griffiths (1988) reported stimulating effects of pig manure vermicomposts on the growth of soybean (*Glycine max*), particularly in terms of increased root lengths, lateral root numbers, and internode lengths of seedlings. In another rooting experiment, vermicomposts improved the establishment of vanilla (*Vanilla planifolia*) cuttings better than applications of other growth media such as mixtures of coir pith and sand did (Siddagangaiah et al., 1996). Similar responses in growth were observed for cloves (*Syzygium aromaticum*) and black peppers (*Piper nigrum*) sown into 1:1 mixtures of vermicompost and soil (Thankamani et al., 1996). In these experiments, black-pepper cuttings raised in vermicomposts were significantly taller and had more leaves than those grown in commercial potting mixtures. Increased plant height, a greater number of branches, and the longest taproots occurred on clove plants grown in vermicompost mixtures. Vadiraj et al. (1998) reported enhanced growth and dry matter yields of cardamom (*Electtaria cardamomum*) seedlings grown in vermicomposted forest litter compared with that in other plant growth media tested. Vermicomposts produced from coconut coir increased the yields of onions (*Allium cepa*; Thanunathan, 1997).

Atiyeh et al. (2001b) demonstrated that vermicompost produced from pig manure substituted into MM360 at a range of concentrations increased the growth of vegetable and ornamental seedlings, especially at low vermicompost substitution rates, although all the nutrients needed by the crops were available to the plants. However, substitution of higher percentages of a range of vermicomposts into the MM360 did not always improve plant growth, possibly because of high salt contents or other adverse factors. They demonstrated in some experiments that as little as 5% of vermicompost substituted into 95% MM360 was enough to produce growth responses of test crops. Significant increases in the growth of tomato seedlings occurred in 10% vermicompost produced from pig waste substituted into 90% MM360 commercial medium compared with plants grown in 100% MM 360, 100% peat–perlite mixture, or 100% coir–perlite mixture when plants received all needed nutrients (Atiyeh et al., 2000b). Substitution of 10, 20, and 50 vermicompost into 90, 80, or 50 MM360 stimulated plant growth significantly, independent of nutrient supply, with significant increases in plant height and root and shoot biomass. Atiyeh et al. (2000b) reported that substitution of 10 or 50% pig manure vermicompost into 90 or 50% MM360 increased dry weights of tomato seedlings significantly compared with those grown in 100% MM360. They reported that the highest marketable fruit yields were obtained in a mixture of 80% MM360 and 20% vermicompost, and lower concentrations of vermicomposts (less than 50%) into MM360 usually produced greater growth effects than the substitution of larger amounts. Substitution of 20% vermicompost into 80% MM360 resulted in 12% higher tomato fruit weights than those grown in MM360. Substitutions of 10, 20, and 40% vermicompost into 90, 80, and 60% MM360 reduced the proportions of nonmarketable fruits significantly and produced larger tomato fruits (Figure 11.4). Mixtures containing 25 and 50% pig manure vermicomposts substituted into 75 and 25% MM360 increased rates of tomato seedling growth and produced greater increases in seedling growth with a 5% substitution of pig manure into 95% MM360, when all needed inorganic nutrients were supplied daily (Atiyeh et al., 2001b).

More recently, the substitution of 30 or 40% pig manure vermicomposts into 20 or 60% MM360 produced more vegetative growth and flowering of marigolds in mixtures containing 40% pig manure vermicompost (Atiyeh et al., 2002). Pig manure vermicompost ranging from 20 to 90% substituted into 80 to 10% MM360 produced marigold plants with significantly more root growth. Peppers produced greatest fruit yields when grown in mixtures containing 40% food waste vermicomposts and 60% MM360 (Arancon et al., 2004a). With petunias as test plants, most flowers were produced in MM360 substituted with 40% food waste vermicompost and 60% MM360 or 40% paper waste vermicompost and 60% MM360. When cow manure vermicompost was tested, petunias produced more flowers in mixtures substituted with 20% vermicompost and 80% MM360 (Arancon et al., unpublished data). In similar experiments, increased yields of tomatoes or peppers or amount of flowering of marigold were not correlated with available mineral-N or microbial biomass in the growth media when all plants were provided with needed nutrients (Atiyeh et al., 2002). It is possible that a number of growth-enhancing factors resulting from the introduction of lower concentrations of food waste vermi-compost into MM360 might have caused the growth changes. Such factors could include improvements in the physical structure of the container medium, increases in enzymatic activities, increased numbers of beneficial microorganisms, or production of biologically active plant growth-influencing substances such as plant growth regulators and humic acids. (See section on plant growth regulator production in vermicomposts.)

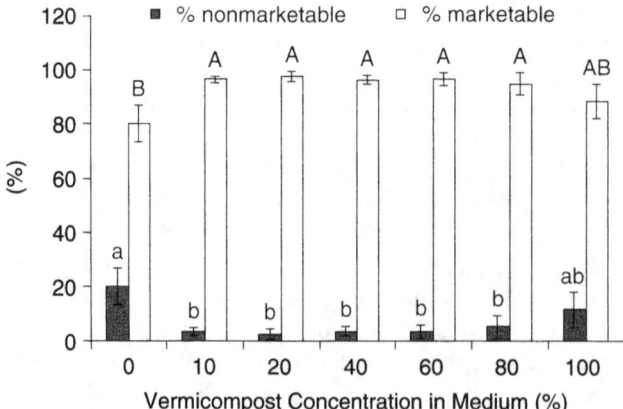

**FIGURE 11.4** Percentage marketable and nonmarketable yields of tomatoes (mean ± standard error) grown in the greenhouse in a range of mixtures of vermicompost and a commercial medium Metro-Mix 360 (MM 360, with all necessary nutrients supplied). Columns followed by same letters are significantly different at $P = 0.05$. (From Atiyeh et al., 2000c. With permission.)

## Effects of Vermicomposts on Growth of Field Crops

A number of field experiments at OSU have reported positive effects of low application rates of vermicomposts on crop growth. Application rates of vermicomposts were correlated with substitution rates that improved growth on crops in greenhouse experiments. Cabbages grown in compressed blocks made from pig waste vermicompost in the greenhouse and transplanted to the field were larger and more mature at harvest than those grown in a commercial blocking material (Edwards and Burrows, 1988). In a field experiment in which cassava peel mixed with guava leaves and vermicomposts produced from poultry droppings were applied to field crops, Mba (1983) reported higher shoot biomass and increased seed yields of cowpea. Masciandro et al. (1997) investigated the effects of soil applications of vermicomposts produced from sewage sludge compared with applications of humic materials extracted from vermicomposts. They reported greater growth of garden cress (*Lepidium sativum*) treated with vermicomposts than with no vermicompost applications. Soil analyses after the vermicompost applications showed marked improvements in the overall physical and biochemical properties of the soil. Surface application of a vermicompost derived from grape residues spread under grape vines and then covered with a straw and paper mulch increased yields of a grape variety Pinot Noir by 55% compared with yields from mulch alone (Buckerfield and Webster, 1998). Both bunch weights and bunch numbers increased, with no losses in grape flavor. In an experiment at a second site, vermicompost applications from animal manures applied under straw mulch increased yields of Chardonnay grapes by up to 35%. Vermicompost applications tended to have greater effects on yields when they were applied under mulches than when applied uncovered to the soil surface, possibly because vermicompost might have degraded on exposure to sun and air.

Venkatesh et al. (1997) reported that yields of Thompson Seedless grapes were significantly higher when vermicomposts were applied. In experiments at OSU, Seyval grapes produced higher marketable yields, more fruit clusters per vine, and bigger berry sizes in response to applications of food waste and paper waste vermicomposts at 2.5 or 5 t/ha (supplemented with inorganic fertilizers; Arancon et al., unpublished data). Vadiraj et al. (1998) compared application rates of vermicomposts of 5 t/ha up to 25 t/ha in 5 t/ha (dry weight basis) increments on the growth of three varieties of coriander. The responses to the vermicompost applications differed for all three coriander varieties tested, and maximum herbage dry weights occurred 60 d after sowing. The varieties RCr-41, Bulgarian, and Sakalespur Local attained highest yields at vermicompost applications rates of 15, 10 to 25, and 20 t/ha, respectively, 60 d after sowing.

Some field experiments have involved amending soils with vermicomposts in conjunction or combination with conventional fertilization programs. Vermicomposts applied at 12 t/ha to field soils together with 100 or 75% of the recommended application rate of inorganic fertilizers increased yields of okra (*Abelmoschus esculentus* Moench) significantly (Ushakumari et al., 1999). Gangadharan and Gopinath (2000) reported in gladiolus significantly increased plant height, number of leaves, leaf area, leaf area indices, fresh weight of whole plant, number of days to spike emergence, length of spikes, width of spikes, length of rachis, number of florets per spike, diameter of corms, and fresh weights of corms after applications of vermicompost combined with 80% of the recommended rate of inorganic fertilizers. Amending soils with vermicomposts, at 2 kg/plant, together with 75% of the recommended rate of inorganic fertilizers promoted shoot production of bananas (Athani et al., 1999). Vermicompost applications to field soils combined with 50% of the recommended inorganic fertilizers increased the yields of tomatoes compared with soils treated with 100% of the recommended inorganic fertilizers only (Kolte et al., 1999). Increased wheat yields were obtained from the residual fertility in soils treated with 50% vermicompost and 50% inorganic fertilizers the previous year (Desai et al., 2000). Vasanthi and Kumaraswamy (1999) reported increased yields of rice after amending soils with vermicomposts at 5 or 10 t/ha, supplemented with recommended application rates of inorganic fertilizer. A lower application rate of 2 t/ha vermicomposts plus the recommended amounts of inorganic fertilizers increased tomato yields to a level similar to those of tomatoes in soils treated with 4 t/ha vermicomposts and 50% of the recommended rates of inorganic fertilizers (Patil et al., 1998). Potatoes produced highest marketable yields after amending soil with 75% of the recommended rate of inorganic fertilizers and 2.5 t/ha vermicomposts (Mrinal et al., 1998). Sunflower (*Helianthus annuus*) yielded more flowers in response to soil treatments with 50% of the recommended application rate of an inorganic fertilizer combined with 5 or 10 t/ha of vermicomposts (Devi et al., 1998). Increased yields of pea (*Pisum sativum*) were recorded after amending soils with 100% of the recommended application rate of inorganic fertilizers in combination with 10 t/ha vermicomposts produced from animal manures (Ramachandra et al., 1998). Zende et al. (1998) reported increased yields of sugarcane after amending soils with vermicomposts applied at 5 t/ha together with 100% of the recommended application rate of inorganic fertilizers. Mulberry (*Muros* spp.) growth increased after amending soils with vermicomposts at 10 t/ha together with 100% of the recommended application rates of inorganic fertilizers (Murakar et al., 1998). Flowering of China aster (*Callistephus chinensis* L.) increased when grown in soils amended with 10 t/ha vermicomposts produced from animal manures, together with 100% of the recommended application rate of inorganic fertilizers (Nethra et al., 1999).

Arancon et al. (2002a) reported significantly increased growth and yields of field tomatoes (*Lycopersicon esculentum*) and peppers (*Capsicum anuum grossum*) when vermicomposts produced commercially from cattle manure, food waste, or recycled paper waste were applied to field plots at 20 and 10 t/ha in 1999 and 10 and 5 t/ha in 2000, compared with those receiving equivalent amounts of inorganic fertilizer. The N, P, and K available from vermicomposts in the first year were 33% of their total N, P, and K contents. Vermicomposts from food waste and recycled paper were applied at 10 or 5 t/ha in 2000 to strawberries (*Fragaria* spp.) grown in field soils in a matted-row system. All the vermicompost-treated plots were supplemented with inorganic fertilizers to equalize the available N levels in all plots at transplanting. Yields were obtained from both the mother plants and runners. The marketable strawberry yields in plots treated with vermicompost (plus fertilizers to balance N inputs) were consistently and significantly higher than those obtained from plots treated with inorganic fertilizers only. Although all plots had similar nutrient inputs, shoot weights, leaf areas, and marketable fruit, yields of pepper plants grown in vermicomposts (plus fertilizers) plots were significantly higher than those of plants grown in plots receiving inorganic fertilizer only. Leaf areas, numbers of strawberry suckers, numbers of flowers, shoot weights, and marketable fruit yields of strawberries increased significantly in response to vermicompost applications (plus fertilizers) compared with those from strawberries that received inorganic fertilizers only (Arancon et al., 2004b; Figure 11.5).

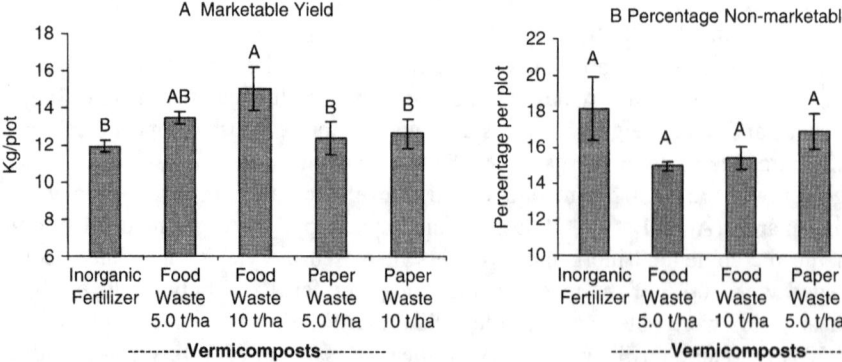

**FIGURE 11.5** Marketable and percentage of nonmarketable yields (mean ± standard error) of strawberries at Piketon, OH. Columns with the same letters are not significantly different at $P = 0.05$. (From Arancon, N.Q. et al., 2004b. *Bioresour. Technol.* With permission.)

## PHYSICOCHEMICAL CHANGES IN SOILS IN RESPONSE TO VERMICOMPOST APPLICATIONS

Improvements in the growth and yields of crops grown in soilless potting media in greenhouses or in field soils substituted or amended with vermicomposts can be attributed to several factors. First, vermicomposts contribute to improvements in physicochemical and biological characteristics of the greenhouse planting media or field soils, which favor better plant growth. For instance, substituting pig-manure-based vermicomposts into a soilless commercial bedding plant medium (MM 360) decreased total porosity, percentage air space, pH, and ammonium concentrations significantly, whereas bulk density, container capacity, electrical conductivity, overall microbial activity, and nitrate concentrations increased significantly in response to increasing substitutions of vermicomposts into MM 360 (Atiyeh et al., 2001b). The changes in pH reported contrasted with those from the work of Tyler et al. (1993), who reported increases in substrate pH in response to increasing concentrations of composted turkey litter added to a plant container medium. Electrical conductivity, an index of salinity, increased linearly in response to increasing concentrations of pig manure vermicomposts in planting media mixtures (Atiyeh et al., 2001b). Klock (1997) reported that the electrical conductivity (salinity) of planting media substituted with vermicomposts increased by 1.3 to 2.8 times over untreated controls. Because most of the mineral N in vermicomposts is usually in the nitrate form (Atiyeh et al., 2001b; Orozco et al., 1996; Benetiz, 1999), it is not surprising that the amount of nitrates in a planting medium increased with the increasing vermicompost concentrations (Atiyeh et al., 2001b).

Reddy and Reddy (1999) reported significant increases in micronutrients in field soils after vermicompost applications compared with those in soils treated with farm manures. Others reported that amounts of soil N also increased significantly after incorporating vermicomposts into soils (Sreenivas et al., 2000; Kale et al., 1992; Nethra et al., 1999) and the amounts of available P and K also increased (Venkatesh et al., 1998). Field experiments at OSU (Arancon et al., 2003a) demonstrated that soils that were treated with vermicomposts and supplemented with inorganic fertilizers and planted with tomatoes had amounts of total N, orthophosphates, dehydrogenase enzyme activity, and microbial biomass usually higher than those that received equivalent amounts of inorganic fertilizers only. In similar field experiments in which pepper was planted, there was more microbial biomass N and orthophosphates in soils to which vermicomposts had been applied than in those that had received inorganic fertilizers only. In soils planted with strawberries, at the end of the growth cycle, the amounts of total extractable N, microbial biomass N, and dissolved organic N in all treatments were statistically similar, but there were more orthophosphates in soils that received vermicomposts than in soils treated with inorganic fertilizers only. Maheswarappa et al. (1999) reported increased organic C, higher pH, decreased bulk density, improved soil porosity and water-holding capacity, and increased microbial populations and dehydrogenase activity of soils in response to vermicompost applications. Therefore, amendment of soils with vermicomposts can produce extensive physicochemical changes in soils.

## PLANT GROWTH REGULATOR PRODUCTION IN VERMICOMPOSTS

A substantial body of evidence demonstrates that microorganisms, including bacteria, fungi, yeasts, actinomycetes and algae, are capable of producing plant growth hormones (PGHs), such as auxins, gibberellins, cytokinins, ethylene, and abscisic acid, in appreciable quantities (Arshad and Frankenberger, 1993; Frankenberger and Arshad, 1995) and humic substances, such as humic and fulvic acids, which might act as plant growth regulators (PGRs; Atiyeh et al., 2001; Canellas et al., 2000; Chapter 4). Many of the microorganisms that are common in the rhizospheres of plants can produce such plant growth-regulating substances. For instance, Barea et al. (1976) reported that of 50 bacterial isolates obtained from the rhizosphere of various plants, 86% could produce auxins, 58% gibberellins, and 90% kinetin-like substances. There have been many studies on the production of plant growth-regulating substances by mixed microbial populations in soil, but relatively few investigations into the availability of these substances to plants, their persistence and fate in soils, or their effects on plant growth (Arshad and Frankenberger 1993).

Some researchers have shown that PGRs can be taken up by plants from soil in sufficient quantities to influence plant growth. For instance, Kucey (1983) showed that auxins produced in soil by *Azospirillum brasilense* could affect the growth of graminaceous plants. There is increasing evidence that microbially produced gibberellins in soils can influence plant growth and development (Mahmoud et al., 1984; Arshad and Frankenberger, 1993). Increased vigor of seedlings has been attributed to microbial production of cytokinins by *Arthrobacter* and *Bacillus* spp. in soils (Jagnow, 1987).

Because the process of vermicomposting increases microbial diversity and activity dramatically, it is possible that some effects of vermicomposts on plant germination, growth, flowering and yields could be explained by PGHs produced by interactions between microorganisms and earthworms. Gavrilov (1963) first suggested that earthworms might produce PGRs. The presence of plant growth-regulating substances in the tissues of *Aporrectodea caliginosa*, *Lumbricus rubellus*, and *Eisenia fetida* was confirmed by Nielson (1965), who isolated indole substances from earthworms and reported increases in growth of peas by earthworm extracts. He also extracted from *Aporrectodea longa*, *Lumbricus terrestris*, and *Dendrobaena rubidus* a similar substance that stimulated plant growth, but his experiments did not exclude the possibility of PGRs that he found being produced by microorganisms living in the earthworm guts and tissues. Graff and Makeschin (1980) tested the effects of substances produced by *L. terrestris*, *A. caliginosa*, and *E. fetida* on the dry matter production of ryegrass. They added eluates from pots containing earthworms to pots not containing earthworms. They concluded that plant growth-influencing substances (PGIs) were released into the soil by all three species, but did not speculate further on the nature of these substances.

Tomati et al. (1983, 1987, 1988, 1990), Grappelli et al. (1987), and Tomati and Galli (1995) tested vermicomposts produced from organic wastes as media for growing ornamental plants and mushrooms. They concluded that the growth increases that occurred in all of their plant growth experiments were too high to be explained purely on the basis of the nutrient contents of the vermicomposts. Moreover, the growth changes they observed included responses such as stimulation of rooting, dwarfing, time of flowering, and lengthening of internodes. They compared the growth of *Petunia*, *Begonia*, and *Coleus* after adding aqueous vermicompost extracts to that after adding auxins, gibberellins, and cytokinins to soil and concluded that the data suggested that hormonal effects were produced by earthworm activity, and this conclusion was supported by the high levels of cytokinins and auxins they found in the vermicomposts. In laboratory experiments involving large earthworm populations, Krishnamoorthy and Vajrabhiah (1986) found that seven species of earthworms increased dramatically the production of cytokinins and auxins in organic wastes. They also demonstrated significant positive correlations ($r = 0.97$) between earthworm populations and the levels of cytokinins and auxins in 10 field soils and concluded that levels of earthworm activity were linked strongly with hormone production. They reported that auxins and cytokinins produced through earthworm activity could persist in soils for up to 10 weeks, although they degraded in a few hours if exposed to sunlight.

Edwards and Burrows (1988) reported that the increases in growth of 28 ornamentals and vegetables in plant growth media produced by the processing of organic wastes by the earthworm *E. fetida* were much higher than in commercially available plant growth media and was too high to be explained solely by the influence of earthworm activity on plant nutrient quality and availability. They found that the growth of ornamentals was influenced significantly even when the earthworm-processed organic wastes were diluted 20:1 with other suitable materials and the nutrient content balanced that of comparable media. Moreover, growth patterns of the plants, which included leaf development, stem and root elongation, as well as flowering by biennial ornamental plants, in the first season of growth indicated that some biological factor other than nutrients was active. Scott (1988) reported that the growth of hardy ornamentals, *Chaemocyparis lawsonian*, *Elaeagnus pungens, Cuppressocypari leylandii, Pyracantha* spp., *Cotoneaster conspicus,* and *Viburnum bod-nantense* increased significantly after addition of low levels of earthworm-worked organic wastes to the growth media even when nutrients in the two media were balanced.

Even under conditions of adequate mineral nutrition, humic substances can have positive effects on plant growth (Chen and Aviad, 1990; Chapter 4). For instance, applications of humic substances to soils increased the dry matter yields of corn and oat seedlings (Lee and Bartlett, 1976; Albuzio et al., 1994); numbers and lengths of tobacco roots (Mylonas and McCants, 1980); dry weights of shoots, roots, and nodules of soybean, peanut, and clover plants (Tan and Tantiwiramanond, 1983); vegetative growth of chicory plants (Valdrighi et al., 1996); and induced shoot and root formation in tropical crops grown in tissue culture (Goenadi and Sudharama, 1995).

During the past decade, the biological activity of humic substances, particularly those derived from earthworm feces on plant growth, has been investigated extensively (Dell'Agnola and Nardi, 1987; Nardi et al., 1988; Muscolo et al., 1993, 1996, 1999). Dell'Agnola and Nardi (1987) reported hormone-like effects of depolycondensed humic fractions obtained from the feces of the earthworms *Allolobophora rosea* and *A. caliginosa*. Treatment of carrots cells with humic substances obtained from the feces of the earthworm *A. rosea* increased their growth and induced morphological changes similar to those induced by auxins (Muscolo et al., 1999). It seems very likely that vermicomposts, which consist of an amalgam of humified earthworm feces and organic matter, can stimulate plant growth considerably beyond that by mineral nutrients because of the effects of humic substances and hormones present in the vermicomposts.

Humates can be extracted from organic materials such as vermicomposts by a classic acid/alkali fractionation technique (Valdrighi et al., 1996) yielding ca. 4 g humic acids/kg vermicompost (Atiyeh et al., 2002). Typical plant growth responses obtained after treating plants with humic substances are increased plant growth in response to increasing concentrations of humic substances, but usually a decrease in growth in response to higher concentrations of the humic materials (Atiyeh et al., 2000; Figure 11.6). This pattern of growth response closely resembles those from substitutions of increasing vermicompost rates into soilless growth media to tomatoes with all needed nutrients supplied (Atiyeh et al., 2002; Figure 11.4).

Such stimulatory effects on plant growth of humic substances at low concentrations has been explained by various theories, including a direct action on the plants that is hormonal in nature, or an indirect action on the metabolism of soil microorganisms, the dynamics of soil nutrients, iron

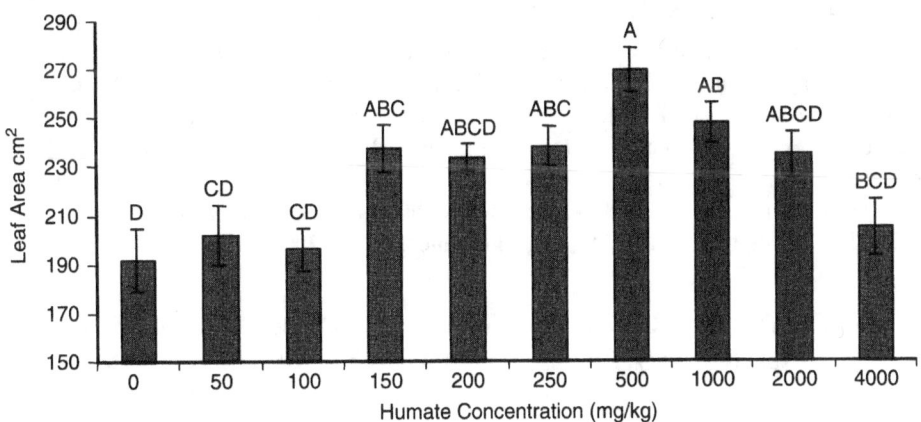

**FIGURE 11.6** Effects of humic acid extracts (mean ± standard error) from pig manure vermicompost applied to a soilless potting medium at different concentrations on tomato leaf area (with all needed nutrients supplied). Columns with same letters are not significantly different at $P = 0.05$. (From Atiyeh et al., 2002. With permission.)

chelation, and soil physical conditions (Cacco and Dell'Agnola, 1984; Nardi et al., 1988; Albuzio et al., 1989; Casenave de Sanfilippo et al., 1990; Chen and Aviad, 1990; Muscolo et al., 1993, 1996, 1999). However, none of these hypotheses provide a completely satisfactory explanation of how humic substances act as PGRs to influence plant growth.

The possibility that hormones might become adsorbed on humic fractions so that any growth response by a PGR is a combined hormone–humic one was given by Edwards (1998) and Atiyeh et al. (2001) and has since been convincingly confirmed by research by Canella et al. (2000), who by structural analysis identified exchangeable auxin groups attached to humic acids in extracts from cattle manure vermicompost. The humic acid–auxin extracts enhanced root elongation, lateral root emergence, and plasma membrane $H^+$-ATPase activity of maize roots. In field experiments Arancon et al. (2002) reported similar increases in growth and yield of tomatoes, peppers and strawberries in response to vermicompost application, and the contribution of nutrients in the significant increases of growth and yield of the field crops was eliminated as a possibility because all treatments were supplemented with inorganic fertilizers to equalize initial nutrient contents of the soil. In a recent work at the OSU, the effects of commercially available humic acids, indole acetic acids (IAAs), and combinations of these materials have been shown to stimulate the effects of humates extracted from vermicomposts.

Vermicomposts contain very rich and diverse microbial populations (Edwards, 1983). Their applications to soils can add diversity and activity to the indigenous soil microorganism populations, thereby resulting in much larger, richer, and diverse soil microbial populations. Some microorganisms such as vesicular arbuscular mycorrhizae can form mutualistic relationships in plant rhizospheres by acting as root extensions; this increases the capacity of plants to utilize soil moisture and nutrients and at the same time benefit from plant root exudates. Other by-products of microbial activities known to promote plant growth include production of antibiotics, microbial disease antagonists, and plant growth-promoting substances such as hormones and humates. Research reports have shown that some microorganisms such as pseudomonads are antagonistic to plant pathogens probably by competing for resources, and by this mechanism microbial competition can induce resistance to plant diseases. Such effects were demonstrated in field experiments at OSU on the suppression of *Verticillium* wilt on strawberries, powdery mildew and *Phomopsis* on grapes in the field, and suppression of *Pythium* and *Rhizoctonia* on cucumbers and radishes in the laboratory (Chaoui et al., 2002).

Bioassays in our laboratory have clearly demonstrated the presence of PGIs in the form of hormones in aqueous extracts from vermicomposts as well as dry insoluble humic acids obtained in base extracts from vermicomposts. These extracts produced significant positive effects on plant growth when applied to soilless plant growth media in relatively small amounts. These positive effects were very similar to those produced by PGH treatments. This can explain why in greenhouse studies (Atiyeh 2000a, 2000b, 2001) vermicomposts added to commercial soilless growth media increased germination, flowering, and fruiting compared with control plant growth media despite all needed nutrients being available in all treatments. The presence of such PGRs in the form of complexes of humates and hormones can help explain the increase in crop growth and yields stimulated by as little as 2.5 t/ha in the field (Arancon et al., 2003a).

## EFFECTS OF VERMICOMPOSTS ON THE INCIDENCE OF PLANT PARASITIC NEMATODES, DISEASES, AND ARTHROPOD PESTS

### INTRODUCTION

Vermicomposts have excellent physicochemical and biological properties, which combine to enhance plant growth, yields, and resistance to diseases. Recent research at OSU and elsewhere has demonstrated significant suppression of plant parasitic nematodes, plant pathogens, and arthropod pests.

## Vermicomposts in Suppression of Plant Parasitic Nematode Population

Extensive scientific literature demonstrates that additions of organic matter to soils sometimes tend to decrease populations of plant parasitic nematodes (Addabdo, 1995; Akhtar and Malik, 2000); a review of 212 scientific papers has discussed the effects of various types of organic amendments on plant parasitic nematode populations (Akhtar, 2000). For instance, many researchers have reported that applications of traditional thermophilic composts can suppress plant parasitic nematode populations (McSorley and Gallaher, 1995; Gutpa, 1997; Miller, 2001). Other studies demonstrated the effectiveness of coffee compost in the control of the nematode *Meloidogyne javanica* on tomatoes (Zambolim et al., 1996), and Chen et al. (2000) demonstrated decreases in production of eggs by the nematode *M. hapla* when brewery compost was added to soils. However, such results were usually obtained in response to large field application rates, which might not be economical for nematode control.

There are relatively few reports of vermicomposts suppressing populations of plant parasitic nematodes. Swathi et al. (1998) demonstrated that 1.0 kg (dry weight)/m$^2$ vermicompost suppressed attacks of *M. incognita* on tobacco plants, and Morra et al. (1998) reported partial control of *M. incognita* by vermicomposts in a tomato–zucchini rotation. Field applications of vermicomposts decreased the number of galls and egg masses of *M. javanica* (Ribeiro et al., 1998). These effects might be due to vermicomposts increasing populations of fungi, including nematode-trapping and cyst parasitic fungi, which are major enemies of plant parasitic nematodes (Kerry, 1988). Vermicompost field applications significantly suppressed plant parasitic nematode populations in soils planted with tomatoes, peppers, strawberries, and grapes (Arancon et al., 2002a, 2002b; Figure 11.7). We speculate that this suppression might be due not only to fungal parasite predators of nematodes but also to increased competition between plant parasitic nematodes and fungivorous and bacterivorous nematodes, resulting from increased availability of food sources for the latter two groups from the vermicompost treatments.

Predator–prey interactions that diminish the populations of plant parasitic nematodes provide another feasible explanation. Vermicomposts probably increase the numbers of omnivorous nematodes or predatory arthropods that prey selectively on plant parasitic nematodes. For instance, according to Bilgrami (1996), the mite *Hypoaspis calcuttaensis* preys voraciously on plant parasitic nematodes, predaceous nematodes, and saprophagous nematodes. Additionally, rhizobacteria can colonize roots and kill plant parasitic nematodes by producing specific enzymes and toxins (Siddiqui and Mahmood, 1999).

## Suppression of Plant Diseases by Vermicomposts

Extensive literature demonstrates the suppression of plant diseases by various organic amendments (Lazarovits et al., 2000; Fikre et al., 2001; Ramamoorthy et al., 2000; Bettiol et al., 1997, 2000; Somasekhara et al., 2000; Rajan and Sarma, 2000; Blok et al., 2000; Shiau et al., 1999; Arafa and Mohamed, 1999; Goudar et al., 1998; Narayanaswamy et al., 1998; Raguchander et al., 1998; Hooda and Srivastava, 1998; Velandia et al., 1998; Dixon et al., 1998; Ehteshamul et al., 1998; Lima et al., 1997; Panneerselvam and Saravanamuthu, 1996; Ara et al., 1996; Karthikeyan and Karunanithi, 1996; Sanudo and Molina Valero, 1995; Dutta and Hegde, 1995; Diyora and Khandar, 1995; Nam et al., 1988; Kannaiyan, 1987; Chapter 5) and traditional thermophilic composts (Huelsman and Edwards, 1998; Goldstein, 1998; Jaworska et al., 1998; Hoitink et al., 1993, 1997) and even by aqueous extracts or vermicompost aqueous extracts or teas (Scheuerell and Mahaffee, 2002). Various mechanisms have been suggested for this pathogen suppression, but most are microbe based, because it has been demonstrated that suppressiveness is lost when the planting media are sterilized (Chaoui et al., 2002). Toyota et al. (1995) suggested two mechanisms for disease suppression by microorganisms. General suppression might be based on a competition for nutrients and energy and involves the total microflora, and this is considered to be the major source of

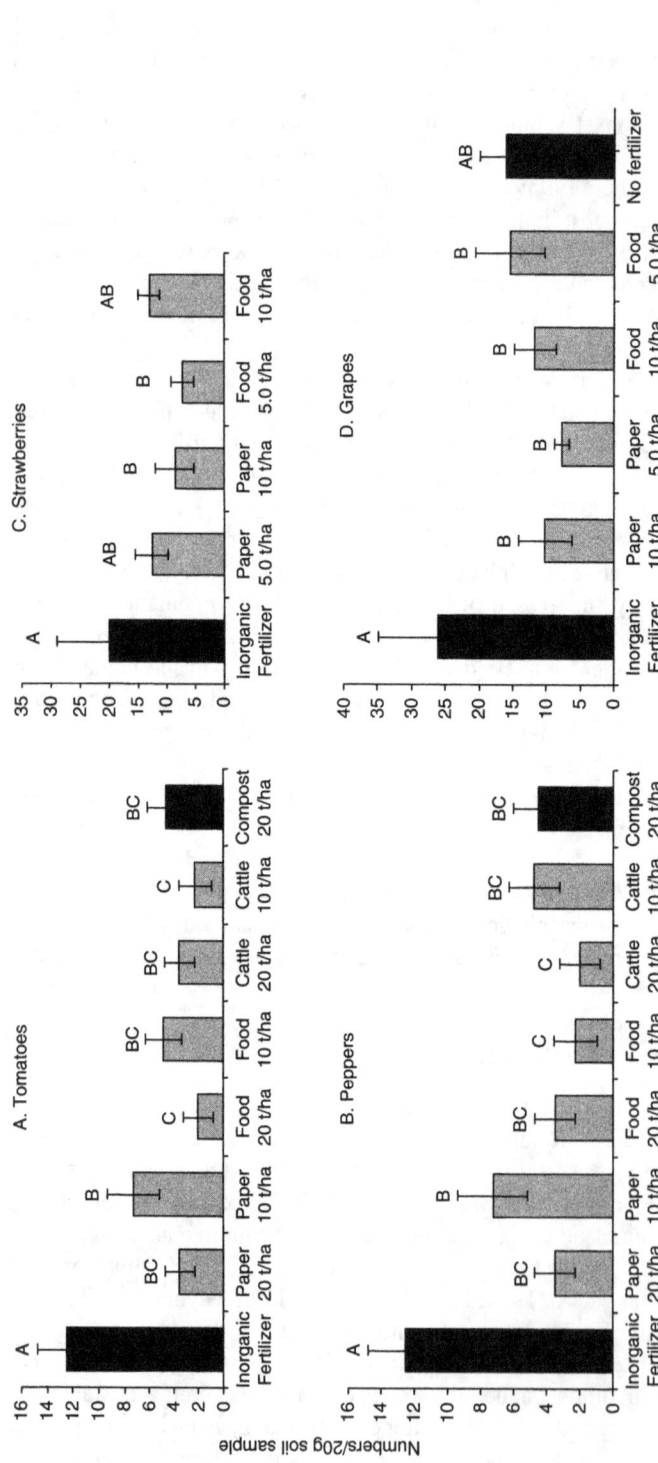

**FIGURE 11.7** Numbers (mean ± standard error) of plant parasitic nematodes in inorganic fertilizer-treated, vermicompost-treated, compost-treated, and unfertilized soils planted with tomatoes (A), peppers (B), strawberries (C), and grapes (D). Columns followed by the same letters are significantly different from $P = 0.05$. (From Arancon et al. 2002. With permission.)

widespread fungistasis in soil. More specific suppression refers to particular antagonistic microorganisms. Toyota et al. (1995) suggested that both mechanisms could occur after vermicompost applications, and because vermicomposts have larger microbial populations and diversity than traditional composts, they might have greater potential to suppress plant pathogens.

The presence of large populations and greater diversity of microorganisms in vermicomposts than in traditional composts has been associated not only with nutrient availability and the production of PGRs in vermicomposts but also with the ability of vermicomposts to suppress plant diseases. Traditional composting is a thermophilic process that kills some microorganisms and promotes specific microbial activity selectively, whereas vermicomposting is a nonthermophilic method and promotes greatly increased activity by a wide range of microorganisms. Considerable research evidence has demonstrated that there is a much greater microbial activity and biodiversity in vermicomposts than in traditional thermophilic composts (Edwards, 1998).

Specific plant pathogens suppressed by traditional thermophilic composts include *Fusarium* (Liping et al., 2001; Cotxarrera et al., 2001; Harender et al., 1997), *Pythium* (Chen et al., 1987), *Gaeumannomyces graminis*, and *Plasmidiophora brassicae* (Pitt et al., 1998) *Phytophthora* (Hoitink and Kuter, 1986; Pitt et al., 1998), and *Rhizoctonia* (Kuter et al., 1983).

Reports on plant pathogen suppression by vermicomposts include suppression of *Plasmodiophora brassicae, Phytophthora nicotianae* (tomato late blight), and *Fusarium lycopersici* (tomato *Fusarium* wilt) by vermicomposts (Nakamura, 1996). Szczech (1993) and Szczech et al. (1999) reported suppression of *Fusarium lycopersici* and *P. nicotianae* on tomatoes by vermicomposts. Rodriguez et al. (2000) demonstrated suppression of fungal diseases caused by *Rhizoctonia solani, Phytophthora drechsleri,* and *F. oxysporum* on gerbera by the incorporating vermicomposts into the growth media. Orlikowski (1999) described a sporulation reduction of the pathogen *P. cryptogea* after treatment with vermicomposts. Studies by Nakasone et al. (1999) showed that aqueous extracts of vermicomposts inhibited mycelial growth of *B. cinerea, Scleretonia sclerotiorum, C. rolfsii, Rhizoctonia solani,* and *F. oxysporum*. At present, many commercial solutions prepared from aqueous extracts of composts are marketed as teas to control plant diseases (Scheuerell and Mahaffee, 2002).

In recent laboratory and greenhouse research at OSU, we have demonstrated significant suppression of *Pythium* and *Rhizoctonia* (Figure 11.8) by substituting low rates (10 to 30%) of vermicompost into horticultural bedding mixtures (Chaoui et al., 2002) and suppression by low rates of application (2.5 t/ha) of vermicomposts on *Verticillium* wilt on strawberries and *Phomopsis* and powdery mildew (*Sphaerotheca fulginae*) on grapes in the field (Figure 11.9; Edwards et al., 2003) compared with grapes grown with recommended rates of inorganic fertilizer. All treatments received the same amounts of total nutrients.

## Suppression of Insect and Mite Attacks by Vermicomposts

Various reports demonstrate that field applications of various types of organic matter, including traditional thermophilic composts, can suppress attacks by insect pests such as aphids and scales (Culliney and Pimentel, 1986; Costello and Altieri, 1995; Yardim and Edwards, 1998; Huelsman et al., 2000; Eigenbrode and Pimentel, 1988). For instance, organic fertilizers (Biradar et al., 1998) suppressed corn insect pests such as aphids (Morales et al., 2001) and European corn borer (Phelan et al., 1996), sucking insect pests (Rao, 2002), brinjal shoot and fruit borer (Sudhakar et al., 1998), and corn insects (Morales et al., 2001). These are discussed in detail in Chapter 7.

More recently, there have been some scattered reports of the suppression of insect pests by vermicomposts. Biradar et al. (1998) reported clear associations between the amounts of vermicomposts in the growth medium in which *Leucaena leucocephala* was grown and the degree of infestation by the psyllid *Heteropsylla cubana*. Rao et al. (2001) reported decreased incidence of the leaf miner *Aproaerema modicella* on peanuts in response to field treatments with vermicomposts. Rao et al. (2001) reported lower overall pest densities of peanuts leaf miner (*A. modicella*) in plots

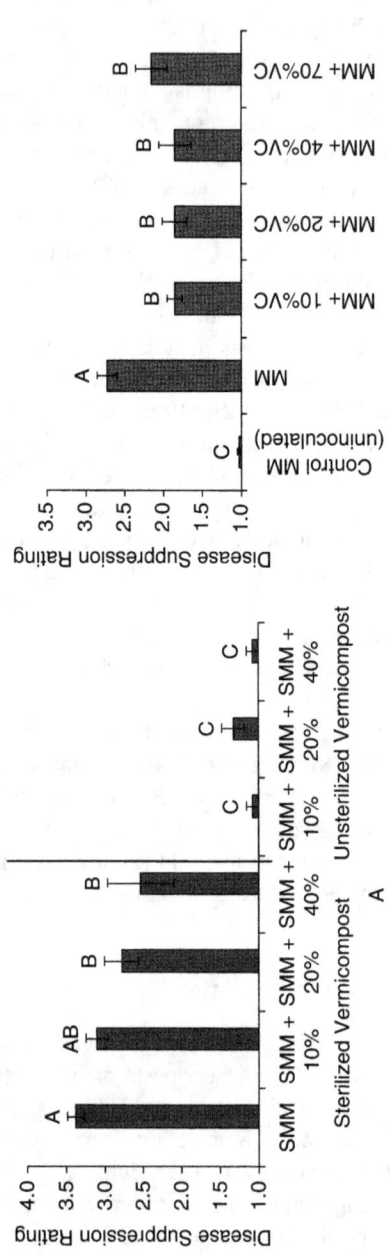

**FIGURE 11.8** (A) *Pythium* symptom suppression rating (mean ± standard error) in cucumber. (B) *Rhizoctonia* symptom suppression rating (mean ± standard error) in radish. Seedlings were planted in a soilless medium MM 360 with varying amounts of vermicompost (VC), inoculated with 1:4000 dilution *Pythium*. SMM is sterilized MM 360. The disease scale is rated 1 (symptomless) to 5 (severe). Columns followed by same letters are not significantly significant at *P* = 0.05. (From Chaoui, H. et al. 2002. *Proc. Brighton Crop Prot. Conf.: Pests Dis.* 8B-3:711–716. With permission.)

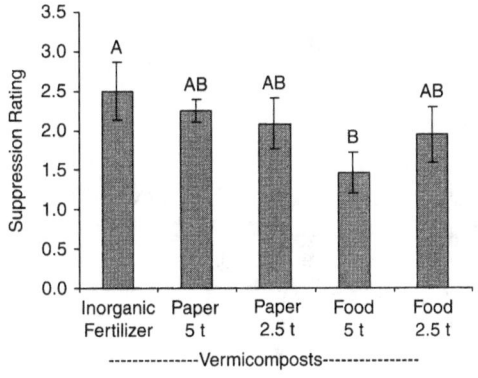 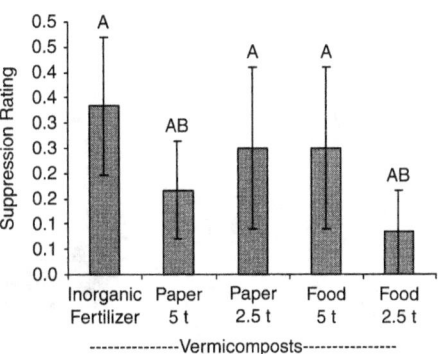

**FIGURE 11.9** (Left) *Phomopsis* symptom suppression rating (mean ± standard error) by paper and food waste vermicomposts applied to field grapes at 2.5 or 5.0 t/ha, compared with that of grapes grown with recommended rates of inorganic fertilizer. (Right) Powdery mildew (*Sphaerotheca fulginae*) symptom suppression rating (mean ± standard error) by paper and food waste vermicomposts applied to grapes at 2.5 or 5.0 t/ha. All treatments received the same amounts of macronutrients. Columns followed by the same letters are not significantly different at $P = 0.05$.

treated with vermicomposts, and Ramesh (2000) reported decreased attacks by sucking pests in response to vermicomposts application. Rao (2002) reported a substantial decrease in attacks by a jassid (*Empoasca verri*) and an aphid (*Aphis craccivora*) and changed predator populations in response to field applications of vermicomposts. Our recent greenhouse experiments demonstrated significant suppression of aphids, mealy bugs (Figure 11.10a), and white cabbage caterpillars by vermicomposts (20 and 40% vermicomposts substitution into 80 and 60% MM 360, respectively) to tomatoes, peppers, and cabbages. There were significant dry weight reductions on tomato, pepper, and cabbage seedlings that did not receive vermicompost treatments after infestations of aphids, mealy bugs (Figure 11.10b), and cabbage white caterpillars compared with those not treated with vermicomposts. Cabbage seedlings grown in 100% MM360 lost more leaf area than those grown in the same media containing 20 or 40% vermicompost after cabbage white caterpillar infestations (Arancon et al., 2003b).

The mechanisms of arthropod pest suppression are not clear, but the availability or balance of nutrient elements can change plant morphology and physiology in ways that might significantly influence plant susceptibility or resistance to pests. Plant characteristics affected by mineral nutrition include growth patterns, onset of senescence, thickness and degree of lignification of the epidermal cells, sugar concentrations in the apoplast, amino-N in phloem sap, and levels of secondary plant compounds (Patriquin et al., 1995). High N levels promoted rapid development of aphids (Dixon, 1969) and resulted in larger lepidopteran larvae (House, 1965). Some studies have shown that total plant N can be correlated positively with the extent of pest insect feeding and growth (Fox and Macauley, 1977). However, inadequate N nutrition can also enhance consumption by insects of plant fluids and tissues and thus promote even more damage to crop plants (Hamilton and Moran, 1980). Products of N metabolism, such as amino acids, have also been correlated with increased insect pest attacks on plants. For instance, high levels of amino acids stimulated growth and fecundity of several herbivorous insects (Prestidge and McNeill, 1983). Adequate levels of N at certain growth stages of a crop might be critical in influencing the extent of insect pest incidence and damage. Variations in the nutritional status of plant tissues other than N, such as P or Mg, might induce resistance or increased susceptibility to insect attacks due to deficiencies of K, Ca, Bo, Zn, and Si (Patriquin, 1995). Based on these data, it seems probable that a balanced nutrition for plants with maximum pest resistance could be achieved by amending vermicomposts into planting media or field soils.

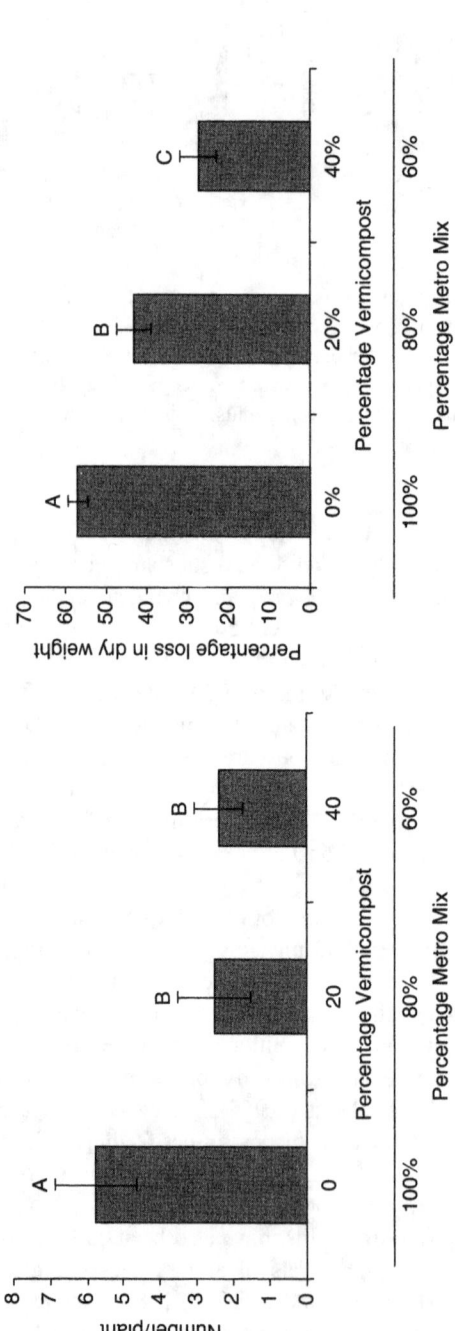

**FIGURE 11.10** (Left) Mealy bug infestations (mean ± standard error) on tomatoes with different amounts of vermicomposts into a soilless medium (MM 360). (Right) Dry weight reductions (mean ± standard error) due to mealy bugs on tomatoes substituted with different amounts of vermicomposts added to a soilless medium (MM 360). Columns with the same letters are not significantly different at $P = 0.05$.

Interaction between vermicomposts and attacks of crops by arthropods, pathogen, and nematodes is relatively new research area, and there is an urgent need to evaluate the effects of such interactions on pest incidence on a range of valuable crops. This would establish critical greenhouse or vermicompost field application rates needed to provide effective suppression as well as a better understanding of the mechanisms involved in this suppression.

## REFERENCES

Addabdo, T.D. (1995) The nematicidal effect of organic amendments: a review of the literature 1982–1994. *Nematologia Mediterranea* 23:299–305.

Adu, J.K., and Oades, J.M. (1978) Utilization of organic materials in soil aggregates by bacteria and fungi. *Soil Biol. Biochem.* 10:117–122.

Aichberger, R. von (1914) Studies on the nutrition of earthworms. *Kleinwelt* 6:53–58, 69–72, 85–88.

Akhtar, M. (2000) Approaches to biological control of nematode pests by natural products and enemies. *J. Crop Prod.* 3: 367–395.

Akhtar, M., and Malik, A. (2000) Role of organic amendments and soil organisms in the biological control of plant parasitic nematodes: a review. *Bioresour. Technol.* 74:35–47.

Albanell, E., Plaixats, J., and Cabrero, T. (1988) Chemical changes during vermicomposting (*Eisenia fetida*) of sheep manure mixed with cotton industrial wastes. *Biol. Fertil. Soils* 6:266–269.

Albuzio, A., Nardi, S., and Gulli, A. (1989) Plant growth regulator activity of small molecular size humic fractions. *Sci. Total Environ.* 347:199–207.

Albuzio, A., Concheri, G., Nardi, S., and Dell'Agnola, G. (1994) Effect of humic fractions of different molecular size on the development of oat seedlings grown in varied nutritional conditions. In Senesi, N., and T.M. Miano (Eds.), *Humic Substances in the Global Environment and Implications on Human Health*. Elsevier Science, Amsterdam, pp. 199–204.

Aldag, R., and Graff, O. (1975) N-Fraktionen in Regenwurmlösung und deren Ursprungsboden. *Pedobiologia* 15:151–153.

Anderson, J.M., Proctor, J., and Vallack, H.W. (1983) Ecological studies in four contrasting lowland rain forests in Gunung Mulu National Park, Sarawak. III. Decomposition processes and nutrient loss from leaf litter. *J. Ecol.* 71:503–527.

Appelhot, M. (Ed.) (1981) *Workshop on the Role of Earthworms in the Stabilization of Organic Residues.* Vol. 1. Beech Leaf Press, Kalamazoo, MI.

Ara, J., Ehteshamul, H.S., Sultana, V., Qasim, R., and Ghaffar, F. (1996) Effect of Sargassum seaweed and microbial antagonists in the control of root rot disease of sunflower. *Pak. J. Bot.* 28(2):219–223.

Arafa, M.K., and Mohamed, E.I. (1999) Soybean seed borne fungi and their control. 2. Effect of soil amendments on the incidence of *Fusarium* root rot and chlamydospores germination. *Egypt. J. Agric. Res.* 77:97–111.

Arancon, N.Q., Edwards, C.A., and Lee, S. (2002) Management of plant parasitic nematode populations by use of vermicomposts. *Proc. Brighton Crop Prot. Conf.: Pests Dis.* 8B-2:705–716.

Arancon, N.Q., Edwards, C.A., Bierman, P., Metzger, J., Lee, S., and Welch, C. (2003a) Applications of vermicomposts to tomatoes and peppers grown in the field and strawberries grown unger high plastic tunnels. In *Proceedings of the International Earthworm Symposium*, Cardiff, Wales, September 2002.

Arancon, N.Q., Galvis, P.A., Edwards C.A., and Yardim, E. (2003b) The trophic diversity of nematode communities in soils treated with vermicomposts. *Pedobiologia* 47:736–740.

Arancon, N.Q., Atiyeh, R.M., Edwards, C.A., and Metzger, J.D. (2004a) Effects of vermicomposts produced from food waste on greenhouse peppers. *Bioresour. Technol.* 93:139–143.

Arancon, N.Q., Edwards, C.A., Bierman, P., Welch, C., and Metzger, J.D. (2004b) The influence of vermicompost applications to strawberries: Part 1. Effects on growth and yield. *Bioresour. Technol.* 93:145–153.

Arancon, N.Q., Edwards, C.A., and Bierman, P. (in press) The influence of vermicompost applications to strawberries. Part 2. Changes in soil microbiological, chemical and physical properties. *Bioresour. Technol.*

Arshad, M., and Frankenberger W.T., Jr. (1993) Microbial production of plant growth regulators. In Metting, F.B., Jr. (Ed.), *Soil Microbial Ecology: Applications in Agricultural and Environmental Management.* Marcel Dekker, New York, p. 307.

Athani, S.I., Hulamanai, N.C., and Shirol, A.M. (1999) Effect of vermicomposts on the maturity and yield of banana. *S. Indian Hortic.* 47(1–6):4–7.

Atiyeh, R.M., Dominguez, J., Subler S., and Edwards, C.A. (2000a) Changes in biochemical properties of cow manure processed by earthworms (*Eisenia andreii*) and their effects on plant growth. *Pedobiologia* 44:709–724.

Atiyeh, R.M., Edwards, C.A, Subler, S., and Metzger, J.D. (2000b) Earthworm processed organic wastes as components of horticultural potting media for growing marigolds and vegetable seedlings. *Compost Sci. Util.* 8:215–233.

Atiyeh, R.M., Arancon, N.Q., Edwards, C.A., and Metzger, J.D. (2000c) Influence of earthworm-processed pig manure on the growth and yield of greenhouse tomatoes. *Bioresour. Technol.* 75:175–180.

Atiyeh, R.M., Arancon, N.Q., Edwards, C.A., and Metzger, J.D. (2001a) The influence of earthworm-processed pig manure on the growth and productivity of marigolds. *Bioresour. Technol.* 81:103–108.

Atiyeh, R.M., Edwards, C.A., Subler, S., and Metzger, J.D. (2001b) Pig manure vermicomposts as a component of a horticultural bedding plant medium: effects on physicochemical properties and plant growth. *Bioresour. Technol.* 78:11–20.

Atiyeh, R.M., Lee, S., Edwards, C.A., Arancon, N.Q., and Metzger, J.D. (2002) The influence of humic acids derived from earthworms-processed organic wastes on plant growth. *Bioresour. Technol.* 84:7–14.

Atlavinyte, O. (1975) *Ecology of Earthworms and Their Effect on the Fertility of Soils in the Lithuanian SSR.* Mokslas Publishers, Vilnius.

Atlavinyte, O., and Pociene, C. (1973) The effect of earthworms and their activity on the amount of algae in the soil. *Pedobiologia* 13:445–455.

Barea, J.M., Navarro, E., and Montana, E. (1976) Production of plant growth regulators by rhizosphere phosphate-solubilizing bacteria. *J. Appl. Bacteriol.* 40:129–134.

Barley, K.P. (1961) The abundance of earthworms in agricultural land and their possible significance in agriculture. *Adv. Agron.* 13:249–269.

Barley, K.P., and Kleinig, C.R. (1964) The occupation of newly irrigated lands by earthworms. *Aust. J. Sci.* 26:290.

Barois, I. (1992) Mucus production and microbial activity in the gut of two species of *Amynthus* (Megascolecidae) from cold and warm tropical climates. *Soil Biol. Biochem.* 24:1507–1510.

Barois, I., and Lavelle, P. (1986) Changes in respiration rate and some physicochemical properties of a tropical soil during transit through *Pontoscolex corethrurus* (Glossoscolecidae, Oligochaeta). *Soil Biol. Biochem.* 18:539–541.

Barois, I., Verdier, B., and Kaiser, P. (1987) Influence of the tropical earthworm *Pontoscolex corethrurus* (Glossoscolecidae) on the fixation and mineralization of nitrogen. In Bonvicini-Pagliai, A.M., and Omodeo, P. (Eds.), *On Earthworms.* Mucchi Editore, Modena, pp. 151–158.

Bassalik, K. (1913) On silicate decomposition by soil bacteria. *Z. Gärungs-physiol.* 2:1–32.

Baweja, K.D. (1939) Studies of the soil fauna with special reference to the recolonisation of sterilised soil. *J. Anim. Ecol.* 8:120–161.

Bernier, N., and Ponge, J.F. (1994) Humus form dynamics during the sylvogenetic cycle in a mountain spruce forest. *Soil Biol. Biochem.* 26:183–220.

Bettiol, W., Migheli, Q., and Garibaldi, A. (1997) Control, with organic matter, of cucumber damping off caused by *Pythium ultimum* Trow. *Pesquisa Agropecuaria Brasileira* 32(1):57–61.

Bettiol, W., Migheli, Q., and Garibaldi, A. (2000) Control of *Pythium* damping-off of cucumber with composted cattle manure. *Fitopatologia Brasileira* 1:84–87.

Beyer, L., Blume, H.P., and Irmler, U. (1991) The humus of "parabraunerde" (Orthic Luvisol) under *Fagus sylvatica* L., and *Quercus robur* L., and its modification in 25 years. *Ann. Sci. Forest.* 48:267–278.

Bhandari, G.S., Randhawa, N.S., and Maskin, M.S. (1967) On the polysaccharide content of earthworm casts. *Curr. Sci. (Bangalore)* 36:519–520.

Bhatnagar, T. (1975) Lombriciens et humification: Un apect nouveau de l'incorporation microbienne d'azote induite par les vers de terre. In Kilbertus, G. Reisinger, O., Mourey, A., and Cancela de Fonseca, J.A. (Eds.), *Biodegredation et Humification.* Pierron, Sarreguemines, pp. 169–182.

Bilgrami, A.L. (1996) Evaluation of the predation abilities of the mite *Hypoaspis calcuttaensis*, predaceous on plant and soil nematodes. *Fundam. Appl. Nematol.* 20:96–98.

Biradar, A.P., Sunita, N.D., Teggelli, R.G., and S.B. Devaranavaddgi. 1998. Effect of vermicompost on the incidence of subabul psyllid. *Insect Environ.* 4:55–56.

Blair, J.M., Parmelee, R.W., and Lavelle, P. (1994) Influences of earthworms on biogeochemistry. In Hendrix, P.F. (Ed.), *Earthworm Ecology and Biogeography in North America*. Lewis Publishers, Chelsea, pp. 127–158.

Blair, J.M., Allen, M.F., Parmelee, R.W. et al. (1995) Changes in soil N pools in response to earthworm population manipulations under different agroecosystem treatments. *Soil Biol. Biochem.* 29:361–367.

Blok, W.J., Lamers, J.G., Termoshuizen, A.J., and Bollen, G.J. (2000) Control of soil-borne plant pathogens by incorporating fresh organic amendments followed by tarping. *Phytopathology* 90:253–259.

Bohlen, P.J., Edwards, W.M., and Edwards, C.A. (1995) Earthworm community structure and diversity in experimental agricultural watersheds in northeastern Ohio. *Plant Soil* 170:233–239.

Bolton, P.J., and Phillipson, J. (1976) Burrowing, feeding, egestion and energy budgets of *Allolobophora rosea* (Savigny) (Lumbricidae). *Oecologia* 23:225–245.

Bouché, M.B. (1971) Relations entre les structures spatiales et fonctionelles des ecosystems illustrees par le rôle pedobiologique des vers de terre. In Desson, P. (Ed.), *La Vie dans les Sols, Aspects Nouveaux, Etudes Experimentiales*. Gauthier-Villars, Paris, France, pp. 187–209.

Brüsewitz, G. (1959) Untersuchungen über den Einfluss des Regenwurms auf Zahl und Leistungen von Mikroorganismen im Boden. *Arch. Microbiol.* 33:52–82.

Buckerfield, J.C., and Webster, K.A. (1998) Worm-worked waste boosts grape yields: prospects for vermicompost use in vineyards. *Aust. N.Z. Wine Ind. J.* 13:73–76.

Buckerfield, J.C., Flavel, T., Lee, K.E., and Webster, K.A. (1999) Vermicomposts in solid and liquid form as plant-growth promoter. *Pedobiolgia* 43:753–759.

Businelli, M., Perucci, P., Patumi, M., and Giusquiani, P.L. (1984) Chemical composition and enzymic activity of some worm casts. *Plant Soil* 80:417–422

Cacco, G., and Dell'Agnola, G. (1984) Plant growth regulator activity of soluble humic complexes. *Can. J. Soil Sci.* 64:225–228.

Canellas, L.P., Olivares, F.L., Okorokova, A.L., Facanha, A.R. (2000) Humic acids isolated from earthworm compost enhance root elongation, lateral root emergence, and plasma H+-ATPase activity in maize roots. *Plant Physiol.* 130:1951–1957.

Casenave de Sanfilippo, E., Arguello, J.A., Abdala, G., and Orioli, G.A. (1990) Content of auxin-, inhibitor- and gibberellin-like substances in humic acids. *Biologia Plantarum* 32:346–351.

Cavender, N.D., Atiyeh, R.M., and Knee, M. (2003) Vermicompost stimulates mycorrhizal colonization of roots of *Sorghum bicolor* at the expense of plant growth. *Pedobiologia* 47:85–90.

Chan, P.L.S., and Griffiths, D.A. (1988) The vermicomposting of pre-treated pig manure. *Biol. Wastes* 24:57–69.

Chaoui, H., Edwards, C.A., Brickner, A., Lee, S., Arancon, N.Q. (2002) Suppression of the plant parasitic diseases: *Pythium* (damping off), *Rhizoctonia* (root rot) and *Verticillium* (wilt) by vermicompost. *Proc. Brighton Crop Prot. Conf.: Pests Dis.* 8B-3:711–716.

Chen, J., Abawi, G.S., and Zuckerman, B.M. (2000) Efficacy of *Bacillus thuringiensis, Paecilomyces marquandii*, and *Streptomyces costaricanus* with and without organic amendment against *Meloidogyne hapla* infecting lettuce. *J. Nematol.* 32:70–77.

Chen, W., Hoitink, H.A., Schmitthenner, A.F., and Touvinen, O.H. (1987) The role of microbial activity in suppression of damping off caused by *Pythium ultimum*. *Phytopathology* 78:314–322.

Chen, Y., and T. Aviad. (1990) Effects of humic substances on plant growth. In MacCarthy, P., Clapp, C.E., Malcolm, R.L., and Bloom, P.R. (Eds.), *Humic Substances in Soil and Crop Sciences: Selected Readings*. American Society of Agronomy and Soil Science Society of America, Madison, WI, pp. 161–186.

Christensen, O. (1988) The direct effect of earthworms on nitrogen turnover in cultivated soils. *Ecol. Bull.* 39:41–44.

Clements, R.O., Murray, P.J., and Sturdy, R.G. (1991) The impact of 20 years' absence of earthworms and three levels of nitrogen fertilizer on a grassland soil environment. *Agric. Ecosyst. Environ.* 36:75–86.

Contreras, E. (1980) Studies on the intestinal actinomycete flora of *Eisenia lucens* (Annelida, Oligochaeta). *Pedobiologia* 20:411–416.

Cooke, A. (1983) The effects of fungi on food selection by *Lumbricus terrestris* L. In Satchell, J.E. (Ed.), *Earthworm Ecology.* Chapman & Hall, London, pp. 365–373.

Cooke, A., and Luxton, M. (1980) Effect of microbes on food selection by *Lumbricus terrestris. Rev. Ecol. Biol. Sol.* 17:365–370.

Costello, M.J., and Altieri, M.A. (1995) Abundance, growth rate and parasitism of *Brevicoryne brassicae* and *Myzus persicae* (Homoptera: Aphididae) on broccoli grown in living mulches. *Agric. Ecosyst. Environ.* 52:187–196.

Cotxarrera, L., Trillas-Gay, M.I., Steinberg, C., and Alabouvette, C. (2001) Use of sewage sludge compost and *Trichoderma asperellum* isolates to suppress *Fusarium* wilt of tomato. *Soil Biol. Biochem.* 34:467–476.

Culliney, T.W., and Pimentel, D. (1986) Ecological effects of organic agricultural practices on insect populations. *Agric. Ecosyst. Environ.* 15:253–266.

Curry, J.P., and Bolger, T. (1984) Growth, reproduction and litter and soil consumption by *Lumbricus terrestris* L. in reclaimed peat. *Soil Biol. Biochem.* 16:253–257.

Curry, J.P., and Byrne, D, (1992) The role of earthworms in straw decomposition and nitrogen turnover in arable land in Ireland. *Soil Biol. Biochem.* 24:1409–1412.

Czerwinski, Z., Jakubczyk, H., and Nowak, E. (1974) Analysis of a sheep pasture ecosystem in the Pieniny Mountains (The Carpathians). XII. The effect of earthworms on the pasture soil. *Ekol. Pol.* 22: 635–650.

Daniel, O. (1991) Leaf-litter consumption and assimilation by juveniles of *Lumbricus terrestris* L. (Oligochaeta, Lumbricidae) under different environmental conditions, *Biol. Fertil. Soils* 12:202–208.

Daniel, O, and Anderson, J.M. (1992) Microbial biomass and activity in contrasting soil material after passage through the gut of the earthworm *Lumbricus rubellus* Hoffmeister. *Soil Biol. Biochem.* 24:465–470.

Darwin, C. (1881) *The Formation of Vegetable Mould through the Action of Worms, with Observations of Their Habits.* Murray, London.

Dash, M.C., Binapani, S., Behera, N., and Dei, C. (1984) Gut load and turnover of soil, plant and fungal material by *Drawida calebi,* a tropical earthworm. *Rev. Ecol. Biol.* 21:387–393.

Dash, M.C., Mishra, P.C., and Behera, N. (1979) Fungal feeding by a tropical earthworm. *Trop. Ecol.* 20:9–12.

Dawson, R.C. (1947) Earthworm microbiology and the formation of water-stable aggregates. *Soil Sci.* 69:175–184.

Day, G.M. (1950) The influence of earthworms on soil microorganisms. *Soil Sci.* 69:175–184.

Dell'Agnola, G., and Nardi, S. (1987) Hormone-like effect and enhanced nitrate uptake induced by depolycondensed humic fractions obtained from *Allolobophora rosea* and *A. caliginosa* feces. *Biol. Fertil. Soils* 4:115–118.

Desai, V.R., Sabale, R.N., and Raundal, P.U. (2000) Integrated nitrogen management in wheat-coriander cropping system. *J. Maharashtra Agric. Univ.* 24(3):273–275.

Devi, D., Agarwal, S.K., and Dayal, D. (1998) Response of sunflower (*Helianthus annuus*) to organic manures and fertilizers. *Indian J. Agron.* 43(3):469–473.

Dickschen, F., and Topp, W. (1987) Feeding activities and assimilation efficiencies of *Lumbricus rubellus* (Lumbricidae) on a plant-only diet. *Pedobiologia* 30:31–37.

Dixon, A.F.G. (1969) Quality and availability of food for a Sycamore aphid population. In Watson, A. (Ed.), *Animal Populations in Relation to the Food Sources.* Blackwell Scientific, Oxford.

Dixon, G.R., Walsh, U.F., and Szmidt, R.A. (1998) Suppression of plant pathogens by organic extracts a review. *Acta Horticulturae* 469:383–390.

Diyora, P.K., and Khandar, R.R. (1995) Management of wilt of cumin (*Cuminum cyminum* L.) by organic amendments. *J. Spices Arom. Crops* 4(1):80–81.

Dkhar, M.C., and Mishra, R.R. (1986) Microflora in earthworm casts. *J. Soil Biol. Ecol.* 6:24–31.

Domsche, K.M., and Banse, H.J. (1972) Mykologische Untersuchungen an Regenwurm Exkrementen. *Soil Biol. Biochem.* 4:31–38.

Doube, B.M., Stevens, P.M., Davoren, C.W., and Ryder, M.H. (1994) Earthworms and the introduction and management of beneficial soil microorganisms. In Pankhurst, C.E, Doube, B.M., Gupta, U.V.S.R., and Grace, P.R. (Eds.), *Soil Biota: Management in Sustainable Farming Systems.* CSIRO, Melbourne, pp. 32–41.

Dutta, P.K., and Hegde, R.K. (1995) Effect of organic amendments on the suppression of *Phytophthora palmivora* Butler causing black pepper wilt. *Plant Hlth.* 1:56–60.

Edwards, C.A. (1983) Utilization of earthworm composts as plant growth media. In Tomati, U. and Grappelli, A. (Eds.), *International Symposium on Agricultural and Environmental Prospects in Earthworms.* Rome, Italy, pp. 57–62.

Edwards, C.A. (1988) Breakdown of animal, vegetable and industrial organic wastes by earthworms. In Edwards, C.A., and Neuhauser, E.F. (Eds.), *Earthworms: Waste and Environmental Management.* SPB Academic Publishers, The Hague, the Netherlands, 21–31.

Edwards, C.A. (Ed.) (1998) *Earthworm Ecology.* CRC Press, Boca Raton, FL, 389 pp.

Edwards, C.A., and Bohlen, P.J. (1996) *Biology and Ecology of Earthworms*, 3rd ed. Chapman & Hall, London, 426 pp.

Edwards, C.A., and Burrows, I. (1988) The potential of earthworm composts as plant growth media. In Neuhauser, C.A. (Ed.), *Earthworms in Environmental and Waste Management.* SPB Academic Publishing, The Hague, the Netherlands, pp. 211–220.

Edwards, C.A., and Fletcher, K.E. (1988) Interactions between earthworms and microorganisms in organic-matter breakdown. *Agric. Ecosyst. Environ.* 24:235–247.

Edwards, C.A., and Heath, G.W. (1963) The role of soil animals in breakdown of leaf material. In Doeksen, J., and van der Drift, J. (Eds.), *Soil Organisms.* North Holland, Amsterdam, 76–80.

Edwards, C.A., and Neuhauser, E.F. (1988) *Earthworms in Waste and Environmental Management.* SPB Academic Publishing, The Hague, the Netherlands.

Edwards, C.A., Bohlen, P.J., Linden, D.R., and Subler, S. (1995) Earthworms in agroecosystems. In Hendrix, P.F. (Ed.), *Earthworm Ecology and Biogeography in North America.* Lewis Publishers, Boca Raton, FL, pp. 185–213.

Ehteshamul, H.S., Zaki, M.J., Vahidy, A.A., Abdul, G., and Ghaffar, A. (1998) Effect of organic amendments on the efficacy of *Pseudomona aeruginosa* in the control of root rot disease of sunflower. *Pak. J. Bot.* 30(1):45–50.

Eigenbrode, S.D., and Pimentel, D. (1988) Effects of manure and chemical fertilizers on plant quality and insect pest populations on collards. *Agric. Ecosyst. Environ.* 20:109–125.

Elliot, P.W., Knight, D., and Anderson, J.M. (1990) Denitrification in earthworm casts and soil from pasture under different fertilizer and drainage regimes. *Soil Biol. Biochem.* 22:601–605.

Elvira, C., Goicoechea, M., Sampedro, L., Mato, S., and Nogales, R. (1996) Bioconversion of solid paper-pulp mill sludge earthworms. *Bioresour. Technol.* 57:173–177.

Ferrière, G. (1980) Fonctions des lombriciens. VII. Une méthode d'analyse de la matière organique végétale ingérée. *Pedobiologia* 20:263–273.

Ferrière, G., and Bouché, M.B. (1985) Première mesure écophysiologique d'un débit d'azote de *Nicodrilus longus* (Ude) (Lumbricidae Oligochaeta) dans la prairie de Citeaux. *CR Acad. Sci.* 301:789–794.

Fikre, H., Sandhu, K.S., and Singh, P.P. 2001. Management of white rot pea through organic amendments and fungicides. *Plant Dis. Res.* 16(2):193.

Flack, F., and Hartenstein, R. (1984) Growth of the earthworm *Eisenia foetida* on microorganisms and cellulose. *Soil Biol. Biochem.* 16:491–495.

Fox, L.R., and Macauley B.J. (1977) Insect grazing on eucalyptus in response to variation in leaf tannins and nitrogen. *Oecologia* 29:145–162.

Frankenberger, W.T., Jr., and Arshad, M. (1995) *Phytohormones in Soils: Microbial Production and Function.* Marcel Dekker, New York.

Gangadhavan, J.D., and Gopinath, G. (2000) Effect of organic and inorganic fertilizers on growth, flowering, and quality of gladiolus cv. white prosperity. *Karnataka J. Agric. Sci.* 13(2):401–405.

Gange, A. (1993) Translocation of mycorrhizal fungi by earthworms during early succession. *Soil Biol. Biochem.* 25:1021–1026.

Garcia, C., Ceccanti, B., Masciandaro, G., and Hernandez, T. (1995) Phosphatase and β-glucosidase activities in humic substances from animal wastes. *Bioresour. Technol.* 53:79–87.

Gavrilov, K. (1963) Earthworms, producers of biologically active substances. *Zh. Obshch. Biol.* 24:149–154.

Ghabbour, S.L. (1966) Earthworms in agriculture: A modern evaluation. *Rev. Ecol. Biol. Soc.* 111(2):259–271.

Ghilarov, M.S. (1963) On the interrelations between soil dwelling invertebrates and soil microorganisms. In Doeksen, J., and van der Drift, J. (Eds.), *Soil Organisms.* North Holland, Amsterdam, pp. 255–259.

Goenadi, D.H., and Sudharama, I.M. (1995) Shoot initiation by humic acids of selected tropical crops grown in tissue culture. *Plant Cell Rep.* 15:59–62.

Goldstein, J. (1998) Compost suppresses disease in the lab and on the fields. *BioCycle* 39(11):62–64.

Goudar, S.B., Srikant, K., and Kulkarni, S. (1998) Effect of organic amendments on *Fusarium udum*, the causal agent of wilt pigeon pea. *Karnataka J. Agric. Sci.* 11:690–692.

Graff, O. (1971) Stikstoff, Phosphor und Kalium in der Regenwurmlösung aut der Wiesenversuchsfläche des Sollingprojektes. *Ann. Zool. Ecol. Anim. Spec. Publ.* 4:503–512.

Graff, O., and Makeschin, F. (1980) Beeinflussung des Ertrags von Weidelgras (*Lolium multiflorum*) durch Ausscheidungen von Regenwurmen dreier verschiedener Arten. *Pedobiologia* 20:176–180.

Grappelli, A., Galli, E., and Tomati, U. (1987) Earthworm casting effect on *Agaricus bisporus* fructification. *Agrochimica* 21:457–462.

Gutpa, M.C., and Kumar, S. (1997) Efficacy of certain organic amendments and nematicides against *Tylenchorhyncus* spp. and *Helicotylenchus* spp. in soil. *Indian J. Nematol.* 27:139–142.

Haimi, J., and Huhta, V. (1990) Effects of earthworms on decomposition processes in raw humus forest soil: a microcosm study. *Biol. Fertil. Soils* 10:178–183.

Hameed, R., Bouché, M.B., and Cortez, J. (1994) Etudes *in situ* des transferts d'azote d'otigine lombricienne (*Lumbricus terrestris* L.) vers les plantes. *Soil Biol. Biochem.* 26:495–501.

Hamilton, W.D., and Moran, N. (1980) Low nutritive quality as a defense against herbivores. *J. Theor. Biol.* 86:247–254.

Hand, P., and Hayes, W.A. (1988) The vermicomposting of cow slurry. In Edwards, C.A., and Neuhauser, E.F. (Eds.), *Earthworms in Waste and Environmental Management*. SPB Academic Publishing, The Hague, the Netherlands, pp. 49–67.

Harender, R., Kapoor, I.J., and Raj, H. (1997) Possible management of *Fusarium* wilt of tomato by soil amendments with composts. *Indian Phytopathol.* 50:387–395.

Harinikumar, K.M., and Bagyaraj, D.J. (1994) Potential of earthworms, ants, millipedes, and termites for dissemination of vesicular-arbuscular mycorrhizal fungi in soil. *Biol. Fertil. Soils* 18:115–118.

Harinikumar, K.M., Bagyaraj, D.J., and Kale, R.D. (1991) Vesicular arbuscular mycorrhizal propagules in earthworm casts. In Veeresh, G.K., Rajagopal, D., and Viraktamath, C.A. (Eds.), *Advances in Management and Conservation of Soil Fauna*. Oxford & IBH, New Delhi, pp. 605–610.

Harmsen, G., and van Schreven, D. (1955) Mineralisation of organic nitrogen in soil. *Adv. Agron.* 7:299–398.

Hartenstein, R. (1982) Soil macroinvertebrates, aldehyde oxidase, catalase, cellulase and peroxidase. *Soil Biol. Biochem.* 15:51–54.

Heath, G.W., Arnold, M.K., and Edwards, C.A. (1966) Studies in leaf litter breakdown. I. Breakdown rates among leaves of different species. *Pedobiologia* 6:1–12.

Hendriksen, N.B. (1991) Gut load and food-retention time in the earthworms *Lumbricus festivus* and *L. castaneus*: A field study. *Biol. Fertil. Soils* 11:170–173.

Hendrix, P.F. (Ed.) (1995) *Earthworm Ecology and Biogeography in North America*. Lewis Publishers, Boca Raton, FL, 244 pp.

Hendrix, P.F., Crossley, D.A., Jr., and Coleman, D.C. (1987) Carbon dynamics in soil microbes and fauna in conventional and no-till ecosystems. *INTECOL Bull.* 15:590–663.

Hirst, J.M., Storey, I.F., Ward, W.C., and Wilcox, H.G. (1955) The origin of apple scab epidemics in the Wisbech area in 1953 and 1954. *Plant Pathol.* 4:91.

Hoffman, J.A., and Purdy, L.H. (1964) Germination of dwarf bunt (*Tilletia controversa*) teliospores after ingestion by earthworms. *Phytopathology* 54:878–879.

Hoitink, H.A., and Kuter, G.A. (1986) Effects of composts in growth media on soil-borne pathogens. In Chen, Y., and Avnimelech, Y. (Eds.), *The Role of Organic Matter in Modern Agriculture*. Martinus Nijhoff, Dordrecht, pp. 289–306.

Hoitink, H.A., Boehm, M.J., and Hadar, Y. (1993) Mechanisms of suppression of soilborne plant pathogens in compost-amended substrates. In Hoitink, A.J., and Keener, H. (Eds.), *Science and Engineering of Composting: Design, Environmental, Microbiological and Utilization Aspects*. Renaissance Publications, Worthington, OH, pp. 601–621.

Hoitink, H.A., Stone, A.G., and Han, D.Y. (1997) Suppression of plant diseases by compost. *HortScience* 32:184–187.

Hooda, K.S., and Srivastava, M.P. (1998) Impact of neem coated urea and potash on the incidence of rice blast. *Plant Dis. Res.* 13:28–34.

House, H.L. (1965) Effects of low nutrient content of food, and nutrient imbalance on the feeding and nutrition of phytophagous larva *Celerio euphobiae L.* (Lep. Sphingidae). *Can. Entomol.* 97:62–68.

Huelsman, M.F., and Edwards, C.A. (1998) Management of disease in cucumbers (*Cucumis sativus*) and peppers (*Capsicum annum*) by using composts. *BCPC Conf.: Pests Dis.* 8d-15:881–886.

Huelsman, M.F., Edwards, C.A., Lawrence, J., and Clarke, H. (2000) A study of the effect of soil nutrient levels on the incidence of insect pests and predators in Jamaican sweet potato (*Ipomoea batatas*) and callaloo (*Amaranthus*). *BCPC Conf.: Pests Dis.* 3:895–900.

Huss, M.J. (1989) Dispersal of cellular slime molds by two soil invertebrates. *Mycologia* 81:677–682.

Hutchinson, S.A., and Kamel, M. (1956) The effect of earthworms on dispersal of soil fungi. *J. Soil. Sci.* 7(2): 213–218.

Jagnow, G. (1987) Inoculation of cereal crops and forage grasses with nitrogen-fixing rhizosphere bacteria. Possible causes of success and failure with regard to yield response — a review. *Z. Pflanzenernaehr. Bodenk.* 150:361–368.

James, S.W. (1991) Soil, nitrogen, phosphorus, and organic matter processing by earthworms in tallgrass prairie. *Ecology* 72:2101–2109.

Jaworska, M., Ropek, D., Glen, K., and Kopec, M. (1998) The influence of different organic fertilization on wholesomeness of mountain meadow grasses. *Progr. Plant Prot.* 38:624–626.

Jolly, J.M., Lappin-Scott, H.M., Anderson, J.M., and Clegg, C.D. (1993) Scanning electron microscopy of two earthworms: *Lumbricus terrestris* and *Octolasion cyaneum*. *Microb. Ecol.* 26:235–245.

Kale, R.D., Mallesh, B.C., Bano, K., Bagyaraj, D.J, and Kretzschmar, A. (1992) Influence of vermicomposts application on the available macronutrients and selected microbial populations in a paddy field. *4th Int. Symp. Earthworm Ecol.* 24:1317–1320.

Kannaiyan, S. (1987) Studies on the biological control of sheath blight disease in rice (India). *Int. Cong. Plant Prot.* 1987.

Karthikeyan, A., and Karunanithi, K. (1996) Influence of organic amendments on the intensity of *Fusarium* wilt of banana. *Plant Dis. Res.* 11:180–181.

Kaushal, B.R., Bisht, S.B.S., and Kalia, S. (1994) Effect of diet on cast production by the megascolecid earthworm *Amynthas alexandri* in laboratory culture. *Biol. Fertil. Soils* 17:14–17.

Keogh, R.G. (1979) Lumbricid earthworm activities and nutrient cycling in pasture ecosystems. In Crosby, T.K., and Pottinger, R.P. (Eds.), *Proceedings of the 2nd Australian Conference on Grassland Invertebrate Ecology*. Government Printer, Wellington, pp. 49–51.

Kerry, B. (1988) Fungal parasites of cyst nematodes. In Edwards, C.A., Stinner, B.R., Stinner, D., and Rabatin, S. *Biological Interactions in Soil*. Elsevier, Amsterdam, pp. 293–306.

Khambata, S.R., and Bhatt, J.V. (1957) A contribution to the study of the intestinal microflora of Indian earthworms. *Arch. Mikrobiol.* 28:69–80.

Klock, K.A. (1997) Growth of salt sensitive bedding plants in media amended with composted urban wastes. *Compost Sci. Util.* 5:55–59.

Knight, D., Elliot, P.W., Anderson, J.M., and Scholefield, P. (1992) The role of earthworms in managed, permanent pastures in Devon, England. *Soil Biol. Biochem.* 24:1511–1517.

Knollenberg, R.W., Merritt, R.W., and Lawson, D.L. (1985) Consumption of leaf litter by *Lumbricus terrestris* (Oligochaeta) in a Michigan woodland floodplain *Am. Midl. Nat.* 113:1–6.

Kolte, U.M., Patil, A.S., and Tumbarbe, A.D. (1999) Response of tomato crop to different modes of nutrient input and irrigation. *J. Maharashtra Agric. Univ.* 14(1):4–8.

Kozlovskaya, L.S., and Zhdannikova, E.N. (1961) Joint action of earthworms and microflora in forest soils. *Dokl. Akad. Nauk. SSSR* 139:470–473.

Kretzschmar, A. (1984) Vermicomposting and humification by earthworms in relation to soil fertility: Research priorities for their management and utilization. In *International Conference in Waste Management*, Cambridge, U.K.

Krishnamoorthy, R.V., and Vajrabhiah, S.N. (1986) Biological activity of earthworm casts: An assessment of plant growth promoter levels in casts. *Proc. Indian Acad. Sci. (Anim. Sci.)* 95:341–351.

Krištfek, V., Pizl, V., and Szabo, I.M. (1990) Composition of the intestinal streptomycete community of earthworms (Lumbricidae). In Lessel, V. (Ed.), *Microbiology in Poecilotherms*. Elsevier, Amsterdam, pp. 137–140.

Krištfek, V., Ravasz, K., and Pizl, V. (1992) Changes in density of bacteria and microfungi during gut transit in *Lumbricus rubellus* and *Aporrectodea caliginosa* (Oligochaeta: Lumbricidae). *Soil Biol. Biochem.* 24:1499–1500.

Krištfek, V., Ravasz, K., and Pizl, V. (1993) Actinomycete communities in earthworm guts and surrounding soil. *Pedobiologia* 37:379–384.

Kubiena, W.L. (1955) Animal activity in soils as a decisive factor in establishment of humus forms. In Kevan, D.K. Mc E. (Ed.), *Soil Zoology.* Butterworths, London, pp. 73–82.

Kucey, R.M.N. (1983) Phosphate-solubilizing bacteria and fungi in various cultivated and virgin. Alberta soils. *Can. J. Soil Sci.* 63:671–678.

Kuter, G.A., Nelson, G.B., Hoitink, H.A., and Madden, L.V. (1983) Fungal population in container media amended with composted hardwood bark suppressive and conductive to *Rhizoctonia* damping-off. *Phytopathology* 73:1450–1456.

Lavelle, P. (1978) Les vers de terre de la savane de Lamto (Cote d'Ivoire). Peuplements, populations et fonctions de l'ecosysteme. *Publ. Lab. Zool. E.N.S.* 12:1–301.

Lavelle, P. (1988) Earthworms and the soil system. *Biol. Fertil. Soil* 6:237–251.

Lavelle, P. (1992) Conservation of soil fertility in low-input agricultural systems of the humid tropics by manipulating earthworm communities (macrofauna project). European Economic Community Project No. TS2-0292-F (EDB).

Lavelle, P., and Martin, A. (1992) Small- and large-scale effects of endogeic earthworms on soil organic matter dynamics in soils of the humid tropics. *Soil Biol. Biochem.* 24:1491–1498.

Lavelle, P., Blanchart, E., and Martin, A. (1993) Impact of soil fauna on the properties of soils in the humid tropics. In Sanchez, P.A., and Lal, R. (Eds.), *Myths and Science of Soils in the Tropics.* SSSA Spec. Publ. 29, Soil Science Society of America, Madison, WI, pp. 157–185.

Lavelle, P., Melendez, G., Pashanashi, B., and Schaefer, R. (1992) Nitrogen mineralization and reorganization in casts of the geophagous tropical earthworm *Pontoscolex corethrurus* (Glossoscolecidae). *Biol. Fertil. Soils* 14:49–53.

Lazarovits, G., Tenuta, M., Conn, K.L., Gullino, M.L., Katan, J., and Matta, A. (2000) Utilization of high nitrogen and swine manure amendments for control of soil-borne diseases: Efficacy and mode of action. *Acta Horticulturae* 532:59–64.

Lee, K.E. (1983) The influence of earthworms and termites on soil nitrogen cycling. In Lebrun, Ph., Andre, H.M., and de Medts, A. et al. (Eds.), *New Trends in Soil Biology.* Proceedings of the VIII International Colloquium of Soil Zoology, Louvain-la-Neuve, Belgium, pp. 35–48.

Lee, K.E. (1985) *Earthworms: Their Ecology and Relationships with Soils and Land Use.* Academic Press, Sydney.

Lee, Y.S., and Bartlett, R.J. (1976) Stimulation of plant growth by humic substances. *J. Am. Soc. Soil Sci.* 40:876–879.

Lima, M.R., May, L.L., and Maccari, A. (1997) Incidence of fungal diseases in potato under several crop systems. *Revista do Setor de Ciencias Agrarias* 16:95–98.

Liping, M.A., Qiao, X., Gao, F., and Hao, B. (2001) Control of sweet pepper *Fusarium* wilt with compost extract and its mechanism. *Yingyong Yin Huanjing Shengwu Xnebao* 7:84–87.

Loquet, M., Bhatnagar, T., Bouche, M.B., and Rouelle, J. (1977) Essai d'estimation del'influence ecologique des lombrices sur les microorganisms. *Pedobiologia* 17:400–417.

MacKay, A.D., and Kladivko, E.J. (1985) Earthworms and the rate of breakdown of soybean and maize residues in soil. *Soil Biol. Biochem.* 17:851–857.

Madge, D.S. (1969) Field and laboratory studies on the activities of two species of tropical earthworms. *Pedobiologia* 9:188–214.

Madsen, E.L., and Alexander, M. (1982) Transport of *Rhizobium* and *Pseudomonas* through soil. *Soil Sci. Soc. Am. J.* 46:557–560.

Mahesewarappa, H.P., Nanjappa, H.V., and Hegde, M.R. 1999. Influence of organic manures on yield of arrowroot, soil physico-chemical and biological properties when grown as intercrop in coconut garden. *Ann. Agric. Res.* 20:318–323.

Mahmoud, S.A.Z., Ramadam, E.M., Thabet, F.M., and Khater, T. (1984) Production of plant growth promoting substance by rhizosphere organisms. *Zeutrbl. Mikrobiol.* 139:227–232.

Mariaglieti, K. (1979) On the community structure of the gut microbiota of *Eisenia lucens* (Annelida, Oligochaeta). *Pedobiologia* 19:213–220.

Marinissen, J.Y.C., and Dexter, A.R. (1990) Mechanisms of stabilization of earthworm casts and artificial casts. *Biol. Fertil. Soils* 9:163–167.

Martin, A. (1991) Short- and long-term effects of the endogenic earthworm *Millsonia anomala* (Omodeo) (Megascolecidae, Oligochaeta) of tropical savannas on soil organic matter. *Biol. Fertil. Soils* 11: 234–238.

Martin, A., Balesdent, J., and Mariotti, A. (1991) Earthworm diet related to soil organic matter dynamics through $^{13}$C measurements. *Oecologia* 91:23–29.

Martin, A., Mariotti, A., Balesdent, J., and Lavelle, P. (1992) Soil organic matter assimilation by a geophagous tropical earthworm based on carbon-13 measurements. *Ecology* 73:118–128.

Masciandaro, G., Ceccanti, B., and Garcia, C. 1997. Soil agroecological management: Fertirrigation and vermicompost treatments. *Bioresour. Technol.* 59:199–206.

Mba, C.C. (1983) Utilization of *Eudrilus eugeniae* for disposal of cassava peel. In Satchell, J.E. (Ed.), *Earthworm Ecology: From Darwin to Vermiculture*. Chapman & Hall, London, pp. 315–321.

McIlveen, W.D., and Cole, H., Jr. (1976) Spore dispersal of Endogonadaceae by worms, ants, wasps, and birds. *Can. J. Bot.* 54:1486–1489.

McSorley, R., and Gallaher, R.N. 1995. Effect of yard waste compost on plant parasitic nematode densities in vegetable crops. *J. Nematol.* 27:245–249.

Miles, H.B. (1963) Soil protozoa and earthworm nutrition. *Soil Sci.* 95:407–409.

Miller, P.M. (2001) Reducing field populations of several plant-parasitic nematodes by leaf mold composts and some other additives. *Plant Dis. Rep.* 61:328–331.

Morales, H., Perfecto, I., and Ferguson, B. (2001) Traditional fertilization and its effect on corn insect populations in the Guatemalan highlands. *Agric. Ecosyst. Environ.* 84:145–155.

Morra, L., Palumbo, A.D., Bilotto, M., Ovieno, P., and Picascia, S. (1998) Soil solarization: Organic fertilization and grafting contribute to build an integrated production system in a tomato-zucchini sequence. *Colture-Protette (Italy)* 27:63–70.

Mrinal, S., Rajkhowa, D.J., and Saikia, M. (1998) Effect of planting density and vermicomposts on yield of potato raised from seedling tubers. *J. Indian Potato Assoc.* 25:141–142.

Muller, P.E. (1878) Nogle Undersogelser af Skovjord. *Tidsskr. Landoko.* 4:259–283.

Muller, P.E. (1884) Studier over Skovjord. II. Am Muld og Mor i Egeskove og paa Heder. *Tidsskr. Skovbrug* 7:1–232.

Murarkar, S.R., Tayade, A.S., Bodhade, S.N., and Ulemale, R.B. (1998) Effect of vermicomposts on mulberry leaf yield. *J. Soil Crops* 8:85–87.

Muscolo, A., Bovalo, F., Gionfriddo, F., and Nardi, S. (1999) Earthworm humic matter produces auxin-like effects on *Daucus carota* cell growth and nitrate metabolism. *Soil Biol. Biochem.* 31:1303–1311.

Muscolo, A., Felici, M., Concheri, G., and Nardi, S. (1993) Effect of earthworm humic substances on esterase and peroxidase activity during growth of leaf explants of *Nicotiana plumbaginifolia*. *Biol. Fertil. Soils* 15:127–131.

Muscolo, A., Panuccio, M.R., Abenavoli, M.R., Concheri, G., and Nardi, S. (1996) Effect of molecular complexity and acidity of earthworm faeces humic fractions on glutamate dehydrogenase, glutamine synthetase, and phosphenolpyruvate carboxylase in *Daucus carota* $\alpha$ II cell. *Biol. Fertil. Soils* 22:83–88.

Mylonas, V.A., and Maccants, C.B. (1980) Effects of humic and fulvic acids on growth of tobacco. I. Root initiation and elongation. *Plant Soil* 54:485–490.

Nakamura, Y. (1996) Interactions between earthworms and microorganisms in biological control of plant pathogens. *Farm. Jpn.* 30:37–43.

Nakasone, A.K., Bettiol, W., and de Souza, R.M. (1999) The effect of water extracts of organic matter on plant pathogens. *Summa Phytopathologica* 25:330–335.

Nam, C.G., Jee, H.J., and Kim, C.H. (1988) Studies in biological control of *Phytophthora* blight of red pepper. 2. Enhancement of antagonistic activity by soil amendment with organic wastes. *Kor. J. Plant Pathol.* 4:313–318.

Narayanaswamy, H., Syamrao, J., Kumar, M.D., Karigowda, C., and Ravindra, H. (1998) Management of damping off disease in FVC tobacco nursery through organic amendments. *Tob. Res.* 24:106–108.

Nardi, S., Arnoldi, G., and Dell'Agnola, G. (1988) Release of hormone-like activities from *Alloborophora rosea* and *Alloborophora caliginosa* feces. *J. Soil Sci.* 68:563–657.

Nethra, N.N., Jayaprasad, K.V., and Kale, R.D. (1999) China aster (*Callistephus chinensis* (L.) Ness) cultivation using vermicomposts as organic amendment. *Crop Res. Hissar* 17:209–215.

Neuhauser, E.F., and Hartenstein, R. (1978) Reactivity of macroinvertebrate peroxidases with lignins and lignin model compounds. *Soil Biol. Biochem.* 10:341–342.

Neuhauser, E.F., Kaplan, D.L., Malecki, M.R., and Hartenstein, R. (1980) Materials supporting weight gain by the earthworm E. *foetida* in waste conversion systems. *Agric. Wastes* 2:43–60.

Nielson, G.E., and Hole, F.E. (1964) Earthworms and the development of coprogenous A1 horizons in forest soils of Wisconsin. *Soil Sci. Soc. Am. Proc.* 28:426–430.

Nielson, R.L. (1965) Presence of plant growth substances in earthworms demonstrated by paper chromatography and the Went pea test. *Nature Lond.* 208:1113–1114.

Nowak, E. (1975) Population density of earthworms and some elements of their production in several grassland environments. *Ekol. Pol.* 23:459–491.

O'Brien, B.J., and Stout, J.D. (1978) Movement and turnover of soil organic matter as indicated by carbon isotope measurements. *Soil Biol. Biochem.* 10:309–317.

Orlikowski, L.B. (1999) Vermicompost extract in the control of some soil borne pathogens. *Int. Symp. Crop Prot.* 64:405–410.

Orlov, D.S., and Biryukova, O.N. (1996) Humic substances of vermicomposts. *Agrokhimiya* 12:60–67.

Orozco, S.H., Cegarra, J., Trujillo, L.M., and Roig, A. (1996) Vermicomposting of coffee pulp using the earthworm *Eisenia fetida*: Effects on C and N contents and the availability of nutrients. *Biol. Fertil. Soils.* 22:162–166.

Panneerselvam, A., and Saravanamuthu, R. (1996) Studies on the saprophytic survival of *Fusarium moniliforme* J Sheld in soil under treatment of oil cakes. *Indian J. Agric. Res.* 30(1):12–16.

Parle, J.N. (1959) Activities of microorganisms in soil and influence of these on soil fauna. Ph.D. Thesis, University of London, London.

Parle, J.N. (1963a) Microorganisms in the intestines of earthworms. *Gen. Microbiol.* 31:1–13.

Parle, J.N. (1963b) A microbiological study *of* earthworm casts. *Gen. Microbiol.* 31:13–23.

Parmalee, R.W., Bohlen, P.J., and Edwards, C.A. (1995). Analysis of nematode trophic structure in agroecosystems: Functional groups versus high resolution taxonomy. In Collins, H.P., Robertson, G.P., and Klug, M.J. (Eds.), *The Significance and Regulation of Soil Biodiversity*. Kluwer Academic Publishers, Dordrecht, pp. 203–207.

Parmelee, R.W., and Crossley, D.A., Jr. (1988) Earthworm production and role in the nitrogen cycle of a no-tillage agroecosystem on the Georgia piedmont. *Pedobiologia* 32:351–361.

Parmelee, R.W., Beare, M.H., Cheng, W. et al. (1990) Earthworms and enchytraeids in conventional and no-tillage agroecosystems: A biocide approach to assess their role in organic matter breakdown. *Biol. Fertil. Soils* 10:1–10.

Patil, M.P, Humani, N.C., Athani, S.I., and Patil, M.G. (1998) Response of new tomato genotype Megha to integrated nutrient management. *Adv. Agric. Res. India* 9:39–42.

Patriquin, D.G., Baines, D., and Abboud, A. (1995) Diseases, pests and soil fertility. *Soil Manage. Sust. Agric.* 161–174.

Phelan, P.L., Norris, K.H., and Mason, J.F. (1996) Soil management history and host preference by *Ostrinia nubilalis*: Evidence for plant mineral balance mediating insect–plant interactions. *Environ. Entomol.* 25:1329–1336.

Piearce, T.G. (1978) Gut contents of some lumbricid earthworms. *Pedobiologia* 18:153–157.

Piearce, T.G., and Phillips, M.J. (1980) The fate of ciliates in the earthworm gut: An *in vitro* study. *Microb. Ecol.* 5:313–320.

Pitt, D., Tilston, E.L., Groenhof, A.C., and Szmidt, R.A. (1998) Recycled organic materials (ROM) in the control of plant disease. *Acta Horticulturae* 469:391–403.

Ponge, J. F. (1991) Succession of fungi and fauna during decomposition of needles in a small area of Scots pine litter. *Plant Soil* 138:99–114.

Ponomareva, S.I. (1962) Soil macro and microorganisms and their role in increasing fertility. *Vtoraya Zoologischeskaya Konfereniya Litovskoi SSR* 97–99.

Potter, D.A., Powell, A.J., and Smith, M.S. (1990) Degradation of turfgrass thatch by earthworms (Oligochaeta: Lumbricidae) and other soil invertebrates. I. *Econ. Ent.* 83:205–211.

Prestidge, R.A., and McNeil, S. (1983) The role of nitrogen in the ecology of grassland Auchenorruncha. In Lee, J.A., McNeil, S., and Rorison, I.H. (Eds.), *Nitrogen as an Ecological Factor*. Blackwell Scientific, Oxford, pp. 257–282.

Rabatin, S.C., and Stinner, B.R. (1989) The significance of vesicular-arbuscular mycorrhizal faunal-soil-macroinvertebrate interactions in agroecosystems. *Agric. Ecosyst. Environ.* 27:195–204.

Raguchander, T., Rajappan, K., and Samiyappan, R. (1998) Influence of biocontrol agents and organic amendments on soybean root rot. *Int. J. Trop. Agric.* 16:247–252.

Rajan, P.P., and Sarma, Y.R. (2000) Effect of organic soil amendments and chemical fertilizers on foot rot pathogen (*Phytophthora capsici*) of black pepper (*Piper nigrum*). *Cent. Conf. Spices Arom. Plants* 249–253.

Ramachandra, R., Reddy, M.A.N., Reddy, Y.T.N., Reddy, N.S., Anjanappa, M., and Reddy, R. (1998) Effect of organic and inorganic sources of NPK on growth and yield of pea (*Pisum sativum*). *Legume Res.* 21:57–60.

Ramamoorthy, V., Alice, D., Meena, B., Muthusamy, M., and Seetharaman, K. (2000) Biological management of *Sclerotium* wilt of jasmine. *Indian J. Plant Prot.* 28:102–104.

Ramesh, P. (2000) Effects of vermicomposts and vermicomposting on damage by sucking pests to ground nut (*Arachis hypogea*). *Indian J. Agric. Sci.* 70:334.

Rao, K.R. (2002) Induce host plant resistance in the management of sucking insect pests of groundnut. *Ann. Plant Prot. Sci.* 10:45–50.

Rao, K.R., Rao, P.A, and Rao, K.T. (2001) Influence of fertilizers and manures on populations of coccinellid beetles and spiders in a groundnut ecosystem. *Ann. Plant Prot. Sci.* 9:43–46.

Ravasz, K., Zicsi, A., Contreras, E. et al. (1986) Über die Darmaktinomyceten Gemeinschaften einiger Regenwurmarten. *Opusc. 2001* 22:85–102.

Ravasz, K., Zicsi, A., Contreras, E., and Szabo, I.M. (1987) Comparative bacteriological analyses of the faecal matter of different earthworm species In Pagliai, A.M.B., and Omodeo, P., *On Earthworms*. Mucchi Editore, Modena, Italy, pp. 389–399.

Raw, F. (1962) Studies of earthworm populations in orchards. I. Leaf burial in apple orchards. *Ann. Appl. Biol.* 50:389–404.

Reddell, P., and Spain, A.V. (1991) Transmission of. infective *Frankia* (Actiriomycetales) propagules in casts of the endogenic earthworm *Pontoscolex corethrurus* (Oligochaeta: Glossoscolecidae). *Soil Biol. Biochem.* 23:775–778.

Reddy, B.G., and Reddy, M.S. (1999) Effect of integrated nutrient management on soil available micro nutrients in maize-soybean cropping system. *J. Res. ANGRAU* 27:24–28.

Reynolds, J.W. (1998) The status of earthworm biogeography, diversity and taxonomy in North America revisited with glimpses into the future. In Edwards, C.A. (Ed.), *Earthworm Ecology*. CRC Press, Boca Raton, FL, pp. 14–34.

Ribeiro, C.F., Mizobutsi E.H., Silva D.G., Pereira J.C.R., Zambolim, L. (1998) Control of *Meloidognye javanica* on lettuce with organic amendments. *Fitopatologia Brasileira* 23:42–44.

Rodríguez, J.A., Zavaleta, E., Sanchez, P., and Gonzalez, H. (2000) The effect of vermicomposts on plant nutrition, yield and incidence of root and crown rot of gerbera (*Gerbera jamesonii H. Bolus*). *Fitopatologia* 35:66–79.

Rosswall, T., and Paustian, K. (1984) Cycling of nitrogen in modern agricultural systems. *Plant Soil* 76:3–21.

Rouelle, J. (1983) Introduction of an amoeba and *Rhizobium japonicum* into the gut of *Eisenia fetida* (Sav.) and *Lumbrucus terrestris* L. In Satchell, J.E. (Ed.), *Earthworm Ecology: From Darwin to Vermiculture*. Chapman & Hall, New York, pp. 375–381.

Ruz-Jerez, B.E., Ball, P.R., and Tillman, R.W. (1992) Laboratory assessment of nutrient release from a pasture soil receiving grass or clover residues, in the presence or absence of *Lumbricus rubellus* or *Eisenia fetida*. *Soil Biol. Biochem.* 24:1529–1534.

Sanudo, S.B., and Molina Valero, L.A. (1995) Cultural control of black leg (*Gaeumannomyces graminis*) in wheat variety ICA Yacuanquer in Nariño. *ASCOLFI Inform.* 21:34–35.

Satchell, J.E. (1958) Earthworm biology and soil fertility. *Soil Fertil.* 21:209–219.

Satchell, J.E. (1963) Nitrogen turnover by a woodland population of *Lumbricus terres*. In Doeksen, J., and van der Drift, J. (Eds.), *Soil Organisms*. North Holland, Amsterdam, pp. 60–66.

Satchell, J.E. (1967) Lumbricidae. In Burgess, A., and Raw, F. (Eds.), *Soil Biology*. Academic Press, London, pp. 259–322.

Scheu, S. (1987) The role of substrate-feeding earthworms (Lumbricidae) for bioturbation in a beechwood soil. *Oecologia* 72:192–196.

Scheu, S. (1991) Mucus excretion and carbon turnover of endogenic earthworms. *Biol. Fertil. Soils* 12:217–220.

Scheu, S. (1994) There is an earthworm-mobilizable nitrogen pool in soil. *Pedobiologia* 38:243–249.

Scheu, S., and Wolters, V. (1991a) Influence of fragmentation and bioturbation on the decomposition of carbon-14-labelled beech leaf litter. *Soil Biol. Biochem.* 23:1029–1034.

Scheu, S., and Wolters, V. (1991b) Buffering of the effect of acid rain on decomposition of carbon-14-labelled beech leaf litter by saprophagous invertebrates. *Biol. Fertil. Soils* 13:285–289.

Scheurell, S., and Mahafee, W. (2002) Compost tea: Principles and prospects for plant disease control. *Compost Sci. Utiliz.* 10(4):313–338.

Scott, M.A. (1988) The use of worm-digested animal waste as a supplement to peat in loamless composts for hardy nursery stock. In *Earthworms in Environmental and Waste Management.* SPB Academic Publishing, The Hague, the Netherlands.

Senesi, N., Saiz-Jimenez, C., and Miano, T.M. (1992) Spectroscopic characterization of metal-humic acid-like complexes of earthworm-composted organic wastes. *Sci. Total Environ.* 117/118:111–120.

Shaw, C., and Pawluk, S. (1986) Faecal microbiology of *Octolasion tyrtaeum, Apporectodea turgida* and *Lumbricus terrestris* and its relation to carbon budgets of three artificial soils. *Pedobiologia* 29:377–389.

Shiau, F.L., Chung, W.C., Huang, J.W., and Huang, H.C. (1999) Organic amendment of commercial culture media for improving control of *Rhizoctonia* damping-off cabbage. 1999. *Can. J. Plant Pathol.* 21:368–374.

Shi-wei, Z., and H. Fuzhen. (1991) The nitrogen uptake efficiency from $^{15}N$ labelled chemical fertilizer in the presence of earthworm manure (cast). In Veeresh, G.K., Rajagopal, D., and Viraktamath, C.A. (Eds.), *Advances in Management and Conservation of Soil Fauna.* Oxford & IBH, New Delhi, pp. 539–542.

Siddagangaiah, B.A., Vadiraj, M.R., Sudharshan M.R., and Krishnakumar, V. (1996) Standardization of rooting media for propagation of vanilla (*Vanilla planifolia* Andr.). *J. Spices Arom. Crops* 5:131–133.

Siddiqui, Z.A., and Mahmood, I. (1999) Role of the bacteria in the management of plant parasitic nematodes: A review. *Bioresour. Technol.* 69:167–179.

Somasekhara, Y.M., Anilkumar, T.B., and Siddaramaiah A.L. (2000) Effect of organic amendments and fungicides on population of *Fusarium udum* and their interaction with *Trichoderma* spp. *Karnataka J. Agric. Sci.* 13:752–756.

Spiers, G.A., Gagnon, D., Nason, G.E. et al. (1986) Effects and importance of indigenous earthworms on decomposition and nutrient cycling in coastal forest ecosystems. *Can. J. For. Res.* 16:983–989.

Sreenivas, C., Muralidhar, S., and Rao, M.S. (2000) Vermicomposts: A viable component of IPNSS in nitrogen nutrition of ridge gourd. *Ann. Agric. Res.* 21:108–113.

Stephens, P.M., Davoren, C.W., Ryder, M.H., and Doube, B.M. (1993) Influence of the lumbricid earthworm *Aporrectodea trapezoides* on the colonization of wheat (*Triticum aestivum* cv. Spear) roots by *Pseudomonas corrugata* strain 2140R in soil. *Soil Biol. Biochem.* 25:1719–1724.

Stockdill, S.M.J. (1966) The effect of earthworms on pastures. *Proc. N.Z. Ecol. Soc.* 13:68–74.

Striganova, B.R., Marfenina, O.E., and Ponomarenko, V.A. (1989) Some aspects of the effect of earthworms on soil fungi. *Biol. Bull. Acad. Sci. USSR* 15:460–463.

Subler, S., Edwards, C.A., and Metzger, J. (1998) Comparing vermicomposts and composts. *BioCycle* 39:63–66.

Sudhakar K., Punnaiah K.C., and Krishnayya P.V. (1998) Influence of organic and inorganic fertilizers and certain insecticides on the incidence of shoot and fruit borer, *Leucinodes orbonalis* Guen, infesting brinjal. *J. Entomol. Res.* 22:283–286.

Sugi, Y., and Tanaka, M. (1978) Number and biomass of earthworm populations. In Kiraj, T., Onoand, Y., and Hosokawa, T. (Eds.), *Biological Production in a Warm-Temperate Evergreen Oak Forest of Japan.* J.I.P.B. Synthesis No. 18, University of Tokyo Press, Tokyo, pp. 171–178.

Svensson, B.H., Bostrom, U., and Klemedston, L. (1986) Potential for higher rates of denitrification in earthworm casts than in the surrounding soil. *Biol. Fertil. Soils* 2:147–149.

Swathi, P., Rao, K.T., and Rao, P.A. (1998) Studies on control of root-knot nematode *Meloidogyne incognita* in tobacco miniseries. *Tob. Res.* 1:26–30.

Syers, J.K., Sharpley, A.N., and Keeney, D.R. (1979) Cycling of nitrogen by surface-casting earthworms in a pasture ecosystem. *Soil Biol. Biochem.* 11:181–185.

Szabó, I.M., Prauser, H., Bodnar, G. et al. (1990) The indigenous intestinal bacteria of soil arthropods and worms. In Lessel, R. (Ed.), *Microbiology in Poecilotherms.* Elsevier, Amsterdam, pp. 109–117.

Szczech, M. (1999) Suppressiveness of vermicompost against *Fusarium* wilt of tomato. *J. Phytopathol.* 147:155–161.

Szczech, M., Randomanski, W., Brzeski, M.W., Sindinska, U., and Kotavski, J.F. (1993) Suppressive effect of a commercial earthworm compost on some root infecting pathogens of cabbage and tomatoes. *Biol. Agric. Hortic.* 10:47–52.

Tan, K.H., and Tantiwiramanond, D. (1983) Effect of humic acids on nodulation and dry matter production of soybean, peanut, and clover. *Soil Sci. Soc. Am. J.* 47:1121–1124.

Thankamani, C.K., Sivaraman, K., and Kandiannan, K. (1996) Response of clove (*Syzygium aromaticum* (L.) Merr. & Perry) seedlings and black pepper (*Piper nigrum* L.) cuttings to propagating media under nursery conditions. *J. Spices Arom. Crops* 5:99–104.

Thanunathan, K., Natarajan, S., Senthilkumar, R., and Arulmurugan, K. (1997) Effect of different sources of organic amendments on growth and yield of onion in mine spoil. *Madras Agric. J.* 84:382–384.

Thompson, L., Thomas, C.D., Radley, J.M.A. et al. (1993) The effects of earthworms and snails in a simple plant community. *Oecologia* 95:171–178.

Tiwari, S.C., and Mishra, R.R. (1993) Fungal abundance and diversity in earth- worm casts and in uningested soil. *Biol. Fertil. Soils* 16:131–134.

Tiwari, S.C., Tiwari, B.K., and Mishra, R.R. (1989) Microbial populations, enzyme activities and nitrogen-phosphorus-potassium enrichment in earthworm casts and in the surrounding soil of a pineapple plantation. *Biol. Fertil. Soils* 8:178–182.

Tomati, U., and Galli, E. (1995) Earthworms, soil fertility and plant productivity. *Proc. Int. Coll. Soil Zool. Acta Zool. Fenn.* 196:11–14.

Tomati, U., Galli, E., Grapppelli, A., and Dihena G. (1990) Effect of earthworm casts on protein synthesis in radish (*Raphanus sativum*) and lettuce (*Lactuca sativa*) seedlings. *Biol. Fertil. Soil* 9:288–289.

Tomati, U., Grappelli, A., and Galli, E. (1983) Fertility factors in earthworm humus. In *Proceedings of the International Symposium on Agricultural and Environmental Prospects in Earthworm Farming*. Publication Ministero della Ricerca Scientifica e Technologia, Rome, pp. 49–56.

Tomati, U., Grappelli, A., and Galli, E. (1987) The presence of growth regulators in earthworm-worked wastes. *In* Bonvicini Paglioi, A.M., and P. Omodeo (Eds.), *On Earthworms: Proceedings of International Symposium on Earthworms: Selected Symposia and Monographs*, Unione Zoologica Italiana, Mucchi, Modena, Italy, pp. 423–435.

Tomati, U., Grappelli, A., and Galli, E. (1988) The hormone-like effect of earthworm casts on plant growth. *Biol. Fertil. Soils* 5:288–294.

Toyota, K.M., Kitamura, M., and Kimura, M. (1995) Suppression of *Fusarium oxysporum* f. sp. raphani PEG-4 in soil following colonization by other *Fusarium* spp. *Soil Biol. Biochem.* 27:41–46.

Trigo, D., and Lavelle, P. (1993) Changes in respiration rate and some physiochemical properties of soil during gut transit through *Allolobophora molleri* (Lumbricidae, Oligochaeta). *Biol. Fertil. Soils* 15:185–188.

Tyler, H.H., Warren, S.L., Bilderback, T.E., and Fonteno, W.C. (1993) Composted turkey litter: I. Effect on physical and chemical properties of pine bark substrate. *J. Environ. Hortic.* 11:131–136.

Urquhart, A.T. (1887) On the work of earthworms in New Zealand. *Trans. N.Z. Inst.* 19:119–123.

Ushakumari, K., Prabhakumari, P., and Padmaja, P. (1999) Efficiency of vermicomposts on growth and yield of summer crop okra (*Abelmoschus esculentus* Moench). *J. Trop. Agric.* 37:87–88.

Vadiraj, B.A., and Siddagangaiah, S., and Potty, N. (1998) Response of coriander (*Coriandrum sativum* L.) cultivars to graded levels of vermicomposts. *J. Spices Arom. Crops* 7:141–143.

Valdrighi, M.M., Pera, A., Agnolucci, M., Frassinetti, S., Lunardi, D., and Vallini, G. (1996) Effects of compost-derived humic acids on vegetable biomass production and microbial growth within a plant (*Cichorium intybus*)-soil system: A comparative study. *Agric. Ecosyst. Environ.* 58:133–144.

van der Bruel, W.E. (1964) Le sol, la pedofauna et les applications de pesticides. *Ann. Gembl.* 70:81–101.

Vasanthi, D., and Kumaraswamy, K. (1999) Efficacy of vermicomposts to improve soil fertility and rice yield. *J. Indian Soc. Soil Sci.* 47:268–272.

Velandia, J., Galindo, R.P., and de Moreno, C.A. (1998) Poultry manure evaluation in the control of *Plasmodiophora brassicae* in cabbage. *Agronomia Colombiana* 15(1):1–6.

Venkatesh, P.P.B., Patil, P.B., Patil, C.V., and Giraddi, R.S. (1998) Effect of *in situ* vermiculture and vermicomposts on availability and plant concentration of major nutrients in grapes. *Karnataka J. Agric. Sci.* 11:117–121.

Venkatesh, P.P.B., Sudhirkumar, K., and Kotikal, Y.K. (1997) Influence of *in situ* vermiculture and vermicompost on yield and yield attributes of grapes. *Adv. Agric. Res. India* 8:53–56.

Wasawo, D.P.S., and Visser, S.A. (1959) Swampworms and tussock mounds in the swamps of Teso, Uganda. *E. Afr. Agric. J.* 25:86–90.

Went, J.C. (1963) Influence of earthworms on the number of bacteria in the soil. In Doeksen, J., and van der Drift, J. (Eds.), *Soil Organisms.* North Holland, Amsterdam, pp. 260–265.

Werner, M., and Cuevas, R. (1996) Vermiculture in Cuba. *BioCycle* 37:61–62.

Wilson, D.P., and Carlile, W.R. (1989) Plant growth in potting media containing worm-worked duck waste. *Acta Horticulturae* 238:205–220.

Witkamp, M. (1966) Decomposition of leaf litter in relation to environment, microflora and microbial respiration. *Ecology* 47:194–201.

Wittich, W. (1953) Untersuchungen über den Verlauf der Streuzersetzung auf einem Boden mit Regenwurmtätigkeit. *Schrift Reige forstl. Fak. Univ. Göttingen* 9:7–33.

Wolters, V., and Joergensen, R.G. (1992) Microbial carbon turnover in beech forest soils worked by *Aporrectodea caliginosa* (Savigny) (Oligochaeta: Lumbricidae). *Soil Biol. Biochem.* 24:171–177.

Yardim, E.N., and Edwards, C.A. (1998) The effects of chemical pest, disease and weed management practices on the trophic structure of nematode populations in tomato agroecosystems. *Appl. Soil Ecol.* 7:137–147.

Zambolim, L., Santos, M.A., Becker, W.F., and Chaves, G.M. (1996) Agrowaste soil amendments for the control of *Meloidogyne javanica* on tomato. *Fitopatologia Brasileira* 21:250–253.

Zende, G.K., Ruikar, S.K., and Joshi, S.N. (1998) Effect of application of vermicomposts along with chemical fertilizers on sugar cane yield and juice quality. *Indian Sugar* 48:357–369.

Zicsi, A. (1983) Earthworm ecology in deciduous forests in central and southeast Europe. In Satchell, J.E., *Earthworm Ecology: from Darwin to Vermiculture.* Chapman & Hall, London, pp. 171–178.

Zrazhevskii, A.I. (1957) *Dozhdevye chervi kak fakto plodorodiya lesnykj pochv,* Akademie Nauk Ukr. SSSR, Kiev.

# Index

## A

For Product Safety Concerns and Information please contact our EU
representative GPSR@taylorandfrancis.com
Taylor & Francis Verlag GmbH, Kaufingerstraße 24, 80331 München, Germany